Mechanical Properties and Performance of Engineering Ceramics and Composites III

Mechanical Properties and Performance of Engineering Ceramics and Composites III

*A Collection of Papers Presented at the 31st International Conference on Advanced Ceramics and Composites
January 21–26, 2007
Daytona Beach, Florida*

Editor
Edgar Lara-Curzio

Volume Editors
Jonathan Salem
Dongming Zhu

The American Ceramic Society

BICENTENNIAL
1807
WILEY
2007
BICENTENNIAL

WILEY-INTERSCIENCE
A John Wiley & Sons, Inc., Publication

Copyright © 2008 by The American Ceramic Society. All rights reserved.

Published by John Wiley & Sons, Inc., Hoboken, New Jersey.
Published simultaneously in Canada.

No part of this publication may be reproduced, stored in a retrieval system, or transmitted in any form or by any means, electronic, mechanical, photocopying, recording, scanning, or otherwise, except as permitted under Section 107 or 108 of the 1976 United States Copyright Act, without either the prior written permission of the Publisher, or authorization through payment of the appropriate per-copy fee to the Copyright Clearance Center, Inc., 222 Rosewood Drive, Danvers, MA 01923, (978) 750-8400, fax (978) 750-4470, or on the web at www.copyright.com. Requests to the Publisher for permission should be addressed to the Permissions Department, John Wiley & Sons, Inc., 111 River Street, Hoboken, NJ 07030, (201) 748-6011, fax (201) 748-6008, or online at http://www.wiley.com/go/permission.

Limit of Liability/Disclaimer of Warranty: While the publisher and author have used their best efforts in preparing this book, they make no representations or warranties with respect to the accuracy or completeness of the contents of this book and specifically disclaim any implied warranties of merchantability or fitness for a particular purpose. No warranty may be created or extended by sales representatives or written sales materials. The advice and strategies contained herein may not be suitable for your situation. You should consult with a professional where appropriate. Neither the publisher nor author shall be liable for any loss of profit or any other commercial damages, including but not limited to special, incidental, consequential, or other damages.

For general information on our other products and services or for technical support, please contact our Customer Care Department within the United States at (800) 762-2974, outside the United States at (317) 572-3993 or fax (317) 572-4002.

Wiley also publishes its books in a variety of electronic formats. Some content that appears in print may not be available in electronic format. For information about Wiley products, visit our web site at www.wiley.com.

Wiley Bicentennial Logo: Richard J. Pacifico

Library of Congress Cataloging-in-Publication Data is available.

ISBN 978-0-470-19633-5

Printed in the United States of America.

10 9 8 7 6 5 4 3 2 1

Contents

Preface   xi

Introduction   xiii

PROCESSING

Synthesis and Characterization of $Ba_3Co_2Fe_{24}O_{41}$ and $Ba_3Co_{0.9}Cu_{1.1}Fe_{24}O_{41}$ Nanopowders and Their Application as Radar Absorbing Materials   3
   Valeska da Rocha Caffarena, Magali Silveira Pinho, Jefferson Leixas Capitaneo, Tsuneharu Ogasawara, and Pedro Augusto de Souza Lopes Cosentino

Microwave Enhanced Anisotropic Grain Growth in Lanthanum Hexa Aluminate-Alumina Composites   15
   Z. Negahdari, T. Gerdes, and M. Willert-Porada

Oxyfuel Combustion Using Perovskite Membranes   23
   E.M. Pfaff and M. Zwick

Synthesis of SiC Nanofibers with Graphite Powders   33
   Andrew Ritts, Qingsong Yu, and Hao Li

Influence of Additional Elements on Densification Behavior of Zirconia Base Amorphous Powders   41
   Tatsuo Kumagai

SILICON-BASED CERAMICS

The Intergranular Microstructure of Silicon Nitride Based Ceramics   55
   L.K.L. Falk, N. Schneider, Y. Menke, and S. Hampshire

Ultrafine Powders Doped With Aluminium In SiCN System 65
Vincent Salles, Sylvie Foucaud, and Paul Goursat

PROPERTIES OF MONOLITHIC CERAMICS

Modulus and Hardness of Nanocrystalline Silicon Carbide as Functions of Grain Size 79
Suraj C. Zunjarrao, Abhishek K. Singh and Raman P. Singh

Stoichiometric Constraint for Dislocation Loop Growth in Silicon Carbide 91
Sosuke Kondo, Yutai Katoh, and Akira Kohyama

Effects of Si:SiC Ratio and SiC Grain Size on Properties of RBSC 101
S. Salamone, P. Karandikar, A. Marshall, D.D. Marchant, and M. Sennett

Electrical Properties of AlN-SiC Solid Solutions with Additions of Al and C 111
Ryota Kobayashi, Junichi Tatami, Toru Wakihara, Katsutoshi Komeya, and Takeshi Meguro

FIBER-REINFORCED CMCS

Effects of Frequency on Fatigue Behavior of an Oxide-Oxide Ceramic Composite at 1200°C 119
G. Hetrick, M.B. Ruggles-Wrenn, and S.S. Baek

Post Creep/Dwell Fatigue Testing of MI SiC/SiC Composites 135
G. Ojard, A. Calomino, G. Morscher, Y. Gowayed, U. Santhosh, J. Ahmad, R. Miller, and R. John

Time-Dependent Response of MI SiC/SiC Composites Part 1: Standard Samples 145
G. Ojard, Y. Gowayed, J. Chen, U. Santhosh, J. Ahmad, R. Miller, and R. John

Time-Dependent Response of MI SiC/SiC Composites Part 2: Samples with Holes 155
Y. Gowayed, G. Ojard, J. Chen, R. Miller, U. Santhosh, J. Ahmad, and R. John

Effects of Environment on Creep Behavior of an Oxide-Oxide Ceramic Composite with ±45° Fiber Orientation at 1200°C 163
G. T. Siegert, M. B. Ruggles-Wrenn, and S. S. Baek

Assessments of Life Limiting Behavior in Interlaminar Shear for Hi-Nic SiC/SiC Ceramic Matrix Composite at Elevated Temperature 179
Sung R. Choi, Robert W. Kowalik, Donald J. Alexander, and Narottam P. Bansal

Architectural Design of Preforms and Their Effects on Mechanical Property of High Temperature Composites 191
Jae Yeol Lee, Tae Jin Kang, and Joon-Hyung Byun

Design Factor Using a SiC/SiC Composites for Core Component of Gas Cooled Fast Reactor. 2: Thermal Stress 199
Jae-Kwang Lee and Masayuki Naganuma

Development of Novel Fabrication Process for Highly-Dense & Porous SiC/SiC Composites with Excellent Mechanical Properties 207
Kazuya Shimoda, Joon-Soon Park, Tatsuya Hinoki, and Akira Kohyama

Effects of Interface Layer and Matrix Microstructure on the Tensile Properties of Unidirectional SiC/SiC Composites 213
Masaki Kotani, Toshio Ogasawara, Hiroshi Hatta, and Takashi Isikawa

Tensile Properties of Advanced SiC/SiC Composites for Nuclear Control Rod Applications 223
Takashi Nozawa, Edgar Lara-Curzio, Yutai Katoh, and Robert J. Shinavski

PARTICULATE REINFORCED AND LAMINATED COMPOSITES

Influence of the Architecture on the Mechanical Performances of Alumina-Mullite and Alumina-Mullite-Zirconia Ceramic Laminates 237
Alessandra Costabile and Vincenzo M. Sglavo

Fabrication of Novel Alumina Composites Reinforced by SiC Nano-Particles and Multi-Walled Carbon Nanotubes 245
Kaleem Ahmad and Wei Pan

Effect of Carbon Additions and B_4C Particle Size on the Microstructure and Properties of B_4C – TiB_2 Composites 257
R.C. McCuiston, J.C. LaSalvia, and B. Moser

Electro-Conductive ZrO_2-NbC-TiN Composites Using NbC Nanopowder Made By Carbo-Thermal Reaction 269
S. Salehi, J. Verhelst, O. Van der Biest, and J. Vleugels

High Temperature Strength Retention of Aluminum Boron Carbide (AlBC) Composites 277
Aleksander J. Pyzik, Robert A. Newman, and Sharon Allen

ENVIRONMENTAL EFFECTS

Corrosion Resistance of Ceramics in Sulfuric Acid Environments at High Temperature 289
C.A. Lewinsohn, H. Anderson, M. Wilson, and A. Johnson

Analyzing Irradiation-Induced Creep of Silicon Carbide 297
Yutai Katoh, Lance Snead, and Stas Golubov

Physico-Chemical Reactivity of Ceramic Composite Materials at 307
High Temperature: Vaporization and Reactivity with Carbon of
Borosilicate Glass
Sebastien Wery and Francis Teyssandier

Irradiation Effects on the Microstructure and Mechanical Properties 319
of Silicon Carbide
Magalie Menard, Marion Le Flem, Lionel Gelebart, Isabelle Monnet,
Virginie Basini, and Michel Boussuge

Oxidation of ZrB_2-SiC: Comparison of Furnace Heated Coupons and 327
Self-Heated Ribbon Specimens
S.N. Karlsdottir, J.W. Halloran, F. Monteverde, and A. Bellosi

The Role of Fluorine in Glass Formation in the Ca-Si-Al-O-N System 337
Amir Reza Hanifi, Annaik Genson, Michael J. Pomeroy, and Stuart Hampshire

Wetting and Reaction Characteristics of Al_2O_3/SiC Composite 347
Refractories by Molten Aluminum and Aluminum Alloy
James G. Hemrick, Jing Xu, Klaus-Markus Peters, Xingbo Liu, and Ever Barbero

NDE AND TEST METHODS

Evaluation of Oxidation Protection Testing Methods on Ultra-High 361
Temperature Ceramic Coatings for Carbon-Carbon Oxidation
Resistance
Erica L. Corral, Alicia A. Ayala, and Ronald E. Loehman

Nondestructive Evaluation of Silicon-Nitride Ceramic Valves from 371
Engine Duration Test
J.G. Sun, J.S. Trethewey, N.S.L. Phillips, N.N. Vanderspiegel, and J.A. Jensen

Model of Constrained Sintering 379
Kais Hbaieb and Brian Cotterell

FRACTURE

Study of Factors Affecting the Lengths of Surface Cracks in 391
Silicon Nitride Introduced by Vickers Indentation
Hiroyuki Miyazaki, Hideki Hyuga, Yu-ichi Yoshizawa, Kiyoshi Hirao,
and Tatsuki Ohji

Strength Recovery Behavior of Machined Alumina by Crack Healing 399
Kotoji Ando, Wataru Nakao, Koji Takahashi, and Toshio Osada

Modeling Crack Bifurcation in Laminar Ceramics 411
K. Hbaieb, R.M. McMeeking, and F.F. Lange

Delayed Failure of Silicon Carbide Fibers in Static Fatigue at 423
Intermediate Temperatures (500-800°C) in Air
    W. Gauthier and J. Lamon

Fracture-Toughness Test of Silicon Nitrides with Different 433
Microstructures Using Vickers Indentation
    Hiroyuki Miyazaki, Hideki Hyuga, Yu-ichi Yoshizawa, Kiyoshi Hirao, and
    Tatsuki Ohji

Self-Crack-Healing Ability of Alumina/ SiC Nanocomposite 443
Fabricated by Self-Propagating High-Temperature Synthesis
    Wataru Nakao, Yasuyuki Tsutagawa, Koji Takahashi, and Kotoji Ando

Through-Life Reliability Management of Structural Ceramic 449
Components Using Crack-Healing and Proof Test
    Kotoji Ando, Masato Ono, Wataru Nakao, and Koji Takahashi

JOINING AND BRAZING

Joining Methods for Ceramic, Compact, Microchannel Heat 463
Exchangers
    C.A. Lewinsohn, J. Cutts, M. Wilson, and H. Anderson

Glass-To-Metal (GTM) Seal Development Using Finite Element 469
Analysis: Assessment of Material Models and Design Changes
    Rajan Tandon, Michael K. Neilsen, Timothy C. Jones, and James F. Mahoney

Integrative Design with Ceramics: Optimization Strategies for 479
Ceramic/Metal Joints
    A. Bezold, H.R. Maier, and E.M. Pfaff

Diffusion Bonding of Silicon Carbide for MEMS-LDI Applications 491
    Michael C. Halbig, Mrityunjay Singh, Tarah Shpargel, and James D. Kiser

Effect of Residual Stress on Fracture Behavior in Mechanical Test 503
for Evaluating Shear Strength of Ceramic Composite Joint
    Hisashi Serizawa, Kazuaki Katayama, Charles A. Lewinsohn, Mrityunjay Singh,
    and Hidekazu Murakawa

Author Index 513

Preface

This volume contains papers presented at the *Symposium on Processing, Properties and Performance of Engineering Ceramics and Composites* during the 31st International Conference & Exposition on Advanced Ceramics & Composites held on January 21-26, 2007 at Daytona Beach, Florida. These papers from researchers in 12 different countries, address core fundamentals as well as timely topics on the science and technology of ceramics and ceramic composites in the best tradition of the International Conference & Exposition on Advanced Ceramics & Composites.

The papers in this volume are organized in the following sections:

- Processing of Ceramics and Composites
- Silicon-Based Ceramics
- Properties of Monolithic Ceramics
- Fibers and Interfaces
- Fiber-Reinforced Ceramic Matrix Composites
- Particulate-Reinforced and Laminated Ceramics
- Environmental Effects
- NDE, Test Methods and Modeling
- Fracture
- Joining & Brazing

The International Conference & Exposition on Advanced Ceramics & Composites is truly the premier international forum to present information on emerging ceramic technologies and on the processing, properties, behavior and application of structural ceramics, functional ceramics and ceramic composites. There are many reasons for the great success of this meeting, including its venue (Daytona Beach for the first time after 30 meetings in Cocoa Beach), its size (the collision frequency and mean-free path is perfect to foster interactions among the attendees), its schedule (there are not many places that beat Florida in January), and the dedication of many volunteers and the staff of The American Ceramic Society. In particular, we thank the attendees, the authors, the session chairs and session organizers, as well as those who helped us review the manuscripts contained in this volume.

EDGAR LARA-CURZIO
Oak Ridge National Laboratory
Oak Ridge, Tennessee

Introduction

2007 represented another year of growth for the International Conference on Advanced Ceramics and Composites, held in Daytona Beach, Florida on January 21-26, 2007 and organized by the Engineering Ceramics Division (ECD) in conjunction with the Electronics Division (ED) of The American Ceramic Society (ACerS). This continued growth clearly demonstrates the meetings leadership role as a forum for dissemination and collaboration regarding ceramic materials. 2007 was also the first year that the meeting venue changed from Cocoa Beach, where it was originally held in 1977, to Daytona Beach so that more attendees and exhibitors could be accommodated. Although the thought of changing the venue created considerable angst for many regular attendees, the change was a great success with 1252 attendees from 42 countries. The leadership role in the venue change was played by Edgar Lara-Curzio and the ECD's Executive Committee, and the membership is indebted for their effort in establishing an excellent venue.

The 31st International Conference on Advanced Ceramics and Composites meeting hosted 740 presentations on topics ranging from ceramic nanomaterials to structural reliability of ceramic components, demonstrating the linkage between materials science developments at the atomic level and macro level structural applications. The conference was organized into the following symposia and focused sessions:

- Processing, Properties and Performance of Engineering Ceramics and Composites
- Advanced Ceramic Coatings for Structural, Environmental and Functional Applications
- Solid Oxide Fuel Cells (SOFC): Materials, Science and Technology
- Ceramic Armor
- Bioceramics and Biocomposites
- Thermoelectric Materials for Power Conversion Applications
- Nanostructured Materials and Nanotechnology: Development and Applications
- Advanced Processing and Manufacturing Technologies for Structural and Multifunctional Materials and Systems (APMT)
- Porous Ceramics: Novel Developments and Applications

- Advanced Dielectric, Piezoelectric and Ferroelectric Materials
- Transparent Electronic Ceramics
- Electroceramic Materials for Sensors
- Geopolymers

The papers that were submitted and accepted from the meeting after a peer review process were organized into 8 issues of the 2007 Ceramic Engineering & Science Proceedings (CESP); Volume 28, Issues 2-9, 2007 as outlined below:

- Mechanical Properties and Performance of Engineering Ceramics and Composites III, CESP Volume 28, Issue 2
- Advanced Ceramic Coatings and Interfaces II, CESP, Volume 28, Issue 3
- Advances in Solid Oxide Fuel Cells III, CESP, Volume 28, Issue 4
- Advances in Ceramic Armor III, CESP, Volume 28, Issue 5
- Nanostructured Materials and Nanotechnology, CESP, Volume 28, Issue 6
- Advanced Processing and Manufacturing Technologies for Structural and Multifunctional Materials, CESP, Volume 28, Issue 7
- Advances in Electronic Ceramics, CESP, Volume 28, Issue 8
- Developments in Porous, Biological and Geopolymer Ceramics, CESP, Volume 28, Issue 9

The organization of the Daytona Beach meeting and the publication of these proceedings were possible thanks to the professional staff of The American Ceramic Society and the tireless dedication of many Engineering Ceramics Division and Electronics Division members. We would especially like to express our sincere thanks to the symposia organizers, session chairs, presenters and conference attendees, for their efforts and enthusiastic participation in the vibrant and cutting-edge conference.

ACerS and the ECD invite you to attend the 32nd International Conference on Advanced Ceramics and Composites (http://www.ceramics.org/meetings/daytona2008) January 27–February 1, 2008 in Daytona Beach, Florida.

JONATHAN SALEM AND DONGMING ZHU, Volume Editors
NASA Glenn Research Center
Cleveland, Ohio

Processing

SYNTHESIS AND CHARACTERIZATION OF $Ba_3Co_2Fe_{24}O_{41}$ AND $Ba_3Co_{0.9}Cu_{1.1}Fe_{24}O_{41}$ NANOPOWDERS AND THEIR APPLICATION AS RADAR ABSORBING MATERIALS

Valeska da Rocha Caffarena [1], Magali Silveira Pinho [2], Jefferson Leixas Capitaneo [3], Tsuneharu Ogasawara [3], Pedro Augusto de Souza Lopes Cosentino [4]

[1] Brazilian Petroleum S.A. – PETROBRAS (GE-MC/TDM)
Av. Almirante Barroso, 81, 32° andar, Centro, Zip Code 20031-004,
Rio de Janeiro, RJ, Brazil, valeskac@petrobras.com.br [2] Brazilian Navy Research Institute (IPqM)
Rua Ipiru 2, Praia da Bica, Ilha do Governador,
Rio de Janeiro, RJ, Brazil, Zip Code 21931-090, magalipinho@yahoo.com.br
[3] Dep. of Metallurgical and Materials Engineering, COPPE-PEMM/UFRJ, PO Box 68505, Zip Code 21941-972, Rio de Janeiro, RJ, Brazil, jeff@metalmat.ufrj.br, ogasawat@metalmat.ufrj.br
[4] Materials Laboratory of the Brazilian Army Technological Center, 28705 Americas Av, Rio de Janeiro, RJ, Brazil, Zip Code 23020-470, pac@ctex.eb.br

ABSTRACT

There has been a growing and widespread interest in the development of radar absorbing materials (RAM) to reduce the radar signatures of navy platforms. Z-type barium hexaferrite is one of the most complex compounds in the family of hexagonal ferrites that due to its good magnetic properties, it is a promising candidate to be used as RAM.

In this work, the nanosized Z-type barium hexaferrite powders ($Ba_3Co_2Fe_{24}O_{41}$ and $Ba_3Co_{0.9}Cu_{1.1}Fe_{24}O_{41}$) were synthesized at 950 °C by the citrate sol-gel process to be used as RAM in polychloroprene (CR) matrices. X-ray diffraction and X-ray fluorescence (XRD and XRF, respectively) were used to characterize these materials. Magnetic properties of the Z-type barium hexaferrites were also evaluated by using the vibrating sample magnetometer (VSM). The Cu^{2+} ions were incorporated into the structure of $Ba_3Co_2Fe_{24}O_{41}$ and consequently, low temperature sintering and good magnetic properties were achieved. The microwave reflectivity levels (dB) of the Z-type barium hexaferrite:polychloroprene composites were determined for the frequency range 8.0 - 16.0 GHz. The permittivity (e) and permeability (m) values were measured by using the transmission/reflection (T/R) method in a waveguide medium. The nanocomposite 80:20 of $Ba_3Co_{0.9}Cu_{1.1}Fe_{24}O_{41}$:CR, 3.0 mm thick, showed the best performance as RAM for the X-band, with a microwave absorption of 99.50 % (reflectivity of - 22.5 dB) in 9.5 GHz, which can be attributed to the increase in the magnetic properties due to the Cu addition.

INTRODUCTION

Radar absorbing materials (RAM) play an important role on the stealth technology, which corresponds to the invisibility of military platforms to the different systems of detection (radar, acoustic, infrared, etc), by suppressing microwaves reflected from metallic structures and so, reducing the radar signatures of the targets.

Recently, a great deal of attention is devoted to hexagonal ferrites as microwave materials for the 1-100 GHz band and a wide range of chemical methods have been used to obtain ultrafine particles. All of these methods require a low-temperature process in order to control the particles' size, as a first synthesis stage. In an attempt either to promote the formation of Z-type

Synthesis and Characterization of $Ba_3Co_2Fe_{24}O_{41}$ and $Ba_3Co_{0.9}Cu_{1.1}Fe_{24}O_{41}$ Nanopowders

ferrite and to improve the magnetic properties, we report the synthesis of these nanostructured materials by the sol–gel–citrate precursor method [1, 2]. As magnetic materials, the barium hexaferrites are not generally replaced by any other magnetic material because they are relatively inexpensive, stable and have a wide range of technological applications. Barium hexaferrites have been classified according to their structures, into five main classes: $BaFe_{12}O_{19}$ (M-type), $BaMe_2Fe_{16}O_{27}$ (W-type), $Ba_2Me_2Fe_{28}O_{46}$ (X-type), $Ba_2Me_2Fe_{12}O_{22}$ (Y-type) and $Ba_3Me_2Fe_{24}O_{41}$ (Z-type), where Me represents a divalent ion from the first transition series.

Z-type barium hexaferrite is a promising material for applications as RAM in the frequency of GHz, which require high permeability, great resistivity and good chemical and thermal stabilities. $(Co-Cu)_2Z$ barium hexaferrite is a new type of soft magnetic compound, which presents these characteristics and a ferromagnetic resonance in the GHz frequency, being useful for inductor cores or in UHF communications, in the microwave region [3, 4].

In the conventional ceramic method, a high sintering temperature is necessary to obtain this Z-type hexaferrite due to the complex crystalline structure.

By using chemical methods, the calcination temperature can be reduced and the introduction of metallic ions makes possible the use of these ferrites as microwave absorbers in different frequency ranges, simply by varying the degree of substitution [5]. In this work, the citrate sol-gel process under inert atmosphere was used to obtain $Ba_3Co_2Fe_{24}O_{41}$ and $Ba_3Co_{0.9}Cu_{1.1}Fe_{24}O_{41}$ nanopowders. The introduction of Cu^{2+} ions in the structure of $Ba_3Co_2Fe_{24}O_{41}$ can also reduce the sintering temperature, because it can acts as a flux, due to its low melting point (1084.62 °C) [6].

Composites of $Ba_3Co_2Fe_{24}O_{41}$ and $Ba_3Co_{0.9}Cu_{1.1}Fe_{24}O_{41}$ with polychloroprene (CR) were obtained for the microwave absorption measurements, for the frequency range: 8.0 – 12 GHz (X-band) and 12 – 16 GHz (Ku-band). The magnetic properties of these materials are largely dependent of the sample microstructure.

X-ray diffraction (XRD), X-ray fluorescence (XRF), thermal analyses (TGA/DTA), and the vibrating sample magnetometer (VSM) were used to characterize the synthesized material.

The microwave measurements were based on the transmission/reflection method (T/R) using rectangular waveguides as the confining medium for the samples [7].

EXPERIMENTAL

Nanosized $(Co-Cu)_2Z$ structured powders were synthesized by the citrate precursor method using reagent grade $Fe(NO_3)_3.9H_2O$, $Ba(NO_3)_2$, monohydrate citric acid, $Co(NO_3)_2.6H_2O$ and $Cu(NO_3)_2.3H_2O$ in stoichiometric molar ratios to obtain $Ba_3Co_2Fe_{24}O_{41}$ and $Ba_3Co_{0.9}Cu_{1.1}Fe_{24}O_{41}$ hexaferrites. The solids were weighing and placing then into appropriate closed vessels subjected to a super dry nitrogen atmosphere to obtain the precursor solutions [8]. Distilled water was added under agitation, until total dissolution of solids.

The solutions were then transferred to a previously evacuated flask and mixed under super dry nitrogen operating as a reflux condenser, with intensive stirring. The resulting mixture was heated to 80 °C to complete the reaction under reflux, in order to keep the inert atmosphere and to allow subsequent additions of ammonium hydroxide (NH_4OH), added drop wise into the solution to render it neutral or slightly alkaline (pH 7.0 - 8.0), for subsequent precipitation of the organo metallic complex [8].

Predried ethanol was previously added drop wise under vigorous stirring into the reaction mixture, to promote the precipitation of a complex citrate gel of barium, iron, copper and cobalt.

Drying at 60 °C, leaving behind the desired solid phase, the remaining aqueous solution was eliminated. The ideal temperature for the citrate gel decomposition was determined by thermogravimetric analysis (TGA) and differential thermal analysis (DTA). Based on the results of the thermal analyses, the batch of dried solid was calcined inside a muffle furnace. TGA and DTA measurements were carried out in a TA Instruments SDT-2960. The experiments were carried out in static air, using platinum crucibles between 20 and 1,000 °C, with a heating rate of 10 °C min^{-1}.

The calcination was performed using the following heating schedule: 2 °C/min up to 410 °C, establishing a plateau for 1 hour, 10 °C/min up to the final sintering temperature with a residence time of 4 h at the sintering temperature. The material was then cooled to room temperature at a rate of 10 °C/min.

Then, X-ray fluorescence (XRF) measurements were carried out on a Philips model PW 2,400 sequential spectrometer. This quantitative method was used to determine the stoichiometry of the ferrite samples, which were analyzed in the form of a fused bead, using lithium tetraborate flux.

For the powder X-ray diffraction (XRD) analysis, the material was placed on a glass sample holder and spread out to form a thin layer. A Siemens AXS D5005 diffractometer with a dwell time of 1 °/min, in the θ-2θ Bragg-Brentano geometry, was employed.

The magnetic hysteresis loops were obtained using the vibrating sample magnetometer VSM 4,500 PAR.

The Z-type barium hexaferrite structure is illustrated in Figure 1.

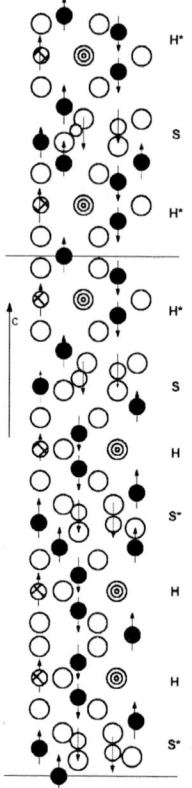

Synthesis and Characterization of Ba$_3$Co$_2$Fe$_{24}$O$_{41}$ and Ba$_3$Co$_{0.9}$Cu$_{1.1}$Fe$_{24}$O$_{41}$ Nanopowders

Figure 1: Z-type barium hexaferrite structure, where: S = spinel block, H = hexagonal block, ○ = cation in tetrahedral sites, ● = cation in octahedral sites, ↓↑ = magnetic momentum direction, ○ = O^{2-}, ◎ = Ba^{2+}, ⊘ = hexahedral sites.

In order to obtain the composites for the measurements of the microwave absorbing properties, the powders were mixed with polychloroprene (CR), resulting in the composition 80:20 (wt. %, ferrite:polychloroprene). The processing was carried out in a Berstorff two roll mill, at room temperature, with velocities of 22 and 25 rpm (back and forward). Vulcanized samples with 8.0 x 4.0 cm and 3.0 mm thick were obtained by compression moulding in a hydraulic press at 150 °C and 6.7 MPa. The vulcanization times were determined by the data obtained in the Monsanto Rheometer TM100 [7]. The dispersion of the magnetic particles in CR was evaluated by SEM, using a ZEISS DSM 940A microscope and the morphological study was performed by a Topometrix II® atomic force microscope.

Synthesis and Characterization of $Ba_3Co_2Fe_{24}O_{41}$ and $Ba_3Co_{0.9}Cu_{1.1}Fe_{24}O_{41}$ Nanopowders

The microwave measurements conducted in this work were based on the Transmission/Reflection method (T/R) using rectangular waveguides as the confining medium for the samples. The microwave absorption of the sheet composites was illustrated by variations of reflectivity (dB) versus frequency (GHz), using the HP 8510 network analyzer system. The materials were analyzed for the frequency range from 8.0 to 16.0 GHz (X-Ku bands) [7; 9-13].

RESULTS AND DISCUSSION

Figure 2 illustrates TGA and DTA curves of the precursor gel. From the DTA curve analysis, no significant weight loss occurred above 400 °C, indicating that any remanent organic material had already been completely eliminated [9]. Based on this result, the gel was calcined at 410 °C.

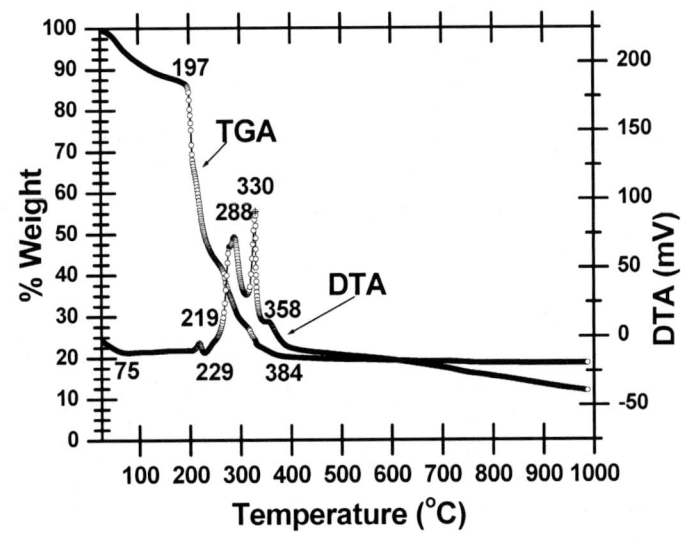

Figure 2: TG-DTA curves for the Z-type precursor gel of $Ba_3Co_{0.9}Cu_{1.1}Fe_{24}O_{41}$ powder.

The thermal behavior of the organic precursor showed that the weight losses from 25 to 197 °C and from 197 to 384 °C were respectively illustrated by an endothermic and four exothermic changes in the DTA curve. The evaporation of water was observed for temperatures below 75°C.

The decomposition of the unreacted citric acid should occur between 150°C - 229°C. The organic complex citrate decomposition occurred between 229°C - 384°C. In total, it resulted in about 82 % loss in weight from the initial temperature up to 400°C.

XRD results in Figure 3 indicate that at 950 °C, the Z-type phase was clearly the majority phase, according to JCPDS 19-0097. However, according to Pullar et al [11]., this material contains small undetectable amounts of Y-type phase ($Ba_2Co_2Fe_{12}O_{22}$), coexisting with the Z-type phase.

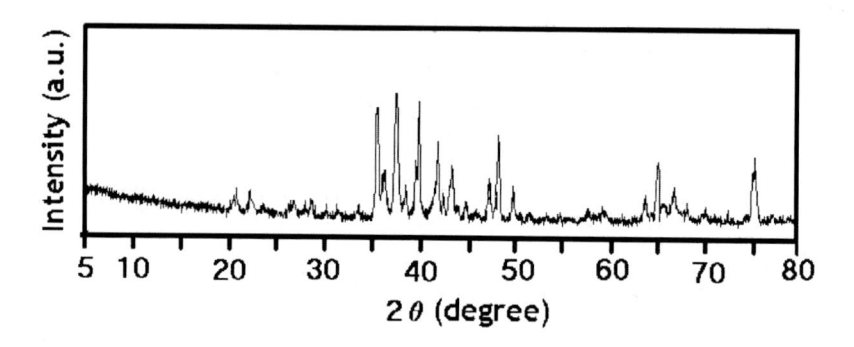

Figure 3: X-Ray diffraction curve for the Z-type powders calcined at 950 °C.

Results from X-ray fluorescence analysis curve for the Z-type powders calcined at 950 °C are shown at Table 1. The theoretic and experimental values are very similar, which indicate the absence of carbon (complete organic precursor elimination) and that the synthesized powders achieved the planned stoichiometry.

Table 1: Chemical compositions of $Ba_3Co_2Fe_{24}O_{41}$ and $Ba_3Co_{0.9}Cu_{1.1}Fe_{24}O_{41}$.

Z-type barium hexaferrite	Chemical Compound							
	BaO		CoO		Fe_2O_3		CuO	
	Theoretic	Experimental	Theoretic	Experimental	Theoretic	Experimental	Theoretic	Experimental
$Ba_3Co_2Fe_{24}O_{41}$	18.2094	18.21	5.9326	5.93	75.8580	75.89	—	
$Ba_3Co_{0.9}Cu_{1.1}Fe_{24}O_{41}$	18.1729	18.17	2.6643	2.66	75.7059	75.71	3.4569	3.46

The hysteresis curves for Z-type barium hexaferrites fired at 950 °C are illustrated in Figure 4. The magnetization curves show typical feature of magnetically soft ferrites. The saturation magnetization M_s was obtained by extrapolating M(1/H)- curves to 1/H = 0, resulting in the value of 57.8 emu/g for $Ba_3Co_{0.9}Cu_{1.1}Fe_{24}O_{41}$ and 52.3 emu/g for $Ba_3Co_2Fe_{24}O_{41}$.

Synthesis and Characterization of $Ba_3Co_2Fe_{24}O_{41}$ and $Ba_3Co_{0.9}Cu_{1.1}Fe_{24}O_{41}$ Nanopowders

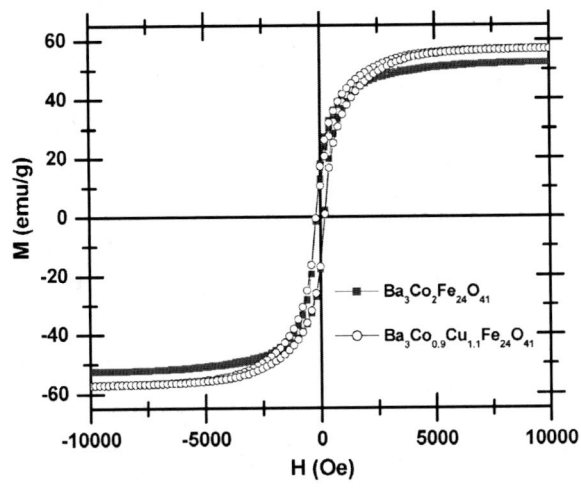

Figure 4: Magnetic hysteresis curves of $Ba_3Co_2Fe_{24}O_{41}$ and $Ba_3Co_{0.9}Cu_{1.1}Fe_{24}O_{41}$ powders.

In the Z-type barium hexaferrites, the metallic ions such as Fe^{3+}, Co^{2+} and Cu^{2+} are located in different sites such as octahedral and tetrahedral ones. Co^{2+} ions show strong magnetocrystalline anisotropy and its substitution by Cu^{2+} ions may result in the improvement of the magnetic properties.

The ionic radius of Cu^{2+} (0.085 nm) is very near to that of Co^{2+} (0.082 nm) and larger than Fe^{3+} (0.067 nm), as show in Table 2.

Table 2: Ionic radius, coordination, magnetic momentum/ion and alignment in Z-type hexaferrite.

Ion	Ionic radius (nm)	Coordination	Magnetic Momentum/ion	Magnetic Momentum alignment
Fe^{3+}/Fe^{2+}	0.067 (0.090)	Tetrahedral/decahedral	5/4	↑↑
Co^{2+}	0.082	Octahedral	3	↑↑
Cu^{2+}	0.085	Octahedral/decahedral	1	↑↓
Ba^{2+}	0.143	Oxygen site	-	-
O^{2-}	0.132	Oxygen site	-	-

The copper ions shall occupy the octahedral sites in the structure of Z-type hexaferrite and substitute partially cobalt ions. The Cu^{2+} ions distort the crystalline field due to their electronic configuration, and this behavior results in an increase of the M_s [5].

Synthesis and Characterization of $Ba_3Co_2Fe_{24}O_{41}$ and $Ba_3Co_{0.9}Cu_{1.1}Fe_{24}O_{41}$ Nanopowders

The introduction of Cu^{2+} ions promotes the partial substitution of the Fe-Fe strong magnetic interations by the weak ones of Co-Fe, as a result it can be observed a decrease of the initial permeability value.

The introduction of Cu^{2+} ions also promotes the reduction of the calcination temperature of the Z-type barium hexaferrite and the probability of the reactions (1), (2) e (3) to occur:

$$Fe^{3+} \rightarrow Fe^{2+} + e^- \qquad (1)$$

$$Cu^{2+} + e^- \rightarrow Cu^+ \qquad (2)$$

$$O^{2-} \rightarrow O + 2\,e^- \qquad (3)$$

Figure 5 shows the AFM micrograph of $Ba_3Co_{0.9}Cu_{1.1}Fe_{24}O_{41}$ magnetic hexagonal particles with nanometric size (230 x 100 nm), while Fig. 6 illustrates the good dispersion of these nanoparticles in polychloroprene matrix (80:20, wt. %).

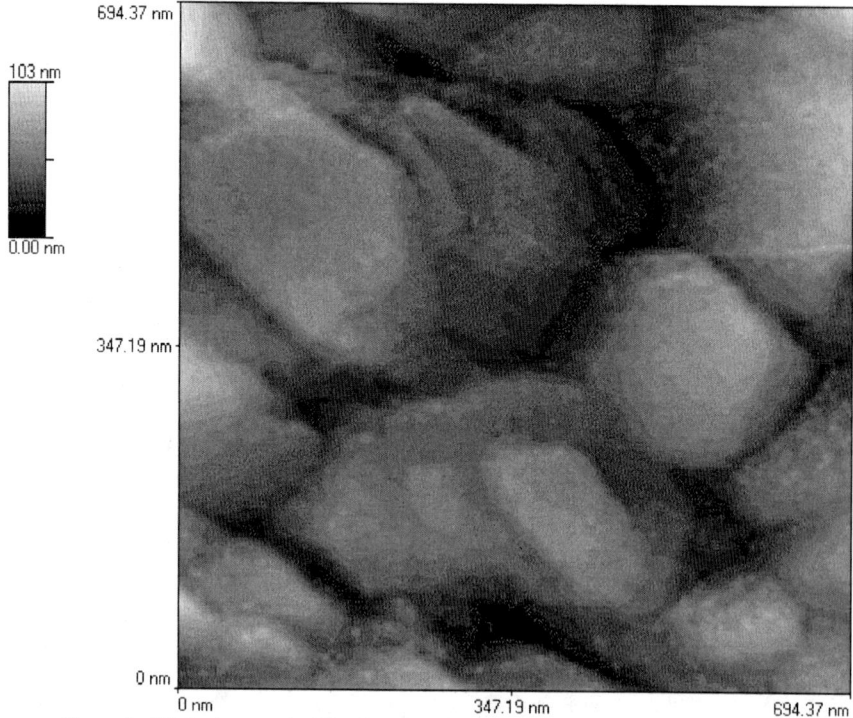

Figure 5: AFM micrograph of $Ba_3Co_{0.9}Cu_{1.1}Fe_{24}O_{41}$ calcined at 950 °C (100,000 X; 200 kV).

Synthesis and Characterization of $Ba_3Co_2Fe_{24}O_{41}$ and $Ba_3Co_{0.9}Cu_{1.1}Fe_{24}O_{41}$ Nanopowders

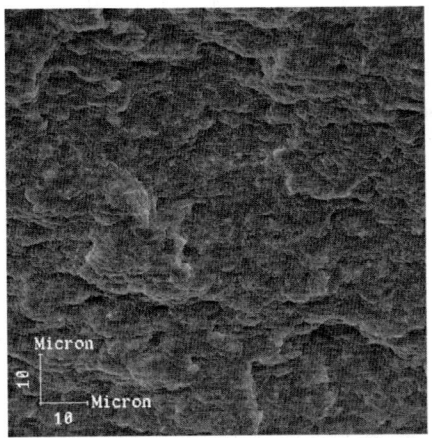

Figure 6: SEM micrograph of $Ba_3Co_{0.9}Cu_{1.1}Fe_{24}O_{41}$:CR, magnification of 1,000.

From the SEM image, it can be seen that rubber mixing is a very good technique for the dispersion of the Z-type hexaferrites particles, in spite of the high weight concentration used (80 %) and the tendency of these particles to form magnetic agglomerates [7].

The effect of Cu addition on the reflectivity measurements can be seen in Fig. 7.

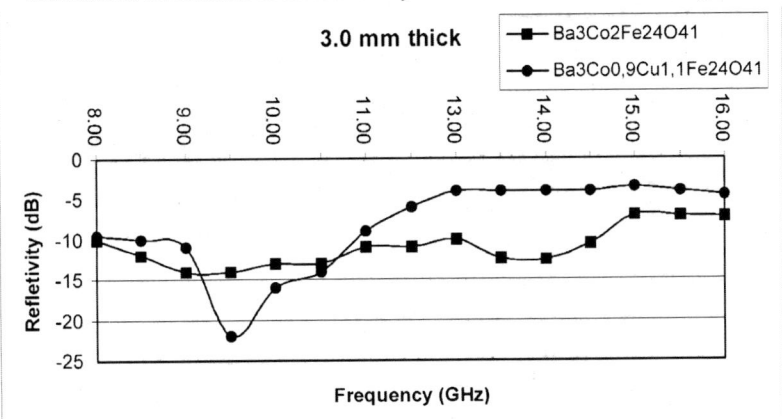

Figure 7: Effect of Cu addition on the reflectivity measurements for the composites 80:20 of $Ba_3Co_2Fe_{24}O_{41}$:CR and $Ba_3Co_{0.9}Cu_{1.1}Fe_{24}O_{41}$:CR, 3.0 mm thick.

The greatest microwave absorption of 99.50 % in 9.5 GHz for the $Ba_3Co_{0.9}Cu_{1.1}Fe_{24}O_{41}$:CR 3.0 mm thick nanocomposite, can be attributed to the addition of Cu, resulting in an increase of the magnetic properties.

CONCLUSION

The citrate sol-gel method promoted the formation of nanocrystalline Z-type hexaferrite at a lower temperature (950 °C) than that employed by the conventional one (1,200 - 1,300 °C). The introduction of Cu^{2+} ions improved the magnetic properties of this ceramic illustrated by the increase of 5.5 emu/g in the Ms. As a result, the nanocomposite 80:20 of $Ba_3Co_{0.9}Cu_{1.1}Fe_{24}O_{41}$:CR 3.0 mm thick, showed the best performance as RAM for the X-band, with a microwave absorption of 99.50 % (reflectivity of - 22.5 dB) in 9.5 GHz.

ACKNOWLEDGEMENTS

The authors thank FAPERJ and CNPq for the financial support.

REFERENCES

[1] D. C., Cramer, *Overview of Technical, Engineering, and Advanced Ceramics*. In: *Engineered Materials Handbook: Ceramics and Glasses*, ASM International, New York, USA, 1998, pp. 16 - 20.

[2] B., Deconihout, C., Pareige, P., Pareige, D., Blavette, A., Menand, "Tomographic Atom Probe New Dimension in Materials Analysis", *Microscopy and Microanalysis*, **5**, 39 - 47 (1999).

[3] H., Zhang, J., Zhou, Z., Yue, P., Wu, Z., Gui, L., Li, "Synthesis of Co_2Z Hexagonal Ferrite with Planar Structure by Gel Self-Propagating Method", *Materials Letters*, **43**, 62 - 65 (2000).

[4] Z., Wang, L., Li, S., Su, Z., Gui, Z., Yue, J., Zhou, "Low-Temperature Sintering and High Frequency Properties of Cu-Modified Co_2Z Hexaferrite", *J. Eur. Cer. Soc.*, **23**, 715 - 720 (2003).

[5] H., Zhang, J., Zhou, L., Li, Z., Yue, Z., Gui, "Dielectric Characteristics of Novel Z-Type Planar Hexaferrite with Cu Modification", *Materials Letters*, **55**, 351 - 355 (2002).

[6] V. R., Caffarena, D. Sc. Thesis, *Study of the Magnetic and Microwave Absorber Properties of Z-Type Barium Hexaferrite Obtained by Citrate Precursor Method*, PEMM-UFRJ, Rio de Janeiro, Brazil, 2004.

[7] M. S., Pinho, M. L., Gregori, R. C. R., Nunes, B.G., Soares, "Performance of Radar Absorbing Materials by Waveguide Measurements for X- and Ku-Band Frequencies", *Eur. Pol. J.*, **38**, 2321 - 2327 (2002).

[8] T., Ogasawara, M. A. S., Oliveira, "Microstructure and Hysteresis Curves of the Barium Hexaferrite from Co-Precipitation by Organic Agent", *J. Magn. Mag. Mat.*, **217**, 147 - 154 (2000).

[9] V. R., Caffarena, T., Ogasawara, J. L., Capitaneo, M. S., Pinho, *Synthesis of (Co-Zn)-Z-Type Barium Hexaferrite for use as Microwave Absorber*. In: *Applied Mineralogy: Developments in Science and Technology*, M. Pechio, F. R. D. Andrade, L. Z. D'Agostino, H. Kahn, L., M.. Sant'Agostino, M. M. M. L. Tassinari (Eds.), Fundação Biblioteca Nacional, São Paulo, Brazil, 2004, pp. 49-52.

[10] D., Austissier, A., Podembski, D. C., Jacquiod, "Microwaves Properties of M and Z Type hexaferrites", *J. Ph.*, **IV C**, 1409-1412 (1997).

[11] R. C., Pullar, "The Manufacture and Characterization of Aligned Fibres of the Ferroxplana Ferrites Co_2Z, 0.67 % CaO-Doped Co_2Z, Co_2Y and Co_2W", *J. Magn. Mag. Mat.*, **186**, 313-325 (1998).

[12] V. R., Caffarena, T., Ogasawara, J. L., Capitaneo, M. S., Pinho, "Magnetic Properties of Z-Type $Ba_3Co_{1.3}Zn_{0.3}Cu_{0.4}Fe_{24}O_{41}$ Nanoparticles", *Materials Chemistry and Physics*, 101, 81 - 86 (2006).

[13] W. B., Weir, "Automatic Measurement System for a Multichannel Digital Tuned Bandpass Filter", *IEEE Trans. Inst. Meas.*, **IM 23**, 140-148 (1974).

MICROWAVE ENHANCED ANISOTROPIC GRAIN GROWTH IN LANTHANUM HEXA ALUMINATE-ALUMINA COMPOSITES

Z. Negahdari, T. Gerdes, M. Willert-Porada
Chair of Materials Processing
Faculty of Applied Science, University of Bayreuth
D-95447 Bayreuth, Germany

ABSTRACT

This paper summarizes results of a study to control the microstructure evolution in alumina/ lanthanum hexa-aluminate composites obtained by application of microwave assisted sintering (2.54 GHz) to commercial grade α–alumina with different concentration of $LaAlO_3$ as additive. Based on measurement of phase compositions, porosity, grain growth and densification elongated grain growth of α–alumina and La-aluminate segregation was investigated. By comparison with results of conventional sintering, evidence is found for an influence of the microwave field on transport phenomena related to formation of La-beta alumina and on the enhancement of elongated alumina grain growth.

INTRODUCTION

Lanthanum-hexaaluminate, $LaAl_{11}O_{18}$, abbreviated as LHA, is used as a coating to adjust bonding between an oxide fiber and alumina matrix[1,2] and also for platelet in-situ reinforcement in Al_2O_3-ceramic matrix composites[3-5]. Formation of $LaAl_{11}O_{18}$, with a magnetoplumbite structure, by the solid state reaction of $LaAlO_3$ and Al_2O_3, is very slow and requires higher temperatures as compared to alumina itself and hours of dwelling at sintering temperature.[6]

Enhancement of mass transport and solid-state reaction rates during microwave heating and processing, which is broadly called as "microwave effect", results in the modification of ceramics microstructures and development of materials with unique combinations of properties.[7,8] Utilization of the "microwave effect" for strengthening of polycrystalline alumina ceramics using La-aluminate additives is a new approach.[9] Lanthanum has a strong potential to modify equiaxed alumina grains to anisotropic elongated grains. This elongated morphology is believed to be due to preferential segregation of ions to basal planes (0001) in alumina grain boundaries.[10] In situ enhanced anisotropic growth of elongated alumina grains during sintering of $LaAlO_3$ doped alumina is expected to toughen the ceramic matrix. In addition, an increase in the mechanical strength could be achieved because of grain growth inhibition in the Al_2O_3-matrix by grain boundary pinning from segregation of lanthanum aluminates in grain boundaries.

EXPERIMENTAL PROCEDURE

Commercially available alumina A16 (ACC-ALCOA) with 99.8% purity, BET surface area 8–11 m^2/g and the average particle size 0.3 μm and lanthanum aluminate ($LaAlO_3$, Alfa Aesar) with 99.9% purity, BET surface area 5 m^2/g were used as starting materials. The mixtures of powders adjusted by weight to Al_2O_3 /2.8 vol.% $LaAl_{11}O_{18}$ and Al_2O_3 /28 vol.% $LaAl_{11}O_{18}$ and PVA dissolved in H_2O (3 wt.% *of powder*) as binder were dispersed in ethanol and stirred for approximately 5 h. The solvent was removed by evaporation, and the powder mixture was further dried with an in house build microwave rotary evaporator at ambient pressure, followed by vacuum drying at 80°C and 10 mbar. The dried soft agglomerated powders were crushed and screened through a 45 mesh sieve. Cylindrical pellets of 13 mm in diameter and 0.8 mm height were prepared by uni-axial cold pressing at 300 MPa.

Before sintering, the binder was removed by heating at 2°C/min to 800°C and dwelling for 3 h in a conventional furnace. The average relative density of the pellets after binder removal was ~60% theoretical density, as determined from the dimensions and weight of at least 20 samples.

Samples were sintered conventionally in air (Nabertherm HT 08/17 High Temperature furnace heated with $MoSi_2$ heating elements) at various temperatures, ranging from 1500° to 1600°C for 1h. Microwave Sintering was carried out in a in-house built multimode 6 kW, 2.45 GHz microwave furnace. The heating and cooling rate for both conventional and microwave sintering was maintained at 15°C/min. The heating rate of the oxide ceramic parts in the microwave furnace is unknown below 900°C, due to utilization of microwave absorbing SiC-rods for preheating the oxide ceramic to a temperature, where sufficient microwave absorption of the oxide ceramic is achieved.

The input power for microwave heating was adjusted manually to control the sintering temperature and heating rate. The samples were placed in an insulating casket made from aluminosilicate fibers board. Figure 1 shows the design of the insulating casket for microwave sintering. The top cover, equipped with SiC rods, was used as a microwave absorber. The SiC rods were used to preheat the alumina samples, because alumina is a poor absorber up to ~900°C, when microwave radiation with 2.45 GHz frequency is used. As soon as the necessary temperature is reached, the top cover of the casket was replaced by an aluminosilicate fiber board with no additional heating elements. All temperature measurements in the microwave furnace were obtained by means of an infrared pyrometer (Keller HCW, PZ20), with an emissivity value for the samples set 0.6. As view port, a hole of 1 cm diameter was drilled into the top cover of the casket and covered with quartz to avoid heat loss by convection. Since it is difficult to determine the accuracy of temperature measurement with the infrared pyrometer, all results related to microstructure and phase composition are compared according to relative density (ratio of bulk density to theoretical density) of the ceramic.

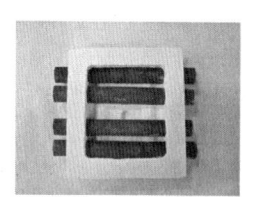

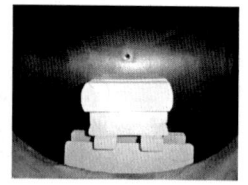

Fig.1. Microwave transparent casket designed for microwave assisted sintering.

The bulk density was measured via the Archimedes method, using water as immersion liquid. The theoretical density values for Al_2O_3/2.8 vol.% $LaAl_{11}O_{18}$: 3.985 g/cm^3 and Al_2O_3/28 vol.% $LaAl_{11}O_{18}$ composite: 4.037 and g/cm^3, respectively, are estimated by the rule of mixture from the true density of each compound (Al_2O_3: 3.98 g/cm^3, $LaAl_{11}O_{18}$: 4.17 g/cm^3). Microstructure observations using scanning electron microscopy (SEM, JEOL, JSM-840) were made on cross sections of polished and thermally etched (in air at 100°C below the sintering temperature for 20 min) samples. Mercury porosimetry was used to measure changes in pore size distribution of conventional and microwave sintered samples. X-ray diffractometry (XRD, Philips X'Pert PW3040) and backscattered electron (BSE) imaging in a scanning electron microscope (SEM) were used to characterize the phase composition and elemental distribution of the composites.

RESULTS AND DISCUSSION

In Figure 2 a comparison of microstructures for Al_2O_3/2.8 vol.% LHA and Al_2O_3/28 vol.% LHA composites is shown, as function of density and sintering method, for conventional and microwave assisted sintering at various temperatures for 1 h dwell time.

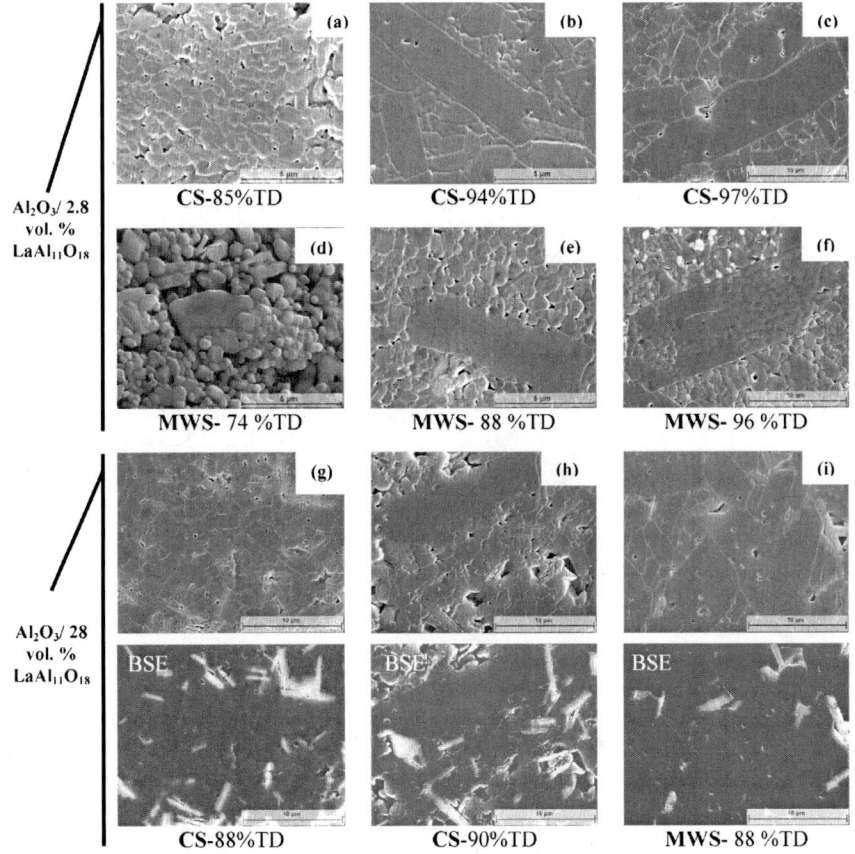

Fig. 2. Microstructure upon conventional (CS) and microwave (MWS) sintering of Al_2O_3/ 2.8 vol.% $LaAl_{11}O_{18}$ composites: CS:(a-c) and MWS:(d-f), and of Al_2O_3/28 vol.% $LaAl_{11}O_{18}$ composites: CS (g), (h) and MWS:(i).

The solid state reaction between LaAlO$_3$ and alumina yields platelet grains of the ß-alumina structure lanthanum hexa-aluminate, visible in BSE-mode as the lighter phase because of the increased content of the heavy element lanthanum. It is clearly visible, that LHA formation and elongated alumina grain growth start at much lower density of around 75% theoretical density in the microwave sintered Al$_2$O$_3$/2.8 vol.% LHA composite as compared with conventional sintering. The change in the aspect ratio at higher densities is characteristic of the increase in grain boundary diffusion at the late intermediate to final stages of densification. At the higher LaAlO$_3$ content of 28vol% besides elongated grains much more magnetoplumbite La-hexa aluminat platelets are visible as compared to the 2.8 vol%-LaAlO$_3$-Al$_2$O$_3$-composite. Elongated grains embedded in a matrix made up from small alumina grains are formed upon conventional sintering at 90% TD, whereas no small grains are left in the microwave sintered sample already at 88%TD.

In the reaction between LaAlO$_3$ and Al$_2$O$_3$, two possible diffusion mechanisms can be considered. It may be La^{3+} or Al^{3+} diffusion, accompanied by oxygen exchange through the gas phase.[6] Based on experimental observations,[9, 11-12] microwave electric fields could induce a nonlinear driving force for charge transport, which in oxide ceramics with a high ion mobility could influence reaction kinetics by enhancing mass transport rates in solid-state reaction. Because formation of La-hexa-aluminate needs only a small amount of La^{3+}-ions to be incorporated into the alumina lattice, LHA-formation could be a very sensitive solid state reaction to detect a "microwave effect", due to the dielectric properties of lanthanum aluminate grains. Enhancement of LHA formation upon microwave sintering would be seen by a decreased relative peak intensity ratio of the planes (104) of alumina to (017) of LHA, as in conventionally sintered samples of comparable density (Figure 3).

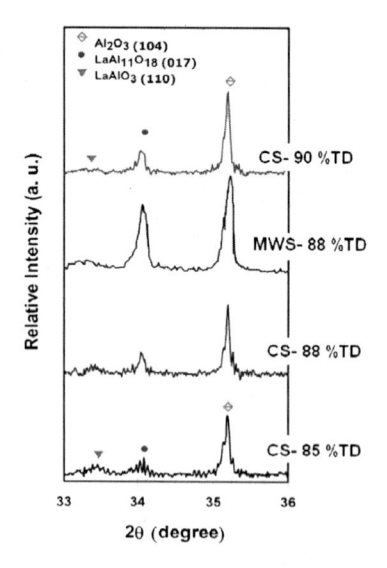

	$I_{Al_2O_3(104)} / I_{LaAl_{11}O_{18}(017)}$
CS- 85 % TD	4.5
CS- 88 % TD	3.7
CS- 90 % TD	3.5
MWS- 88 % TD	1.4

Fig. 3. Diffraction patterns of the region from 33° to 36° 2θ for conventional (CS) and microwave (MWS) sintered Al$_2$O$_3$/ 28 vol.% LaAl$_{11}$O$_{18}$ composites as function of density. JCPDS cards used for identification were aluminum oxide (46-1212), lanthanum aluminum oxide (33-0699), and aluminum lanthanum oxide (31-0022).

Microwave Enhanced Anisotropic Grain Growth

The pore-size distribution of the microwave and conventional sintered Al$_2$O$_3$/2.8 vol.% LHA is illustrated in Figure 4. The total porosity and the average pore size decrease, while the sintering temperatures (or density) increases.

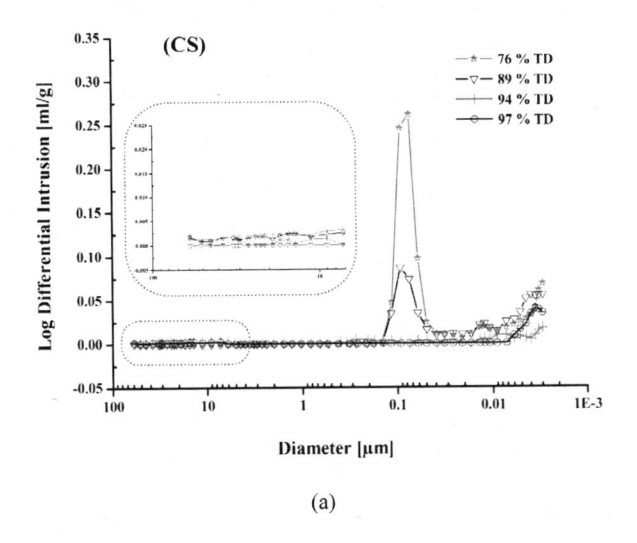

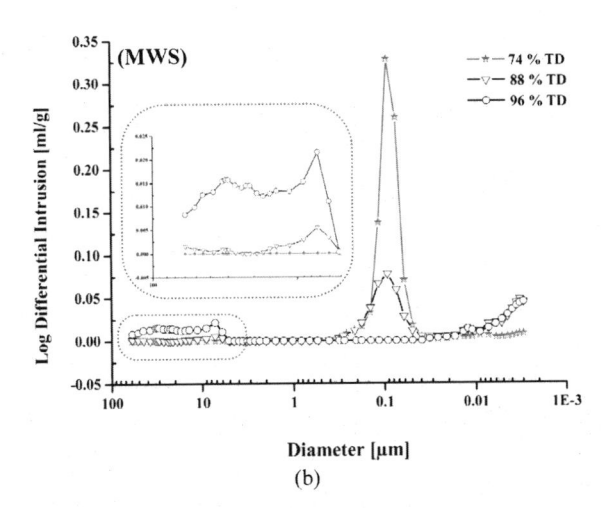

(a)

(b)

Fig. 4. Comparison of pore-size distribution for (a) conventional, (b) microwave sintered Al$_2$O$_3$/2.8 vol. % LHA composite as function of density.

It is evident that the pore coalescence and pore removal occur at different density levels upon microwave sintering as compared with conventional sintering. The originally mono-modal distribution of pore-size remains as such upon CS up to 97%TD, with the average size shifted from 100 nm to <10 nm and a significant total pore volume reduction. In case of MWS, the distribution changes at 88%TD and 96%TD into a bi-modal, and in addition to small open pores of $\Phi \approx 5$ nm some larger open pores at $\Phi \approx 7$ μm appear. For the CS-samples this population of pores has a very small volume, close to the detection limit, as visible in the insert of Figure 4a.

The enlargement of pore channels, determined by mercury porosimetry for the samples sintered by microwave to 88%TD and 96%TD suggests the development of void spaces in the late intermediate to final stages of sintering.

Pores migrate through the microstructure by random walk. This process will increase the probability of pore coalescence, driven by reduction of interfacial energy if no other processes, like e.g., solid state reactions with a significant change in the shape of grains occur. General knowledge of the coalescence of pores indicates, that the mechanism proceeds with grain-boundary motion at the late intermediate to final stages of densification: i.e., at the stage, in which the coarsening kinetics of the grains dominates the densification process.[13-17] The development of this structure and the formation of pores with mean pore size of 10 μm during microwave sintering at much lower density level as compared to CS are supposed to be an indirect consequence of enhanced anisotropic alumina grain growth in lanthanum beta alumina-alumina composites. Pores can grow by different mechanisms: surface diffusion, particle size distribution effects, particle coalescence, phase transformation, and evaporation/condensation.[18] Particle coalescence is the most probable mechanism in the case of the alumina/lanthanum-ß Alumina composites.

CONCLUSIONS

It has been demonstrated that upon microwave sintering enhancement of a solid state reaction and anisotropic grain growth in alumina/ in-situ-formed lanthanum hexa-aluminate composites occur. The influence of microwave fields on pore coalescence and pore removal at the late intermediate to final stages of sintering in Alumina-LHA composites has been also confirmed.

Since the remarkable influence of microwave sintering on the microstructures of alumina- lanthanum beta alumina composites opens up a way to alumina ceramics not obtained upon conventional sintering, the aim of future work will be the evaluation of mechanical properties of microwave sintered alumina matrix composites.

ACKNOWLEDGEMENTS

The financial support of Bavarian Science Foundation within the doctoral grant DPA-52/05 is gratefully acknowledged.

REFERENCES

[1] P. E. D. Morgan and D. B. Marshall, "Functional interfaces for oxide/oxide composites," Mat. Sci. Eng. A, **162**, 15–25 (1993).

[2] B.Saruhan, I.R. Abothu, and S. Komarneni, "Crystallization of electrostatically seeded lanthanum hexaluminate films on polycrystalline oxide fibers," J. Am. Ceram. Soc. **83** [12], 3172-78 (2000).

[3] P. l.. Chen and I. W. Chen, "In-Situ Alumina/Aluminate Platelet Composites," J. Am. Ceram. Soc. **75** [9], 2610-12 (1992).

[4] M. Yasuoka, K. Hirao, M. E. Brito, S. Kanzaki, "High-Strength and High-Fracture-Toughness Ceramics in the Al_2O_3- $LaAl_{11}O_{18}$ Systems, " J. Am. Ceram. Soc. **78** [7], 1853-56 (1995).

[5] Y.Wu, Y. Zhang, X. X. Huang, J. K. Guo, " In-situ growth of needlelike $LaAl_{11}O_{18}$ for reinforcement of alumina composites," Ceram. Int., **27** [8], 903-906 (2001).

[6] R. C. Ropp and B. Carroll, "Solid-state Kinetics of $LaAl_{11}O_{18}$," J. Am. Ceram. Soc. **63** [7-8], 416 -419 (1980).

[7] M. Willert-Porada, "Novel aspects of microwave processing of ceramics materials," 501-506 in *Ceramic Processing Science and technology* Edited by H. Hausner, G. L. Messing, and S. I. Hirano, Am. Ceram. Soc., Westerville, OH, Ceram. Trans. **51** (1995).

[8] J. H. Booske, R. F. Cooper, and S. A. Freeman, "Microwave enhanced reaction kinetics in ceramics," Mat. Res. Innovat., **1**, 77-84(1997)

[9] M. Willert-Porada, Z. Negahdari, T. Gerdes, A. Muller, and M. Paneerselvam, "Microwave-material interaction in oxide ceramics with different additives," Advanced in Science and Technology, **45**, 845-850 (2006).

[10] D. Amutha Rani, Y. Yoshizawa, K. Hirao, and Y. Yamauchi, "Effect of Rare-Earth Dopants on Mechanical Properties of Alumina," J. Am. Ceram. Soc., **87** [2], 289–92 (2004).

[11] S. Vodegel, C. Hannappel , and M.Willert-Porada, " Gefügeeinstellung beim Mikrowellen-Sintern von Al_2O_3 ," Metall, **48** [3], 206-210 (1994).

[12] S. Vodegel, *Mikrowellensintern von Aluminiumoxide*, Fortscher. Ber. VDI Reihe 3, Nr. 354 (1994), ISBN 3-18-335403-9.

[13] N. Shinohara and M. Okumiya, "Morphological Changes in Process-Related Large Pores of Granular Compacted and Sintered Alumina," J. Am. Ceram. Soc., **83** [7], 1633–40 (2000).

[14] W. D. Kingery and B. Francois, "The Sintering of Crystalline Oxides, I. Interactions between Grain Boundaries and Pores," 471–98 in *Sintering and Related Phenomena* Edited by G. C. Kuczynski, N. A. Hooton, and G. N. Gibbon. Gordon and Breach, New York (1967).

[15] F. F. Lange, "Powder Processing Science and Technology for Increased Reliability," J. Am. Ceram. Soc., **72** [1], 3–15 (1989).

[16] F. F. Lange and B. J. Kellett, "Thermodynamics of Densification: II, Grain Growth in Porous Compacts and Relation to Densification," J. Am. Ceram. Soc., **72** [5] , 35–41 (1989).

[17] G. Rossi and J. E. Burke, "Influence of Additives on the Microstructure of Sintered Al_2O_3," J. Am. Ceram. Soc., **56** [12], 654-59 (1973).

[18] O. J. Whittemore and J. J. Sipe, "Pore growth during the initial stages of sintering ceramics," Powder Technology, **9** [4], 159-164 (1974).

OXYFUEL COMBUSTION USING PEROVSKITE MEMBRANES

E.M. Pfaff, M. Zwick
Aachen University
Nizzaallee 32
D-52072 Aachen, Germany

ABSTRACT

As part of world wide activities a cooperative research project OXYCOAL-AC at Aachen University (Germany) is aiming at a CO_2-emmision-free coal combustion process for power generation. In the proposed process the coal dust is burned by a mixture of purified re-circulated CO_2 and oxygen. The exhaust gas consists in this case mainly of CO_2 and H_2O which can be easily condensed out so that nearly pure CO_2 remains. All of the CO_2 above the re-circulated amount is then taken out of the process and can be, i.e., liquefied for a long-term storage in the underground of geological formations to prevent its emission in the atmosphere. The core component of this process is the high temperature oxygen transport membrane (OTM) that makes the whole concept economically attractive.
The ceramic partner has to consider all aspects of the process chain respectively as the membrane is the weakest link in this chain the process has also to be adapted to chemical and physical requirements from the material site. Perovskite mixed conductors should have a high ionic conductivity and have to meet the requirements in a power plant.
A main task is to join the ceramic components gas tight against 20 bars to high temperature resistant steel. In order to reduce and control thermal and mechanical stresses special designs have to be calculated by FEM. Brazing alternatives have to be compared to force fitting for tube arrangements and honeycomb structures.

INTRODUCTION

In the course of actions to reduce CO_2-emissions of fossil burning power plants additionally to increasing the efficiency by using higher steam temperatures so called Oxyfuel-Processes are developed intensively. Aim is the CO_2-free coal combustion for power generation. This will be enabled by coal combustion with pure oxygen in a nitrogen free atmosphere. The burning exhausts consist in this case mainly of CO_2 and H_2O which can be easily condensed out so that nearly pure CO_2 remains which can be used or liquefied for a long-term storage in the underground of geological formations.
Cryogenic air fractionation is industrial manageable and estimates for a power plant process a loss of efficiency in the order of 5 – 10 % points[1]. Hellfritsch et al.[2] presented 2004 a brown coal fired oxyfuel-process using cryogenic oxygen with an efficiency factor of 40.6 %. Building a pilot plant with a thermal power of 30 MW Vattenfall Europe AG will establish for the first time a cryogenic oxygen generation at an industrial scale. But it was pointed out that furthermore a major R&D effort is necessary.
Oxygen generation by oxygen transport membranes OTM is subject of intensive research activities because this technology promises a loss of efficiency in the order of 2–5% points only[1]. OTM operate at high temperatures and make them ideally suited for direct integration with coal gasification plants[3]. In the cooperative European research project AZEP amongst others the

companies Alstom Power and Norsk Hydro investigated the oxygen generation by OTM for gas turbines. EnCap another European cooperative research project investigates the supply of oxygen by membranes for coal gasification power plants as well as for oxyfuel processes.

OXYCOAL-AC

The German cooperative project OXYCOAL-AC accompanied by industrial partners like RWE Power AG, E.ON Energie AG, Siemens AG Power Generation and Linde AG has the objective to develop a CO_2-emission-free coal burning process for energy generation. In the proposed process the coal dust is burned by a mixture of purified re-circulated CO_2 and oxygen whereas the oxygen will be generated at temperatures up to 850 °C by a ceramic membrane from a compressed air stream of about 20 bars, see figure 1.
The core component of this process is the high temperature oxygen transport membrane (OTM) which is built by Perovskite mixed ionic/electric conductors[4]. The advantages of this process are on the one hand an easy separation of minor components out of the exhaust gas and very low NO_x-emissions on the other hand the oxygen generation by OTM is thermodynamically convenient. Also a re-fitting of older plants is possible.
In a multidisciplinary effort the whole process is under construction including ceramic membranes and optimization of burners as well as the firing process, development of turbo components and an all-embracing process control. For this six research institutes of Aachen University mechanical engineering department are working together with different tasks.

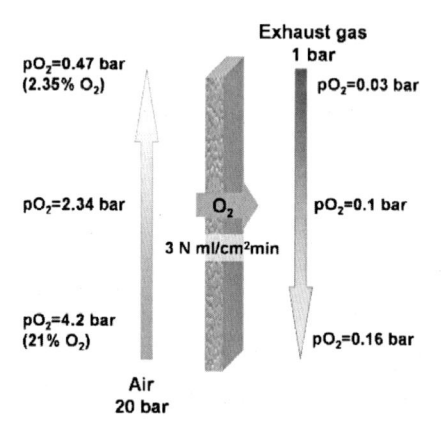

Figure 1. Schematic diagram of the oxygen enrichment

The main facilities are schematically presented in figure 2 and comprised a combustion chamber, the OTM-module, an air compressor combined with a nitrogen turbine, a turbo component to circulate the exhaust gas and to charge the OTM-module, hot gas filtration and the steam generator. Of cause are in such a process adequate static and dynamic process simulations essential for an optimal integration and developing of the components.
The combustor in a 100kW coal dust combustion facility and the burner were modified for the coal combustion in CO_2/O_2-atmosphere. Because of the higher molar mass of CO_2 compared to

N_2 a longer flame with lower flame velocity occurred at comparable conditions. The combustion is measured and simulated[5].

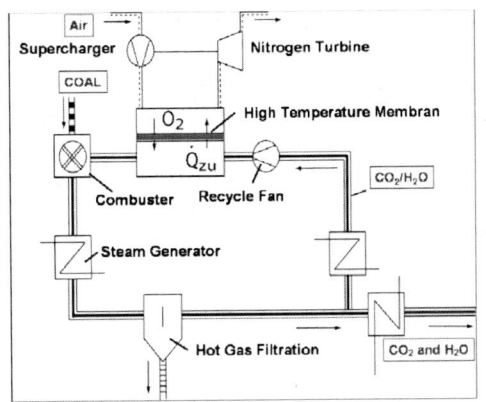

Figure 2. Schematic presentation of the OXYCOAL-AC process

Based on the use of an OTM-module and the linked recirculation of exhaust gas partials new turbomachinery are necessary[6]. In particular the combination of mass flow, temperature und pressure conditions enforces a new design or at least modification of commercial available machines.

OTM-MODULE
A main target of the described research activities is the design and buildup of a membrane module based on available materials and element shapes. No material development is planed. In principle planar and tubular executions are possible and have to be compared concerning feasibility, reliability and costs. Perovskite mixed ionic/electronic conductors (MIEC) should have a high ionic conductivity at the above mentioned conditions and have to resist the requirements in a power plant. As the oxygen flux increase with decreasing wall thickness thin film technologies are taken into consideration. Based on a porous support with an intermediate layer a dense membrane layer of about 25 – 100 µm is coated i.e. by slurry deposition. On the membrane an additional layer for surface optimization can be deposed. Supports can be extruded or isostatic pressed as tubes or cast as plates and will than co-fired with the deposed layers. Also honeycombs are possible. Alternatives are monolithic dense components of thin wall thickness because of their higher resistance against erosion. Furthermore a thin wall is beneficial for heat flux because of the low heat conductivity of perovskite materials which additionally decreases with porosity.
This leads over to other aspects in designing a reliable working OTM-module. Main tasks of the engineering point of view are the analysis and optimization of force flow, heat flow and mass flow as well as load stresses, joint stresses and residual stresses. In a first concept a module with one side closed tubes was favored, see figure 3. Tubes are easy to produce, to be coated and to be jointed. A design study of Vente[7] comes to the conclusion that a tubular system is the optimal choice for all considered conditions. Calculations indicate that multi-channel monoliths

(honeycombs) and tube-and-plate concepts do not have major advantages in terms of specific area over a simple single-hole-tube concept.

Because the perovskite tubes are brazed an upright configuration is used to reduce tensile stresses and creep caused in the weight of the tubes. The tubes are fixed only on one side on a base plate. Therefore each tube is brazed on a joining socket and then welded by electron beam into the base plate. Base plate and socket are of alloy 602CA which was found as best material for brazing as stated below.

The exhaust gas will be discharged through a metallic inside tube to the bottom of the perovskite tube and streams back to the outgas floor. The compressed air streams on the outside of the perovskite tubes and can be directed by support plates. Based on simulations the optimal length of the tubes was fixed to 500 mm taking into consideration the oxygen gradient along the length, streaming and back pressure behavior and other aspects.

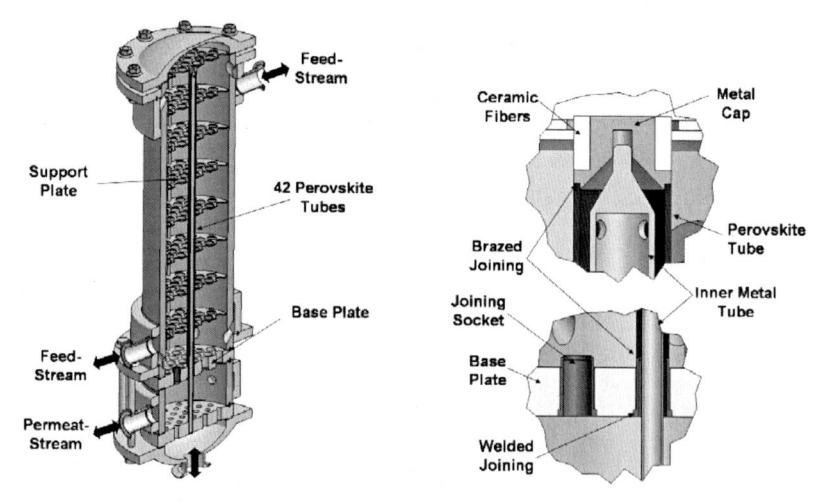

Figure 3: Schematic presentation of the tubular concept

PEROVSKITE MIEC MATERIALS

A lot of mixed ionic/electronic conducting materials in particular perovskite type ceramics were investigated. The market of such materials is increasing. Most experience exists for electrodes in SOFCs but also as membrane rector for partial oxidation of methane, syngas reactors and so on. In the center of investigation is the understanding and improvement of ion conductivity under oxidation or reducing atmospheres. The multiplicity of literature is difficult to compare because of different measuring conditions. For same materials different results were found possibly because of not defined microstructures[8] and surfaces. A techno-economical evaluation by Bredesen and Sogge[9] postulate a necessary oxygen flux of 10 N ml/(cm^2 min) for an economically use in oxygen separation. Fluxes of this order of magnitude are reported in lab experiments, usually with He carrier gas was measured. Our own requirement was defined with a flux of 3 N ml/(cm² min) at above described conditions.

Depending on the different applications also mechanical and thermal material data are very important for the designer and here is a gap of knowledge. For stress FEM calculations the coefficient of thermal expansion and Young's Modulus are to be known as well as the tthermal conductivity. For long term reliability studies also high temperature strength, crack propagation and creep behavior must be investigated. Because crystal size changes with oxygen partial pressure (chemical strain) measurements have to be done finally under realistic conditions and must be done with samples cut of real components not with special produced laboratory samples. Because of this necessity a characterization chain was established to prove different materials and components.

Of main interest for use in oxygen separation techniques are materials $ABO_{3-\delta}$ with A = (Sr, Ba, Ca, La) and B = (Co, Mn, Fe)[10,11]. For this project $Sr_{0.5}Ca_{0.5}Mn_{0.8}Fe_{0.2}O_{3-\delta}$ and $Ba_{0.5}Sr_{0.5}Co_{0.8}Fe_{0.2}O_{3-\delta}$ based tubes were provided by HITK Hermsdorf (Germany) and were tested. BSCF has a better ionic conductivity of about 3 N ml/(cm² min) for a dense tube with 1mm thickness but is not stable under CO_2-conditions and can be used only under special operating requirements. SCMF is more stable at least in short term applications but has a one order smaller conductivity and can be used as a thin membrane layer on a porous support only. As mentioned above also mechanical behavior has to be taken into consideration. Particularly with regard to brazing process high strength are required. It is obvious that a porous tube has a lower strength than a dense one. This low strength can become critical at room temperature because the superposed joint stresses have a maximum value at this temperature. Total porosity, pore size distribution and wall thickness influence not only strength but also oxygen flux through the wall so that an optimization must be found. Up to now we have not found the ideal commercial material and component respectively for our process conditions. In table 1 some material data of tested components are summarized.

Table 1. Material data of selected perovskite materials, measured on commercial components

	mechanical 20 °C					thermal 800°C		
	ρ	$ρ_{th}$	$σ_{4pb}$*	m	E	CTE	λ	cp
	[g/ccm]	[g/ccm]	[MPa]		[GPa]	[10^{-6}/K]	[W/mK]	[J/gK]
SCMF	3.84	5.12	33	6.1	30-70	15.8	2.2	0.84
BSCF	5.70	5.80	62	6.3	60-90	18.2	1.8	0.71

*calculated from Ring tests

The Bending strength was calculated from ring tests using Weibull theory. This test arrangement, figure 4, is ideal for samples out of tubes and can be used easily for high temperature tests, too. No special sample manufacturing is necessary. The tested BSCF sample was cut of a dense tube, the SCMF sample of a 20% open porous support tube.

The BSCF material was tested concerning creep behavior at temperatures of 850, 900 and 950 °C under a constant load of 50N, see figure 5. Though the resulting stress in the ceramic is more than the double stress in later application, a reducing application temperature to 800 °C makes sense.

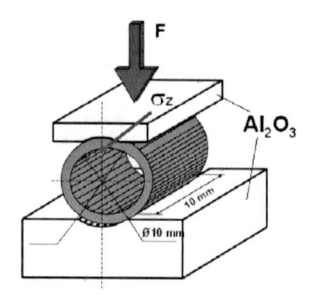

Figure 4. Ring test

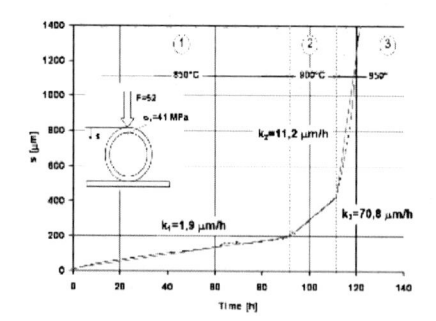

Figure 5. Creep behavior of a BSCF tube at different temperatures

JOINING TECHNIQUES

For the gas tight joining of ceramic components to metal housing principally two alternatives are possible: force fitting with a seal element and brazing. For force fittings in OTM the seal elements must withstand high temperatures and must retain enough elasticity. An allowable leakage must be defined by system optimization. For SOFC brazing methods are in use and intensively investigated[12]. Active metal brazing can be used for a wide range of material combination but needs high vacuum for processing. These conditions decompose OTM-materials and are not usable. An alternative is reactive air brazing (RAB) with a braze alloy of Ag and CuO as reactive component[13].

Another viable alternative used for SOFC is glass bonding. An adaptation of the coefficient of thermal expansion is solved up to CTE = $11.5 \cdot 10^{-6}$/K for the use of zirconia membranes[14]. Actual activities are concentrated on long term stability because interconnect steels react with glass components[15].

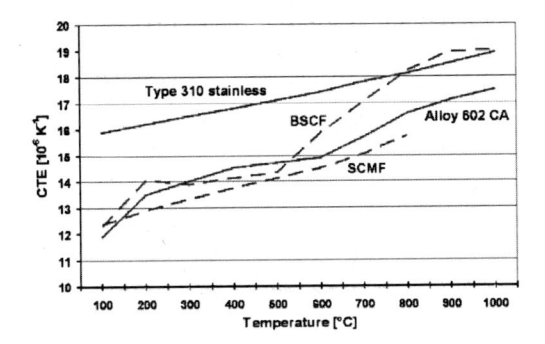

Oxyfuel Combustion Using Perovskite Membranes

Figure 6: Temperature dependence of CTE for perovskite and steel alternatives

The main requirement in OTM-systems is to handle the high CTE-values of joint materials, see figure 6. The CTE curves as function of temperature of alloy 602CA and BSCF show an acceptable conformity at temperatures below 550 °C. The mismatch at higher temperature is not critical if using RAB joints because the plasticity of the brazing alloy at these temperatures can reduce stresses. FEM calculations show maximum principle stresses at room temperature in the order of 100 MPa for a tube with 10mm outer diameter and 1mm wall thickness. This stresses will be reduced by heating up again on working temperature of 800 °C to 17 MPa caused by softening and plastic deformation of brazing alloy.

A difference between FEM calculations and ring test results could be explained by assumptions in theoretical simulation model. One of these assumptions is a single brazing layer between components without respect of diffusion zones at the interfaces. Also the braze alloy is for a time being treated as pure Ag without consideration of CuO influence.

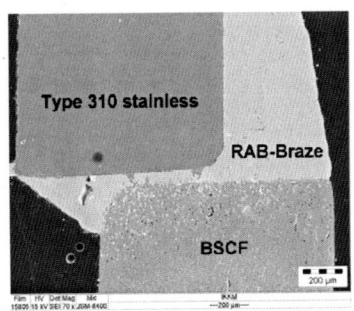

Figure 7. SEM picture of a BSCF-310steel joint

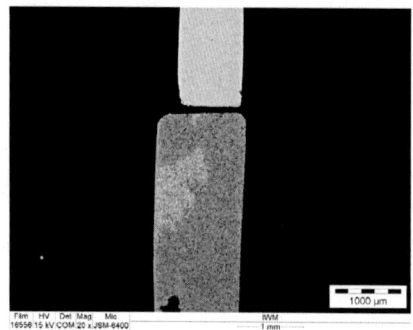

Figure 8. Porous support absorb brazing alloy

Although figure 6 shows a higher mismatch in CTE between perovskite materials and Type 310 stainless steel also such joints can be realized, see figure 7. Difficulties occur for porous support tubes (SCMF). The pores absorb the brazing alloy and transfer it from the joining zone into the structure by capillary forces, figure 8. Therefore glass sealing could be the better alternative but an adaptation of CTE differences is much more difficult then for SOFC and the long term stability is similarly to investigate as described for metal joints before. The adaptation of glass sealing in our research concentrates currently on a composite glass ceramic based on $BaO-B_2O_3$-silicate-glass with fillers like glass ceramic in the zinc-aluminum-silicate system. First results show sufficient short term strength at room temperature but have to be carried on. It is to remember that porous perovskite tubes have very low strength so that joining stresses are difficult to handle. On the other hand the decreasing of strength with temperature in ceramic will be compensated by an increasing plasticity of brazing materials.

CONCLUSIONS
For a realization of industrial OTM-reactors a lot of facts must be considered. The whole process chain of the respective application has to be optimized and to compensate weakness of ceramic materials. For the proposed Oxyfuel process OTM materials are not yet good enough and need to be improved not only concerning oxygen conductivity but also concerning their mechanical, thermal and chemical behavior and must be tested as early as possible under real conditions. The multidisciplinary effort as started in this project must be expanded and is a mutual challenge for material scientists and system engineers.

ACKNOWLEDGEMENT
This work is supported by German Federal Ministry of Economics and Technology.

REFERENCES
[1] T. Griffin, A. Bill, S.G. Sundkvist, J.L. Marion, N. Nsakala, "CO_2 Control Technologies: Alstom Power Approach", 6th Int. Conference on Greenhouse Gas Control Technologies (GHGT-6), Kyoto, October 2002
[2] S. Hellfritsch, P.G. Gilli, N. Jentsch, "Concept for a Lignite-fired Power Plant Based on the Optimised Oxyfuel Process with CO_2-Recovery", VGB Powertech 8/2004

[3]A.K. Anand, C.S. Cook, J.C. Cormann, A.R. Smith, "New Technology Trends for Imroved IGCC System Performance", Transactions of the ASM. Journal of Engineering for Gas Turbines and Power, **118** [4] 732-6 (1996)

[4]S. Engels, M. Modigell, "Development of a zero emission coal-fired power plant by means of mixed ion conductiong high temperature membranes", Desalination **199** 291-92 (2006)

[5]D. Toporov, S. Tschunko, J. Erfurt, R. Kneer, "Modelling of Oxycoal Combustion in a small Scale Test Facility", 7th Eur. Conf. of Industrial Furnaces and Boilers IFUB 18.-.21.04.2006, Porto, Portugal

[6]J.-Ch. Haag, A. Hildebrandt, H. Hönen, M. Assadi, R. Kneer, "Turbomachinery Simulation in Design Point and Part-Load Operation for Advanced CO_2 Capture Power Plant Cycles", to be presented at ASME Turbo Expo 14.-17.05.2007, Montreal, Canada

[7]J.F. Vente, W.G. Haije, R.IJpelaan, F.T. Rusting, "On the full-scale module design of an air separation unit using mixed ionic electronic conducting membranes", J. of Membrane Sci. **278** 66-71 (2006)

[8]Y.L. Zhang, L. Yang, A. Ponnusamy, K. Jacobson, K. Salama, "Effect of microstructure on oxygen permeation in $SrCo_{0.8}FeO_{3-\sigma}$", J.of Mater. Science **34** 1367-72 (1999)

[9]R. Bredesen, J. Sogge, "A technical and economic assessment of membrane reactors for hydrogen and syngas production" in: Seminar on Ecological Applications of Innovative Membrane Technology in the Chemical Industry" (1996)

[10] J.F. Vente, W.G. Haije, Z.S. Rak, "Performance of functional Perovskite Membranes for Oxygen Production", J. Membrane Sci. **276** 178-84 (2006)

[11]Y. Teraoka, H. Shimokawa, Ch.Y. Kang, H. Kusaba, K. Sasaki, "Fe-based perovskite-type oxides as excellent oxygen-permeable and reduction-tolerant materials", Solid State Ionics **177** 2245-48 (2006)

[12]R.M. do Nascimento, A.E. Martinelle, A.J.A. Buschinelli, "Review Article: Recent Advances in Metal-Ceramic-Brazing", Cerâmica **49** 178-198 (2003)

[13]K.S. Weil, J.S. Hardy, "Development of a New Ceramic-to-Metal Brazing Technique for Oxygen Separation/Generation Applications" Proc. 16th Annual Conf. on Fossil Energy Materials (2002)

[14]Jeffrey W. Fergus, "Sealants for solid oxide fuel cells", J. Power Sources **147** 46-57 (2005)

[15]P.Batfalsky, V.A.C. Haanappel, J.Malzbender, N.H. Menzler, V.Shemet, I.C. Vinke, R.W. Steinbrech, "Chemical interaction between glass-ceramic sealants ant interconnect steels in SOFC stacks", J. Power Sources **155** 128-137 (2006)

SYNTHESIS OF SiC NANOFIBERS WITH GRAPHITE POWDERS

Andrew Ritts[1], Qingsong Yu[1], and Hao Li[2]
[1]Chemical Engineering Department, [2]Mechanical & Aerospace Engineering Department
University of Missouri-Columbia
Columbia, Mo 65211

ABSTRACT

Nanotubes and nanofibers have been used as composite reinforcing materials because of their outstanding mechanical properties, and enhanced thermal properties. SiC nanofibers with excellent mechanical properties are especially preferred in harsh environments. SiC nanofibers were fabricated using graphite powders and Ni catalysts in a hot wall chemical vapor deposition (CVD) chamber. Our experiments show that the dimension and structures of SiC nanofibers strongly depend on the Ni catalyst, reaction time, and the carbon source used. Sufficient Ni catalysts will facilitate the growth of SiC nanofibers with uniform diameters while SiC nanofibers were found shorter with less Ni catalyst or with large and nonuniform diameters when no Ni nanoparticles exist. Increased reaction time, from 7 to 15 hours, is generally required for extending the growth of high quality SiC nanofibers. The graphite particle (325 mesh, <44μm) appears to be a much better carbon source for the growth of SiC nanofibers compared with graphite particles with size of 1-15 μm. SiC particles, several microns in size, were also found as a by-product in the as-fabricated samples. Future work will investigate the mechanisms behind some of the above observations and tailor experimental conditions to synthesize SiC nanofibers with desired microstructure and properties.

INTRODUCTION

Silicon carbide (SiC) is one of the most important ceramic materials in the form of powders, molded shapes, and thin films.[1] It has wide industrial applications due to its excellent mechanical properties, high thermal and electrical conductivity, excellent oxidation resistance, and it has potential application as a functional ceramic or a high temperature semiconductor as well.[2,3] Nanotubes and nanofibers are expected to be good composite reinforcing materials because of their outstanding mechanical properties, enhanced thermal properties, and outstanding electrical properties.[4] SiC nanofibers with excellent mechanical properties are especially preferred in harsh environments, such as high temperature and highly corrosive environments. Nanoscale SiC fibers have important potential applications in nanoelectronics, field emission devices and nanocomposites.[5] Methods to fabricate SiC nanofibers, such as carbon nanotubes confined reactions,[6,7] thermal plasma synthesis,[3] and template/catalyst-free processes[8] have been reported. The primary carbon source studied previously include carbon nanotubes and graphite powders.[6-8] Using carbon nanotubes as precursors as compared to graphite is expensive. Thermal plasma synthesis and template/catalyst free processes have many steps and produce poor quality and few SiC nanofibers. Despite of the growth method, previous research suggested several primary growth mechanisms of SiC nanofibers: vapor-solid (VS) without metal catalyst, vapor-solid-liquid (VLS) mechanisms, direct conversion from CNTs using CNTs as a template. The objective of the present study is to analyze the fundamental processing-structure relationship of the SiC nanofibers fabricated using inexpensive graphite powders, silica powders, silicon powders, and metal catalysts. In our experiments, we expect that the VS and VLS are the two primary possible growth mechanisms for the growth of SiC nanofibers. It is hoped that this method using inexpensive precursors may lead to mass production of high quality SiC nanofibers with well controlled structures and advanced properties.

EXPERIMENTAL

Silica, with 0.5-10 mm particle size and 99% purity (Sigma Aldrich, WI) and silicon with 325 mesh (less than 44 microns) and 99% purity (Sigma Aldrich), were mixed with a 2:1 molar ratio and then put in a high purity alumina crucible and leveled, as shown in Figure 1. Then graphite particles are added in the crucible and leveled. The crucible was put inside a hot wall CVD chamber with 100 sccm Ar/H_2 (5%H_2) gas. The temperature was raised to 1500 °C at 150 °C/hour and hold for either 7 or 15 hours. After the reaction, the temperature was cooled down with the Ar/H_2 gas on before reaching room temperature.

Two graphite powders, with 325 mesh and 99.99% purity (Sigma Aldrich) and with particle size ranging from 1 micron to 15 microns and a purity of 99.9995% (Alfa Aesar, MA) were used as the carbon sources. Graphite with large size will be denoted with "large graphite particles" and graphite with smaller size will be denoted as "small graphite particles" for simplicity. Nickel (II) nitrate hexahydrate was used to prepare nickel nitrate ($Ni(NO_3)_2$) solutions. Two concentrations (0.063 mole/liter and 0.0063 molar/liter) have been used and they are denoted with "high concentration nickel nitrate solution" and "low concentration nickel nitrate solution" in the present paper. When catalysts are used, carbon source materials was soaked and sonicated in nickel nitrate solutions for 30 minutes at room temperature and was filtered/dried over night. The dried graphite was put on top of the Si/SiO_2 mixture for ease of separation after the reaction. When nickel nitrate was used, the H_2 in Ar/H_2 gas was used to reduce the nickel nitrate to nickel nanoparticles. The as-received samples were characterized with SEM (Scanning Electron Microscopy), EDS (Energy Eispersive Spectrometer), and XRD (X-Ray Diffraction). In addition, efforts have been made to eliminate the impurities, such as graphite, Ni catalyst, and silica shells. The general procedures are: 1. oxidation at 700°C for 10 hours to oxidize graphite, 2. etching with 6 mole/liter HCl for 30 minutes to dissolve the Ni catalyst, and 3. etching with 1% HF for 30 minutes to dissolve silica layers. The treated samples were also characterized afterwards.

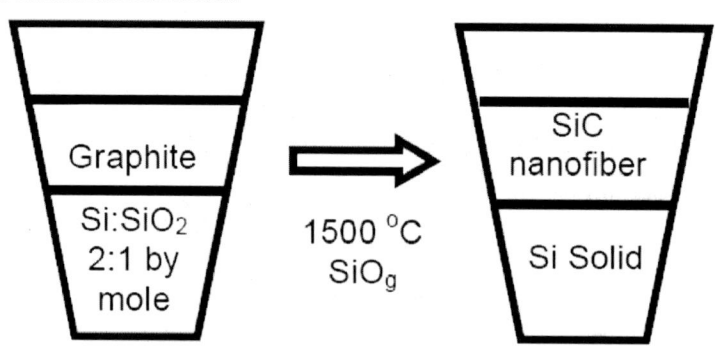

Figure 1. Conversion of graphite powders to SiC nanofibers using SiO_g at 1500°C with Ar/H_2 flow at atmospheric pressure in a high purity alumina crucible.

RESULTS AND DISCUSSION

The X-Ray diffraction shows that the as-received samples are β-SiC. Our previous TEM data on SiC nanofibers fabricated with carbon nanotubes shows that most of SiC nanofibers have a silica layer. It is very likely that the SiC nanofibers fabricated using graphite particles also have the silica layer, which is supported by our EDS data. TEM characterization on the present samples will be conducted in the future. In general for all the experiments, there are three distinguishable layers which can be easily separated manually after the experiments. The top layer is SiC nanofibers or particles with few impurities. The top layer is gray when no catalyst is applied and is lightly blue when Ni catalysts are used. The SEM images were taken from this part if not specified otherwise. The middle layers are also composed SiC nanofibers, but with more by products or impurities based on our SEM study (not shown here). This middle layer is typically gray and/or green. The bottom layer is mostly the left over Si/SiO$_2$ mixture with very small amount of SiC. Such three layers were also found in synthesis of SiC nanofibers using carbon nanotubes as the carbon source.[7] The following characterization will focus on the top layers containing relatively high quality SiC nanofibers.

Many reactions are thought to occur during the fabrication process of SiC. When using graphite as a precursor we believe that the first reaction is a direct conversion reaction which produces CO_g shown in equation 1. Nanofiber growth is believed to be either by VS or VLS growth mechanisms as shown in equation 2. It is also believed that the CO_{2g} produced in equations 1 and 2 will react with the graphite to produce more CO_g allowing more nanofibers to grow without requiring the replacement reaction shown in equation 1.

$$SiO_g + 2C(graphite) \rightarrow SiC(particles) + CO_{2g} \quad (1)$$

$$SiO_g + 3CO_g \rightarrow SiC(nanofibers) + 2CO_{2g} \quad (2)$$

$$CO_{2g} + C(graphite) \rightarrow 2CO_g \quad (3)$$

Figure 2. SEM images as-received SiC nanofibers synthesized at 1500°C for 7 hours using large graphite particles without catalyst.

Synthesis of SiC Nanofibers with Graphite Powders

Figure 3. Comparison of SiC nanofiber fabricated using graphite particles treated with low (left) and high (right) concentration nickel solution for 7 hours.

Figure 2 shows SEM images of SiC nanofibers fabricated with large graphite particles without Ni catalyst at 1500°C for 7 hours. The yield of this reaction is low and few SiC nanofibers could be identified in the top layers using SEM. The diameters of these nanofibers range from 40 to 250 nm. Previous research indicates that VS mechanism result in large diameter distribution of SiC nanofibers. For this experiment, the only metal catalyst for VLS growth may come from the impurities if there is any catalyst. It is expected the VLS mechanism is significantly impeded due to the limitation of catalyst source. Large microparticles are found in the sample and EDS results show that the surface composition are primarily Si and C. These microparticles will need to be removed in order to obtain pure SiC nanofibers.

Figure 3 depict the microstructures of SiC nanofibers fabricated with large graphite particles treated with nickel nitrate solution. Compared to those fabricated without catalyst in Figure 2, more SiC nanofibers were fabricated when graphite particles are treated with nickel nitrate solutions. It appears that the presence of Ni catalyst in the graphite significantly promote the growth of SiC nanofibers. Samples with lower Ni catalyst concentration (0.0063 M) have much shorter lengths with a maximum length of 10 µm, while samples with higher catalyst concentration (0.063 M) had longer lengths over 40 µm. Both samples with Ni catalyst have clusters of the SiC nanofiber and micro sized particles distributed throughout the sample. The diameter distribution is much smaller 40-100 nm when catalyst is used. It is interesting that the lower concentration of nickel nitrate treated graphite particles resulted in SiC nanofibers with shorter length and large diameters, average diameters of ~80 nm. The higher concentration Ni catalyst samples have average diameters of ~60 nm. Typically the diameter of nanofibers grown with a VLS mechanism is approximately proportional to the catalyst size. The VLS mechanisms are evident based on our SEM and EDS results for both low concentration and high concentration experiments. This observation may indicate that the lower concentration of nickel nitrate result in bigger Ni nanoparticles or that with less catalyst more SiC is created through VS mechanism widening the fiber diameter. We have no explanation for this observation at this point if this is indeed true. It is expected it take a longer time to form Ni nanoparticles with lower concentration nickel nitrate solution, which may indicate less time for SiC nanofiber growth considering the total experimental time is the same. The less growth time may yield shorter length, consistent with our observation. When there are insufficient Ni catalysts in the case of low concentration nickel nitrate solution, extra SiO_g and CO may or may not react on the surface of the SiC nanofibers, which may make it thicker. Such reaction may happen with an increased local CO_g concentration as shown by Tang et al.[7]

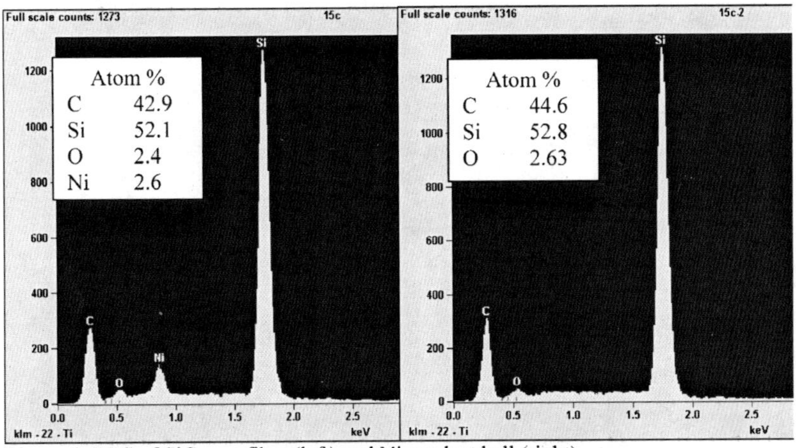

Figure 4. EDS of SiC nanofiber (left) and Ni catalyst ball (right)

Left panel — Atom %:
C 42.9
Si 52.1
O 2.4
Ni 2.6

Right panel — Atom %:
C 44.6
Si 52.8
O 2.63

In both cases, we cannot conclusively exclude the possibility of VS growth mechanism even though we do believe the VS growth was significantly impeded in the case of high concentration nickel nitrate solution. In the case of low concentration nickel nitrate solution, SiC nanofibers may grow with a VS mechanism and have larger diameter and larger diameter distribution, consistent with our observation. Another major difference between the two images in Figure 3 is that the SiC nanofibers grown with more catalysts are flexible, which may indicate high strength, while the SiC nanofibers grown with less catalysts are straight, more like the SiC whiskers. Yang[9] et al and Wong[10] et al suggested that SiC nanofiber strength are inversely proportional to its diameter square root and may reach ~24 GPa with diameter of 100 nm. Although this assumption matches experimental data very well, we have to consider the defects which can also significantly degrade the nanofiber strength. Our previous TEM data suggested that nanofibers grown with diameter over 50 nm with VLS mechanism have very few defects and also have oxide layer outside. Figure 4 shows EDS data of SiC nanofibers and SiC nanofibers and catalysts, which also indicate the presence of oxide. In addition Ni catalyst is evident when EDS was taken in an area with catalyst.

Longer growth time, 15 hours, was also used for experiments using low concentration nickel nitrate solution. The general observation is that longer reaction time yield more and longer SiC nanofibers. The SiC nanofibers diameters seem not be influenced significantly by the reaction time. SiC microparticles were observed.

Figure 5. SiC nanofiber from small graphite particles at 1500°C for 7 hours.

Figure 6. SiC nanofiber from small graphite with different Ni catalyst mixing method.

Figure 5 shows SEM images of SiC nanofibers produced using smaller graphite particles with catalyst prepared using high concentration of nickel nitrate solution. EDS confirmed the particles and fibers fabricated to have Si, C, and O. There are significantly fewer SiC nanofibers synthesized using small graphite particles compared to the case using large graphite particles. The following reasons may contribute to such low yield of SiC nanofibers fabrication using small graphite particles. Mixing the nickel nitrate solution with the small graphite particles was difficult due to lack of wetability of the graphite when using the same method as the larger graphite based on observation with naked eyes. The decreased wetability could be due to the increased surface energy of smaller particles used. Another mixing method was used to with the intention to improve the yield and the results are shown in figure 6, but there were still less SiC nanofibers compared to those using large graphite particles. This mixing method involves hand stirring of the nickel nitrate solution (rather than sonication) and repeatedly submerging small graphite particles in the nickel nitrate solution. It is also possible that, smaller graphite particles have much larger surface areas which consume most of the SiO_g vapor that was the source for SiC nanofiber growth, and then resulted in the low yield.

During both high and low catalyst concentrations the SiC nanofiber were found close to each other and form kind of clusters, but not agglomeration. It is very likely that this is due to localized catalyst on the graphite particles. Figure 7 depicts the situation where localized catalyst, which promote the growth of SiC nanofiber clusters. We speculate this is due to the low wettability of water on graphite. Due to the low wettability, nickel nitrate solution tends not to wet/cover the graphite and then form droplets, which resulted in the Ni catalyst as shown in figure 7.

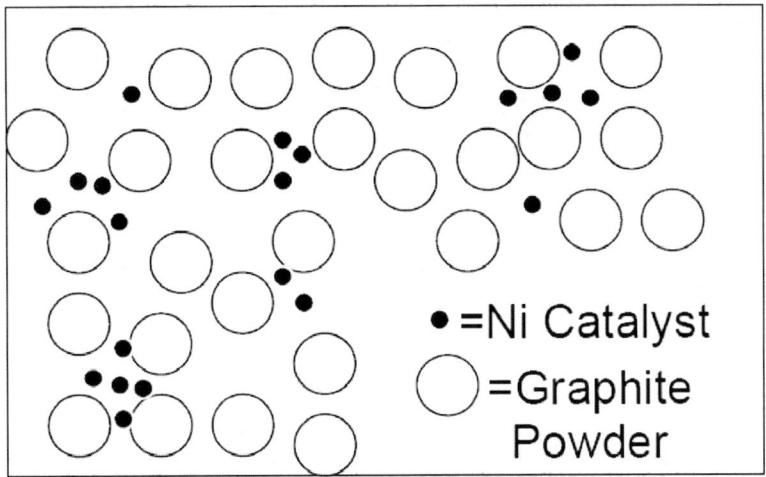

*Ni catalysts were scaled up for illustration.
Figure 7. Depiction of localized Ni catalyst.

Removal of the unwanted microparticles and other impurities are essential for the applications of SiC nanofibers. However the method including oxidation and HCl and HF etching as described in experimental section is not effective to remove the microparticles. One of the fundamental reasons is that the microparticles are very likely SiC or at least have shell of SiC, which could not be easily oxidized while protecting the SiC nanofibers. We are continuing making efforts to solve this critical problem. Figure 8 shows typical SEM images of a sample after such treatment.

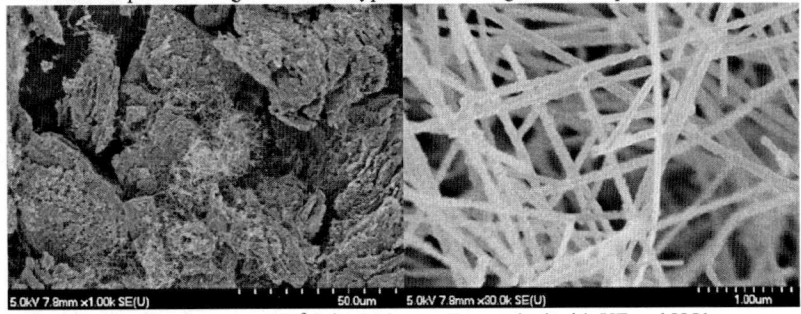

Figure 8. SiC nanofiber baked at 800°C for 15 hours, then etched with HF and HCl

CONCLUSION AND FUTURE WORK

Our experimental results indicate that Ni catalyst concentration, graphite particle size, and reaction time play important roles during the synthesis of SiC nanofibers. The diameters of SiC nanofibers range from 40-250 nm without assistance of Ni catalyst, while SiC nanofibers have a much narrower diameter distribution of 40-100 nm when Ni catalysts were used. Higher or sufficient Ni catalyst concentration and prolonged reacting time yield more and longer SiC nanofibers. Larger graphite particles of average size smaller than 44 microns was found to be a better precursor compared to small graphite particles in terms of promoting SiC nanofiber growth and controlling the dimension. One of the greatest challenges is how to purify these SiC nanofibers. In the future, we will continue to investigate the mechanisms behind some of the above observations and tailor experimental conditions to synthesize SiC nanofibers with desired microstructure and properties, including diameter, length, defects, yield, and color.

ACKNOWLEDGEMENTS

This work was supported by National Science Foundation under the award No. CMMI-0620906 and award No. CMMI- 0556150 and the GAANN Fellowship program.

REFERENCES

[1] H.O. Pierson, Handbook of Refractory Carbides and Nitrides, William Andrew, Noyes, 137 (1996).
[2] K. Krnel, Z. Stadler, and T. Kosmač, Preparation and Properties of C/C–SiC Nano-Composites, *J. Eur. Ceram. Soc.*, **27**, 2-3, 1211-6 (2007).
[3] L Tong, and R G Reddy, Thermal Plasma Synthesis of SiC Nano-Powders/Nano-Fibers *Mater. Res. Bull.*, **41**, 2303-2310 (2006).
[4] O. Breuer, and U. Sundararaj, Big Returns from Small Fibers: A Review of Polymer/Carbon Nanotube Composites, *Polym. Compos.*, **25**, 6, 630-41 (2004).
[5] W. Yang, H. Araki, S. Thaveethavorn, H. Suzuki, and T. Noda, In Situ Synthesis and Characterization of Pure SiC Nanowires on Silicon Wafer *Appl. Surf. Sci.*, **241**, 236 (2005).
[6] H. Dai, E.W. Wong, Y.Z. Liu, S. Fan, and C.M. Lieber, Synthesis and Characterization of Carbide Nanorods, *Nature*, **375**, 769 (1995).
[7] C.C. Tang, S.S. Fan, H.Y. Dang, J.H. Zhao, C. Zhang, P. Li, and Q. Gu, Growth of SiC Nanorods Prepared by Carbon Nanotubes-Confined Reaction *J. Cryst. Growth*, **210**, 595 (2000).
[8] G.W. Meng, L.D. Zhang, C.M. Mo, S.Y. Zhang, Y. Qin, S.P. Feng, and H.J. Li, Preparation of β-SiC Nanorods with and without Amorphous SiO2 Wrapping Layers, *J. Mater. Res.*, **13**, 2533 (1998).
[9] W. Yang, H. Araki, C. Tang, S. Thaveethavorn, A. Kohyama, H. Suzuki, and T. Noda, "Single-Crystal SiC Nanowires with a Thin Carbon Coating for Stronger and Tougher Ceramic Composites," *Adv. Mater.*, **17**, 1519-23, (2005).
[10] E. W. Wong, P. E. Sheehan, and C. M. Lieber, "Nanobeam Mechanics: Elasticity, Strength, and Toughness of Nanorods and Nanotubes," *Science*, **227**, 1971-75, (1997).

INFLUENCE OF ADDITIONAL ELEMENTS ON DENSIFICATION BEHAVIOR OF ZIRCONIA BASE AMORPHOUS POWDERS

Tatsuo Kumagai
Department of Mechanical Engineering, National Defense Academy
1-10-20 Hashirimizu, Yokosuka-shi
239-8686, JAPAN

ABSTRACT
Thermal characteristics and densification behaviors of mechanically milled amorphous powders of yttria partially stabilized zirconia (ZrO_2-3mol%Y_2O_3: 3YZ) and 3YZ with additives such as B, SiO_2 and Al_2O_3 have been investigated. In all the present amorphous powders, irrespective of additives, the temperatures of the end of exothermic peak (T_f) in DTA traces, i.e., finishing temperature of crystallization, are lower than the corresponding onset temperatures of rapid densification (T_{rd}), and thus crystallization (mainly tetragonal zirconia solid solution (t-ZrO_{2ss})) has already occurred sufficiently prior to beginning of densification. The T_f and T_{rd} values for the amorphous 3YZ-B powder are considerably lower than those for the amorphous 3YZ-SiO_2 powder. In both cases, however, the T_f values are close to the corresponding T_{rd} values. In addition, the large slope of the density-temperature (D-T) curve for the amorphous 3YZ-B powder is quite similar to that for the amorphous 3YZ-SiO_2 powder. In the case of the amorphous 3YZ and 3YZ-Al_2O_3 powders, the T_f values are not close to and relatively lower than the corresponding T_{rd} values. Moreover, the slopes of the D-T curves are considerably smaller than those for the amorphous 3YZ-B and -SiO_2 powders. Faster densification in the amorphous 3YZ-B and -SiO_2 powders seems to be strongly related to the presence of residual B (or B_2O_3) and SiO_2 base amorphous phases after t-ZrO_{2ss} crystallization. In contrast, slower densification in the amorphous 3YZ and 3YZ-Al_2O_3 powders is considered to be due to the disappearances of the amorphous phases.

INTRODUCTION
Partially stabilized zirconia (PSZ) is well known to be a structural ceramic material showing superior fracture toughness.[1-4] Recently, Kimura et al.[5,6] reported that further improvements in mechanical properties such as strength, toughness and workability could be achieved in nanoscale crystalline ZrO_2-20mol%Al_2O_3 and yttria partially stabilized zirconia (ZrO_2-3mol%Y_2O_3: 3YZ)-20mol%Al_2O_3 bulk ceramics fabricated by consolidation of mechanically milled amorphous powders using electrical discharge consolidation system. However, they concluded that densification of the amorphous powders proceeds via viscous flow of the amorphous phases without conformation of maintenance of the amorphous structure during sintering. Our recent study[7] revealed that the crystallization temperature of the amorphous 3YZ-20mol%Al_2O_3 powder measured by thermal analysis is considerably lower than the starting temperature of densification and, moreover, apparent amorphous phase could not be recognized in the full dense nanoscale crystalline bulks by transmission electron microscopy.

In this study, in order to investigate the densification behavior, consolidation experiments were conducted for both as-received crystalline 3YZ and mechanically milled amorphous 3YZ powders. In addition, the effects of ternary elements of B, Si and Al on the thermal characteristics and densification behaviors were also investigated.

EXPERIMENTAL

Initial compositions of mechanically milled powders used in this study, except silica added powder, whose composition is measured after mechanical milling, are listed in Table 1. As-received 3YZ (98%, High Purity Chemicals Co., Ltd.) and 20mol% boron (97%, Hermann C. Starck Inc.) added 3YZ powders (3YZ-20B) were mechanically milled by a mono type planetary ball mill system (P-6, Fritch). In these cases, pot and balls (5mm in diameter) made of PSZ were used in order to prevent contamination. A 20mol%Al_2O_3 (99.99%, High Purity Chemicals Co., Ltd.) added 3YZ (3YZ-20Al_2O_3) powder was attrition milled using a rotation-arm reaction ball milling system.[5] In this case, tank, agitating arms and balls (5mm in diameter) made of PSZ were used for the same reason mentioned above. Ball mill of a 3YZ powder initially containing 5mol%CeO_2 (99.9%, High Purity Chemicals Co., Ltd.) by a twin type planetary ball mill system (P-7, Fritch) using agate pots and PSZ balls (5mm in diameter) produces an amorphous 3YZ-26mol%SiO_2-3mol%CeO_2 (3YZ-26SiO_2) powder, because of the abrasion of the soft agate pots by the hard PSZ balls. Mechanical milling treatments are carried out at rotational speeds of $7s^{-1}$ for 3YZ and 3YZ-20B, $5s^{-1}$ for 3YZ-Al_2O_3 and $10s^{-1}$ for 3YZ-26SiO_2 in an argon atmosphere. The ball to powder weight ratios are 10:1 for 3YZ, 3YZ-20B and 3YZ-26SiO_2, and 4:1 for 3YZ-Al_2O_3.

Densification of the obtained powders was conducted using an electrical discharge consolidation system equipped with a vacuum system.[5] This consolidation system can monitor the temperature inside a carbon die, applied stress, current, voltage and displacement in real-time. Each powder (about 1g) packed into a graphite die with inner and outer diameters of 10 and 45mm, respectively, was heated at a rate of about 7 to 8K/s to a certain temperature under the constant applied stress of 150MPa in a vacuum chamber (10Pa). The total period of time of heating and subsequent isothermal heat-treating was fixed at 900s (15min).

Thermal characteristics of the mechanically milled powders were evaluated by differential thermal analysis (DTA, DTA-30, Shimadzu Seisakusho Co., Ltd.). DTA scans were conducted at heating rates of 0.08 to 1.67K/s (5 to 100K/min) in a flowing argon atmosphere. Phase structures of the powders and consolidated bulks were determined by X-ray diffractometry (XRD, M03XHF22, Mac Science Co., Ltd.) using CuKα radiation operated at 50kV and 32mA. The 2-theta values of crystalline peaks were calibrated by the external standard method comprising a pure Si single crystal for calculation of lattice parameters. Microstructures were examined by optical microscopy, field emission scanning electron microscopy (FE-SEM; S-4500, Hitachi Co.,

Table 1 Powder Compositions

Sample Designation	Powder Composition
3YZ	97mol%ZrO_2-3mol%Y_2O_3
3YZ-2B	98mol%3YZ-2mol%B
3YZ-20B	80mol%3YZ-20mol%B
3YZ-26SiO_2*	71mol%3YZ-26mol%SiO_2-3mol%CeO_2*
3YZ-20Al_2O_3	80mol%3YZ-20mol%Al_2O_3

All are initial compositions except 3YZ-26SiO_2*; the composition of 3YZ-26SiO_2* is obtained after mechanical milling.

Ltd.) and field emission transmission electron microscopy (FE-TEM; HF-2000, Hitachi Co., Ltd) operated at 200kV. For SEM observations, the bulk samples polished down to 0.5μm by diamond paste for optical microscopy were coated with carbon, followed by thermal etching at 1323K for 600s in evacuated quartz ampoules, and then coated with carbon. FE-TEM images were obtained from the edges of the mechanically milled powders.

RESULTS AND DISCUSSION
Fabrication of mechanically milled amorphous powders

Fig.1 shows the change in XRD profile of the as-received crystalline 3YZ (cry-3YZ) powder without any additives during mechanical milling. Crystalline peaks corresponding to tetragonal zirconia solid solution (t-ZrO_{2ss}) decrease with increasing the period of milling time. Instead, two broad peaks corresponding to amorphous phase appear around 30 and 50 degrees. As shown in Fig.2, similar amorphous patterns could be observed in the mechanical milled 3YZ powders containing boron, silica and alumina. As shown in a typical high-resolution TEM image obtained from the mechanically milled 3YZ-20Al$_2$O$_3$ powder (Fig.3), periodic fringe contrast could not be recognized in a major part of the figure, indicating the formation of the amorphous phase at the atomic level.

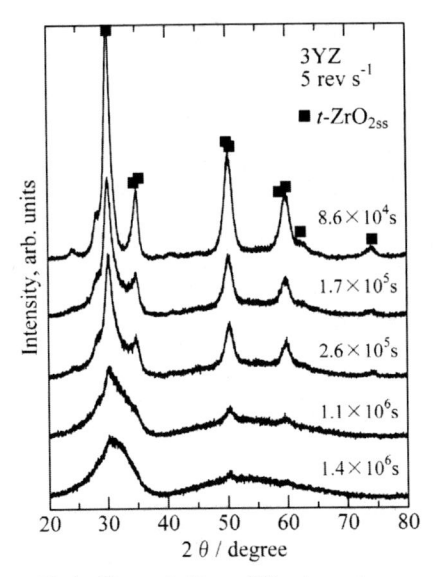

Fig.1 Change in X-ray diffraction patterns of ZrO$_2$-3mol%Y$_2$O$_3$ (3YZ) powder during mechanical milling.

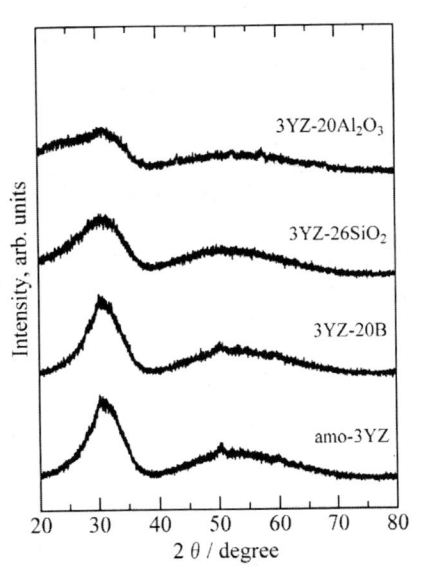

Fig.2 X-ray diffraction patterns of mechanically milled amorphous powders of 3YZ (amo-3YZ), and 3YZ containing 20mol%B (3YZ-20B), 26mol%SiO$_2$ (3YZ-26SiO$_2$) and 20mol%Al$_2$O$_3$ (3YZ-20Al$_2$O$_3$).

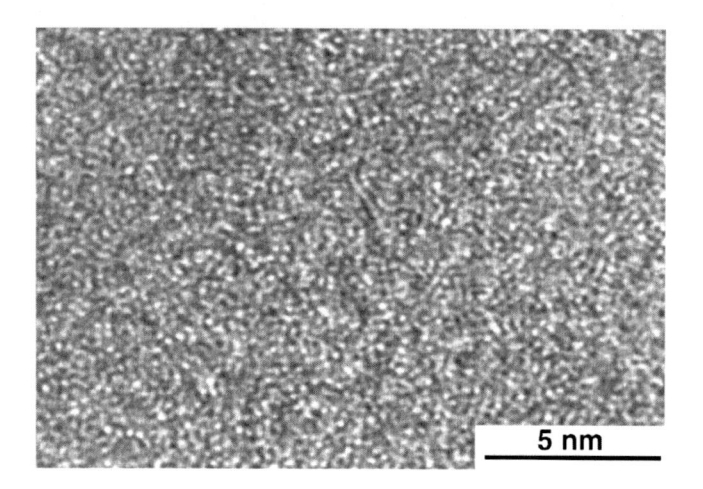

Fig.3 HRTEM image from amorphous 3YZ-20Al$_2$O$_3$ powder.

Consolidation of amorphous and crystalline 3YZ powders

Prior to consolidation experiments, thermal characteristics of the amorphous 3YZ (amo-3YZ) powder were examined by DTA at heating rates of 5, 10, 20 and 50K/min to about 1173K. As shown in Fig. 4, each DTA curve reveals a sharp exothermic peak, which corresponds to crystallization of zirconia solid solution. These crystallization temperatures increase slightly from 554-592 to 604-627K with increasing heating rate from 5 to 50K/min, indicating the heating rate dependence of the crystallization temperature.

Densification processes for the amo-3YZ (open circle) and cry-3YZ (solid circle) powders are shown in Fig. 5. In both cases, changes in relative density could not be recognized at temperatures below about 900K. As the sintering temperature increases above about 900K, the values of relative density increase sharply from 0.76 to 1.0 for considerably short periods of time (about 130s). Figs. 6 and 7 show the optical

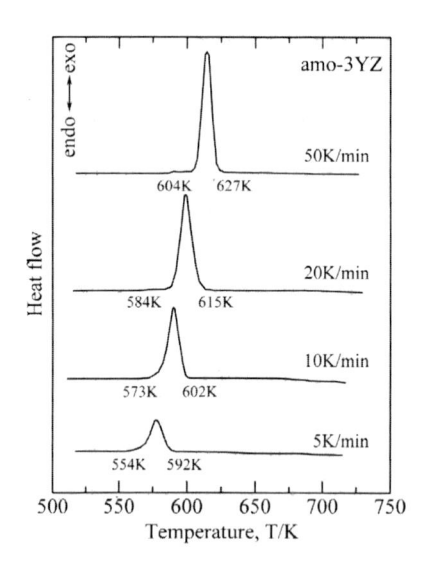

Fig.4 Differential thermal analysis (DTA) curves for amo-3YZ powder at different heating rates.

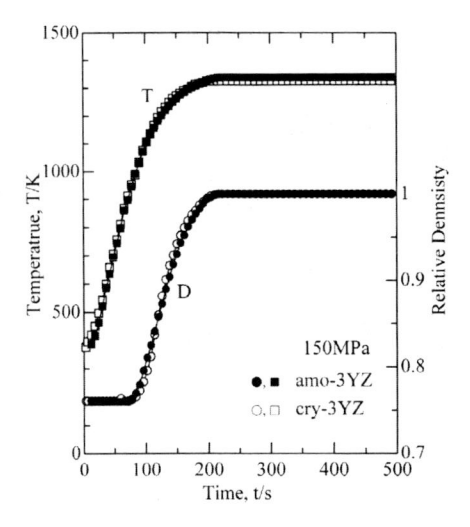

Fig.5 Relative density and temperature as a function of time during electrical discharge consolidation of mechanically milled amo-3YZ (solid circle and square) and as-received crystalline 3YZ (cry-3YZ, open circle and square) powders under constant applied stress of 150MPa.

microstructure and the corresponding SEM image of the full dense bulk obtained from the amo-3YZ powder. These photos indicate that the full dense bulk has no apparent microcrack and

Fig.6 Optical microstructure of 3YZ bulk consolidated at 1338K. Vickers indentation at the central portion was obtained with a load of 49N for 30s.

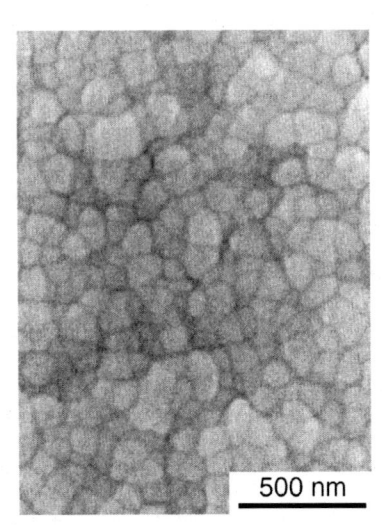

Fig.7 SEM image of 3YZ bulk consolidated at 1338K.

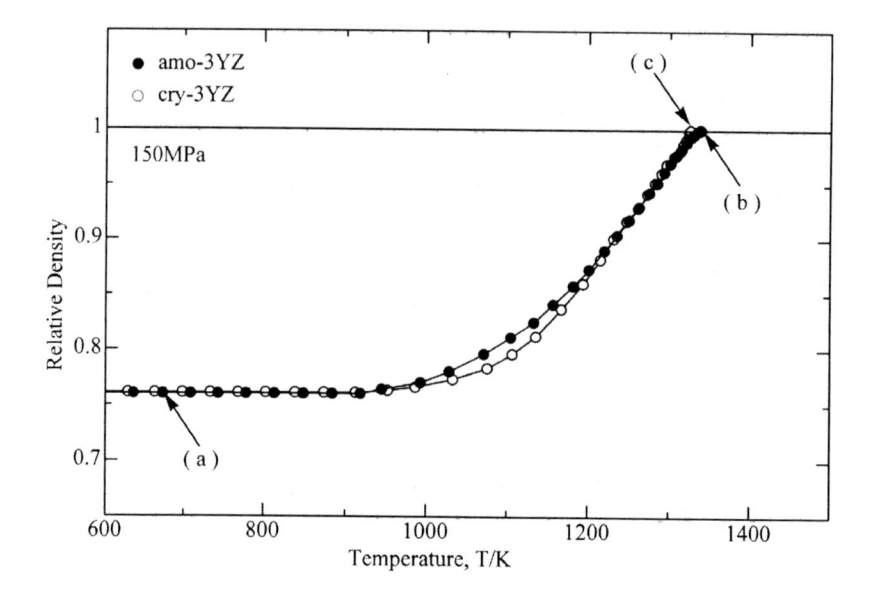

Fig.8 Change in relative density as a function of temperature during consolidation of mechanically milled amo-3YZ (solid circle) and as-received cry-3YZ (open circle) powders under constant applied stress of 150MPa.

void, and consists of fine grains with an average diameter of about 80nm. Fig.8 shows change in relative density as a function of sintering temperature for the amo-3YZ (open circle) and cry-3YZ (solid circle) compacts. Although the starting amo-3YZ powder consists of amorphous phase, its densification curve is quite similar to that of the cry-3YZ powder. According to the XRD experiments (Fig.9), the amo-3YZ compact with low relative density of 0.7 (marked (a) at 673K in Fig.8) shows the XRD pattern of t-ZrO_{2ss} similar to that of the amo-3YZ compact with high relative density (marked (b) at 1338K in Fig.8), although the split of some peaks arising from tetragonality is ambiguous in the amo-3YZ compact with low relative density, probably due to the smaller crystallite size. In addition, the cry-3YZ compact with high relative density (marked (c) at 1325K in Fig.8) also shows the XRD pattern of t-ZrO_{2ss} similar to those of the amo-3YZ compacts with low and high relative density.

All the results shown in Figs. 4, 5, 8 and 9 clearly indicate that crystallization in the amo-3YZ compact has already occurred sufficiently during heating, at least, to 673K, prior to beginning of densification, and that densification of the amo-3YZ compact proceeds by plastic flow of the crystalline t-ZrO_{2ss} phase instead of viscous flow of the amorphous 3YZ phase.

Consolidation of amorphous 3YZ-B, 3YZ-SiO$_2$ and 3YZ-Al$_2$O$_3$ powders

Thermal characteristics of the 3YZ base amorphous powers were examined by DTA and the results are shown in Fig.10. The additions of B, SiO$_2$ and Al$_2$O$_3$ cause the delay in crystallization, i.e., stabilization of the amorphous structures. Although both B and Si are the constituents of typical network formers [8, 9] (B$_2$O$_3$ and SiO$_2$), the increment of crystallization temperature in the B added 3YZ amorphous powder is about half of that in the SiO$_2$ added amorphous powder. On the other hand, the addition of Al$_2$O$_3$, which is classified as intermediate oxide, [9] raises the crystallization temperature to a value comparable to that corresponding to 3YZ-SiO$_2$.

Fig.11 displays densification behaviors of the 3YZ base amorphous powders. As shown in Fig.11, apparent increase in relative density for 3YZ-20B could not be recognized at temperatures above 898K. Further increase in temperature causes rapid densification as in the case of amo-3YZ (above 903K), but the slope of the density-temperature (*D-T*) curve for 3YZ-20B is considerably larger than that for amo-3YZ. The rapid densification and the large slope of the *D-T* curve, which are quite similar to 3YZ-20B, could be observed for 3YZ-26SiO$_2$, although the onset temperature of rapid densification (T_{rd}) for 3YZ-26SiO$_2$ (1206K) is considerably higher than that for 3YZ-20B

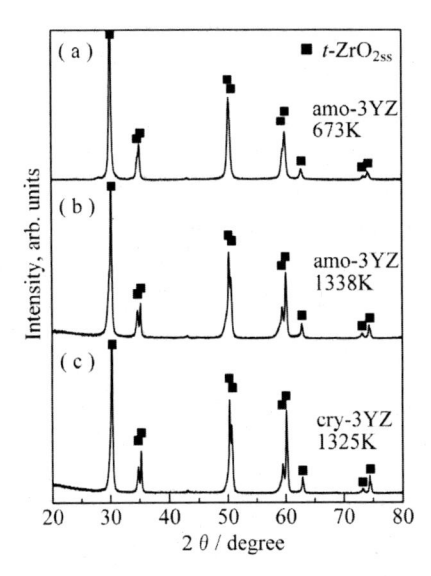

Fig.9 X-ray diffraction patterns for 3YZ compacts consolidated at (a) 673K and (b) 1338K using amo-3YZ powders, and 3YZ compact consolidated at (c) 1325K using as-received cry-3YZ powder. Relative densities of 3YZ compacts for (a), (b) and (c) are marked by arrows in Fig. 8.

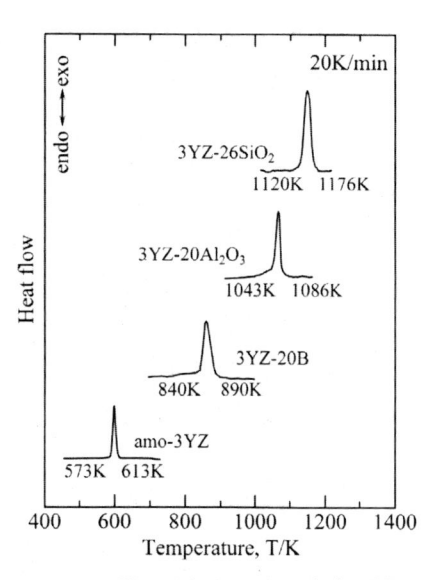

Fig.10 Differential thermal analysis (DTA) curves for amorphous powders of amo-3YZ, 3YZ-2B, 3YZ-20B, 3YZ-26SiO$_2$ and 3YZ-20Al$_2$O$_3$ at constant heating rate of 20K/min.

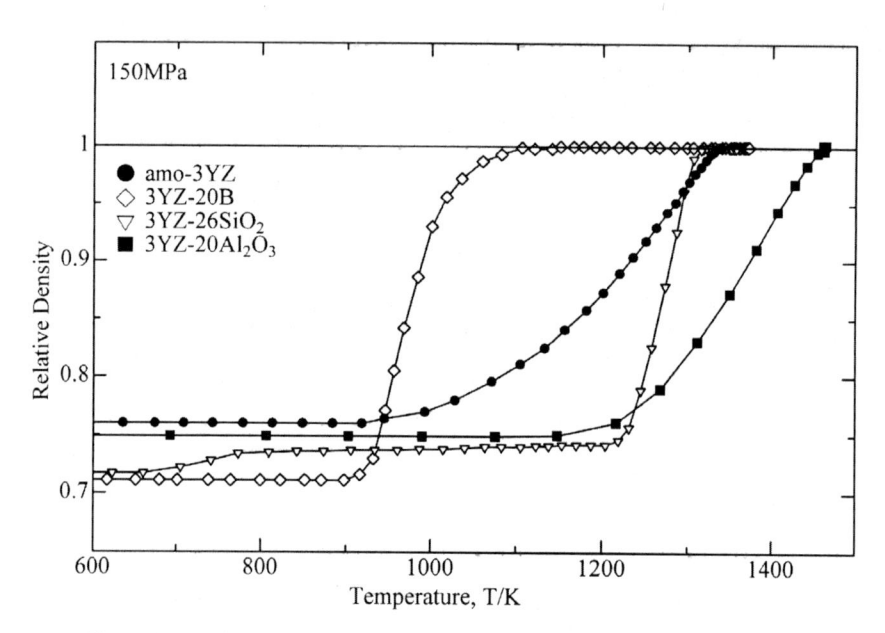

Fig.11 Change in relative density as a function of temperature during consolidation of amorphous powders of amo-3YZ (●), 3YZ-20B (◇), 3YZ-26SiO$_2$ (▽) and 3YZ-20Al$_2$O$_3$ (■) under constant applied stress of 150MPa.

(898K). Interestingly, as shown by DTA analysis (see Fig.10), T_{rd} for 3YZ-26SiO$_2$ and 3YZ-20B are quite similar to the temperatures of the end of exothermic peak (T_f), i.e., finishing temperature of crystallization, for 3YZ-26SiO$_2$ (1176K) and 3YZ-20B (890K), respectively. In the case of Al$_2$O$_3$ addition, T_{rd} (1148K) is different from and higher than T_f (1086K). The same tendency, i.e., $T_{rd} > T_f$, is also observed in amo-3YZ, although T_{rd} for 3YZ-Al$_2$O$_3$ is higher than that for amo-3YZ (903K). Moreover, the small slopes of the D-T curves for 3YZ-Al$_2$O$_3$ and amo-3YZ are similar.

According to XRD analysis (Fig.12), all the 3YZ base bulks consist mainly of t-ZrO$_{2ss}$, although a small amount of monoclinic (m-) ZrO$_{2ss}$, cubic (c-) ZrO$_{2ss}$ and alpha-Al$_2$O$_3$ peaks are also observed in 3YZ-20B, 3YZ-26SiO$_2$ and 3YZ-20Al$_2$O$_3$, respectively. In order to clarify the effect of Boron on the phase structure, the analyzed XRD pattern of 3YZ-2mol%B (3YZ-2B) bulk was also shown in Fig.12. As the B content increases from 2 to 20mol%, the intensities of m-ZrO$_{2ss}$ peaks increase, indicating that B promotes the t- to m-ZrO$_{2ss}$ phase transformation.

Table 2 Lattice Parameters for Dominant Crystalline Phase in Bulk Samples Obtained by Consolidation of As-received Crystalline 3YZ (cry-3YZ) and Mechanically Milled Amorphous 3YZ (amo-3YZ) and 3YZ Base Powders

Sample Designation	Temperature	Phase Structure (**minor phase)	Lattice Parameter (nm)	Tetragonality	Volume (nm^3)
amo-3YZ	1400K	tetragonal	a = 0.510212 c = 0.517724	1.0147	0.13477
cry-3YZ	1349K	tetragonal	a = 0.510105 c = 0.517519	1.0145	0.13466
3YZ-2B	1400K	tetragonal (+ monoclinic**)	a = 0.509456 c = 0.518093	1.0170	0.13447
3YZ-20B	1383K	tetragonal (+ monoclinic**)	a = 0.508897 c = 0.518726	1.0193	0.13434
3YZ-26SiO$_2$	1410K	tetragonal (+ cubic**)	a = 0.509984 c = 0.518141	1.0160	0.13476
3YZ-20Al$_2$O$_3$	1432K	tetragonal	a = 0.510178 c = 0.517693	1.0147	0.13475

Lattice parameters for the dominant crystalline phase (t-ZrO$_{2ss}$) in the consolidated bulk samples are listed in Table II. All values of lattice parameters, tetragonality and volume of unit cell for the crystalline t-ZrO$_{2ss}$ phase in the amo-3YZ bulk are in good agreement with those in the cry-3YZ bulk, indicating no change in chemical composition during mechanical milling. In the case of SiO$_2$ addition, tetragonality of t-ZrO$_{2ss}$ (1.0160) is slightly larger than that in the amo-3YZ bulk (1.0147), but volume of unit cell of t-ZrO$_{2ss}$ (0.13476nm^3) is quite similar to that in the amo-3YZ bulk (0.13477nm^3). Little difference in volume of unit cell suggests little solubility of Si in t-ZrO$_{2ss}$. Since no other crystalline peaks but zirconia solid solution could be recognized in XRD pattern (Fig.12), the 3YZ-26SiO$_2$ bulk is considered to be composed mainly of the crystalline t-ZrO$_{2ss}$ phase and the SiO$_2$ base amorphous phase as reported in our previous study.[10] In the case of B addition, t-ZrO$_{2ss}$ has large tetragonality and small volume of unit cell as compared to those in the amo-3YZ and the 3YZ-26SiO$_2$ bulks. However, since there is little difference in volume of unit cell of t-ZrO$_{2ss}$ between 3YZ-2B and -20B, solubility of B in t-ZrO$_{2ss}$ is also considered to be restricted as in the case of SiO$_2$ addition. Moreover, all the crystalline peaks in the 3YZ-2B and -20B bulks correspond to t- or

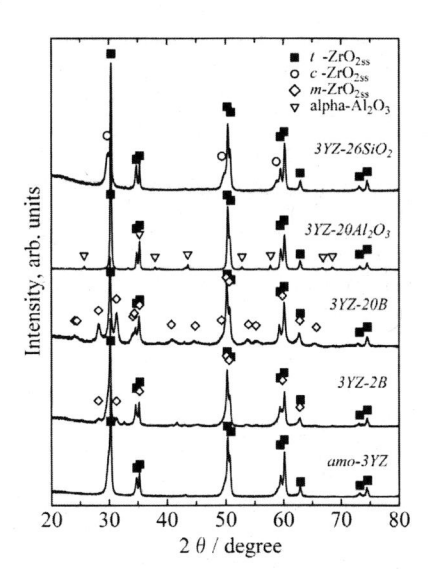

Fig.12 X-ray diffraction patterns of amo-3YZ, 3YZ-2B, 3YZ-20B, 3YZ-26SiO$_2$ and 3YZ-20Al$_2$O$_3$ bulks consolidated at 1400, 1400, 1383, 1410 and 1432K, respectively.

m-ZrO_{2ss}. These results indicate that the B added bulks also consist of the crystalline ZrO_{2ss} and the B base amorphous phases. The large tetragonality and small volume of unit cell of t-ZrO_{2ss} and the promotion of t- to m-ZrO_{2ss} transformation in the 3YZ-2B and -20B bulks are probably due to extraction of Y_2O_3 from t-ZrO_{2ss} by the B (or B_2O_3) base amorphous phase. In the case of Al_2O_3 addition, both of tetragonality (1.0147) and volume of unit cell (0.13475) of t-ZrO_{2ss} are close to those in the amo-3YZ bulk, indicating that Al atoms also hardly be contained in t-ZrO_{2ss}. The absence of the Al_2O_3 base amorphous phase as confirmed by previous TEM microscopy[7] and the presence of crystalline alpha-Al_2O_3 phase (Fig.12) indicate that the 3YZ-20Al_2O_3 bulk sample consists of the crystalline phases of t-ZrO_{2ss} and alpha-Al_2O_3.

All the results mentioned above clearly indicate the difficulty of densification of the 3YZ and 3YZ base amorphous powders below the crystallization temperatures, irrespective of additions of network formers (B and SiO_2) and intermediate oxide (Al_2O_3). The difficulties of viscous flow of the 3YZ and 3YZ base amorphous phases and fabrication of the 3YZ and 3YZ base "amorphous bulks" may be related to absence of the glass transition temperatures of the 3YZ and 3YZ base amorphous phases. Instead, as mentioned above, densification of the amorphous 3YZ powder occurs by plastic flow after full crystallization of t-ZrO_{2ss}. In the cases of B and SiO_2 additions, after crystallization of t-ZrO_{2ss}, which hardly contains B and SiO_2,[11-13] the amorphous phases with low ZrO_2 and high B contents and low ZrO_2 and high SiO_2 contents, respectively, remain and promote rapid densification, resulting in the large slopes of the D-T curves (see Fig.11). On the other hand, in the case of Al_2O_3 addition, the t-ZrO_{2ss} crystals hardly contain Al_2O_3 as in the cases of B and SiO_2 additions, but the remains with low ZrO_2 and high Al_2O_3 contents could not maintain the amorphous structure and transform to alpha-Al_2O_3 crystals. Therefore, the absence of glassy phase after t-ZrO_{2ss} crystallization in 3YZ-20Al_2O_3 causes the small slope of the D-T curve as in the case of amo-3YZ (see Figs.8 and 11).

In the present study, direct TEM observations of the amorphous B (or B_2O_3) and SiO_2 phases coexisting with the crystalline t-ZrO_{2ss} phase have not been carried out. Effects of dispersion morphology and/or volume fraction of these amorphous phases on the densification behaviors are currently under investigation. Apart from the details of microstructure, full densification of amorphous 3YZ powder at low temperature by addition of B seems to be favorable for fabrication of composite bulks composed of 3YZ ceramics and another materials with low full densification temperature (e.g., amorphous TiAl powder).

CONCLUSION

Thermal characteristic and densification behavior of a mechanical milled amorphous yttria partially stabilized zirconia (ZrO_2-3mol%Y_2O_3: amo-3YZ) powder have been investigated. The obtained results are as follows. 1) The onset temperature of rapid densification (T_{rd}, 903K) is considerably higher than the temperature of the end of exothermic peak (T_f, 615K). 2) The compact just before rapid densification shows the crystalline XRD profile, which corresponds to tetragonal ZrO_2 solid solution (t-ZrO_{2ss}), quite similar to that of the full dense bulk. 3) The density-temperature (D-T) curve for the amo-3YZ powder is in good agreement with that for the as-received crystalline 3YZ powder. These results clearly indicate that densification of the amo-3YZ powder proceeds via plastic flow of the crystalline t-ZrO_{2ss} phase rather than viscous flow of the amorphous 3YZ phase.

Effects of ternary element additions (B, SiO_2 and Al_2O_3) to 3YZ on thermal characteristics and densification behaviors have also been investigated. The obtained results are as follows. 1) All of the B, SiO$_2$ and Al$_2$O$_3$ additions cause increase in crystallization temperature, i.e., stabilization of amorphous structure. 2) T_{rd} for 3YZ-B and -SiO$_2$ (898 and 1206K, respectively) are in good agreement with corresponding T_f (890 and 1176K, respectively). In contrast, T_{rd} for 3YZ-Al$_2$O$_3$ (1148K) is higher than corresponding T_f (1086K) as in the case of amo-3YZ. 3) The slopes of the D-T curves for 3YZ-B and -SiO$_2$ are considerably larger than those for amo-3YZ and 3YZ-Al$_2$O$_3$. Based on the XRD results, faster densification in 3YZ-B and -SiO$_2$ seems to be due to the presence of the B (or B$_2$O$_3$) and SiO$_2$ base amorphous phases coexisting with the crystalline ZrO$_{2ss}$ phase. In contrast, slower densification in amo-3YZ and 3YZ-Al$_2$O$_3$ is probably due to the disappearances of amorphous phases.

ACKNOWLEDGMENTS

The author would like to thank Drs. Hiroshi Kimura, Hisao Esaka and Kei Shinozuka, The National Defense Academy, Japan, for the use of some necessary equipment, and to Dr Kazuhiro Hongo, The National Defense Academy, Japan, for technical supports during this investigation.

REFERENCES

[1] R. M. McMeeking and A. G. Evans, "Mechanics of Transformation-Toughening in Brittle Materials," *J. Am. Ceram. Soc.*, **65** [5], 242-246 (1982).

[2] M. Rühle, N. Claussen and A. H. Heuer, "Transformation and Microcrack Toughening as Complementary Processes in ZrO$_2$-Toughened Al$_2$O$_3$" *J. Am. Ceram. Soc.*, **69** [3], 195-197 (1986).

[3] A. H. Heuer, "Transformation Toughening in ZrO$_2$-Containing Ceramics," *J. Am. Ceram. Soc.*, **70** [10], 689-698 (1987).

[4] W. R. Cannon, "Transformation Toughened Ceramics for Structural Applications," *Treatise Mater. Sci. Technol.*, **29**, 195-228 (1989).

[5] H. Kimura, "Non-Equilibrium Powder Processing of Full Density Nanoceramics," pp.55-61 in Proceedings of the 1999 International Conference on Powder Metallurgy & Particulate Materials (PM^{2}TEC'99), *Advances in Powder Metallurgy & Particulate Materials, Part12, Intermetallics, Nanophase Materials & Shape Memory Alloys / Machinability / Structure, Properties & Characterization*, Metal Powder Industries Federation, Princeton, NJ, 1999.

[6] H. Kimura and K. Hongo, "In Progress Nanocrystalline Control Consolidation of the Amorphous ZrO$_2$-20mol%Al$_2$O$_3$ Powder," *Mater. Trans.*, **47** [5], 1374-1379 (2006).

[7] T. Kumagai, K. Hongo and H. Kimura, "Phase Transformation and Densification of an Attrition Milled Amorphous Yttria-Partially Stabilized Zirconia - 20mol% Alumina Powder," *J. Am. Ceram. Soc.*, **87** [4], 644-650 (2004).

[8] R. C. Buchanan and A. Sircar, "Densification of Calcia-Stabilized Zirconia with Borates," *J. Am. Ceram. Soc.*, **66** [2], C20-C21 (1983).

[9] S. Knickerbocker, M. R. Tuzzolo and S. Lawhorne, "Sinterable β -Spodumene Glass-Ceramics," *J. Am. Ceram. Soc.*, **72** [10], 1873-1879 (1989).

[10] T. Kumagai, "Production of Mechanically Alloyed Amorphous Ceramic Powder in ZrO$_2$-SiO$_2$-CeO$_2$-Y$_2$O$_3$ System and its Consolidation by Pressured Sintering," J. Japan Inst. Metals, **69** [8], 769-774 (2005).

[11] T. Ono, M. Kagawa and Y. Syono, "Ultrafine Particles of the ZrO$_2$-SiO$_2$ System Prepared by the Spray-ICP Technique," *J. Mater. Sci.*, **20** [7], 2483-2487 (1985).

[12] V. S. Nagarajan and K. J. Rao, "Crystallization Studies of ZrO_2-SiO_2 Composite Gels," *J. Mater. Sci.*, **24** [6], 2140-2146 (1989).

[13] F. del Monte, W. Larsen and J. D. Mackenzie, "Stabilization of Tetragonal ZrO_2 in ZrO_2-SiO_2 Binary Oxides," *J. Am. Ceram. Soc.*, **83** [3], 628-634 (2000).

Silicon Based Ceramics

THE INTERGRANULAR MICROSTRUCTURE OF SILICON NITRIDE BASED CERAMICS

L.K.L. Falk, N. Schneider[+], Y. Menke[+] and S. Hampshire[+]
Department of Applied Physics, Chalmers University of Technology
SE-412 96 Gothenburg, Sweden
[+]Materials and Surface Science Institute, University of Limerick
Limerick, Ireland

ABSTRACT
The intergranular microstructure of liquid phase sintered silicon nitride based ceramics is a consequence of the oxynitride liquid that forms during sintering. A tailored starting powder composition containing a combination of metal oxide and nitride additives that would form an oxynitride glass-ceramic may result in a liquid phase sintering medium that is readily crystallized after densification leaving a minimum of residual glass. This work is concerned with the incorporation of B-phase, which is a five-component phase, into the intergranular regions of silicon nitride ceramics.

INTRODUCTION
The strong covalent bonding and atomic arrangement of α- and β-Si_3N_4 result in a combination of good inherent mechanical and chemical properties. These structures have, however, extremely low self-diffusivities below their decomposition temperatures. As a consequence, the fabrication of dense Si_3N_4-based ceramic materials requires a sintering aid that promotes densification and inhibits decomposition [1-3]. The sintering additives will in general introduce an intergranular microstructure that will govern the properties of the ceramic material [3-6]. It becomes desirable to control the fraction of intergranular volume and its structure and elemental composition.

This paper will discuss the development of intergranular microstructure in liquid phase sintered Si_3N_4-based ceramics, and the incorporation of oxynitride glass-ceramic phases. In particular, investigations of five-component B-phase will be described. B-phase has often been referred to as Y_2SiAlO_5N, but this composition is now under review.

LIQUID PHASE SINTERED Si_3N_4 CERAMICS
Metal oxide and nitride sintering additives react with the inherent surface silica present on the Si_3N_4 starting powder particles above relevant eutectic temperatures [1,3,7]. This leads to the formation of a liquid phase sintering medium. Some of the Si_3N_4 may also participate in the formation of this liquid and this may lower the eutectic temperature of the system [7,8]. The liquid serves as mass transport medium, and densification is achieved through particle rearrangement and solution/reprecipitation of Si_3N_4 followed by coalescence [7,9]. In addition, secondary crystalline phases may partition from the liquid during the sintering process [10,11]. The remaining part of the liquid phase sintering medium will cool as a glass, see Figures 1 and 2.

A number of the crystalline phases that form in these systems are not stable at the high sintering temperatures used. The volume of residual glass may, then, be reduced through a post-densification heat treatment [1-3,12,13]. A complete crystallization of the glassy pockets may, however, not be achieved. The glass may support hydrostatical stresses that oppose crystallization [14]. Thin intergranular films of residual glass generally remain, even when substantial crystallization of the intergranular volume has been achieved, see Figure 2. These films are rich in oxygen and cations originating from the sintering additives [10,11].

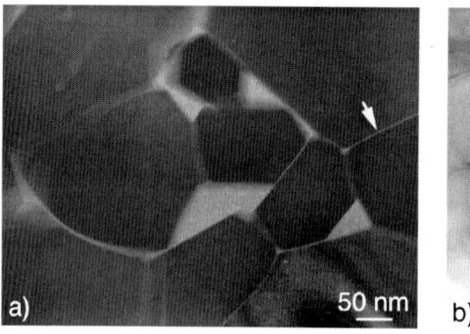

Figure 1: (a) The residual intergranular glass (bright contrast) in a Si_3N_4 ceramic sintered with the addition of Y_2O_3. Thin glassy films (arrowed) are present in the grain boundaries. (b) α'-Si_3N_4 grains and intergranular Sm-melilite (m) in an α-sialon microstructure fabricated with the addition of Sm_2O_3.

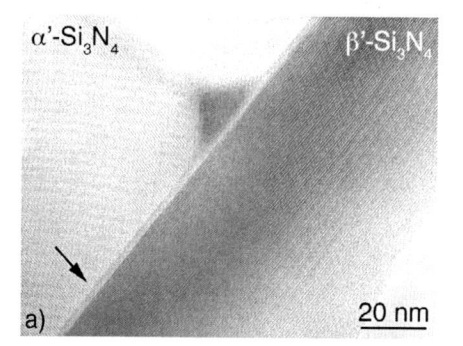

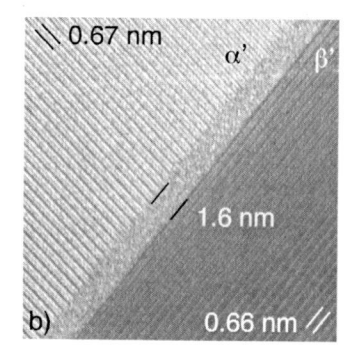

Figure 2: A partly crystallized triple grain junction in a duplex α/β-sialon microstructure fabricated with the addition of Yb_2O_3. Thin films of residual glass are present in the grain boundaries. Figure (b) is a high resolution detail of the arrowed boundary in (a).

The incorporation of liquid phase constituents into solid phases during sintering will result in a reduced volume of residual glass. This may be achieved in two different ways, either through the partitioning of secondary crystalline phases as discussed above, or through the formation of solid solutions based on the α- and β-Si_3N_4 structures. β-sialon (β'-Si_3N_4) forms when silicon in the β-Si_6N_8 structure is substituted by aluminium, and some nitrogen at the same time is replaced by oxygen for the retention of charge neutrality [15,16]. Such anion and cation substitutions in the α-$Si_{12}N_{16}$ structure give, together with the incorporation of an additional interstitial cation, α-sialon (α'-Si_3N_4) [16].

A partly crystallized intergranular microstructure in a Si_3N_4-based ceramic may be viewed as an oxynitride glass-ceramic. A tailored starting powder composition containing a combination of metal oxide and nitride additives that would form an oxynitride glass-ceramic may, hence, result in a liquid phase sintering medium that is readily crystallized after densification leaving a minimum of residual glass. The formation of just one intergranular crystalline oxynitride phase would be of particular interest.

SiAlON B-PHASE GLASS-CERAMICS

A number of investigations have been concerned with the formation glass-ceramics in yttrium and rare earth Si-Al-O-N systems [17-30]. Appropriate combinations of glass composition and crystallisation heat treatment schedule for the formation of just one five-component crystalline phase have been established. These single-phase glass-ceramics contain either B-phase or the related high temperature phase I_w [25-27,29,30].

B-phase glass-ceramics may be fabricated through the nucleation and crystallization heat treatment of nitrogen-rich parent glasses with composition (e/o) 35R:45Si:20Al:83O:17N, where R stands for Er, Yb, Y or a mixture of Y and Yb [25,26,29,30]. The element R determines the degree of crystallization, and the temperature range for single B-phase formation. The crystallization at 1050 °C of an yttrium B-phase parent glass results in a single crystalline phase glass-ceramic with very little residual glass [26]. When the yttrium is fully replaced by ytterbium, J-phase ($Y_4Si_2O_7N_2$) [22,23] forms in addition to B-phase at this crystallization temperature, see Figure 3.

Figure 3: The microstructure after crystallization heat treatment of an ytterbium B-phase parent glass at 1050 °C for 10 h. Elongated B-phase crystal sections (B) and dendritic J-phase (J) are surrounded by a residual glass.

Freely growing B-phase crystals adopt a lenticular shape, see Figure 4. Imaging and electron diffraction in the transmission electron microscope (TEM) has shown that there is a relationship between the crystal shape and the hexagonal B-phase structure (space group $P6_3/m$ [31]). The crystals grow preferentially parallel to the basal plane of the hexagonal lattice so that the axis of the lenticular crystal is in the direction of the c-axis of the lattice [26], see Figure 5.

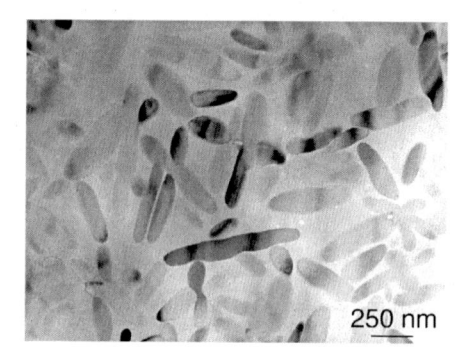

The Intergranular Microstructure of Silicon Nitride Based Ceramics

Figure 4: Lenticular B-phase crystals surrounded by residual glass in an erbium B-phase glass-ceramic crystallized at 1050 °C for 1 h following a nucleation heat treatment at 948 °C for 50 h.

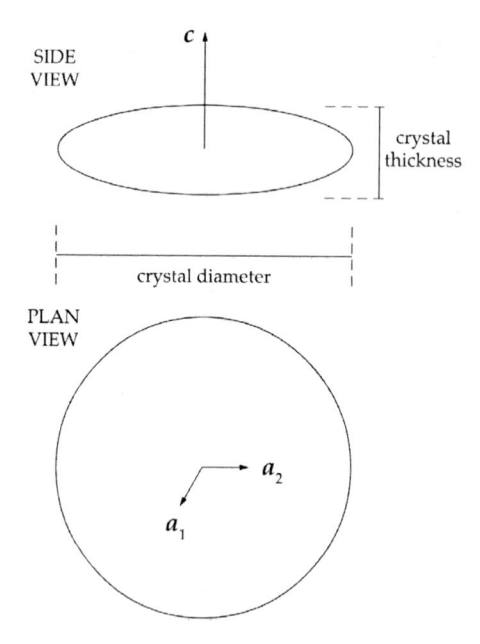

Figure 5: The relationship between the lenticular B-phase crystal shape and the hexagonal lattice. After Young et al. [26].

Point analysis by energy dispersive X-ray spectroscopy (EDX) in the TEM has shown that B-phase takes up an extensive solid solution range, and that there is a clear anti-correlation between the element R and silicon [26,29,30,32,33], see Figure 6. This has been observed also in the related I_w structure [27]. In addition, it was found that a larger R^{3+} cation radius moves the B-phase composition to lower R contents, and as a consequence of the anti-correlation with silicon, the silicon solid solution range goes to higher values. The EDX results lend support to the proposition that the B-phase structure consists of a two-dimensional network of randomly linked $(Si,Al)(O,N)_4$ tetrahedra between layers of R^{3+} cations [31]. In addition, the EDX results suggest that, apart from the random substitution of silicon by aluminium in the $(Si,Al)(O,N)_4$ tetrahedra, a locally increased density in the bi-dimensional network of randomly oriented tetrahedra is associated with an increased density of vacancies in the R^{3+} cation lattice [29,30,32,33]. The wide range of yttrium B-phase composition varies around a cation ratio that may be rationalized to Y:Si:Al = 3:2:1 [26]. This corresponds to a ratio in e/o of 45:40:15.

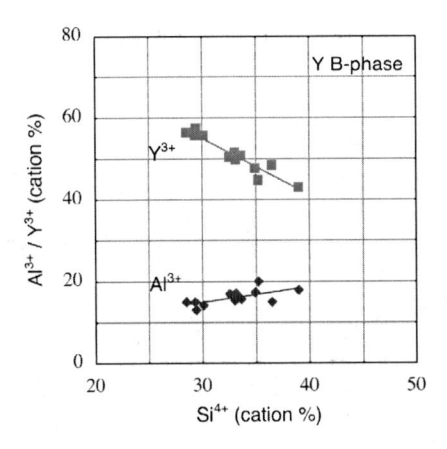

Figure 6: The cation solid solution range of yttrium B-phase after crystallization heat treatment at 1050 °C for 10 h following a nucleation heat treatment at 965 °C for 1 h. Data taken from reference [26].

Si_3N_4 / B-PHASE GLASS MATERIAL

A powdered yttrium B-phase parent glass has been used as the sintering additive to Si_3N_4 in order to produce a Si_3N_4 ceramic with a controlled intergranular microstructure [34]. An appropriate mixture of metal oxides, with the same cation contents as the glass, was also used in order to give a reference material. The additive cation content was (e/o) 35Y:45Si:20Al, and the total additive concentration was 10 wt%. Green bodies were pressureless sintered at 1800 °C for 2 h in a nitrogen atmosphere. Post-sintering heat treatments were carried out with the aim of producing intergranular B-phase from the glass residue. A nucleation heat treatment at 960 °C was followed by a crystallisation heat treatment at 1050 °C. The holding time at the two temperatures was 10 h. These two temperatures have previously been determined as the optimum nucleation and crystallization temperatures for yttrium B-phase [24].

The use of B-phase parent glass as the sintering additive resulted in substantial crystallization of the intergranular volume. Pockets at multi grain junctions contained several smaller crystals with different crystallographic orientations, see Figure 7. The crystals were surrounded by residual glass, and thin glassy films were present also in Si_3N_4 grain boundaries. The presence of many secondary crystals in an intergranular pocket indicates that a number of nuclei formed and grew in the glass pocket during the post-sintering heat treatment.

The intergranular crystals were rich in silicon and yttrium. The yttrium content varied between 18 and 38 e/o, which is lower than the values observed in yttrium B-phase crystals in glass-ceramics [26,34]. These B-phase crystals have yttrium contents that are in the range 38 to 53 e/o after a prolonged crystallization heat treatment at 1050 °C. Some of the analysed intergranular crystals also contained smaller amounts of aluminium (less than 10 e/o). The aluminium content of the crystals was, hence, significantly lower than that of glass-ceramic B-phase. The residual glass surrounding the secondary crystals contained silicon, yttrium and aluminium, and showed significant variations in local cation composition, see Figure 8.

The intergranular microstructure of the Si_3N_4 ceramic fabricated with the addition of a mixture of oxides was virtually the same as that of the Si_3N_4 / B-phase glass material [34].

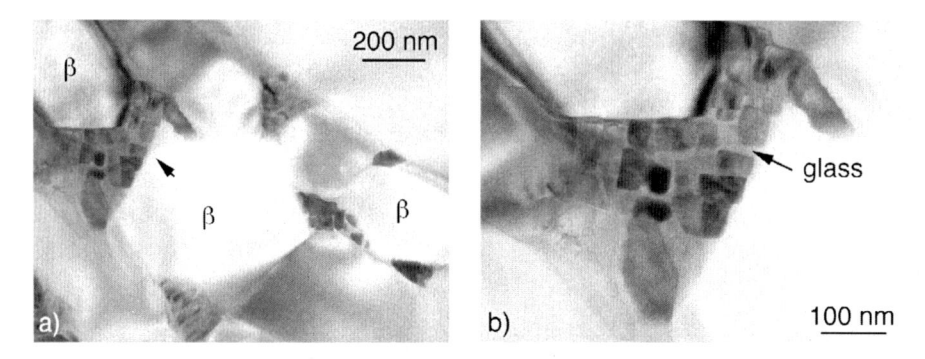

Figure 7: Intergranular pockets with faceted secondary crystals in the material fabricated with the addition of a B-phase parent glass. The image in (b) is a high magnification detail of the pocket arrowed in (a).

Elemental analysis of the intergranular regions confirmed that Si_3N_4, in addition to the inherent surface silica present on the Si_3N_4 starting powder particles, participated in the formation of the liquid phase sintering medium, see Figure 8. As a consequence, the composition of the overall intergranular phase became significantly more silicon-rich than the sintering additive cation composition [34]. The inherent surface silica, which in this case was estimated to 2.6 wt% of the Si_3N_4 starting powder, would shift the sintering additive cation composition from (e/o) 35Y:45Si:20Al to 26Y:59Si:15Al. This means that the relative silicon concentration of the sintering additive cation composition increased at a constant yttrium : aluminium ratio as illustrated by the arrow in the section through the Jänecke prism in Figure 8. A liquid phase sintering medium that incorporated the surface silica only would, hence, not account for the observed compositions of the intergranular regions as illustrated in Figure 8.

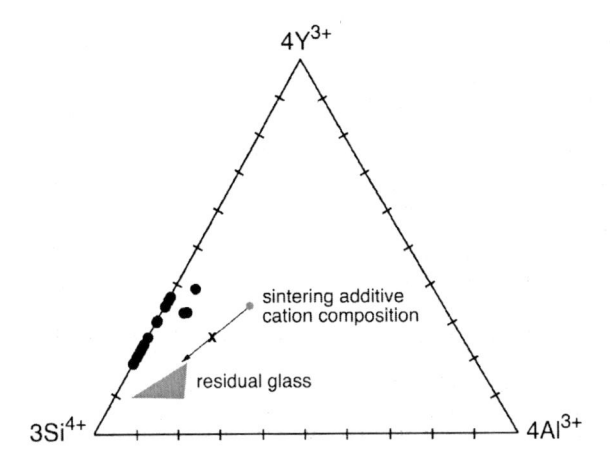

Figure 8: The results from EDX point analyses of the intergranular regions in the Si_3N_4 material fabricated with the addition of a B-phase parent glass. The results are projected onto a section through the Jänecke prism with a constant oxygen : nitrogen ratio (e/o) containing the B-phase parent glass composition (17 e/o nitrogen). Cation compositions of the secondary crystalline phase are marked with black circles. The range of cation composition of the residual glass is also indicated. The sintering additive cation composition including the surface silica present on the Si_3N_4 starting powder is marked by **x**. The arrow indicates the shift in liquid cation composition caused by the Si_3N_4 starting powder.

Analysed β-Si_3N_4 grains did not contain any detectable amounts of aluminium in any of the two microstructures. If, despite this, a dilute β'-Si_3N_4 solid solution had formed, the z-value would be extremely low (< 0.06). This is in contrast to other observations; the addition of Al_2O_3 has generally been observed to result in the formation of β'-Si_3N_4 solid solutions with detectable amounts of aluminium [10,11].

CONCLUDING REMARKS

In sintered silicon nitride ceramics, some of the Si_3N_4 that is dissolved in the liquid phase sintering medium does not reprecipitate as β-Si_3N_4, but remains in the intergranular oxynitride regions. As a consequence, the intergranular microstructure is not only richer in silicon and oxygen than the overall composition of the sintering additives, but contains also an increased amount of nitrogen. This adjustment of the liquid phase composition is inherited by the intergranular structure in the sintered material, and has to be taken into account when preparing tailored starting powder compositions.

ACKNOWLEDGEMENTS
We gratefully acknowledge the provision of financial support for this work from the European Commission under the TMR Research Networks scheme, grant number: FMRX – CT96 – 0038 (DG 12 – ORGS). Drs William Young, Hervé Lemercier and Valerie Peltier-Baron, former post.-docs. at Chalmers and Limerick, are thanked for their contribution to this research. We are grateful for useful discussions with our colleagues in the TMR NEOCERAM network: Prof. D. P. Thompson, University of Newcastle upon Tyne, Prof. R. Harris, University of Durham, Prof. J. P. Descamps, Ecole Polytechnique de Mons, Dr F. Cambier, Belgian Ceramics Research Centre, Prof. P. Goursat, University of Limoges and the late Prof. J.-L. Besson, ENSCI, Limoges.

REFERENCES
[1] F. F. Lange, Silicon Nitride Polyphase Systems: Fabrication, Microstructure and Properties, *Int. Met. Rev.*, **247**, 1-20 (1980).
[2] K. H. Jack, Silicon Nitride, Sialons, and Related Ceramics, *Ceramics and Civilization, Vol. III, High-Technology Ceramics – Past, Present, and Future*, (American Ceramic Society, Columbus, OH, 1986) pp. 259-288.
[3] M. H. Lewis and R. J. Lumby, Nitrogen Ceramics: Liquid Phase Sintering, *Powder Metallurgy*, **26**, 73-81 (1983).
[4] R. Raj, Fundamental Research in Structural Ceramics for Service Near 2000 °C, *J. Am. Ceram. Soc.*, **76**, 2147-2174 (1993).
[5] M. H. Lewis, Sialons and Silicon Nitrides; Microstructural Design and Performance, *Mat. Res. Soc. Symp. Proc. Vol. 287, Silicon Nitride Ceramics - Scientific and Technological Advances*, edited by I.-W. Chen, P. F. Becher, M. Mitomo, G. Petzow and T.-S. Yen (Materials Research Society, Pittsburg, 1983) pp. 159-172.
[6] M. Knutson-Wedel, L. K. L. Falk and T. Ekström, Characterization of Si_3N_4 Ceramics Formed with Different Oxide Additives, *J. Hard Materials*, **3**, 435-445 (1992).
[7] S. Hampshire and K. H. Jack, The Kinetics of Densification and Phase Transformation of Nitrogen Ceramics, *Special Ceramics 7*, edited by D. Taylor and P. Popper (British Ceramic Research Association, Stoke-on-Trent, 1981) pp. 37-49.
[8] S. Hampshire, R. A. L. Drew and K. H. Jack, Oxynitride Glasses, *Phys. Chem. Glasses*, **26**, 182-186 (1985).
[9] W. D. Kingery, Densification during Sintering in the Presence of a Liquid Phase. I. Theory *J. Appl. Phys.*, **30**, 301-306 (1959).
[10] L. K. L. Falk, Imaging and Microanalysis of Liquid Phase Sintered Silicon-based Ceramic Microstructures, *J. Mater. Sci.*, **39**, 6655-6673 (2004).
[11] L. K. L. Falk, Electron Spectroscopic Imaging and Fine Probe EDX Analysis of Liquid Phase Sintered Ceramics, *J. Eur. Ceram. Soc.*, **18**, 2263-2279 (1998).
[12] D. R. Clarke and F. F. Lange, Oxidation of Si_3N_4 Alloys: Relation to Phase Equilibria in the System Si_3N_4-SiO_2-MgO, *J. Am. Ceram. Soc.*, **63**, 586-593 (1980).
[13] L. K. L. Falk and G. L. Dunlop, Crystallisation of the Glassy Phase in a Si_3N_4 Material by Post-Sintering Heat Treatments, *J. Mater. Sci.*, **22**, 4369-4376 (1987).
[14] R. Raj and F. F. Lange, Crystallization of Small Quantities of Glass (or Liquid) Segregated in Grain Boundaries, *Acta Metallurgica*, **29**, 1993-2000 (1981).

[15]M. H. Lewis, B. D. Powell, P. Drew, R. J. Lumby, B. North and A. J. Taylor, The Formation of Single-Phase Si-Al-O-N Ceramics, *J. Mater. Sci.*, **12**, 61-74 (1977).
[16]K. H. Jack, The Characterization of α-Sialons and the α-β Relationships in Sialons and Silicon Nitrides, *Progress in Nitrogen Ceramics*, edited by F. L. Riley (Martinus Nijhoff Publishers, The Hague, 1983) pp. 45-60.
[17]G. Thomas, C. Ahn and J. Weiss, Characterisation and Crystallization of Y-Si-Al-O-N Glass, *J. Am. Ceram. Soc.*, **65**, C185-188 (1982).
[18]G. Leng-Ward and M. H. Lewis, Crystallization in Y-Si-Al-O-N glasses, *J. Mater. Sci. Eng.*, **71**, 101-111 (1985).
[19]T. R. Dinger, R. S. Rai and G. Thomas, Crystallization behaviour of a glass in the Y_2O_3-SiO_2-AlN system, *J. Am. Ceram. Soc.*, **71**, 236-244 (1988).
[20]G. Leng-Ward and M. H. Lewis, Oxynitride Glasses and their Glass-ceramic Derivatives, in *Glasses and Glass-ceramics*, Edited by M. H. Lewis. Chapman and Hall, London, 1990.
[21]J.-L. Besson, D. Billiers, T. Rouxel, P. Goursat, R. Flynn and S. Hampshire, Crystallization and Properties of a Si-Y-Al-O-N Glass-ceramic, *J. Am. Ceram. Soc.*, **76**, C2103-2105 (1993).
[22]S. Hampshire, E. Nestor, R. Flynn, J. L. Besson, T. Rouxel, H. Lemercier, P. Goursat, M. Sabai, D. P. Thompson and K. Liddell, Oxynitride Glasses: Properties and Potential for Crystallization to Glass Ceramics, *J. Euro. Ceram. Soc.*, **14**, 261-268 (1994).
[23]K. Liddell, J. Parmentier, D. P.Thompson, L. Audouin, D. Foster, P. Goursat, N. Schneider, H. Lemercier, S. Hampshire, W. T. Young, L. K. L. Falk, P. R. Bodart, R. K. Harris, G. Massouras, J.-L. Besson, M. Gonon, J.-C Descamps and F. Cambier, Structural Characterisation of New Y-Si-Al-O-N Oxynitride Glass Ceramics, *Key Eng. Materials*, **132-136**, 794-797 (1997).
[24]J. L. Besson, H. Lemercier, T. Rouxel and G. Troillard, Yttrium Sialon Glasses: Nucleation and Crystallization of $Y_{35}Si_{45}Al_{20}O_{83}N_{17}$, *J. Non-Cryst. Sol.*, **211**, 1-21 (1997).
[25]H. Lemercier, R. Ramesh, J.-L. Besson, K. Liddell, D. P. Thompson and S. Hampshire, Preparation of Pure B-Phase Glass-Ceramics in the Yttrium-Sialon System, *Key Eng. Materials*, **132-136**, 814-817 (1997).
[26]W. T. Young, L. K. L. Falk, H. Lemercier, V. Peltier-Baron, Y. Menke and S. Hampshire, The Crystallization of the Yttrium-Sialon Glass: $Y_{15.2}Si_{14.7}Al_{8.7}O_{54.1}N_{7.4}$, *J. Non-Cryst. Sol.*, **270**, 6-19 (2000).
[27]I. MacLaren, L. K. L. Falk, A. Diaz and S. Hampshire, Effect of Composition and Crystallization Temperature on Microstructure of Y- and Er- Si-Al-O-N I_w-Phase Glass-Ceramics, *J. Am. Ceram. Soc.*, **84** [7], 1601-1608 (2001).
[28]M. J. Pomeroy, E. Nestor, R. Ramesh and S. Hampshire, Properties and Crystallization of Rare Earth SiAlON Glasses Containing Mixed Trivalent Modifiers, *J. Am. Ceram. Soc.*, **88**, 875-881 (2005).
[29]Y. Menke, L. K. L. Falk and S. Hampshire, The Crystallization of Er-Si-Al-O-N B-Phase Glass-Ceramics, *J. Mater. Sci.*, **40**, 6499-6512 (2005).
[30]Y. Menke, L. K. L. Falk and S. Hampshire, Effect of Composition on Crystallization of Y/Yb-Si-Al-O-N B-Phase Glasses, *J. Am. Ceram. Soc.*, accepted 2006.
[31]M. F. Gonon, J.-C. Descamps, F. Cambier and D. P. Thompson, Crystal Structure Determination of Y_2SiAlO_5N "B-Phase" by Rietveld Analysis, *Mater. Sci. Forum*, **325-326**, 325-334 (2000).
[32]L. K. L. Falk, Y. Menke and S. Hampshire, The B-phase Solid Solution Range in Yttrium and Rare Earth Sialon Systems, *Silicates Industriels*, **69**, 119-124 (2004).
[33]L. K. L. Falk, Y. Menke and S. Hampshire, The Crystallisation of B-Phase Glass-Ceramics, *Advances in Science and Technology*, **45**, 30-35 (2006).

The Intergranular Microstructure of Silicon Nitride Based Ceramics

[34]L. K. L. Falk, N. Schneider, Y. Menke and S. Hampshire, The Intergranular Oxynitride Microstructure in Silicon Nitride Based Ceramics, *Mater. Sci. Forum*, accepted 2006.

ULTRAFINE POWDERS DOPED WITH ALUMINIUM IN SICN SYSTEM

Vincent Salles, Sylvie Foucaud and Paul Goursat
SPCTS, UMR CNRS 6638, Faculté des Sciences et Techniques
123 Avenue Albert Thomas
87060 Limoges, France

ABSTRACT

Densification of carbide and nitride silicon-based composites usually needs alumina and yttria as sintering aids. An homogeneous microstructure of these materials requires an atomic distribution of Al and Y elements. In the two last decades, studies on polymers, like polycarbosilanes and polysilazanes, led to the synthesis of hybrid precursors. Thus, metallic elements (Al, Y) can be introduced and the chemical composition of multicomponent ceramics can be controlled to atomic level.
A preliminary study on hexamethyldisilazane (HMDS = $Si_2(CH_3)_3NH$), and trimethylaluminium (TMA = $Al(CH_3)_3$), allowed to obtain a silazane with various aluminium contents and a suitable viscosity. The synthesis was carried out at room temperature. The reaction, followed by mass spectrometry and FTIR spectroscopy, was confirmed by methane evolution and formation of Al-N bonds between HMDS and TMA.
Then, a thermal spray-pyrolysis process was set up in order to elaborate pre-alloyed powders. Using an ultrasonic nebulizer, droplets of the aluminosilazane were carried towards the furnace where the organic/inorganic conversion took place. The fabricated powders (80 nm in average diameter) were collected on filter-barriers. Pyrolysis parameters : furnace temperature (1200 and 1400 °C), atmosphere composition (Ar, Ar/NH_3) and gas flow rate (1 and 3 L.min^{-1}), were investigated as well as Si/Al precursor content. The elemental composition of Si/C/N/Al/O powders was determined for the different batches. The morphology and the structural properties were characterized by SEM, FTIR and NMR.

INTRODUCTION

Silicon nitride-based materials are known for their good thermomechanical properties. However their densification needs sintering aids introduction which often induces microstructural heterogeneities.[1-4] To overcome this phenomenon, different ways were explored mainly from the organometallic precursor pyrolysis (polycarbosilane or polysilazane). Whatever the process used, material properties are still low owing to the presence of cracks, porosity or segregated free carbon resulting from the release of gaseous species during thermal treatment.[5]
For all these reasons, the production of dense silicon nitride pieces from nanopowders followed by a sintering step, is a more realistic process. But, conventional methods with powder mixing in liquid media involve the use of dispersing agents and Si_3N_4-based powders don't have equivalent surface groups as oxides such as Al_2O_3 and Y_2O_3. Therefore, powder blends stabilization and final material homogeneity are very difficult to obtain.[6] A new route from pre-alloyed nanopowders (Si/C/N/Al/O) seems to be promising. The main interest of this approach is to favour an atomic scale distribution of sintering aids which should lead to an homogeneous microstructure after densification. Morever, nanostructured materials can be fabricated if an ultrafine multielement powder method is used. Several researchers have worked or are still

working in this area,[7-12] mainly using the laser-pyrolysis process, but their investigations concern mixtures of non mixible precursors. They didn't realize an unique precursor containing metallic elements needed for sintering.

The first papers about the fixation of boron or aluminum on silazanes,[13-17] to obtain directly monolithic ceramics after pyrolysis, have been published in the last two decades. From these results, the synthesis of an hybrid precursor with a suitable viscosity for aerosol formation have been achieved. Then, the thermal conversion was investigated in order to produce pre-alloyed nanopowders containing Si/C/N/Al/(O) elements.

EXPERIMENTAL PROCEDURE
Precursor synthesis

Hexamethyldisilazane (HMDS = $(Si(CH_3)_3)_2NH$) and trimethylaluminum (TMA = $(CH_3)_3Al$), in a 2 mol.L^{-1} solution in hexane were of commercial grade (Aldrich). Handling and reactions were carried out under a controlled atmosphere of argon (99.999 purity). The precursor synthesis is described in a previous paper.[18] HMDS reactant, kept into a glove box under argon, was transfered in the reaction flask. Then, TMA was slowly added at room temperature by a syringe. After the addition of TMA, the solution was stirred for 15 hours at 20 °C. Finally, hexane was separated from the hybrid precursor by distillation at 46 °C, under a partial pressure of 100 mbar for 1 hour. Different Al contents were studied, corresponding to Si/Al molar ratios : 5, 10, 15 and 30.

The aluminosilazane characterizations were performed by mass spectrometry (TA instrument SDT 2960, 1-55 UMA) and Fourier Transform IR spectrometry (Spectrum One FTIR, Perkin-Elmer). All the samples were prepared in a glove box. Before IR analysis a silazane drop was deposited between two KBr windows, and the cell protected from ambient air pollution by a tight film.

Powder synthesis

The conversion of the aluminosilazane into ultrafine powders was performed in a thermal pyrolysis equipment. The aim was to obtain ultrafine powders with high thermal stability and a versatile composition. The aluminosilazane was directly introduced in an aerosol generator device (RBI, Meylan, France). The frequency and the power of the ultrasonic nebulizer were adjusted to obtain the aerosol with a suitable yield (a 1 mL.min^{-1} aerosol rate for a 3 L.min^{-1} carrier gas rate). Its temperature was held to 30 °C by a regulated water circulation. An argon flow was used to carry the fine droplets towards the pyrolysis furnace, equipped with an alumina tube. During pyrolysis, the total pressure was fixed to 1000 mbar. The as-formed particles were finally trapped in two collectors containing filter-barriers made of microporous alumina (pore size # 1 µm). The gaseous species were evacuated by a vacuum pumping system, and by a liquid nitrogen trap. Pyrolysis conditions were studied through three parameters : carrier gas flow rate (1 and 3 L.min^{-1}), furnace temperature (1200-1400 °C) and atmosphere composition (Ar, Ar/NH$_3$). To avoid the entire decomposition of ammonia, this gas was introduced at 700°C in the alumina tube. All the powders were stored into a glove box under argon.

The analytical techniques used for the characterization were : SEM using a Philips XL 30, in which powders were scattered in an ethanol solution, using a drop of the suspension for observation after solvent evaporation, specific surface area measurements (Micromeritics –

ASAP 2010, BET 8 pts, N_2), FTIR spectroscopy (Spectrum One FTIR, Perkin-Elmer, using KBr pellets method), ^{29}Si, ^{27}Al and ^{13}C NMR spectroscopy (Brücker spectrometer, 400 MHz/89 mm, Service de RMN du solide, GDPC, Université de Montpellier II). For elemental analysis (Service central d'Analyse du CNRS, Vernaison, France) : Si and Al were determined by ICP-AES after chemical dissolution, C by combustion to 1800 °C under oxygen, and N by pyrolysis to 3000 °C under He in a graphite crucible.

RESULTS AND DISCUSSION
Precursor synthesis

A gas evolution is visible as soon as TMA is added to HMDS at room temperature. Species corresponding to 15 and 16 m/z are detected by mass spectrometry. It has been shown that the release of ammonia during heat treatment of silazanes starts only above 90-100 °C.[19] Consequently, m/z = 16 and 15 seem to correspond to methane and to a by-product CH_3^+ formed in the quadrupole.

In Figure 1, FTIR spectra for pure HMDS and HMDS after reaction with TMA are reproduced for Si/Al = 30. The product spectrum contains new bands that could be assigned to Al-N bonds. The reaction with TMA induces a decrease of the N-H band (1181 cm^{-1}) which implies that N-H is the silazane reactive group. Taking into account the molecular structure of the starting compounds, the methane formation should result from a reaction between N-H and Al-CH_3 groups.

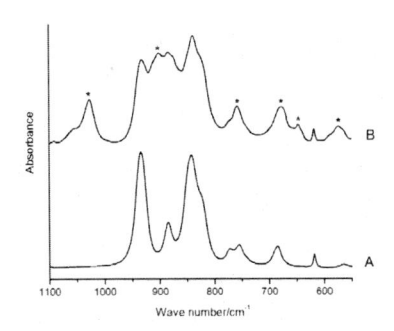

Figure 1. FT-IR spectra of HMDS (A) and the aluminosilazane (B); * Al-N bond

These results prove that aluminium can be fixed on a silazane monomer by direct reaction at room temperature between HMDS and TMA, with a methane release according to reaction (1).

(1)

In addition to previous papers on synthesis of solid polyaluminosilazanes from tetramethyldisilazane and TMA (or dimethylaluminum amide), this study shows that colorless liquid aluminosilazanes can be prepared with Si/Al ratios higher than 5.[17] But the apparent viscosity increases with the Al content in the precursor.

Powder synthesis

A particular nomenclature is used to identify each powder batch (Table 1) which summarizes pyrolysis conditions (temperature, atmosphere, gaseous flow rate) and Si/Al ratio.

Powder batches	Pyrolysis temperature (°C)	Atmosphere	Total flow rate (L.min^{-1})	initial Si/Al molar ratio
12A1L30	1200	Ar	1	30
12A3L30			3	
14A1L30	1400		1	
14A3L30				
14A3L15			3	15
14A3L10				10
14AN3L15(85/15)		Ar/NH$_3$ (85/15 vol.)		15

Table 1. Pyrolysis conditions and nomenclature

Influence of pyrolysis temperature and carrier gas flow rate

Spherical grains with a rather narrow granulometric distribution are observed on SEM micrographs (Figure 2). The particle size seems to be unchanged whatever the decomposition temperature in the studied range. On the contrary, the flow rate variation implies a change of the grain size distribution. The more the flow rate increases the smaller the particles are. In the furnace, the precursor decomposition leads to the formation of gaseous species, and then to solid particles by reaction between these radicals. In these conditions, the growth process of the particles is controlled by the residence time of the gaseous species in the reactional zone, which is related to the carrier gas flow rate.

These changes are confirmed by the specific surface area and the values of equivalent disc diameter calculated with a density of 2.4 measured by helium pycnometry (Table 2). These as-fabricated powders are dense since the calculated diameters are consistent with those of observed particles (Figure 2).

Figure 2. SEM micrographs of powders synthesized at 1200-1400°C with 1-3 L.min^{-1} flow rate : 12A1L30 (A), 12A3L30 (B), 14A1L30 (E) and 14A3L30 (F).

batches	Specific surface area (m^2.g^{-1})	Equivalent disc diameter (nm) [a]
12A1L30	18	139
12A3L30	32	78
14A1L30	20	132
14A3L30	32	78
14A3L15	29	86
14AN3L15(85/15)	69	36

[a] d_{eq} (nm) = 6000 / (d_{th} . S_{BET}) ; with d_{th} = 2.4 and S_{BET} expressed in m^2.g^{-1}.

Table 2. Specific surface area and particle size

The elemental compositions of powders synthesized at 1200 and 1400°C, as well as theoretical (initial mixture) and final (measured) C/N, C/Si, Si/N and Si/Al ratios are shown in Table 3. Pyrolysis temperature in the studied range don't have a significant effect on powder composition.

Whatever the pyrolysis conditions, C appears to be the major element. Flow rate changes, from 1 to 3 L.min^{-1}, induce a decrease of carbon content of about 10 wt.%. Consequently, Si, N and Al contents logically increase. This effect is clearly shown by C/N and C/Si ratios which are reduced to 30 % with an increase from 1 to 3 L.min^{-1}. Resulting from silazane decomposition, carbon containing species are faster evacuated which limits their recombination in solid particles. A parallel behaviour is observed with a laser pyrolysis process, for shorter residence times in the reactional zone, carbon contents are lower for the conversion of HMDS under argon.[20]

At 1400 °C, Si/N values are similar whatever flow rate conditons, but they are slightly higher than theoritical ones (2.0). This can be explained by a decrease in nitrogen content due to transamination reactions which involves ammonia evolution during precursor heating.[21] Nevertheless, this loss is less significant than compositional changes induced by carbon content variations.

The Si/Al ratio changes with the flow rate show that the Al content in the powder increases with the shorter residence time of the precursor species in the pyrolysis furnace.

batches	Composition (wt. %)				Molar ratio							
	Si	C	N	Al	C/N		C/Si		Si/N		Si/Al	
					th.[a]	final	th.[a]	final	th.[a]	final	th.[a]	final
12A1L30	40.3	52.2	11.6	0.3	6.2	5.3	3.1	3.0	2.0	1.7	30.0	129.1
12A3L30	47.5	34.8	15.2	0.5		2.7		1.7		1.6		91.3
14A1L30	37.6	49.7	8.7	0.2		6.6		3.1		2.1		212.5
14A3L30	46.0	40.8	10.7	0.4		4.4		2.1		2.1		100.5

^{a} calculated from quantities of reactants used for the aluminosilazane synthesis.

Table 3. Chemical composition of nanopowders

The X-ray analysis have shown the formation of an amorphous structure with the appearance of two peaks corresponding to β–SiC and graphite carbon.[22] In the Figure 3, ^{29}Si NMR spectra peaks can been attributed to environments such as SiC$_4$ and/or SiC$_2$N$_2$.[23] Other peaks corresponding to SiN$_4$, SiCN$_3$ and SiC$_3$N, have a low intensity. It means that silicon has a carbon rich environment, which is in agreement with the high carbon content in the powders detected by elemental analysis.

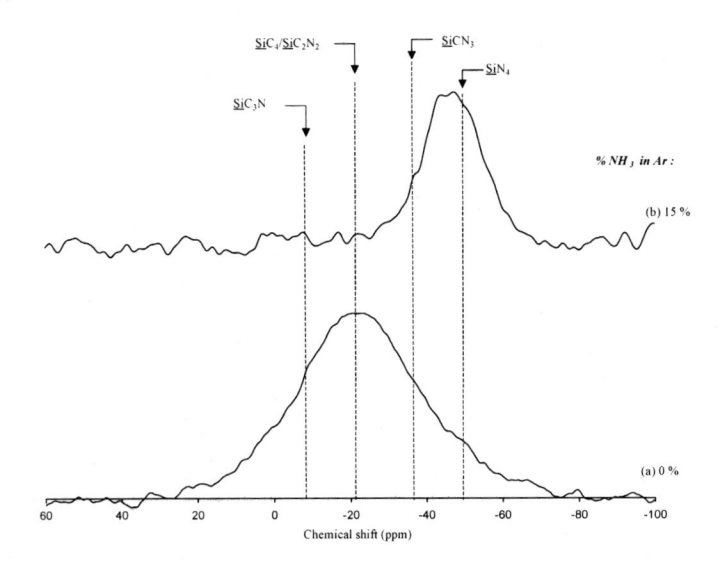

Figure 3. ^{29}Si NMR spectra of powders synthesized without ammonia (a) and with 15 vol. % in argon (b)

Influence of the pyrolysis atmosphere

The presence of high content of free carbon in Si_3N_4/SiC powders has a deleterious effect on the sintering behaviour and also on the thermomechanical properties of the final materials. The efficiency of the pyrolysis under ammonia to reduce the carbon content was demonstrated in several previous investigations.[11, 23, 25] So ammonia was added to argon with different volumic fractions (total flow rate of 3 L.min^{-1}).
Pyrolysis temperature which doesn't influence the final composition and morphology of the powders was fixed at 1400°C to promote the precursor decomposition and thus improve the powder thermal stability. The 3 L.min^{-1} flow rate seems to be the more efficient because it limits the residence time in the reactional zone with a decrease of particle size and a higher carbon loss.

The Table 4 presents the chemical compositions of powders issued from the same precursor pyrolysed under argon or with 15 vol. % of NH_3 in argon.

batches	Composition (wt.%)					Molar ratio			
	Si	C	N	Al	O	C/N	C/Si	Si/N	Si/Al
14A3L15	36.6	45.8	11.9	0.5	-	4.5	2.9	1.5	74.9
14AN3L15 (85/15)	51.6	4.6	37.0	1.2	5.1	0.1	0.2	0.7	41.3

Table 4 : Chemical composition of nanopowders

A comparison of the chemical compositions shows an important decrease of carbon content from 45.8 to 4.6 wt.% with 15 vol. % of ammonia in the pyrolysis atmosphere, and the carbon loss is related to the presence of nitrogen in the pyrolysis atmosphere. According to themodynamical studies, ammonia which is introduced in the furnace at about 700°C isn't entirely decomposed.[26] It could react with gaseous species or radicals resulting from the the decomposition of the aluminosilazane. Several mechanisms involving ammonia are mentionned in literature.[27, 28] The observed nitridation during the thermal conversion of the aluminosilazane could be due to a nucleophilic displacement of carbon in the $Si-CH_3$ groups. Ammonia in the reactional zone influences also the Al content in the synthesized powders. Comparing the chemical composition, ammonia addition induces an increase of Al content. Several competing mechanisms occur in the alumina tube during the thermal conversion.
Al-N bonds (Ei : - 2.9 eV) being less stable than Si-N (Ei : - 3.42 eV) and Si-C (Ei : - 3.12 eV) bonds, the increase of aluminium content could be explained by a stabilization of Al-N bonds in an atmosphere containing ammonia or nitriding species.

The ^{29}Si NMR spectra (Figure 3) are also representative of the nitridation mechanism with changes in the environment of silicon. While the pyrolysis under argon enhances carbon rich environments (SiC_4/SiC_2N_2) about 20 ppm, the pyrolysis under argon/ammonia mixture promotes nitrogen rich environments ($SiCN_3$, SiN_4) about -35 and -50 ppm.

Compositional changes of powders versus pyrolysis parameters (total gas flow rate and the nature of the pyrolysis atmosphere) are summarized in the ternary Si-C-N diagram (Figure 4).

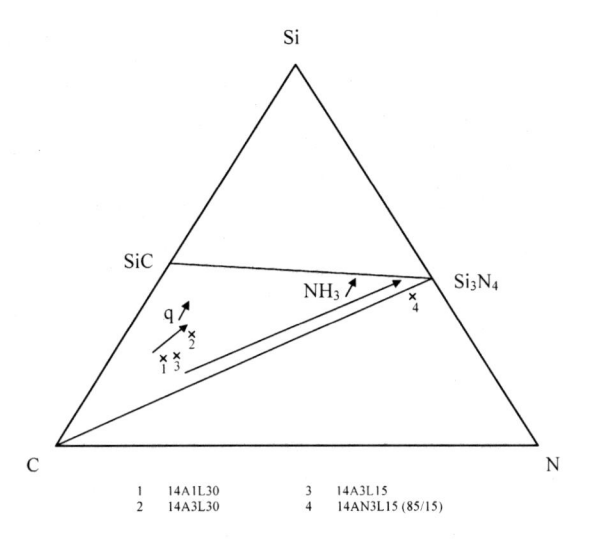

Figure4. Ternary Si-C-N diagram

The carbon amount decreases as the gas flow rate increases but the Si/N ratio remains constant. In order to reach to the composition of the binary Si_3N_4/SiC system, the residence time in the reactive zone should be reduced. But free carbon content is still too high whatever pyrolysis conditions.

Pyrolysis carried out under a pure argon atmosphere (points 1,2 and 3) lead to a mixture of SiC-Si_3N_4 and free carbon. On the contrary, for a pyrolysis under Ar/NH_3 (point 4), nitrogen content increases with the ammonia amount. Ammonia induces a decrease of free carbon but the shift toward the SiC/Si_3N_4 system is difficult to manage. An alternative to reduce carbon content could be the use of a mixed atmosphere like H_2/NH_3 which should allow a better control of the C/N ratio. [29]

CONCLUSION

To improve the microstructural homogeneity and therefore the thermomechanical properties of silicon nitride based composites, the synthesis of pre-alloyed powders from an hybrid precursor seems promising.

A liquid aluminosilazane with a wide range of composition (Si/Al = 5, 10, 15 and 30) and a suitable viscosity can be obtained by the reaction between HMDS and TMA at room temperature. The spray-pyrolysis of this hybrid precursor leads to spherical SiCNAl(O) ultrafine powders with a narrow particle size (80 nm in average diameter). Various compositions in the Si/C/N system can be obtained by changes in experimental parameters (gas flow rate, pyrolysis atmosphere). The addition of ammonia to argon induces a decrease of carbon content and the stabilization of Al-N bonds during the organic / inorganic conversion. Experiments are in progress to explain the nitridation mechanisms and the thermal stability of the as-synthesized powders.

ACKNOWLEDGEMENTS

The authors wish to thank P. Gaveau for the solid NMR analyses (GDPC, Université Montpellier II, France), D. Tétard and E. Laborde for their contribution for the setting of the pyrolysis device (SPCTS, Université de Limoges, France).

REFERENCES

1. G.F. Terwilliger, F.F. Lange, J. Mater. Sc., 1975, **10**, 169-74.
2. C. Greskovich, S. Prochazka and J. H. Rosolowski, in Nitrogen Ceramics, edited by F.L. Riley, 1977, No 23 (Applied Science, Leyden).
3. J.P. Mary, P. Lortholary, P. Goursat and M. Billy, Anvar patent, Fr 7309345, 1973.
4. Th. Chartier, J.L. Besson, P. Goursat, Int. J. High Technology Ceramics, 1986, **2**, 33-45.
5. R. Riedel, G. Passing, H. Schöfelder, R.J. Brook, Nature, 1992, **355**, 714.
6. P. Goursat, D. Bahloul-Hourlier, B. Doucey and m. Mayne, Nanostrucuctured Silicon Based powders and composites, edited by Taylor and Francis, London, 2003, p. 238-263.
7. E. Borsella, S. Botti, R. Fantozzi, R. Alexandrescu, I. Morjan, C. Popescu, T. Dikonimos-Makris, R. Giorgi and S. Enzo, J. Mater. Res., 1992, **7**, 2257.
8. J.S. Haggerty and R.W. Cannon, Laser-induced Chemical Processing, Plenum Press, New-York, 1981, 165-241.
9. M. Cauchetier, O. Croix, M. Luce, M.J. Baraton, T. Merle, P. Quintard, J. Eur. Ceram. Soc., 1991, **8**, 215-19.
10. M. Cauchetier, X. Armand, N. Herlin, M. Mayne, S. Fusil, E. Lefevre, J. Mater. Sc., 1999, **34**, 5257-64.
11. D. Bahloul-Hourlier, B. Doucey, E. Laborde and P. Goursat, J. Mater. Chem., 2001, **11**, 2028-34.
12. M. Mayne., D. Bahloul-Hourlier, B. Doucey, P. Goursat, M. Cauchetier and N. Herlin, J. Eur. Ceram. Soc., 1998, **18**, 1187-94.
13. D. Seyferth, H. Plenio, J. Am. Ceram. Soc., 1990, **73**, 2131-2133.
14. G. D. Soraru, A. Ravagni, R. Campostrini, J. Am. Ceram. Soc., 1991, **74**, 2220-2223.
15. O. Funayama, T. Kato, Y. Tashiro, T. Isoda, J. Am. Ceram. Soc., 1993, **76**, 717-723.
16. Y. Iwamoto, K.-i Kikuta, S.-i. Hirano, J. Mat. Res., 1998, **13**, 353-361.
17. B. Boury, D. Seyferth, Appl. Org. Chem., 1999, **13**, 431-440.
18. V. Salles, S. Foucaud-Raynaud, R. Granet, J.-L. Besson and P. Goursat, Proceeding of Groupe Français de la Céramique, Bordeaux, 2004.
19. N. S. Choong Kwet Yive, PhD Thesis, Université de Montpellier II, 1990.
20. B. Doucey, PhD Thesis, Université de Limoges, 1999.
21. D. Bahloul-Hourlier, B. Doucey, J.-L. Besson and P. Goursat, in Nanostructured silicon-based powders and composites, edited by Taylor and Francis, London, 2003, p. 41.
22. V. Salles, S. Foucaud, E. Laborde, E. Champion and P. Goursat, J. Eur. Ceram. Soc., 2007, **27**, 357-366.

23. R. Dez, F. Ténégal, C. Reynaud, M. Mayne, X. Armand and N. Herlin-Boime, J. Eur. Ceram. Soc., 2002, **22**(16), 2969-2979.
24. A.P. Legrand, J.B. d'Espinose de la Caillerie and Y. El Kortobi, in Nanostructured silicon-based powders and composites, edited by Tylor and Francis, London, 2003, p. 111.
25. M. Birot, J.-P. Pillot and J. Dunoguès, Chem. Rev., 1995, **95**, 1443-1477.
26. G. Soucy, J. W. Juremicz and M.I. Boulos, J. Mater. Sc., 1995, 30, 2008-2018.
27. R.J.P. Corriu, D. Leclercq, P. H. Mutin and A. Vioux, Chemistry of Mater., 1992, 4, 711-716.
28. V. Salles, S. Foucaud, E. Champion and P. Goursat, (paper to be published).
29. M. Cauchetier, E. Musset, M. Luce, N. Herlin, X. Armand and M. Mayne, in Nanostructured silicon-based powders and composites, edited by Taylor and Francis, London, 2003, p. 6.

Properties of Monolithic Ceramics

MODULUS AND HARDNESS OF NANOCRYSTALLINE SILICON CARBIDE AS FUNCTIONS OF GRAIN SIZE

Suraj C. Zunjarrao, Abhishek K. Singh and Raman P. Singh
Mechanics of Advanced Materials Laboratory
School of Mechanical and Aerospace Engineering
Oklahoma State University
Stillwater, Oklahoma 74078
Email: raman.singh@okstate.edu

ABSTRACT
This experimental investigation reports on the mechanical properties of nanocrystalline silicon carbide as a function of the crystallite size. Monolithic SiC samples are fabricated by controlled pyrolysis of a polymer precursor under an inert atmosphere. The degree of crystallization, and the resulting grain size, is controlled by varying the processing conditions. Subsequently, the modulus and hardness are determined using instrumented indentation. Meanwhile the microstructure is characterized using X-ray diffraction. The process yields grain sizes in the range of ~0-12 nm. It is seen that the presence of nanocrystalline domains in amorphous SiC significantly influences the modulus and hardness. A non-linear relationship is observed with optimal mechanical properties for a grain size of 3.5 nm.

INTRODUCTION
Nanocrystalline ceramics are touted to be the key to new super and ultra hard materials. These materials have exhibited promise of unique properties, such as very high hardness and toughness, governed primarily by their microstructure[1-3]. Material microstructure is the most important factor governing the properties of these materials and appropriate tailoring can lead to significantly improved properties. Even common hard ceramic compounds such as silicon carbide can have significantly improved hardness[1]. Although significant work has focused on the improved properties of nanocrystalline materials, especially metals, in the realm of ceramics, most of the existing research has looked on ceramics formed by conventional sintering techniques or, among new techniques, by chemical vapor deposition.

Ceramics derived from preceramic polymers are a fast growing class of advanced materials due to the various advantages they offer, such as low processing temperatures, near net shape fabrication and high ceramic yield. Owing to this, they are being researched for varied applications from disc brakes to nuclear reactor fuels[4]. Since the discovery of polycarbosilane and its subsequent demonstration as a SiC precursor[5], a variety of polymeric precursors to SiC have been formulated and proposed, primarily as binder materials for ceramic powders in the preparation of dense (sintered) ceramic monoliths. Fabrication of ceramics by pyrolysis of preceramic polymers allows the production of highly three-dimensional covalent refractory components that are difficult to fabricate via the traditional powder-processing route[6]. Furthermore, control over the reaction kinetics and microstructure evolution can yield a rich amorphous/nano-crystalline ceramic material[7-9]. Such nano-crystalline ceramics are speculated to show unique properties such as "super hardness" and significantly high toughness[1, 3, 10, 11].

Of the several available SiC polymer precursors, allylhydridopolycarbosilane (AHPCS, Starfire Systems Inc., Malta, New York) is an ultra high purity precursor that yields a near stoichiometric ratio on complete pyrolysis[12]. Its high ceramic yield, relatively low shrinkage and ability to be handled and processed in ambient conditions have attracted wide attention, especially as precursor to SiC fibers and more recently as matrix material[13-16]. Bulk characterization of AHPCS–derived SiC sample, prepared by PIP process, has been reported by Mores et al.[16]. The materials were characterized in terms of density (by immersion method), fracture toughness (by bulk V–notched beam method) and hardness (by bulk Vickers indentation test). Porosity was found to adversely affect the properties, and highest values for fracture toughness and hardness were found to be about 167 MPa.m$^{1/2}$ and 13 GPa, respectively.

There is a need to characterize the properties of these materials in the monolithic domain. In this context, nanoindentation, which is widely used to characterize the mechanical properties at the nano–scale[17], is particularly helpful in determining the true properties of ceramics that have inherent micro–pores. In a recent study, Liao et al.[1] successfully used nanoindentation to determine hardness of super–hard SiC films deposited by thermal plasma chemical vapor deposition. But there is still a need to realize the processing–microstructure–property–application relationship for this organic polycarbosilane.

Currently, nanocrystalline SiC is being pursued for its superior electrical and optical properties[18]. And recently, "superhard" nanocrystalline materials, including SiC materials have gained interest, however, less data exists on their mechanical properties. Vassen et al.[19] reported hardness of up to 27 GPa for bulk–sintered SiC materials with grain sizes as small as 70 nm, while Tymiak et al.[20] reported hardness value of about 37 GPa with grain sizes around 20 nm, for SiC films deposited by hypersonic plasma particle deposition. Liao et al.[1] reported hardness as high as 50 GPa for nanocrystalline SiC films with grain sizes of 10–20 nm, deposited by thermal plasma CVD. Although prime factors influencing the hardness in such materials are nanocrystallinity and grain size, one other factor influencing the hardness in SiC could be the presence of hydrogen. As a result, amorphous SiC can be superhard if free of hydrogen[21]. There is limited data in literature on mechanical characterization of nanocrystalline SiC derived from AHPCS. Considering the increasing interest in the applications of AHPCS and the ease of fabrication of nanocrystalline ceramics from polymer precursor, it would be desirable to explore the mechanical characteristics of AHPCS–derived nanocrystalline SiC.

Material microstructure largely governs the final properties in ceramics, and hence microstructure characterization, as well as, the study of nucleation and crystallization, are vital for understanding structure–property relationships. Work of characterizing the crystal structure of AHPCS–derived SiC and probing its structure–property relations as a function of processing parameters, will help in fundamental understanding of these materials, and is necessary for the use of AHPCS–derived SiC as an effective matrix material or for joining applications.

While very limited literature, if any, exists on microstructural characterization of AHPCS–derived SiC, there is some data on Si–C systems derived by other chemical methods. Mitchell et al.[9] examined the nucleation and crystallization process in Si–C and Si–N–C systems produced by pyrolysis of granulated polymethylsilane and found processing temperature and time to influence the crystal sizes. Nanocrystalline beta–SiC derived from chlorine containing polysilanes/polycarbosilane prepared from poly(chloromethylsilane–co–styrene) was characterized by DTA, TGA, XRD, TEM, mass spectroscopy and infrared spectroscopy by Mitchell et al.[9]. Grain sizes were calculated to be 5–20 nm. Nechanicky et al.[22] reported TGA, FTIR, TEM and XRD studies on alpha–SiC/beta–SiC particulate reinforced composites prepared by PIP using a hyper–branched polymethylsilane (mPMS). Kerdiles et al.[23] used FTIR and HREM to study nano–crystalline SiC in SiC thin films grown by reactive magnetron co–sputtering of SiC and C targets. Hongtao Zhang et al.[24] characterized crystal structure of nanocrystalline SiC by TEM and Raman Spectroscopy in SiC thin films deposited by a plasma–enhanced CVD process. Yongcheng Ying et al.[25] used XRD, TEM and SAED to study the microstructure of nanocrystalline SiC prepared by reacting magnesium silicide (Mg_2Si) and carbon tetra fluoride (CCl_4) in an autoclave and reported crystal sizes of 30–80 nm.

In this paper, we focus on the mechanical property characterization of AHPCS–derived SiC, in terms of hardness and modulus. Ceramics derived from AHPCS are prone to develop cracks and porosity, especially at higher processing temperatures, due to volume shrinkage and evolution of hydrogen gas during pyrolysis[13, 16, 26, 27]. Hence, nanoindentation was used to access the "true" local properties of the material, bypassing the effects of inherent material porosity. Particularly, focus is on correlating these properties with crystallite sizes and degree of crystallinity determined from x–ray diffraction studies. In addition, we study the chemical changes, phase transformations and microstructural changes occurring during the controlled pyrolysis of allylhydridopolycarbosilane, leading to the formation of amorphous/nanocrystalline SiC, under inert atmosphere and at ambient pressure.

MATERIAL PROCESSING

Allylhydridopolycarbosilane, acquired from Starfire Systems Inc., Malta, New York, USA (SMP-10), is an olefin-modified polymer that undergoes pyrolysis when heated under inert atmosphere and yields near stoichiometric SiC. Carefully weighed quantities of liquid AHPCS were heated, under argon atmosphere, from room temperature up to 300°, 500°, 700° and 900°C, respectively. Samples were held at final temperatures for 30 min to ensure thermal equilibrium. A box furnace was fitted with a retort and modified for inert gas pyrolysis up to 900°C, and pyrolysis beyond 900°C was performed in a specially modified high temperature furnace (Model no. F46248, Barnstead International, Dubuque, Iowa, USA). During the initial stages of heating, cross-linking in the polymer is accompanied by loss of volatile components in the precursor which were observed to form yellowish-white deposits on the inner walls of the furnace up to about 700°C. Hence the heating was controlled at the slow rate of 5°C/min, for samples heated up to 700°C, to ensure minimal loss of polymer due to volatilization prior to cross-linking. There is no lose of volatile components beyond this temperature and hence higher heating rates can be safely employed beyond 700°C. Small quantities of amorphous SiC derived from AHPCS pyrolyzed at 900°C were loaded in alumina crucibles and heated to different temperatures of 1150°, 1400° and 1650°C starting from room temperature at 5°C/min and held at final temperature for 30 min, in the high temperature oven under a constant flow of argon. Samples were drawn from the final pyrolyzed product for mechanical and microstructural characterization.

PHYSICAL CHARACTERIZATION

Mass loss of the polymer precursor was carefully monitored by taking precise weight measurements before and after each heating for multiple experiments. Figure 1 shows the ceramic yield obtained as a function of decomposition temperature. The loss in weight is attributed to the loss of low-molecular weight oligomers and hydrogen gas[12]. It should be noted that in our case, volatilization driven mass loss was not limited even with very slow heating rates. Marginal loss in mass was observed beyond 700°C and about 72–74% ceramic yield in the form of amorphous SiC was obtained in the range 900–1650°C. In a separate study[28] on reaction kinetics during the pyrolysis of AHPCS, the polymer pyrolysis was characterized as a three-step process consisting of volatilization, cross-linking and crystallization; and activation energies for the volatilization and cross-linking were determined as 83.1 kJ/mol and 149.7 kJ/mol, respectively, using the mass loss data discussed here.

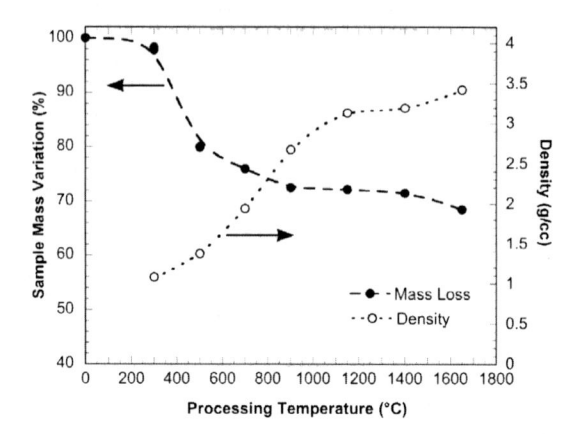

Figure 1. Mass loss and density variation of AHPCS derived SiC as a function of temperature.

Density measurements were performed using helium pycnometry on finely crushed powders obtained by heating the polymer precursor, dried for 2-3 h at 60°C in a drying oven. This method yields true density of inherently porous materials, which cannot be accessed by bulk density measurements. Starting with a liquid AHPCS having a density of 0.997 g/cc (as mentioned by Starfire Systems Inc., USA), a dry and partially cross-linked solid with density about 1.07 g/cc is obtained at 300°C. Further heating results in more cross-linking accompanied by the loss of low molecular weight oligomers and hydrogen gas. As the processing temperatures increase, density is observed to increase steadily until it reaches values that are close to theoretical density for SiC at ~1150°C.

MECHANICAL CHARACTERIZATION
Mechanical properties of SiC derived from AHPCS pyrolyzed to 900°, 1150°, 1400°, and 1650°C were characterized in terms of hardness and elastic modulus using instrumented nanoindentation. Chunks of SiC material from the pyrolysis products were mounted in epoxy (Epoxicure, Buehler, Lake Bluff, Illinois, USA) and polished to a mirror finish using Ecomet 3 polisher (Buehler, USA). Samples were then indented using a sharp Berkovich diamond indenter (Nano Test System, Micro Materials, Wrexham, UK) to peak loads of 50, 75 and 100mN for each material. A total of 10 indentations were performed for every sample. Figure 2 shows typical load–depth curves obtained using peak load of 75mN for SiC pyrolyzed to different temperatures. A change in resistance to indentation offered by the material when processed to different temperatures is quite clear from the figure. While the material processed to 900°C deforms more easily, that processed to 1150°C and 1400°C offers more resistance, which leads to increased hardness and modulus for these materials. Material processed to 1650°C appears to be more complaint. Figures 3 and 4 show the modulus and hardness as a function of processing temperature obtained using three different peak loads of 50, 75 and 100mN.

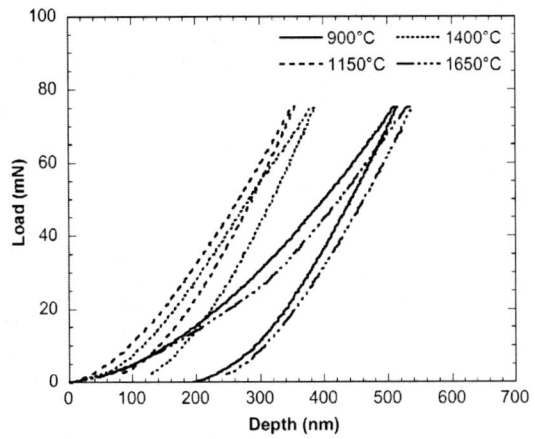

Modulus and Hardness of Nanocrystalline Silicon Carbide as Functions of Grain Size

Figure 2. Typical load–depth curves obtained by nanoindentation using a peak load of 75mN for SiC derived from AHPCS heated to 900° C, 1150°C, 1400°C and 1650°C.

Smooth curves fitted through the data points show an interesting trend of initial increase in modulus and hardness and an eventual drop for material processed at 1650°C. Nominal values of hardness and modulus were obtained for SiC that was pyrolyzed at 900°C. For 50 mN peak load, these values were about 180 GPa for modulus and about 20 GPa for hardness, which are typical for a-SiC.

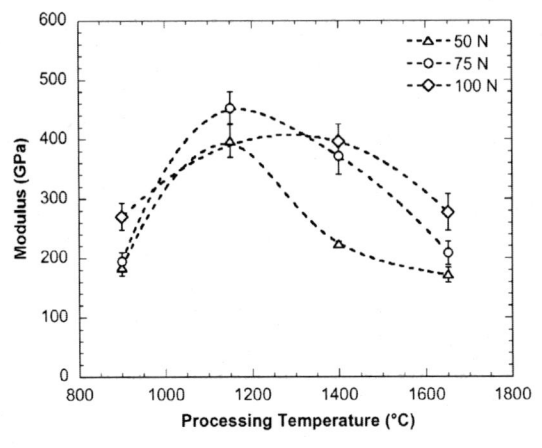

Figure 3. Modulus determined by nanoindentation for SiC derived from AHPCS heated to 900°C, 1150°C, 1400° C and 1650°C.

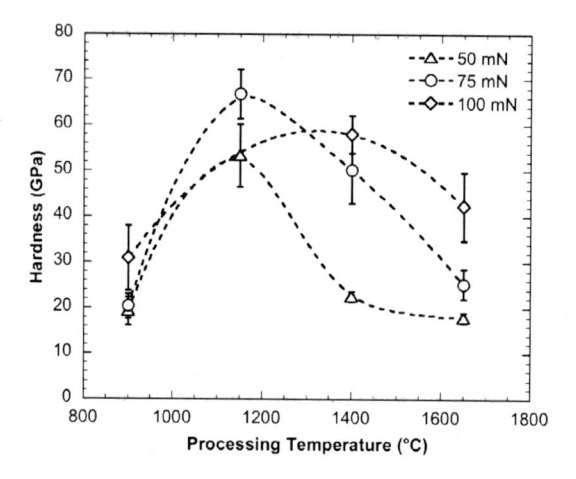

Figure 4. Hardness determined by nanoindentation for SiC derived from AHPCS heated to 900°C, 1150°C, 1400°C and 1650°C.

The values for both hardness and modulus increased SiC processed at 1150° C, and then further dropped down for SiC processed at 1400°C and 1650°C. Highest values of hardness and modulus were observed for the SiC samples processed at 1150 C. For 50 mN peak load, these were 400 GPa for modulus and 54 GPa for hardness. At such high values of hardness, these materials qualify as superhard materials[3]. These results, in conjunction with the crystallite size and degree of crystallinity calculations, make for interesting observations.

MICROSTRUCTURE CHARACTERIZATION

AHPCS is a hyper-branched polymer with structural formula $[R_3SiCH_2-]_a[-SiR_2(CH_2-]_b[=SiR(CH_2-)_{1.5}]_c[=Si(CH_2-)_2]_d$ (where R is H or (-CH$_2$CH=CH$_2$)) that completely decomposes into a network of SiC upon pyrolysis to temperatures above 850°C[16, 29]. The conversion of polymer precursor to SiC was studied using Fourier transform infrared spectroscopy (FTIR) and DTA/TG and reported earlier[30].

X-ray diffraction studies were performed on SiC powders pyrolyzed at 900°, 1150°, 1400° and 1650°C. Powder samples were prepared by wet milling in a planetary ball mill (PM-100, Retsch GmbH, Haan, Germany) for 4h in ethanol and then mounted on glass slide. Powder diffraction patterns were collected using Scintag PAD-X automated diffractometer with a CuKa radiation (λ = 0.1540 nm) using a scanning rate of 0.5 per min and operating at 45 kV and 25 mA. Figure 5 shows the XRD patterns obtained for various samples; data is offset to aid comparison.

Amorphous SiC formed at 900°C shows a greatly diffused peak whereas the peak intensity increases as the processing temperatures increase. Gradual growth of SiC peaks at 2θ values of 35.7°, 60.2° and 72.0° suggests increasing ordering as nano–crystalline domains form and grow in amorphous SiC. It is noted that small peaks for residual tungsten carbide (WC), from the grinding media, are seen in the patterns. Also, even though a peak for WC lies very close to the SiC peak at 35.7°, the prominent peaks at this 2θ are attributed to SiC since the WC peak at 35.6° and 48.3° are expected to be of same intensity according to JCPDS (ICCD 29-1131).

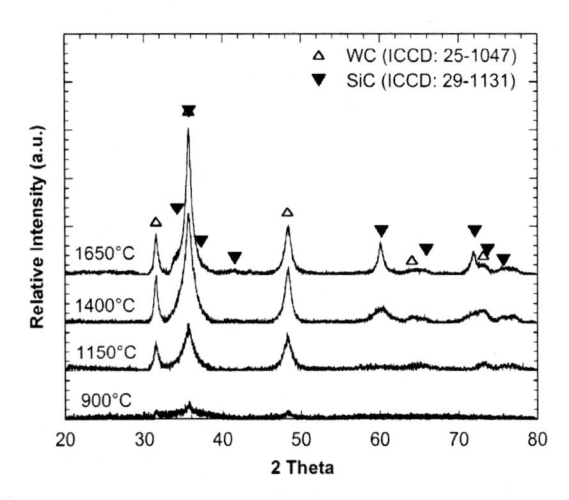

Figure 5. Powder diffraction patterns for SiC derived from AHPCS.

An estimate of the crystallite size was obtained from the peak broadening using the Debye-Scherrer equation[31]. Peak broadening, in terms of full width at half-maximum (FHWM), was determined by fitting the obtained pattern using XFIT program, which uses the pseudo Voigt (PV) and split Pearson (PVII) functions along with a fundamental parameters (FP) convolution approach[32]. Instrument broadening, determined by using NIST-traceable line-width standard LaB6 sample (SRM 696), was accounted for while determining the crystallite sizes at different temperatures. The crystallite sizes were found to be about 3 nm, 6 nm and 11 nm at 1150°, 1400° and 1650°C, respectively. Similar observations, for this material system, were made by Berbon et al.[13].

An estimate of the degree of crystallinity in our SiC samples processed to different temperatures was obtained using an approach similar to that presented by Blanchard and Scwab[33]. The authors determined volumetric degree of crystallinity in amorphous/crystalline Si_3N_4 derived from polysilazane using the area under the XRD patterns. They have successfully demonstrated a good accuracy of their approach by testing standard samples containing known weight fractions of amorphous and crystalline phases. In our case, a preliminary calculation of degree of crystallinity was done using the area under the XRD patterns and assuming SiC obtained at 900°C as 100% amorphous and that obtained at 1650°C as 100% crystalline. Thus, the degree of crystallinity in our samples was determined to be about 45% at 1150°C and about 57% at 1400°C. A more vigorous calculation of the degree of crystallization and validation using standard samples with know weight fractions of amorphous and crystalline phases is being undertaken.

In addition to the XRD patterns, further evidence on the presence of amorphous SiC at 900°C and its polycrystalline nature at higher temperature was seen from electron diffraction patterns. Selected area electron diffraction (SAED) patterns were obtained during transmission electron microscopy (TEM) studies on SiC processed at 900°, 1150°, 1400° and 1650°C and are show in fig. 6. As seen in the figure, greatly diffused concentric rings for SiC processed at 900 C (a) suggests a largely amorphous structure. These rings are seen to gradually become distinct and sharp for SiC processed at higher temperatures (b-d), which is suggestive of growing crystallite size. SAED patterns obtained for SiC processed at 1650°C (d) shows tiny bright specks intermittently along the rings. These are typically seen for nano-sized polycrystalline materials[34].

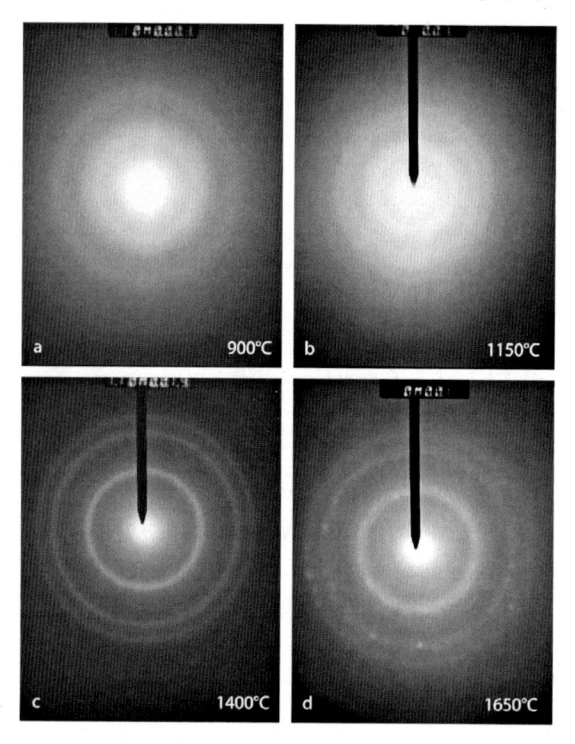

Figure 6. SAED patterns for SiC derived from AHPCS heated to 900°C (a), 1150°C (b), 1400°C (c) and 1650°C (d).

CONCLUSIONS

The fundamental property–structure relationships for SiC derived by inert-gas pyrolysis of allylhydridopolycarbosilane were probed. Onset of crystallization in amorphous SiC was seen close to 1100° C leading to the formation of amorphous/ nanocrystalline SiC with crystallite size, estimated from powder diffraction patterns, found to be in tens of nanometers at 1650°C. Preliminary calculations for the degree of crystallinity were performed using the area under the

XRD patterns. Mechanical property characterization in terms of hardness and modulus were carried out using instrumented nanoindentation to access the local properties and results are viewed in the light of degree of crystallinity and crystallite sizes. The chemical changes, phase transformations and microstructural changes occurring during the controlled pyrolysis of allylhydridopolycarbosilane, leading to the formation of amorphous/nano–crystalline SiC were studied by tracking mass loss, density changes, chemical bonding and evolution of microstructure up to 1650°C. Thus, efforts are made to establish the influence of processing on microstructure and the mechanical properties of AHPCS–derived SiC.

ACKNOWLEDGMENT

The authors would like to thank Wei-Guang Chi, Shanshan Liang and Dr. Sanjay Sampath of the Center for Thermal Spray Research (CTSR), Stony Brook University for their help with helium pycnometry and DTA/TG measurements. The authors would also like to thank the DOE for financial support on this project (Award No. DE-FC07-05ID14673).

REFERENCES

[1] F. Liao, S. Girshick, W. Mook, W. Gerberich and M. Zachariah, "Superhard nanocrystalline silicon carbide films". *Applied Physics Letters*, **86**(17) (2005).
[2] V. Richter and M. Von Ruthendorf, "On hardness and toughness of ultrafine and nanocrystalline hard materials". International Journal of Refractory Metals & Hard Materials, **17**(1-3), 141–152 (1999)
[3] S. Veprek, P. Nesladek, A. Niederhofer, F. Glatz, M. Jilek and M. Sima, "Recent progress in the superhard nanocrystalline composites: towards their industrialization and understanding of the origin of the superhardness". Surface & Coatings Technology, **109**(1-3), 138–147 (1998).
[4] A. K. Singh, S. C. Zunjarrao and R. P. Singh, "Silicon Carbide and Uranium Oxide based Composite Fuel Preparation using Polymer Infiltration and Pyrolysis", Proceedings of ICONE 14, 14th International Conference on Nuclear Engineering, July 17-20, 2006, Miami, USA (2006).
[5] S. Yajima, J. Hayashi, and M. Omori, "Continuous silicon-carbide fiber of high-tensile strength". Journal of Physical Chemistry B(9), 931–934 (1975).
[6] P. Greil, "Polymer derived engineering ceramics". Advanced Engineering Materials, 101(45)(45), 9195–9205 (1997).
[7] J. Wan, M. J. Gasch and A. Mukherjee, "Silicon Nitride/Silicon Carbide Nanocomposites from Polymer Precursor", **23**(4), American Ceramic Society (2002).
[8] G. Danko, R. Silberglitt, P. Colombo, E. Pippel, and J. Woltersdorf, "Comparison of microwave hybrid and conventional heating of preceramic polymers to form silicon carbide and silicon oxycarbide ceramics". Journal of the American Ceramic Society, **83**(7), 1617–1625 (2000).
[9] B. S. Mitchell, H. Zhang, N. Maljkovic, M. Ade, D. Kurtenbach and E. Muller, "Formation of nanocrystalline silicon carbide powder from chlorinecontaining polycarbosilane precursors". Journal of the American Ceramic Society, **82**(8), 2249–2251 (1999).
[10] I. Szlufarska, A. Nakano, and P. Vashishta, "A crossover in the mechanical response of nanocrystalline ceramics". Science, **309**(5736), 911–914 (2005).
[11] Y. Zhao, J. Qian, L. Daemen, C. Pantea, J. Zhang, G. Voronin and T. Zerd, "Enhancement of fracture toughness in nanostructured diamond-sic composites". Applied Physics Letters, **84**(8), 1356–1358 (2004).
[12] L. Interrante, "High yield polycarbosilane precursors to stoichiometric SiC. Synthesis,

pyrolysis and application", **346**, Materials Research Society, Pittsburgh, PA, USA (1994).
[13]M. Z Berbon, D. R. Dietrich, D. B. Marshall and D. Hasselman, "Transverse thermal conductivity of thin c/sic composites fabricated by slurry infiltration and pyrolysis". Journal of the American Ceramic Society, **84**(10), 2229–2234 (2001).
[14]S. Dong, Y. Katoh, A. Kohyama, S. Schwab, and L. Snead, "Microstructural evolution and mechanical performances of sic/sic composites by polymer impregnation/ microwave pyrolysis (pimp) process". Ceramics International, **28**(8), 899–905 (2002).
[15]L. Interrante, J. Jacobs, W. Sherwood and C. Whitmarsh, "Fabrication and properties of fiber- and particulate-reinforced sic matrix composites obtained with (a)hpcs as the matrix source". Key Engineering Materials, 127-131(Pt 1), 271–278 (1997).
[16]K. V. Moraes, and L. V. Interrante, "Processing, fracture toughness, and vickers hardness of allylhydridopolycarbosilane-derived silicon carbide". Journal of the American Ceramic Society, **86**(2), 342–346 (2003)
[17]M. R. VanLandingham, "Review of instrumented indentation". Journal of Research of the National Institute of Standards and Technology, **108**(4), 249–265 (2003).
[18]R. Madar, "Materials science - silicon carbide in contention". NATURE, **430**(7003), 974–975 (2004).
[19]R. Vassen, and D. Stover, "Processing and properties of nanograin silicon carbide". Journal of the American Ceramic Society, **82**(10), 2585–2593 (1999).
[20]M. Hrabovsky, M. Konr´ad, and K. Kopecky, eds., 1999.Vol. 3, Institute of Plasma Physics, Academy of Sciencesof the Czech Republic.
[21]V. Kulikovsky, V. Vorlicek, P. Bohac, A. Kurdyumov and L. Jastrabik, "Mechanical properties of hydrogen-free a-c : Si films". Diamond and Related Materials, 13(4-8), 1350–1355 (2004).
[22]M. Nechanicky, K. Chew, A. Sellinger and R. Laine, "Alpha-silicon carbide/beta-silicon carbide particulate composites via polymer infiltration and pyrolysis (pip) processing using polymethylsilane". Journal of the European Ceramic Society, **20**(4), 441–451 (2000).
[23]S. Kerdiles, R. Rizk, F. Gourbilleau, A. Perez-Rodriguez, B. Garrido, O. Gonzalez-Varona and J. Morante, "Low temperature direct growth of nanocrystalline silicon carbide films". Materials Science and Engineering B - Solid State Materials for Advanced Technology, B, 530–535 (2000).
[24]H. Zhang, and Z. Xu, "Microstructure of nanocrystalline sic films deposited by modified plasma-enhanced chemical vapor deposition". Optical Materials, **20**(3), 177–181 (2002).
[25]Y. Ying, Y. Gu, Z. Li, H. Gu, L. Cheng and Y. Qian, "A simple route to nanocrystalline silicon carbide". Journal of Solid State Chemistry, **177**(11), 4163–4166 (2004).
[26]J. Zheng, and M. Akinc, "Green state joining of sic without applied pressure". Journal of the American Ceramic Society, **84**(11), 2479–2483 (2001).
[27]C. A. Lewinsohn, R. H. Jones, P. Colombo and B. Riccardi, "Silicon carbide-based materials for joining silicon carbide composites for fusion energy applications". Journal of Nuclear Materials, 307-311(2 SUPPL), 1232–1236 (2002).
[28] W. Xiaolin, S. C. Zunjarrao, H. Zhang and R. P. Singh, "Advanced Process Model for Polymer Pyrolysis and Uranium Ceramic Material Processing", Proceedings of ICONE 14, 14th International Conference on Nuclear Engineering, July 17-20, 2006, Miami, USA (2006).
[29]I. Rushkin, Q. Shen, S. Lehman and L. Interrante, "Modification of a hyperbranched hydridopolycarbosilane as a route to new polycarbosilanes". Macromolecules, **30**(11), 3141 – 3146 (1997).
[30]S. C. Zunjarrao, A. K. Singh and R. P. Singh, "Structure-Property Relationships in Polymer

Derived Amorphous/ Nanocrystalline Silicon Carbide for Nuclear Applications", Proceedings of ICONE 14, 14th International Conference on Nuclear Engineering, July 17-20, 2006, Miami, USA (2006).
[31]B. D. Cullity, Elements of X-ray Diffraction. Addison-Wesley Publishing Co. Unc., London (1978)
[32]R. W. Cheary and A. A. Coelho, "Programs xfit and fourya, deposited in ccp14 powder diffraction library, engineering and physical sciences research council, Daresbury Laboratory, Warrington, England (1996).
[33]C. R. Blanchard and S. T. Schwab, "X-ray diffraction analysis of the pyrolytic conversion of perhydropolysilazane into silicon nitride". Journal of the American Ceramic Society, **77**(7), 1729–39 (1994).
[34]D. Williams, and C. B. Carter, Transmission Electron Microscopy: A Textbook for Materials Science. Kluwer Academic / Plenum Publishers, New York (1996).

STOICHIOMETRIC CONSTRAINT FOR DISLOCATION LOOP GROWTH IN SILICON CARBIDE

Sosuke Kondo, Yutai Katoh
Oak Ridge National Laboratory
Material Science and Technology Division
Oak Ridge, TN, 37831 USA

Akira Kohyama
Institute of Advanced Energy
Kyoto University
Gokasho, Uji, Kyoto, 611-0011 Japan

ABSTRACT

A stoichiometric constraint effects on Frank loop growth in β-SiC during irradiation were studied by ion beam irradiation, transmission electron microscopy, and calculation based on a kinetic model. Tilted-ion beam (5.1 MeV Si^{2+}) was irradiated at 1000 °C to induce lattice damage and to implant additional Si interstitial atoms simultaneously. Growth rate of the loops observed within the damaged region appeared to be positively correlated with the deposition rate of Si ions. Analysis based on a kinetic model showed that a small amount of deposited excess Si substantially increased the net flux of Si interstitials, which believed to govern the loop growth rate. The experimental and computational results confirm the stoichiometric constraint effect on Frank loop growth in SiC, and that availability of Si interstitial atoms determines the Frank loop growth rate in most irradiation conditions.

INTRODUCTION

Silicon carbide (SiC) is a refractory ceramic material that exhibits exceptional stability in radiation environments[1]. The radiation-stability of SiC is believed to be related with defect properties and the characteristic development of the microstructure. Recent study of irradiation induced microstructures in high purity β-SiC revealed the temperature and fluence dependent microstructural development[2]. It is likely that isolated vacancies and interstitial clusters are dominant defects in SiC irradiated at low to intermediate temperatures. Predominant growth of interstitial type Frank loops takes place at high temperatures[3]. Progressive microstructural development commences presumably with unstable growth of irradiation induced dislocation loops.

Theoretical models for these microstructural developments in SiC have not been established partly because of complexity of defect evolutions in ceramic compounds. One of the key to understand microstructural development in SiC, which is somewhat ionic compound, is to determine the effect of stoichiometric constraint on defect cluster evolutions. SiC consists of Si and C sublattices, each with different atomic masses and displacement threshold energy, and that results in greater production rate for C Frenkel defects than Si Frenkel defects[4]. Therefore, if the Si/C atomic ratio within Frank loops is forced to maintain stoichiometry, the loop growth rate has to be determined by the supply

Stoichiometric Constraint for Dislocation Loop Growth in Silicon Carbide

of Si interstitials, and in turn that would affect the antisite fractions. The primary objective of present work is to experimentally clarify the stoichiometric constraint effects on Frank loop growth in irradiated SiC using a method of Si-ion irradiation. The effect of Si ion deposition on the loop evolution is also examined by calculation based on a kinetic model and is compared with the experimental result.

IRRADIATION METHOD

Typically, implanted heavy ions lose their kinetic energy mostly through elastic collisions with target atoms in vicinity of the end of the. To avoid the undesirable effects of implanted ions on microstructurul changes[5], it is generally recommended to avoid such regions upon microstructural examination in studies of irradiation damage. However, in this work, deposition of Si ions was taken as an advantage to modify the Si/C ratio of interstitial production during irradiation.

The specimen used were high purity β-SiC poly crystal produced through chemical vapor deposition process (by Rohm & Haas Co., Woburn, MI, USA). The ion-irradiation was carried out at DuET facility[6], Kyoto University. The ion beam (5.1 MeV Si^{2+}) was irradiated from an angle of 36 degrees with respect to the irradiated surface normal to enable a broader distribution in depth of the implanted Si ions. Two samples were irradiated for 1.6 and 4.7 hours (4.1×10^{20} and 1.2×10^{21} ions/m^2 in total, respectively) at 1000 °C. Figure 1 (a) shows the depth profiles of displacement damage rate and the

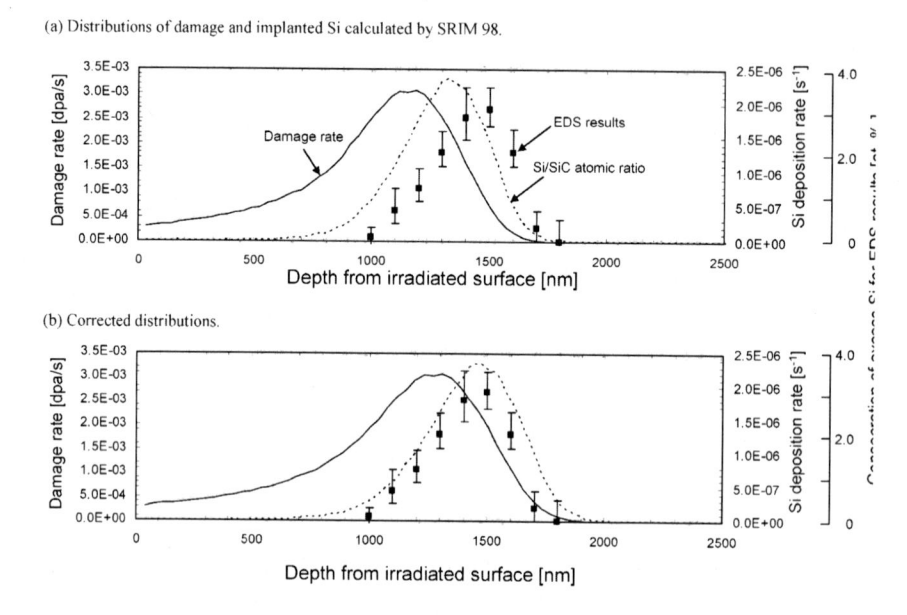

Fig. 1 Depth profile of displacement damage rate and Si deposition rate (deposited Si to the target displacement atomic ratio) in SiC irradiated with 5.1 MeV Si^{2+} ions; (a) calculated by SRIM 98 assuming E_d=35 eV, (b) corrected based on EDS results of Si concentration. Solid squares were obtained from EDS experiment.

Stoichiometric Constraint for Dislocation Loop Growth in Silicon Carbide

deposition rate of implanted Si (expressed by atomic ratio of deposited Si to the target displacement; Si/dpa) calculated by SRIM 98[7]. Results of quantitative excess Si profiling by energy dispersive X-ray spectrometry (EDS) in the implanted specimen are also plotted. The mismatch between SRIM and EDS results observed here is likely to be caused by over-estimated electron stopping power in TRIM calculation[4]. Additionally, crystallographic orientation with beam direction might have affected the ion range. For proper estimation of Si deposition rate at various locations, the TRIM profile was calibrated to fit the EDS profile as shown in Figure 1 (b).

Foils for cross sectional transmission electron microscope (TEM) observations were prepared from the irradiated samples by the focused ion beam method (Micrion JFIB 2100). TEM examination was performed by using JEOL JEM2010 for Frank loop analysis and JEOL JEM2100F for EDS analysis, both of which were operated at 200 keV.

RESULTS OF MICROSTRUCTURAL EXAMINATION

Figure 2 shows the cross sectional TEM image of β-SiC irradiated to 1.2×10^{21} ions/m^2 at 1000 °C, which was taken from near the [110] electron beam direction. Ion-beam has been irradiated from the specimen surface on the left of the image in the direction of an inset arrow. The final distribution of displacement damage and deposited Si are inserted below the TEM image. Edge-on Frank loops formed on specific (111) family plane were shown at the whole damaged area. The loop distribution seems to be well accorded with the calibrated damage profile. Higher magnification images of the loops formed at 4 depths in the sample are shown in Fig. 3. The damage levels are approximately 10 dpa for (a) and (d), and approximately 35 dpa for (b) and (c). The Si/dpa ratio, however, are increased with increasing the depth. In both the damage levels,

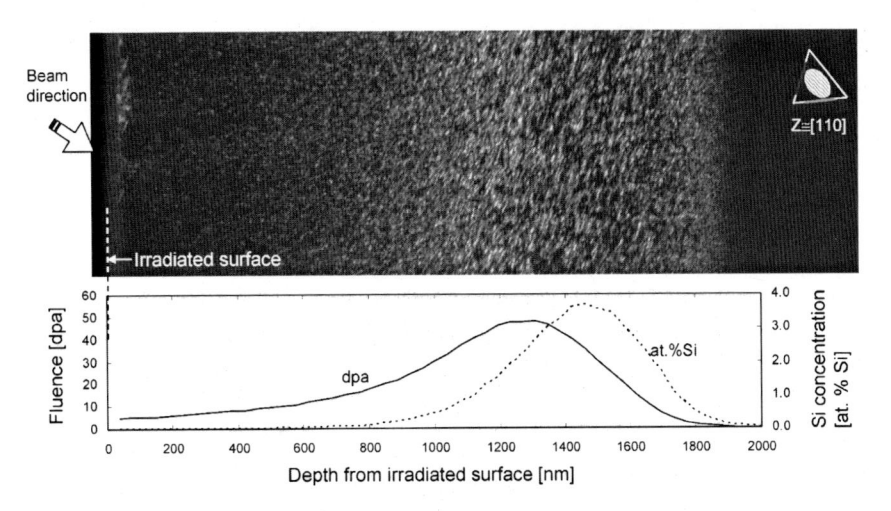

Fig. 2 Cross sectional TEM micrograph showing edge-on Frank loops with Thompson's tetrahedron and the final distribution of the displacement damage and deposited Si for the sample irradiated to 1.2×10^{21} ions/m^2.

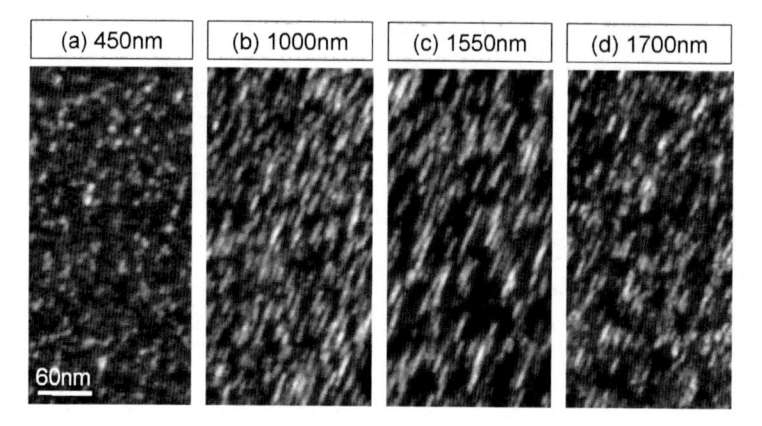

Fig. 3 TEM micrographs of Frank loops formed at (a) 450 nm, (b) 1000 nm, (c) 1550 nm, and (d) 1700 nm, in the sample irradiated to 1.2×10^{21} ions/m^2. Displacement damage and Si/dpa are (a) 10 dpa, 1.5×10^{-4}, (b) 35 dpa, 2.0×10^{-4}, (c) 35 dpa, 1.1×10^{-3}, and (d) 10 dpa, 2.5×10^{-3}, respectively.

the loop size appears to different despite the same damage level. No voids were observed in either of the irradiated specimens.

Mean radius of the loops formed at various depths were plotted with respect to damage rates for both the specimen irradiated to 4.1×10^{20} and 1.2×10^{21} ions/m^2 in Fig. 4. In most cases, a data point was generated from the averaged value of the loops

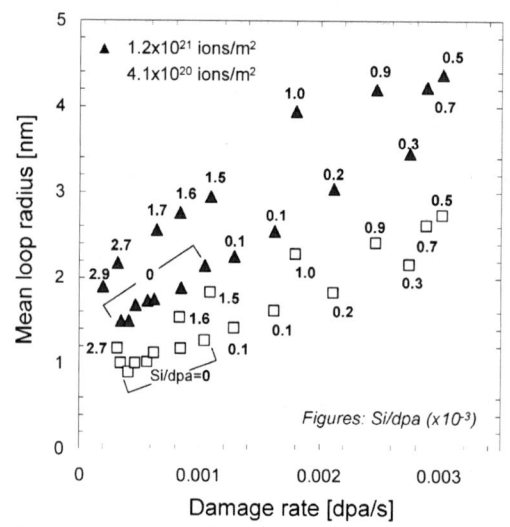

Fig. 4 Experimentally observed mean radius of the Frank loops in SiC irradiated to 4.1×10^{20} ions/m^2 and 1.2×10^{21} ions/m^2 as a function of the displacement damage rate. The data point labels indicate Si deposition to atomic displacement ratio (Si/dpa) divided by 10^{-3}.

observed in the area of 50 nm in depth by 800 nm in width centered at the intended depth. The defects observed in the range between 0 and 150 nm in depth are not included in the results to avoid potential surface effects. The numbers indicated next to the data points are ratios of Si ion deposition to atomic displacement (Si/dpa) at the corresponding depth. Nearly linear trend of loop growth with damage rates were observed below 3×10^{-4} Si/dpa for both irradiated specimens. Loop sizes at higher Si/dpa ratios appear larger than at lower Si/dpa ratios, when compared at comparable damage rates in the same specimen.

Loop growth rates at various depths were roughly estimated from the difference of loop sizes in the two specimens as shown in Fig. 5. Damage rates in relation to the Si/dpa ratio are also shown in the figure, because damage rates largely affect the loop growth rates. It is well-known phenomena in metals and alloys. It is noted that the loop growth rates at 2.8×10^{-3} Si/dpa is approximately three times more than that without Si deposition, when compared at similar damage rates.

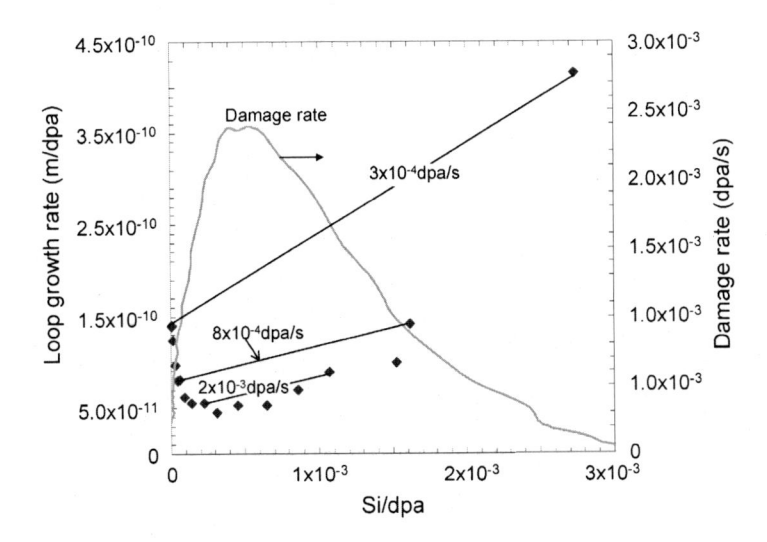

Fig. 5 Loop growth rates as a function of Si deposition to atomic displacement ratio (Si/dpa).

KINETIC MODEL

The experimental results clearly showed the effect of excess Si deposition on loop size, despite the very small Si/dpa ratios. The Si/dpa ratio was only 2.9×10^{-3} at the highest. In an attempt to validate the experimental results obtained, point defect concentrations and loop growth rate were estimated by using simple kinetic models based on rate equations for each defect. Time derivatives for concentrations of interstitial atoms (I_{Si}, I_C), vacancies on both sites (V_{Si}, V_C), and antisite defects (Si_C, C_{Si}) were considered.

Time derivatives of radiation defect concentrations can be written by the following kinetic equations:

Stoichiometric Constraint for Dislocation Loop Growth in Silicon Carbide

$$\frac{dC_{ISi}}{dt} = (\eta f_{Si} + p_{Si})G_{dpa} + \alpha_{Si\ C}^{IC}C_{IC}C_{Si\ C} + \phi_{ISi}^{I-emit}S_{ISi}^{I}$$
$$-C_{ISi}\left[\alpha_{VSi}^{ISi}C_{VSi} + \alpha_{VC}^{ISi}C_{VC} + \alpha_{C\ Si}^{ISi}C_{C\ Si} + D_{ISi}z_{ISi}^{I}S_{ISi}^{I}\right];\quad (1)$$

$$\frac{dC_{IC}}{dt} = \eta f_{C}G_{dpa} + \alpha_{C\ Si}^{ISi}C_{ISi}C_{C\ Si} + \phi_{IC}^{I-emit}S_{IC}^{I}$$
$$-C_{IC}\left[\alpha_{VC}^{IC}C_{VC} + \alpha_{VSi}^{IC}C_{VSi} + \alpha_{Si\ C}^{IC}C_{Si\ C} + D_{IC}z_{IC}^{I}S_{IC}^{I}\right];\quad (2)$$

$$\frac{dC_{VSi}}{dt} = \eta f_{Si}G_{dpa} + \alpha_{C\ Si}^{VC}C_{VC}C_{C\ Si}$$
$$-C_{VSi}\left[\alpha_{VSi}^{ISi}C_{ISi} + \alpha_{VSi}^{IC}C_{IC} + \alpha_{Si\ C}^{VSi}C_{Si\ C} + D_{VSi}z_{VSi}^{I}S_{VSi}^{I}\right];\quad (3)$$

$$\frac{dC_{Si\ C}}{dt} = \alpha_{VC}^{ISi}C_{ISi}C_{VC} - C_{Si\ C}\left(\alpha_{Si\ C}^{IC}C_{IC} + \alpha_{Si\ C}^{VSi}C_{VSi}\right)\quad (4)$$

, where η is a cascade efficiency (assumed η =0.3), p_{Si} is the Si/dpa ratio, G_{dpa} is the displacement damage rate, S^I are the sink strength of loops, and z^I are the bias factors (assumed z^I =1.2 for interstitials, 1 for vacancies). Fraction of Si and C Frenkel pairs to total irradiation-produced defects were calculated by SRIM 98, where 0.32 for Si and 0.68 for C, respectively. The remaining two equations for C_{VC} and $C_{C/Si}$ can be available through changing the subscript Si to C and C to Si. I_{Si} and I_C were treated as under quasi-equilibrium with defects of other types. The last term in equations (1) and (2) are added to compensate for over estimated I_{Si} or I_C influx into loops in defiance of stoichiometric constraint. The I_{Si} or I_C flux for compensation is simply expressed by

$$\phi_{ISi}^{I-emit} = \Delta\phi_{ISi}^{I} - \Delta\phi_{IC}^{I} \quad \text{or} \quad \phi_{IC}^{I-emit} = \Delta\phi_{IC}^{I} - \Delta\phi_{ISi}^{I} \quad (5)$$

, where the $\Delta\phi^I$ are the net defect flux into loops, because these net influxes must have equal value due to stoichiometry constraint. The following defect reactions were included

Table 1. Parameters used for the calculation.

Parameter	Value	Unit	Refs.
Displacement damage rate	1.0 x 10⁻³	dpa/s	Related to experimental condition.
Irradiation temperature	1000	°C	Related to experimental condition.
Initial loop radius	1.4	nm	Obtained from present TEM work.
Loop density	2.4 x 10²²	m⁻³	Obtained from present TEM work.
Displacement fraction for Si	0.32		Calculated by SRIM 98 using E_d= 40eV.
Displacement fraction for C	0.68		Calculated by SRIM 98 using E_d= 20eV.
Migration energy for I_{Si}	1.4	eV	[11]
Migration energy for I_C	0.5	eV	[11]
Migration energy for V_{Si}	3.4	eV	[11]
Migration energy for V_C	3.5	eV	[11]

in the equations:
1) combination of a migrating interstitial atom with a vacancy of the same type (recombination), a vacancy of the other type (antisite production), or an antisite occupied by an atom of the other type (antisite annihilation and interstitial exchange);
2) combination of a migrating vacancy with an antisite occupied by an atom of the same type (antisite annihilation and vacancy exchange);
3) capture of migrating interstitials or vacancies by Frank loops.

Interstitial cluster formations were ignored here, because the observed matured loop microstructures provide much stronger sink strength compared to isolated interstitials. Also, vacancy cluster formation was not treated in the present work. Thermal dissociation of loops and antisites were ignored because of very large formation energy of interstitial atoms[8]. When loop growth is dictated by the Si interstitial net influx, which is expected to be the case in most condition without Si deposition, the I_{Si} flux emitted from loops is $\phi_{ISi}^{I-emit} = 0$. Then radial growth rate of loops is given by

$$\frac{dr^I}{dt} = \frac{\Delta\phi_{ISi}^I}{b} \tag{6}$$

, where b is Burgers vector of the Frank loop formed on (111) family.

The parameters used in the equation are listed in Table. 1. Initial loop density and average size were given by present TEM results the damage level of at approximately 3 dpa. The effects of charge states of defects[9,10] were not considered.

COMPARISON BETWEEN EXPERIMENTAL AND CALCULATION

Figure 6 shows the time dependence of loop growth rate and net Si and C interstitial influxes into loops for the 3×10^{-4} Si/dpa case. The I_{Si} net influx rapidly increases during the initial stage of irradiation, although no significant effects of deposited Si were observed at this Si/dpa level in TEM study. Furthermore, the I_{Si} net influx exceeds the I_C net influx after 3830 sec., and then I_C net influx starts to control the loop growth in turn. After that, loop growth rate is almost saturated at the value of approximately 3×10^{-13} m/s.

Figure 7 shows the simulated loop growth for the several Si/dpa conditions, in which experimental values with same damage rates as the calculation are also plotted. The time required for I_{Si} net influx to excess I_C net influx decreased with increasing Si/dpa ratio. The saturated growth rate, however, is almost independent of Si/dpa ratio except for the case of Si/dpa=0. The growth rate observed in the TEM experiment (8.1×10^{14} m/s for the case of 3×10^{-4} Si/dpa) was about one order of magnitude less than the calculated values. However, the results of calculation indicates that much smaller amount of deposited Si atoms comparing to displacement atoms can largely affect the Frank loop growth rates. It strongly supports that the observed Si/dpa dependence of loop size in our irradiation experiments is stoichiometric constraint effects.

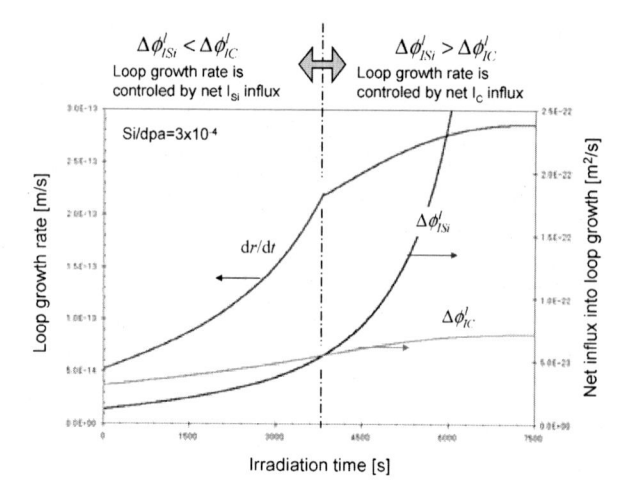

Fig. 6 Excess Si effects on the time dependence of net influx of intestitials and loop growth rate under the condition of Si/dpa=3x10^{-4}.

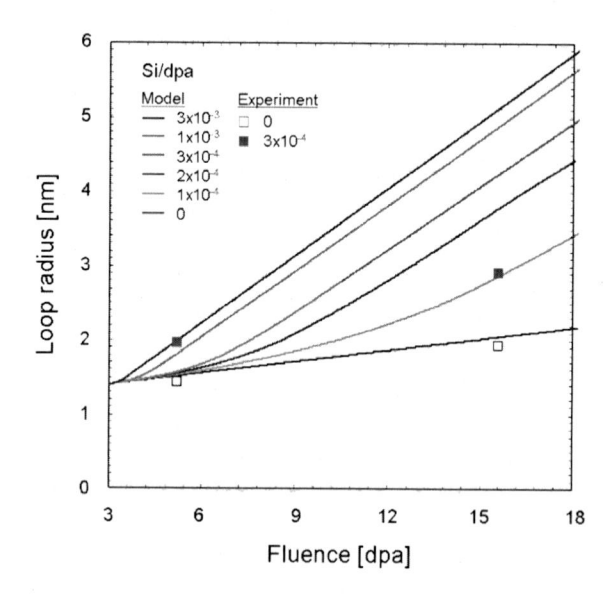

Fig. 7 Comparison of the fluence dependence of the calculated Frank loop radius in β-SiC with experiments under the various Si/dpa conditions.

CONCLUSION

The stoichiometric constraint effects on Frank loop growth in irradiated SiC were studied. Irradiation experiments demonstrated the accelerated growth of Frank loops in the regions simultaneous Si deposition. By calculation based on simple kinetic models, it was shown that a very small amount of excess Si atoms increases the growth rates of Frank loops substantially, and that supply of Si interstitials governs the growth of the irradiation induced loops in SiC in most irradiation environments due to the stoichiometric constraint effect. It is implied that the stoichiometric constraint for defect clustering contributes to the radiation stability of SiC by retarding microstructural development.

REFERENCES

[1] B. Riccardi, L. Giancarli, A. Hasegawa, Y. Katoh, A. Kohyama, R.H. Jones, L.L. Snead, "Issues and advances in SiCf/SiC composites development for fusion reactors," *J. Nucl. Mater.*, **329-333**, 56-65 (2004).

[2] Y. Katoh, N. Hashimoto, S. Kondo, L.L. Snead, A. Kohyama, "Microstructural development in cubic silicon carbide during irradiation at elevated temperatures," *J. Nucl. Mater.*, **351**, Issues 1-3, 228-240 (2006).

[3] T. Yano, H. Miyazaki, M. Akiyoshi, T. Iseki, "X-ray diffractometry and high-resolution electron microscopy of neutron-irradiated SiC to a fluence of 1.9×10^{27} n/m^2," *J. Nucl. Mater.*, **253**, 78-86 (1998).

[4] W. J. Weber, R. E. Williford and K. E. Sickafus, "Total displacement functions for SiC," *J. Nucl. Mater.*, **244**, Issue 3, 205-211 (1997).

[5] S.J. Zinkle, "Microstructure of ion irradiated ceramic insulators," *Nucl. Instr. and Meth. B.*, 91, **234-246** (1994).

[6] A. Kohyama, Y. Katoh, M. Ando, K. Jimbo, "A new Multiple Beams–Material Interaction Research Facility for radiation damage studies in fusion materials," *Fusion Eng. and Des.*, **51-52**, 789-795 (2000).

[7] J.P. Biersack, L.G. Haggmark, "A Monte Carlo computer program for the transport of energetic ions in amorphous targets," *Nucl. Instr. and Meth. B*, **174**, 257 (1980); J. Ziegler, Software and Web site, http://www.SRIM.org.

[8] G. Lucas, L. Pizzagalli, "Structure and stability of irradiation-induced Frenkel pairs in 3C-SiC using first principles calculations," *Nucl. Instr. and Meth. B*, (2006), in press.

[9] S.J. Zinkle and C. Kinoshita. "Defect production in ceramics," *J. Nucl. Mater.* **251**, 200-217 (1997).

[10] A.I Ryazanov, A.V. Klaptsov, A. Kohyama, H. Kishimoto, "Radiation swelling of SiC under neutron irradiation," *J. Nucl. Mater.*, **307-311**, 1107-1111 (2002).

[11] M. Bockstedte, A. Mattausch, O. Pankratov, "Ab initio study of the migration of intrinsic defects in 3C-SiC," *Phys. Rev. B*, **68**, 205201 (2003).

EFFECTS OF SI:SIC RATIO AND SIC GRAIN SIZE ON PROPERTIES OF RBSC

S. Salamone, P. Karandikar, A. Marshall
M Cubed Technologies, Inc.
1 Tralee Industrial Park
Newark, DE 19711

D. D. Marchant
Simula, Inc.
7822 South 46th Street
Phoenix, AZ 85044

M. Sennett
US Army RD&E Command
Natick Soldier RD&E Center
AMSRD-NSC-SS-NS
Kansas Street
Natick, MA 01760

ABSTRACT

Reaction bonded silicon carbide (RBSC) ceramics are typically produced by the reactive infiltration of molten Si into preforms consisting of SiC and carbon. During the infiltration process, the Si and carbon react to form SiC. In this system, silicon to silicon carbide ratio and grain size are important parameters in determining the final physical properties of the composite. Through any number of processing techniques, the final Si:SiC ratio can be systematically changed. Examples include the green density of the starting preform and the carbon content (i.e., SiC to carbon ratio) of the preform formulation. Moreover, the grain size of the final composite can be systematically varied, with starting SiC particle size and carbon content of the preform being the primary process variables. Si to SiC ratio, grain size and other factors work together to dictate the mechanical behavior of an RBSC ceramic. For instance, the Si:SiC ratio of several disparate microstructures may be equivalent but their properties can differ significantly. This can be a result of the degree of interconnectivity of the SiC phase and/or the grain sizes of the various phases. This work presents microstructures and properties of a wide range of RBSC ceramics. For instance, final Si content is varied from less than 10 vol. % to about 30 vol. % and grain size is varied over an order of magnitude (from 12 to 150 microns). Based on the data, correlations are developed.

INTRODUCTION

Reaction bonded silicon carbide (RBSC) composites are produced using the general processing steps outlined in the schematic of Figure 1. The carbon incorporated into the preform reacts with the infiltrating molten Si to yield SiC. This product along with the initial SiC particles are bound together in an interconnected ceramic network filled with residual silicon metal. Once cooled, the resultant component is a dense highly loaded ceramic-metallic composite.

By varying such parameters as particle size, carbon content, forming technique and alloy composition, many different composite microstructures and properties can be realized [1-6]. Combine tailorability with the ability to form complex, net shape geometries [7,8], it is easy to see why reaction bonded systems are suited to a wide range of applications.

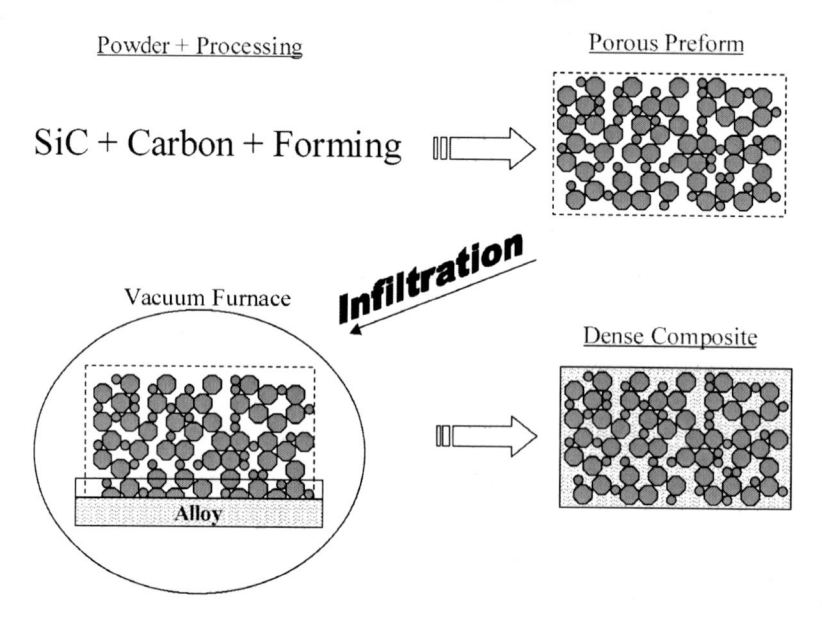

Figure 1: Process Schematic for Reaction Bonded Silicon Carbide

One specific application of RBSC composites is in the area of armor development. The armor industry demands a material with impressive physical properties, low overall cost (raw materials and ease of manufacturing), net shape capability, and the ability to meet rigorous and changing objectives. Pressureless sintered and hot-pressed SiC based materials are often expensive and/or difficult to produce. Thus, RBSC composites which can be cost effectively manufactured are considered viable candidates for armor applications [9].

The ability to manipulate the microstructure and composition, thereby tailoring the physical properties is the key to success of any materials system. The RBSC composites in the present work are an example of how two variables (grain size and Si:SiC ratio) can be used to increase the complexity of the microstructure and enhance properties.

EXPERIMENTAL PROCEDURES

The RBSC composites were produced by the method outlined in Figure 1. The powder consisting of a mixture of 12, 45, and/or 150 μm SiC particles, adjusted depending on the desired microstructure, was combined with the proper carbon content. Preforms were fabricated using these SiC/C mixtures, and were then infiltrated with molten Si to yield a RBSC ceramic.

The physical and mechanical properties of the infiltrated composites were measured using several common techniques summarized in Table I. Several or all of the properties were measured depending on the applicability and effectiveness of the technique to show the desired differences or similarities of the various composites. All the microstructures were characterized by examining polished sections using a LEICA DM 2500 M optical microscope. Phase composition was characterized by quantitative image analysis (QIA) using a Clemex JS-2000 digital scanning system.

Table I: Summary of properties and techniques used to quantify the various composites.

Property	Technique	Standard
Density	Immersion	ASTM B 311
Elastic Modulus	Ultrasonic Pulse Echo	ASTM D 2845
Hardness	Knoop (2kg load)	ASTM C 1236
Flexural Strength	Four Point Bend	ASTM C 1161

RESULTS AND DISCUSSION

Effect of SiC Grain Size

A reduction in SiC grain size typically leads to higher strength RBSC ceramics. However, finer particles tend to pack poorly relative to coarse particles. Thus, finer grain size generally leads to ceramics with lower SiC content and increased Si metal content. In the present study, three RBSC composites with dramatically different grain sizes were examined. The microstructural characteristics spanned an order of magnitude from 12 to 150 microns, so that a wide range of influences may be observed.

In the following photomicrographs, the two phases (SiC and Si) are identified as follows: dark regions are the SiC particles and the lighter regions are comprised of the metallic Si phase. A comparison of the microstructures of all three differing grain size materials at the same magnification is provided in Figure 2. An order of magnitude difference in grain size is clearly evident. In addition, Figure 3 shows the microstructure of the 150 micron-based ceramic at lower magnification. This is necessary to properly depict this very coarse-grained material. A review of the microstructures demonstrates that the 12 μm material is a mono-grain size ceramic. The two coarser materials use particle size distributions to enhance particle packing. These composites will be designated by the maximum particle size in the blend. For example, the 45 μm system contains 12 and 45 μm size particles.

Effects of Si:SiC Ratio and SiC Grain Size on Properties of RBSC

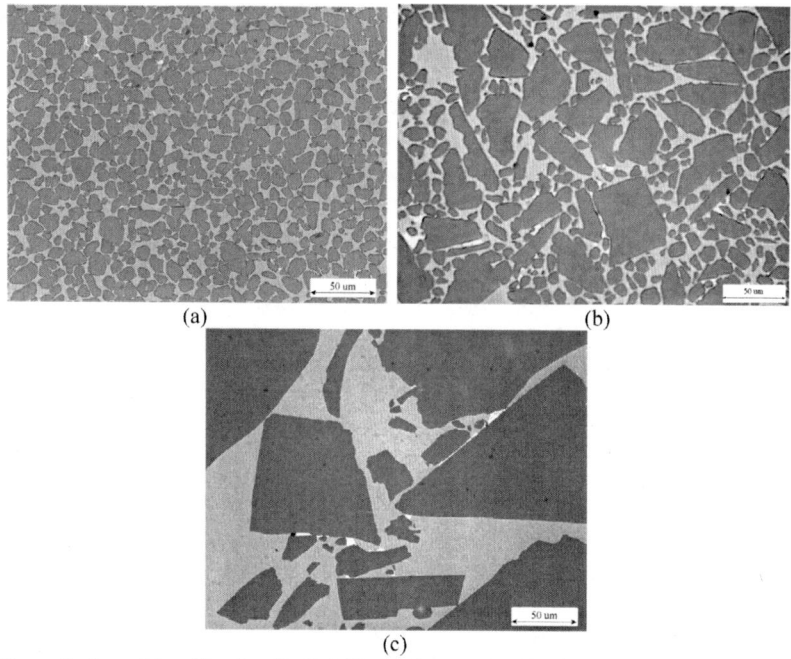

Figure 2: Same Magnification Optical Photomicrographs of (a) 12, (b) 45 and (c) 150 μm Particle Size RBSC.

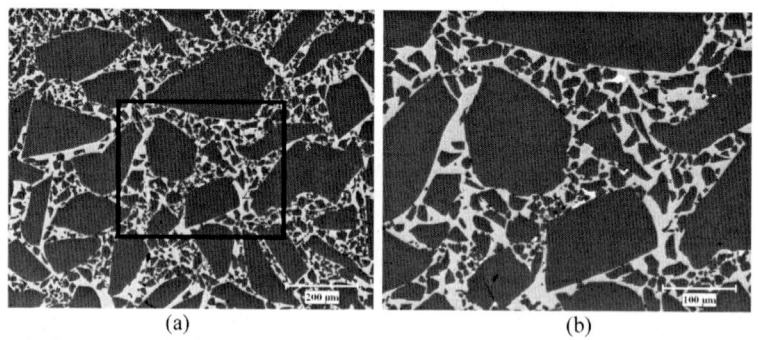

Figure 3: Low Magnification Optical Photomicrographs of 150 μm Particle Size RBSC, (b) is the region inside the box defined in (a).

Results of property measurements are provided in Table II. As expected, the finer grain size has led to higher strength, but lower SiC content (higher Si content) due to less particle packing. Washburn and Coblenz reported a sharp decrease in flexural strength with grain sizes greater than 100 μm, as compared to samples with 12 μm grains. This is clearly evident in the present system where the strength of the 150 μm samples is about half of the 12 μm set. Carroll et. al. found that the room temperature strength of 4-6 μm reaction bonded samples were twice that of samples with a bimodal grain size of 10 & 100 μm [10]. For particles under 100 μm, it was found that samples with two orders of magnitude difference in grain size (.2 to 23-65 μm), but comparable density and SiC content to samples in Table II, that the strengths were not vastly different from one another [6]. All these trends correlate well to what has been found in the present study.

Hardness and Young's modulus track the SiC content, with higher SiC contents leading to stiffer and harder ceramics. Note that there is no hardness value presented for the 150 μm composite. The scale of the microstructure in this composite is greater than that of the indent. This leads to wide variability and abnormally high values, because the indent can be entirely contained in a single SiC grain.

Table II: Effect of SiC Grain Size on Properties of RBSC Ceramics

	RBSC Composites		
SiC Grain Size (μm)	12	45	150
Density (g/cc)	2.96	3.00	3.04
SiC Content (vol. %)	72	76	80
Young's Modulus (GPa)	335	353	375
Flexural Strength (MPa)	360	270	165
Knoop 2 kg Hardness (kg/mm^2)	1184 ± 60	1197 ± 244	---

Effect of Si:SiC Ratio on Microstructure

Increasing the carbon content of the perform formulation prior to infiltration with molten Si allows more SiC to form at the expense of Si. This reaction creates a composite with an increased ceramic phase, as compared to the metallic phase. The additional SiC that forms tends to initiate on the surfaces of the existing SiC grains, in effect, coarsening the grains. Figure 4 provides a series of photomicrographs showing how the addition of carbon to the perform changes the morphology of the grains. The initial large grains are "blocky" with sharp and distinct edges. As the amount of silicon decreases, the large grains become more rounded and they begin to impinge on the neighboring grains. In many of the regions, the small grains become interconnected and are engulfed by the larger grains. These regions transform into dense SiC rich areas. Figures 5 and 6 reveal the same coarsening and impinging behavior in the 12 and 150 μm RBSC systems. The spatial relationships of Figures 5 and 6 allow observation on a more global and local view to the phenomenon, respectively. They also show the ability to engineer a relatively homogeneous structure, in the case of the 12 μm composite, and a heterogeneous microstructure as seen in the 150 μm system.

Effects of Si:SiC Ratio and SiC Grain Size on Properties of RBSC

(a) (b)

(c)

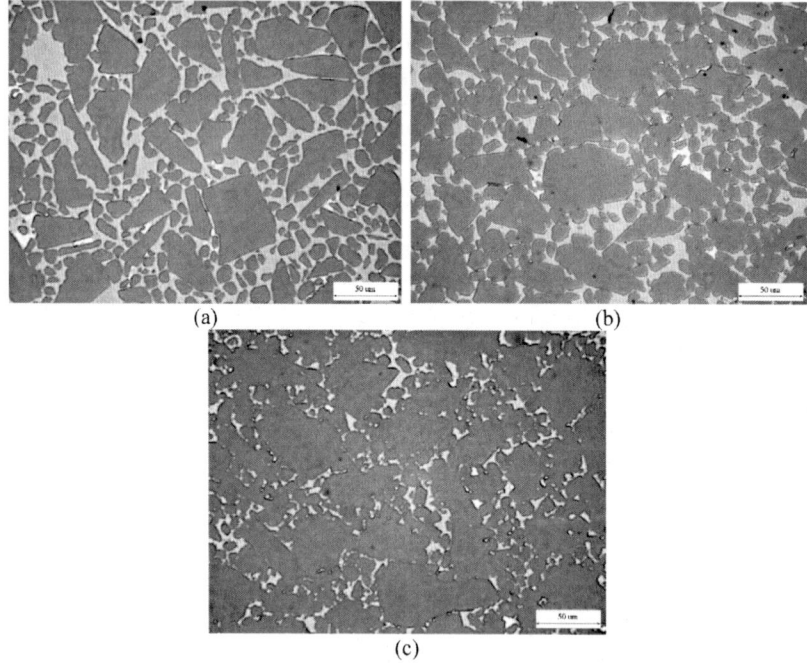

Figure 4: Optical Photomicrographs Showing Effect of Increasing Carbon Content (from a to c) of the Preform in the 45 μm RBSC system. Microstructures are Post-Reactive Infiltration with Molten Silicon.

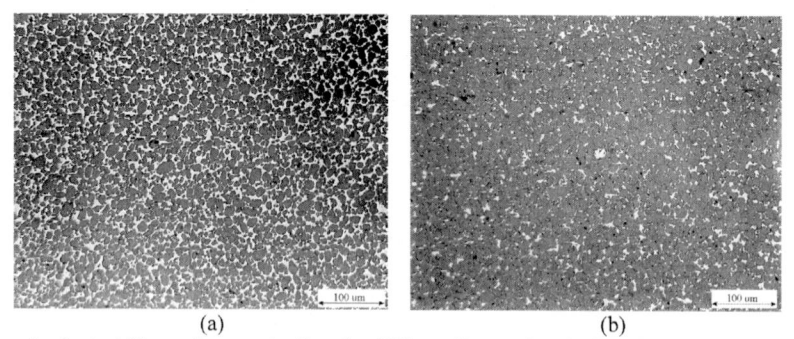

(a) (b)

Figure 5: Optical Photomicrographs Showing Effect of Increasing Carbon Content From (a) to (b) of the Preform in the 12 μm RBSC system.

(a)   (b)

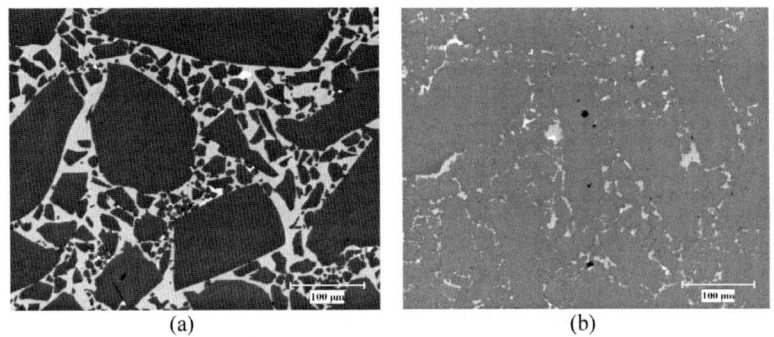

Figure 6: Optical Photomicrographs Showing Effect of Increasing Carbon Content From (a) to (b) of the Preform in the 150 μm RBSC system.

Effect of Si:SiC Ratio on Physical Properties

The result of decreasing the Si:SiC ratio (increasing the ceramic phase) is higher density, higher hardness and higher Young's modulus in the finished composite. This is consistent with the fact that SiC has higher density, modulus and hardness than the respective properties of Si. In the case of all the preforms made with SiC particles, the data in Table III clearly demonstrate that higher carbon content is effective in reducing Si content of the ceramic and in turn enhancing the aforementioned properties. With small changes in the density of the composites, a correspondingly large change in modulus and hardness is evinced. This density/modulus correlation was also seen in biomorphic silicon carbide structures where increased density corresponds to a large increase in elastic modulus [11]. A graph of modulus and density versus volume percent SiC for RBSC composites and constituents (Si metal and SiC) is plotted in Figure 7. As the Si:SiC ratio decreases, changing from pure silicon to pure silicon carbide, the density and modulus increase. For approximately a five percent increase in density, the composite modulus can be enhanced by about five times this value. It is instructive to notice that the modulus values obtained are approaching that of pure SiC.

Table III: Effects of Carbon Additions to Preforms of Varying Particle Size

Preform Type*	SiC Vol%	Density (g/cc)	Young's Modulus (GPa)	Knoop (2 kg) Hardness (kg/mm^2)
12 μm SiC	72	2.96	335	1184
12 μm SiC	84	3.07	387	1397
45 μm SiC	76	3.02	354	1197
45 μm SiC	82	3.05	379	1410
45 μm SiC	88	3.12	407	1536
150 μm SiC	80	3.04	375	----
150 μm SiC	92	3.10	409	----

* Represents maximum particle size in preform blend.

Effects of Si:SiC Ratio and SiC Grain Size on Properties of RBSC

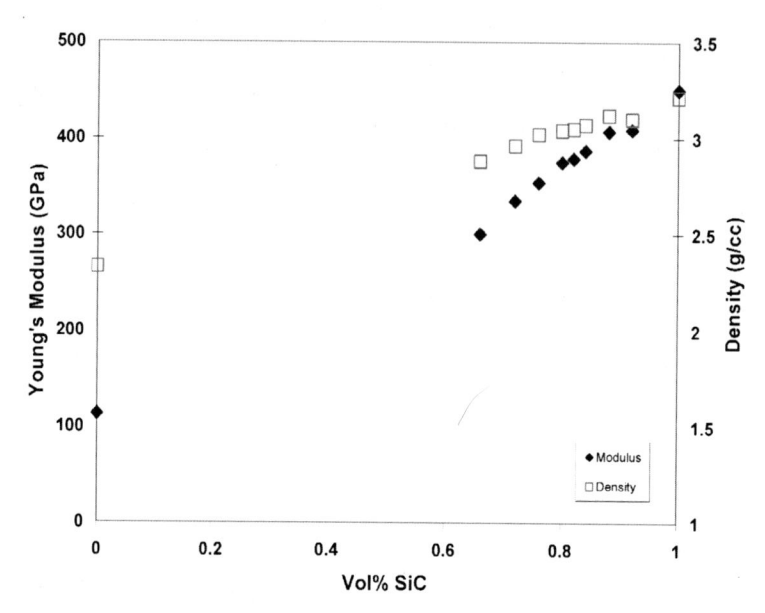

Figure 7: Young's modulus and density as a function of the volume percent of SiC, a comparison of RBSC and constituents.
(Si & SiC properties found in references 12 and 13)

SUMMARY

The ability to tailor the microstructure and physical properties of RBSC composites is one of the main advantages for using these materials. Understanding and influencing the resultant material behavior is critical to using RBSC and will be important when determining specific applications for the composites. In this study, it was found that the grain size had a strong influence on the flexural strength of the composite as compared to the other physical properties. The reduction of residual Si in the composite, via increasing the carbon content, had a much larger effect on the modulus than on the density. As the volume of SiC increased to 92%, the composite modulus (as one would expect) increased close to that of pure SiC. The real value of this work comes from the ability to use the strengths of RBSC composites while approaching the properties of the monolithic material.

ACKNOWLEDGEMENTS

This work was partially supported by the US Army Natick Soldier RD&E Center under contract number W911QY-06-C-0041.

REFERENCES
1. M. E. Washburn and W. S. Coblenz, "Reaction-Formed Ceramics", *Am. Ceram. Soc., Bull.*, **67**, [2] 356-363 (1988).
2. J. M. Fernandez, A. Munoz, A. R. D. Lopez, F. M.V. Feria, A. Dominguez-Rodriguez, M. Singh, "Microstructure-mechanical properties correlation in siliconized silicon carbide ceramics", *Acta Mater.* **51** 3259-3275 (2003).
3. M. Singh and D.R. Behrendt, "Reactive melt infiltration of silicon-niobium alloys in microporous carbons", *J. Mater. Res.*, **9** [7] (1994).
4. U. Paik, HC Park, SC Choi, CG Ha, JW Kim, YG Jung, "Effect of particle dispersion on microstructure and strength of reaction-bonded silicon carbide", *Mater. Sci. Eng.*, A334 267-274 (2002).
5. YM Chiang, R. P. Messner, and C. D. Terwilliger, "Reaction-formed silicon carbide", *Mater. Sci. Eng.*, A144 63-74 (1991).
6. O. P. Chakrabarti, S. Ghosh, and J. Mukerji, "Influence of Grain Size, Free Silicon Content and Temperature on the Strength and Toughness of Reaction-Bonded Silicon Carbide", *Ceramics International*, **20** 283-286 (1994).
7. M. K. Aghajanian, B. N. Morgan, J. R. Singh, J. Mears, R. A. Wolffe, "A New Family of Reaction Bonded Ceramics for Armor Applications", in Ceramic Armor Materials by Design, *Ceramic Transactions*, **134**, J. W. McCauley et al. editors, 527-40 (2002).
8. P. G. Karandikar, M. K. Aghajanian and B. N. Morgan, "Complex, Net-Shape Ceramic Composite Components for Structural, Lithography, Mirror and Armor Applications, *Ceram. Eng. Sci. Proc.*, **24** [4] 561-6 (2003).
9. E. Medvedovski, "Silicon Carbide-Based Ceramics for Ballistic Protection", in Ceramic Armor and Armor Systems, *Ceramic Transactions*, **151**, E. Medvedovski editor, 19-35 (2003).
10. D. F. Carroll, R. E. Tressler, Y. Tsai, and C. Near, "High Temperature Mechanical Properties of Siliconized Silicon Carbide Composites", pp. 775-88, in Materials Science Research, Vol. 20, Tailoring Multiphase and Composite Ceramics, ed. R.E. Tressler, G.L. Messing, C.G. Pantano, and R. E. Newnham. Plenum, New York, 1986.
11. M. Singh and J. A. Salem, "Mechanical properties and microstructure of biomorphic silicon carbide ceramics fabricated from wood precursors", *J. Euro. Ceram. Soc.*, **22**, 2709-2717 (2002).
12. *Metals Handbook: Desk Addition* (ASM International, Metals Park, OH, 1985).
13. *Engineered Materials Handbook, Vol. 4, Ceramics and Glasses*, (ASM International, Metals Park, OH, 1991).

ELECTRICAL PROPERTIES OF AlN-SiC SOLID SOLUTIONS WITH ADDITIONS OF Al AND C

Ryota Kobayashi, Junichi Tatami, Toru Wakihara, Katsutoshi Komeya, Takeshi Meguro
Graduate School of Environmental and Information Sciences, Yokohama National University
79-7, Tokiwadai, Hodogaya-ku
Yokohama, 240-8501, Japan

ABSTRACT

The electrical conductivity of AlN-SiC solid solutions was controlled by addition of Al_4C_3 (Al and C). Powder mixtures of AlN and SiC with small amounts of Al and C (below 3 mol% as Al_4C_3) were consolidated by spark plasma sintering (SPS). Then the consolidated samples were heat-treated to obtain homogeneous AlN-SiC solid solutions. The relative densities of the samples were over 95% of theoretical densities. XRD analysis showed the heat-treated samples were composed of only 2H AlN-SiC solid solutions. While the samples without Al and C were electrical insulators, the samples containing Al and C were electrical conductors. The electrical conductivity of the samples increased with increasing temperature, indicating semiconductive behavior. The heat-treated samples containing 3 mol% Al_4C_3 had the highest electrical conductivity of all samples (100 S/m at 300°C).

INTRODUCTION

Aluminum nitride (AlN) has high thermal conductivity (320 W/mK, comparable with Al) and high electrical resistivity.[1,2] Therefore, AlN has been used as substrate material for power electronics. Silicon carbide (SiC) has high mechanical strength at elevated temperatures (500 MPa at 1500°C), along with excellent oxidation and corrosion resistance. Therefore, SiC is widely used for high-temperature structural applications. In addition, SiC has a very wide bandgap (3.2eV, three times as Si). Thus, SiC is also attractive for high performance semiconductor devices.

In 1978, Cutler et al. showed that AlN and SiC form a 2H solid solution.[3] Furthermore, the properties of the AlN-SiC solid solutions vary with compositions. For example, the thermal conductivities of the solid solutions are minimized near the middle of composition (1/10 of AlN).[4,5] The bandgap of the solid solution changes between 3.0 to 6.2 eV, and is lowest near the middle of composition, which is similar to the trend for thermal conductivity.[6,7] It is also shown that AlN-SiC solid solutions exhibit p-type conduction and their electrical conductivity and Seebeck coefficient increased between 25 and 50 mol% AlN.[8] Therefore, AlN-SiC solid solutions potential use as thermoelectric conversion elements and new wide-bandgap semiconductors. However, controlling the electrical properties of AlN-SiC solid solutions, such as electrical conductivity and conduction types has not been yet possible.

It is well known that SiC semiconductors doped with Al and N exhibit p-type and n-type conductions, respectively. In the case of AlN–SiC solid solutions, Al and C occupy the Si and N sites at a 1:1 ratio, respectively, indicating that AlN-SiC solid solutions are compensated semiconductors. Nevertheless, the AlN–SiC solid solutions fabricated without sintering additives reported in our previous study exhibited p-type conduction.[8] The reason for this might be composition shifts toward Al rich side due to the decomposition of AlN, suggesting the p-type conduction of AlN-SiC solid solutions can be controlled by the addition of small amount of

Al$_4$C$_3$. In this work, we fabricated dense and homogeneous AlN-SiC solid solutions with addition of Al$_4$C$_3$ and investigated their electrical properties.

EXPERIMENTAL PROCEDURE

AlN (Tokuyama Corp., Grade-F, 0.6 μm), SiC (Yakushimadenko Co, OY-20, 0.55 μm), Al (Toyo Aluminum K.K., AC1300), and C (Mitsubishi Chemical Corp., 650B, 0.018 μm) were used as starting powders. Because Al$_4$C$_3$ is easily decomposed by water or oxygen, we used Al and C as the source of Al$_4$C$_3$. The starting powders were wet mixed for 24 hours in ethanol using SiC balls as grinding media. The starting compositions were 50AlN-50SiC (0AC), 50AlN-50SiC-Al$_4$C$_3$ (1AC), and 50AlN-50SiC-3Al$_4$C$_3$ (3AC). The powder mixtures were consolidated by SPS at 2000°C for 10 minutes in 1 atm Ar atmosphere. A uniaxial pressure of 30 MPa was applied during SPS. The SPSed samples were heat-treated to obtain homogeneous AlN-SiC solid solutions at 2200°C for 3 hours in an Ar flow. Packing powders having the same compositions as the samples were used for preventing decomposition of the samples.

The densities of the samples were measured by Archimedes method, and the microstructures were characterized by scanning electron microscopy (SEM). The compositions were analyzed by X-ray fluorescence (XRF), and the Al/Si ratios were calculated. The phases present were identified by X-ray diffraction (XRD). The electrical conductivities were measured by the four-point probe method between room temperature and 300°C.

RESULT AND DISCUSSION

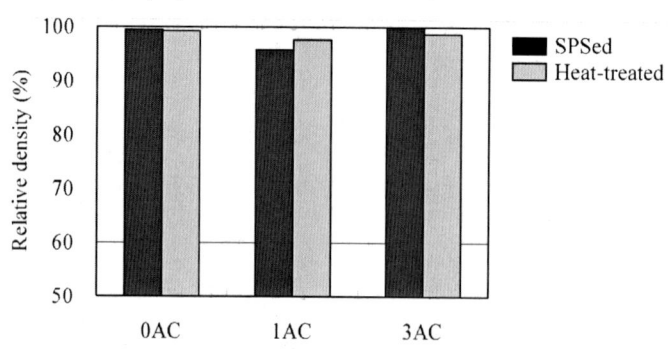

Figure 1 shows the relative densities of the SPSed and heat-treated samples. The theoretical densities of the samples were calculated from the densities of AlN, SiC, Al, and C (AlN: 3.26 g/cm^3, SiC: 3.21 g/cm^3, Al: 2.7 g/cm^3, C: 2.25 g/cm^3). The relative densities of the SPSed samples were over 95% of the theoretical value. In particular, the relative densities of 0AC samples were over 99% of the theoretical value, which is comparable to our previous work.[9] The relative densities of the 0AC and 3AC samples decreased slightly after the heat-treatment. In contrast, the relative density of 1AC sample was slightly increased after the heat-treatment. Even so, the samples had high relative densities before and after the heat-treatment. The microstructures of the samples were changed by addition of Al$_4$C$_3$. Although all the samples had equiaxed textures, the grain sizes of the samples were increased by addition of Al$_4$C$_3$ (0AC sample: 2~3 μm, 1AC and 3AC: >10 μm).

Fig.1 Relative densities of the samples

Figure 2 shows the Al/Si ratios as the compositions of SPSed and heat-treated samples. The Al/Si ratios of 0AC and 1AC samples slightly increased after heat-treatment due to the incorporation of Al from packing powders. However, the Al/Si ratio of 3AC sample slightly decreased after heat-treatment due to the release of Al from the 3AC samples. Even so, the changes in the Al/Si ratios were restrained by the dense textures of the samples.

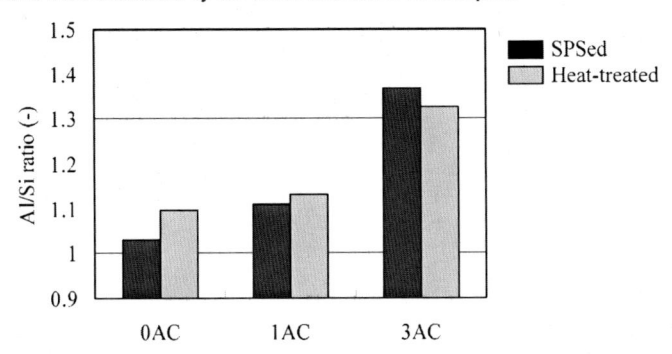

Fig.2 Al/Si ratios of the samples

Figure 3 identifies the phases present in the SPSed and heat-treated samples. The SPSed 0AC and 1AC samples were composed of the mixtures of 2H and 6H AlN-SiC solid solutions. In contrast, the SPSed 3AC sample was composed of only 2H AlN-SiC solid solution. After heat-treatment, the phases present of 0AC and 1AC samples were converted to only 2H AlN-SiC solid solution from the mixture of 2H and 6H solid solutions. This result followed the phase diagram of AlN-SiC system presented by Zangvil et al.[10]

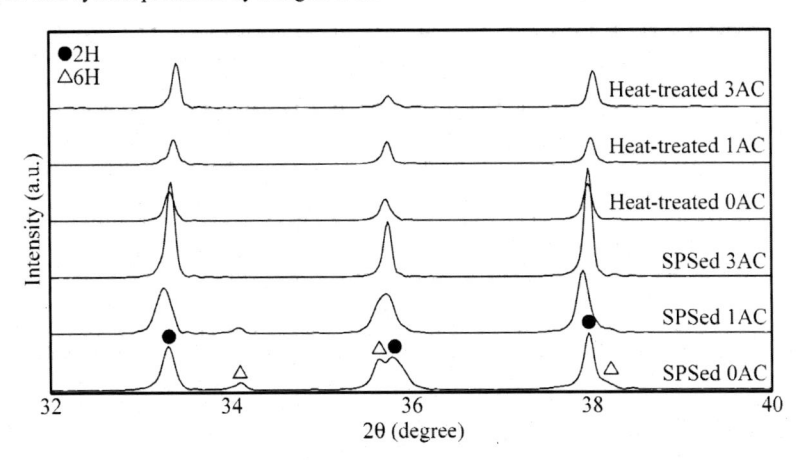

Fig.3 XRD profiles of the samples

Figure 4 shows the electrical conductivities of SPSed and heat-treated samples. To understand the temperature dependence of the electrical conductivity, we plotted inverse of temperature in *x*-axis and electrical conductivity in *y*-axis. The activation energies of carrier formation (E_a) were also calculated from the slope of the curves. Although the samples without Al and C were electrical insulators, the samples containing Al and C were electrical conductors. The electrical conductivity of the 1AC and 3AC samples was found to increase with increasing measurement temperature, indicating not metallic but semiconductive behavior. 3AC samples had the highest electrical conductivity of all samples (SPSed: 38 S/m, heat-treated: 100 S/m, measured at 300°C). 1AC samples had lower electrical conductivity than 3AC samples (SPSed: 1.5 S/m, heat-treated: 30 S/m, measured at 300°C) close to the values for samples fabricated by PLS reported in our previous work.[8] The activation energy for electrical conductivity was found to decrease after heat-treatment. It is believed that the activation energy is affected by the homogeneity of the samples because heat-treated samples are more homogeneous than SPSed samples. The electrical conductivity and activation energy of heat-treated samples might change by the homogeneity of Al and C impurities added.

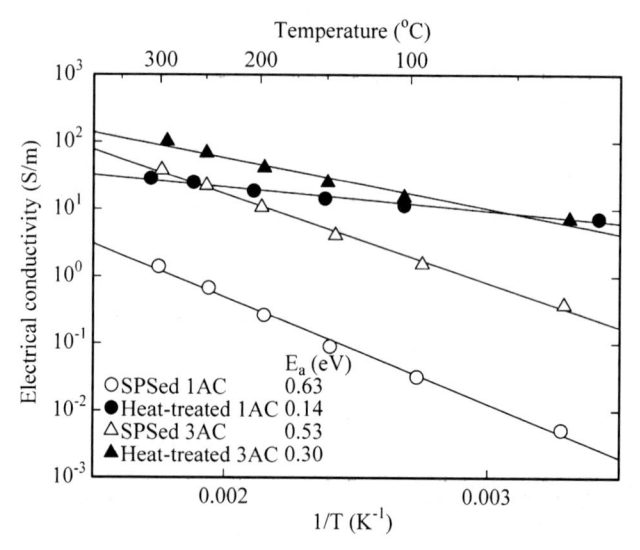

Fig.4 Electrical conductivities of the samples

SUMMARY

The electrical properties of AlN-SiC solid solutions with addition of Al_4C_3 were evaluated. The relative density of the samples was over 95% of theoretical value. The density and composition changed slightly after heat-treatment. The heat-treated samples were composed of only 2H AlN-SiC solid solutions. The electrical conductivity of the samples increased with increasing measurement temperature, indicating semiconductive behavior. Increasing the content of Al_4C_3 enhanced the electrical conductivity of the samples.

REFERENCES

[1] G. A. Slack, "Nonmetaric Crystals with High Thermal Conductivity," *J. Phys. Chem. Solids.*, 34, 321-335 (1973).

[2] G. Long and L. M. Foster, "Aluminum Nitride; a Refractory for Aluminum to 2000°C," *J. Am. Ceram. Soc.*, 42, 53-59 (1959).

[3] I. B. Cutler, P. D. Miller, W. Rafaniello, H. K. Park, D. P. Thompson, K. H. Jack, "New Materials in the Si-C-Al-O-N and Related Systems," *Nature*, 275, 434-435 (1978).

[4] W. Rafaniello, K. Cho and A. V. Virkar, "Fabrication and Characterization of SiC-AlN Alloys," *J. Mater. Sci.*, 16, 3479-3488 (1981).

[5] M. Miura, T. Yogo and S. Hirano, "Mechanical and Thermal Properties of SiC-AlN Ceramics with Modulated Structure," *J. Ceram. Soc. Jpn.*, 101, 1281-1286 (1993).

[6] R. Roucka, J. Tolle, A. V. G. Chizmeshya, P. A. Crozier, C. D. Poweleit, D. J. Smith, I. S. T. Tsong and J. Kouvetakis, "Low-Temperature Epitaxial Growth of the Quaternary Wide Band Gap Semiconductor SiCAlN," *Phys. Rev. Lett.*, 88, 206102 (2002).

[7] J. Tolle, R. Roucka, A. V. G. Chizmeshya, P. A. Crozier, D. J. Smith, I. S. T. Tsong and J. Kouvetakis, "Novel Synthetic Pathways to Wide Bandgap Semiconductors in the Si-C-Al-N System," *Solid State Sci.*, 4, 1509-1519 (2002).

[8] R. Kobayashi, J. Tatami, T. Wakihara, T. Meguro, K. Komeya, "Temperature Dependence of the Electrical Properties and Seebeck Coefficient of AlN-SiC Ceramics," *J. Am. Ceram. Soc.*, 89, 1295-1299 (2006).

[9] R. Kobayashi, J. Tatami, T. Wakihara, K. Komeya, T. Meguro, T. Goto, "Evaluation of Al-Si-C-N Ceramics Fabricated by Spark Plasma Sintering," *Ceram. Trans.*, 194, 273-277 (2006).

[10] A. Zangvil and R. Ruh, "Phase Relationships in the Silicon Carbide-Aluminum Nitride System," *J. Am. Ceram. Soc.*, 71, 884-890 (1988).

Fiber-Reinforced CMCs

EFFECTS OF FREQUENCY ON FATIGUE BEHAVIOR OF AN OXIDE-OXIDE CERAMIC COMPOSITE AT 1200°C

G. Hetrick, M.B. Ruggles-Wrenn[*],
Department of Aeronautics and Astronautics
Air Force Institute of Technology
Wright-Patterson Air Force Base, Ohio 45433-7765, USA

S.S. Baek
Agency for Defense Development
Daejeon, Korea

ABSTRACT

The effect of frequency on fatigue behavior of an oxide-oxide continuous fiber ceramic composite (CFCC) was investigated at 1200°C in laboratory air and in steam environment. The composite consists of a porous alumina matrix reinforced with laminated, woven mullite/alumina (Nextel™720) fibers, has no interface between the fiber and matrix, and relies on the porous matrix for flaw tolerance. Tension-tension fatigue tests were performed at frequencies of 0.1 and 10 Hz for fatigue stresses ranging from 75 to 170 MPa. Fatigue run-out was defined as 10^5 cycles at 0.1 and 1.0 Hz and as 10^6 cycles at 10 Hz. The CFCC exhibited excellent fatigue resistance in laboratory air. The fatigue limit was 170 MPa (88% UTS at 1200°C). The material retained 100% of its tensile strength. In steam, fatigue performance was strongly influenced by loading frequency. Presence of steam significantly degraded the fatigue performance, with the degradation being most pronounced at 0.1 Hz. Fatigue limit and fatigue lifetime decrease with decreasing loading frequency. Composite microstructure, as well as damage and failure mechanisms were investigated. In steam, more coordinated fiber failure is observed at the lowest frequency, while the amount of fiber pullout as well as the damage zone size decrease with decreasing frequency.

INTRODUCTION

Advances in aerospace technologies have raised the demand for structural materials that exhibit superior long-term mechanical properties and retained properties under high temperature, high pressure, and varying environmental factors [1]. Ceramic-matrix composites (CMCs), capable of maintaining excellent strength and fracture toughness at high temperatures, continue to attract attention as candidate materials for such applications. Additionally, the lower densities of CMCs and their higher use temperatures, together with a reduced need for cooling air, allow for improved high-temperature performance when compared to conventional nickel-based superalloys [2]. Advanced reusable space launch vehicles will likely incorporate CMCs in critical propulsion components [3]. Because these applications require exposure to oxidizing environments, the thermodynamic stability and oxidation resistance of CMCs are vital issues.

---

[*] Corresponding author.
The views expressed are those of the authors and do not reflect the official policy or position of the United States Air Force, Department of Defense or the U. S. Government.

Non-oxide fiber/non-oxide matrix composites generally exhibit poor oxidation resistance [4, 5], particularly at intermediate temperatures (~800 °C). The degradation involves oxidation of fibers, fiber coatings, and matrices and is typically accelerated by the presence of moisture [6-8]. Using a non-oxide fiber/oxide matrix or oxide fiber/non-oxide matrix composites generally does not substantially improve the high temperature oxidation resistance [9]. The need for environmentally stable composites motivated the development of CMCs based on environmentally stable oxide constituents [10-18].

It is widely accepted that in order to avoid brittle fracture behavior in CMCs and improve the damage tolerance, a weak fiber/matrix interface is needed, which serves to deflect matrix cracks and to allow subsequent fiber pull-out [19-22]. It has been demonstrated that similar crack-deflecting behavior can be achieved by using a matrix with finely distributed porosity instead of a separate interface between matrix and fibers [23]. This concept has been successfully demonstrated for oxide-oxide composites [10, 14, 18, 24-28]. Resulting oxide/oxide CMCs exhibit damage tolerance combined with inherent oxidation resistance. An extensive review of the mechanisms and mechanical properties of porous-matrix CMCs is given in [29].

In many potential applications, CFCCs will be subject to fatigue loading under a wide range of frequencies. Several studies examined high-temperature fatigue performance of CMCs at loading frequencies ≤ 10 Hz [30-36]. At higher frequencies (ranging from 10 to 375 Hz), a strong effect of loading frequency on fatigue life has been demonstrated for CMCs with weak fiber-matrix interfaces tested at room temperature [37-39]. It was reported that fatigue life decreased sharply as the loading frequency increased. This decrease in fatigue life was attributed to frictional heating and interface and fiber damage. More recently, it has been shown that the room-temperature fatigue life of certain ceramic-matrix composites with a strong fiber-matrix interface shows little dependence on the loading frequency [40]. Vanswijgenhoven et al [41] found that at 1200°C the fatigue limit of a Nicalon-fabric-reinforced CMC was unaffected by the loading frequency, while the number of cycles to failure increased and the time to failure decreased with increase in frequency.

Porous-matrix oxide/oxide CMCs exhibit several behavior trends that are distinctly different from those exhibited by traditional non-oxide CMCs with a fiber-matrix interface. For the non-oxide CMCs, fatigue is significantly more damaging than creep. Zawada et al [42] examined the high-temperature mechanical behavior of a porous matrix Nextel610/ Aluminosilicate composite. Results revealed excellent fatigue performance at 1000 °C. Conversely, creep lives were short, indicating low creep resistance and limiting the use of that CMC to temperatures below 1000 °C. Ruggles-Wrenn et al [43] showed that Nextel™720/Alumina (N720/A) composite exhibits excellent fatigue resistance in laboratory air at 1200°C. The fatigue limit (based on a run-out condition of 10^5 cycles) was 170 MPa (88% UTS at 1200°C). Furthermore, the composite retained 100% of its tensile strength. However, creep loading was found to be considerably more damaging. Creep run-out (defined as 100 h at creep stress) was achieved only at stress levels below 50% UTS. Mehrman et al [44] demonstrated that introduction of a short hold period at the maximum stress into the fatigue cycle significantly degraded the fatigue performance of N720/A composite at 1200°C in air. In steam, superposition of a hold time onto a fatigue cycle resulted in an even more dramatic deterioration of fatigue life, reducing it to the much shorter creep life at a given applied stress. These results suggest that the loading rate plays a significant role in damage development. This study investigates the effects of loading frequency on fatigue behavior of N720/A, an oxide-oxide CMC, at 1200°C in air and in steam environments. Results reveal that the loading frequency has a marked effect on fatigue life, especially in steam. The composite microstructure, as well as damage and failure mechanisms are discussed.

EXPERIMENTAL PROCEDURE

The material studied was Nextel™720/Alumina (N720/A), a commercially available oxide-oxide ceramic composite (COI Ceramics, San Diego, CA), consisting of a porous alumina matrix reinforced with Nextel™720 fibers. The composite was supplied in a form of 2.8 mm thick plates, comprised of 12 0°/90° woven layers, with a density of ~2.77 g/cm^3 and a fiber volume of approximately 45%. Matrix porosity was ~24%. The fiber fabric was infiltrated with the matrix in a sol-gel process. The laminate was dried with a "vacuum bag" technique under low pressure and low temperature, then pressureless sintered [45]. No coating was applied to the fibers. The damage tolerance of the N720/A composite is enabled by a porous matrix. Representative micrograph of the untested material is given in Fig. 1(a), which shows 0° and 90° fiber tows as well as numerous matrix cracks. In the case of the as-processed material, most are shrinkage cracks formed during processing rather than matrix cracks generated during loading. Porous nature of the matrix is seen in Fig. 1(b).

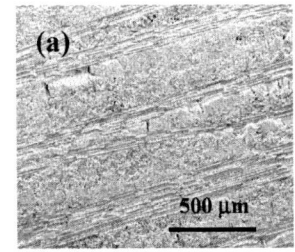

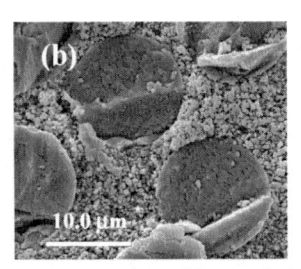

Fig. 1. As-received material: (a) overview, (b) porous nature of the matrix is evident.

A servocontrolled MTS mechanical testing machine equipped with hydraulic water-cooled collet grips, a compact two-zone resistance-heated furnace, and two temperature controllers was used in all tests. An MTS TestStar digital controller was employed for input signal generation and data acquisition. Strain measurement was accomplished with an MTS high-temperature air-cooled uniaxial extensometer. For elevated temperature testing, thermocouples were bonded to the specimens using alumina cement (Zircar) to calibrate the furnace on a periodic basis. The furnace controller (using a non-contacting thermocouple exposed to the ambient environment near the test specimen) was adjusted to determine the power setting needed to achieve the desired temperature of the test specimen. The determined power setting was then used in actual tests. The power setting for testing in steam was determined by placing the specimen instrumented with thermocouples in steam environment and repeating the furnace calibration procedure. Thermocouples were not bonded to the test specimens after the furnace was calibrated. Tests in steam environment employed an alumina susceptor (tube with end caps), which fits inside the furnace. The specimen gage section is located inside the susceptor, with the ends of the specimen passing through slots in the susceptor. Steam is introduced into the susceptor (through a feeding tube) in a continuous stream with a slightly positive pressure, expelling the dry air and creating steam environment inside the susceptor.

Fracture surfaces of failed specimens were examined using SEM (FEI Quanta 200 HV) as well as an optical microscope (Zeiss Discovery V12). The SEM specimens were carbon coated.

All tests were performed at 1200°C. Each specimen was heated to 1200°C in 25 min, and held at 1200°C for additional 15 min prior to testing. Dog bone shaped specimens of 152 mm total length with a 10-mm-wide gage section were used in all tests. Tensile tests were performed in displacement control with a constant rate of 0.05 mm/s at 1200°C in laboratory air. The effects of frequency on the fatigue behavior were evaluated in tension-tension fatigue tests conducted at the frequencies of 0.1 and 10 Hz at 1200°C, in laboratory air and in steam environments. All fatigue experiments were carried out in load control with the ratio R (minimum to maximum stress) of 0.05. Fatigue run-out was defined as 10^5 cycles at 0.1 Hz, and as 10^6 cycles at 10 Hz. The 10^5 cycle count value represents the number of loading cycles expected in aerospace applications at that temperature. Fatigue run-out limits were defined as the highest stress level, for which run-out was achieved. Cyclic stress-strain data were recorded throughout each test. Thus stiffness degradation, changes in hysteresis energy density (HED), as well as strain accumulation with fatigue cycles and/or time could be examined. All specimens that achieved run-out were subjected to tensile test to failure at 1200°C to determine the retained strength and stiffness.

RESULTS AND DISCUSSION
Monotonic Tension

Tensile results obtained at 1200°C were consistent with those reported earlier [43, 46]. The ultimate tensile strength (UTS) was 190 MPa, elastic modulus, 76 GPa, and failure strain, 0.38 %. It is worthy of note that in all tests reported herein, the failure occurred within the gage section of the extensometer.

Tension-Tension Fatigue

Tension-tension fatigue tests were conducted at 0.1 and 10 Hz at 1200°C in air and in steam. Results are summarized in Table I and presented in Fig. 2 as the stress vs time to failure curves. Fatigue results at 1.0 Hz from prior work [43] are included for comparison. Data in Table I show that the loading frequency has little effect on fatigue performance in air. At 0.1 and 1.0 Hz the fatigue limit in air was 170 MPa (88% UTS at 1200°C). This fatigue limit is based on the run-out condition of 10^5 cycles, approximate number of loading cycles expected in aerospace applications at 1200°C. It is recognized that a more rigorous run-out condition could have resulted in a lower fatigue limit. Because the fatigue performance was expected to improve with increasing loading frequency, no tests were conducted in air at the frequency of 10 Hz.

Presence of steam causes noticeable degradation in fatigue performance. At all loading frequencies investigated, the fatigue limits obtained in steam are significantly lower than those obtained in air. Moreover, in steam the influence of the loading frequency on fatigue life becomes dramatic. In air, the high 170 MPa fatigue limit was obtained at both 0.1 and 1.0 Hz. In steam, the best fatigue performance was obtained at 10 Hz. Yet, even at 10 Hz the in-steam fatigue limit is only 150 MPa (78% UTS at 1200°C), noticeably lower than what could be expected in air. At 1.0 Hz, the fatigue limit drops to 125 MPa (69% UTS at 1200°C). As the frequency decreases by another order of magnitude, the fatigue performance deteriorates drastically. At 0.1 Hz, run-out was not achieved even at the low stress level of 75 MPa (39%

UTS at 1200°C). These results suggest that the time under stress is the critical parameter in determining the cyclic lifetime in steam.

Table I. Summary of fatigue results for the N720/A composite at 1200°C. Results of fatigue tests at 1.0 Hz from prior work [43] are included for comparison.

Test Environment	Max Stress (MPa)	Cycles to Failure	Time to Failure (h)	Failure Strain (%)
Fatigue at 0.1 Hz				
Laboratory Air	170	100,017 [a]	278 [a]	1.93 [a]
Steam	75	56,093	156	3.35
Steam	100	17,498	48.6	1.80
Steam	125	1,850	5.14	1.15
Steam	150	75	0.21	0.67
Steam	170	12	0.03	0.53
Fatigue at 1.0 Hz				
Laboratory Air [b]	100	120,199 [a]	33.4 [a]	0.63 [a]
Laboratory Air [b]	125	146,392 [a]	40.7 [a]	1.14 [a]
Laboratory Air [b]	150	167,473 [a]	46.5 [a]	1.66 [a]
Laboratory Air [b]	170	109,436 [a]	30.4 [a]	2.25 [a]
Steam [b]	100	100,780 [a]	28.0 [a]	0.71 [a]
Steam [b]	125	166,326 [a]	46.2 [a]	1.08 [a]
Steam [b]	150	11,782	3.27	1.12
Steam [b]	170	202	0.06	0.81
Fatigue at 10 Hz				
Steam	150	1,000,010 [a]	27.8 [a]	0.77 [a]
Steam	170	11,387	0.32	1.03

[a] run-out, [b] Data from Ruggles-Wrenn et al [43].

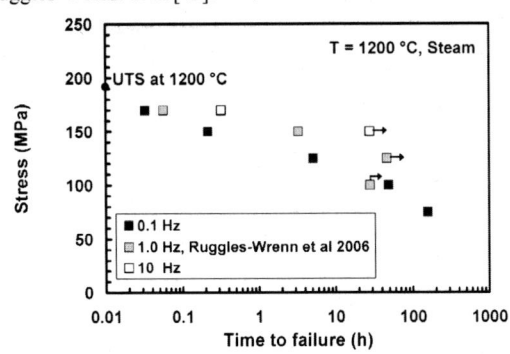

Fig. 2. Fatigue S-N curves for Nextel™720/alumina ceramic composite at 1200°C in steam. Fatigue data at 1.0 Hz from Ruggles-Wrenn et al [43] are included for comparison.

Evolution of the hysteresis response of N720/A with fatigue cycles is typified in Fig. 3, which shows hysteresis stress-strain loops for tests conducted in steam at various loading

Effects of Frequency on Fatigue Behavior of an Oxide-Oxide Ceramic Composite at 1200°C

frequencies. In all tests, regardless of the frequency, the most extensive damage occurs on the first cycle, where considerable permanent strain is seen upon unloading. Afterwards hysteresis loops stabilize quickly. Ratcheting, defined as progressive increase in accumulated strain with increasing number of cycles, continues throughout the test. Effects of loading frequency and test environment on hysteresis response are illustrated in Figs. 4 (a) and (b), respectively. It is seen that the permanent strain produced during the first cycle increases with decreasing frequency, suggesting that time-dependent deformation mechanism (akin to that in creep) is active. It is also seen that larger permanent strain is produced in steam than in air.

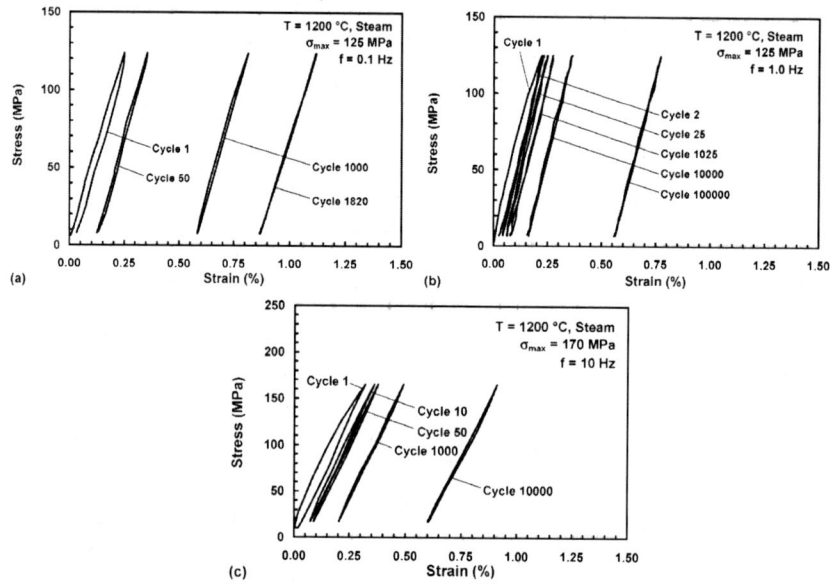

Fig. 3. Evolution of stress-strain hysteresis response with cycles in steam: (a) at 0.1 Hz and 125 MPa, (b) at 1.0 Hz and 125 MPa, data from Ruggles-Wrenn et al [43], (c) at 10 Hz and 170 MPa.

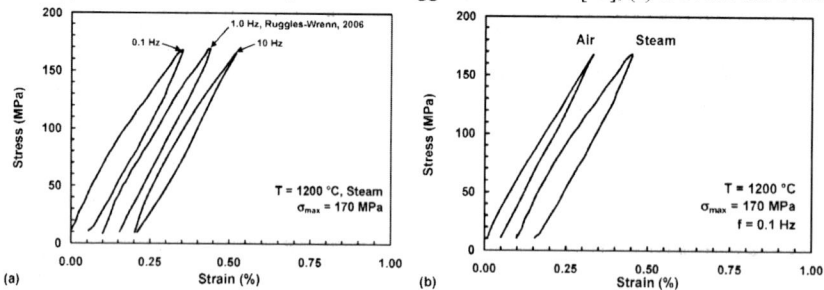

Fig. 4. The stress-strain response of N720/A composite at 1200°C: (a) in steam at three different loading frequencies, data at 1.0 Hz from Ruggles-Wrenn et al [43], and (b) at 0.1 Hz in air and in steam. Curves shifted by 0.1% for clarity.

Effects of Frequency on Fatigue Behavior of an Oxide-Oxide Ceramic Composite at 1200°C

Of importance in cyclic fatigue is the reduction in stiffness (hysteresis modulus determined from the maximum and minimum stress-strain data points during a load cycle), reflecting the damage development during fatigue cycling. Change in modulus is shown in Fig. 5, where normalized modulus (i. e. modulus normalized by the modulus obtained in the second cycle) is plotted vs fatigue cycles. The first cycle was not used for normalization because of the large permanent strain offset upon unloading. It is noteworthy that although all in-air tests achieved run-out, a decrease in normalized modulus with cycling was still observed (Fig. 5(a)). Modulus loss increased with increasing fatigue stress level, but was fairly independent of the loading frequency. Modulus loss was 18 % in the 170 MPa test conducted at 0.1 Hz and 17% in the 170 MPa test at 1.0 Hz.

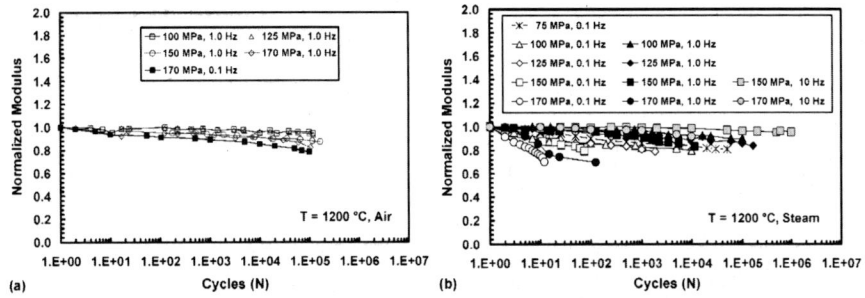

Fig. 5. Normalized modulus vs fatigue cycles at 1200°C (a) in laboratory air and (b) in steam. Fatigue data at 1.0 Hz from Ruggles-Wrenn et al [43] are included for comparison.

Changes in normalized modulus as well as the influence of loading rate on modulus evolution become more pronounced in steam (Fig. 5(b)). While in air the reduction in normalized modulus was limited to 18%, in steam the normalized modulus loss reached 30% (170 MPa tests at 0.1 and at 1.0 Hz). As in air, in steam modulus loss increases with increasing fatigue stress level, but is not significantly affected by increase in frequency from 0.1 to 1.0 Hz. However, increase in frequency by another order of magnitude to 10 Hz has a noticeable effect. At 10 Hz, the reduction in normalized modulus was only 5% in the 150 MPa test, which achieved a run-out, and 8% in the 170 MPa test. Continuous decrease in modulus observed both in air and in steam suggests progressive damage with continued cycling. Because the fatigue damage is still evolving at 10^5 cycles (10^6 cycles at 10 Hz), the 10^5 (10^6 at 10 Hz) fatigue limit does not meet the criteria of a true endurance fatigue limit proposed by Sorensen et al [47] and may not be a true endurance fatigue limit.

Maximum Cyclic Strain

Maximum cyclic strains as functions of cycle number for fatigue tests conducted at 1200°C in air and in steam are presented in Figs. 6 (a) and (b), respectively. It is seen that ratcheting takes place in all tests conducted at 1200°C. In laboratory air the rate of strain accumulation increases with increasing fatigue stress level. On the other hand, the loading frequency appears to have little effect on strain accumulation rate. For the fatigue stress of 170 MPa, the evolution of strain with cycles observed at 0.1 Hz is similar to that at 1.0 Hz. Note that all tests conducted in air achieved fatigue run-out. As seen in Figs. 6 (a) and (b), strains accumulated in steam are considerably lower than those accumulated in air at the same fatigue stress and loading frequency. Generally, lower strain accumulation with cycling indicates that less damage has occurred, and that it is mostly limited to some additional matrix cracking.

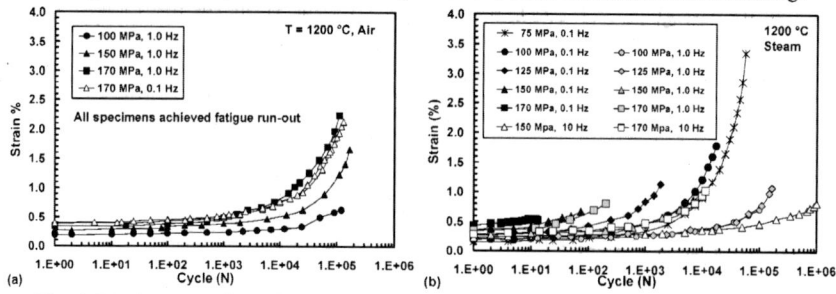

Fig. 6. Maximum strain vs fatigue cycles at 1200°C: (a) in laboratory air and (b) in steam. Fatigue data at 1.0 Hz from Ruggles-Wrenn et al [43] are included for comparison.

However, lower accumulated strains observed in steam invariably correspond to shorter fatigue lives. In this case lower accumulated strains are more likely due to early bundle failures leading to specimen failure. In steam at a given loading frequency, the accumulated strain increases with decreasing fatigue stress. This trend is particularly pronounced at the loading frequency of 0.1 Hz. The 170 MPa test failed after 12 cycles accumulating only 0.53% strain, while the 75 MPa test survived 56093 cycles and accumulated a much larger strain of 3.35%. Specimens with longer cyclic lives also exhibited larger amounts of fiber pullout, which accounts for larger accumulated strains. In steam, the evolution of maximum strain with cycles is strongly influenced by the loading frequency. For a given fatigue stress level, the rate of strain accumulation increases with decreasing frequency. In the case of the 150 MPa tests, specimen cycled at 0.1 Hz accumulated 0.59% strain during the first 50 cycles, while those tested at 1.0 and 10 Hz accumulated 0.32 and 0.25% strain, respectively. However, higher rate of strain accumulation does not necessarily translate into higher failure strain. The decrease in loading frequency causes a dramatic decrease in fatigue life, hence allowing much less time for strain accumulation.

HED

The hysteresis energy density (HED) behavior obtained at 0.1 and 1.0 Hz in air and in steam is shown in Figs. 7 (a) and (b), respectively. At 10 Hz only minimal hysteresis was observed, the HED values quickly became negligibly small and were therefore not included in Fig. 7. The HED values shown in Fig. 7 are fairly small, with the average of ~10 kJ/m^3. Most traditional composites with interfaces and classical fiber debonding typically produce HED values ≥80 kJ/m^3 when fatigued above the proportional limit. It is seen in Fig. 7 that the HED exhibits a significant decrease within the first 10 cycles. From this cycle number on there appears to be only a slight decrease in HED with continued cycling. The initial HED values obtained at 0.1 Hz are somewhat higher than those obtained at 1.0 Hz. However, the stabilized HED values are fairly independent of the loading frequency. The presence of steam also appears to have little influence on the HED behavior at the loading frequencies investigated. In conventional

Effects of Frequency on Fatigue Behavior of an Oxide-Oxide Ceramic Composite at 1200°C

composites with interfaces, a decrease in HED with fatigue cycling is generally attributed to degradation of interfacial shear resistance at the fiber matrix interface. For most brittle matrix composites, it was also observed [20] that continuous damage development, such as matrix cracking and fiber/matrix debonding, in a cyclically loaded specimen may have a significant effect on the HED behavior.

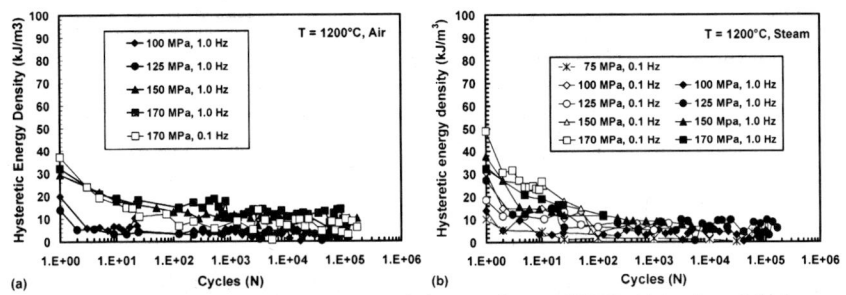

Fig. 7. Hysteresis energy density (HED) vs fatigue cycles at 1200°C: (a) in air and (b) in steam. Fatigue data at 1.0 Hz from Ruggles-Wrenn et al [43] are included for comparison.

Table II. Retained properties of the N720/A specimens subjected to prior fatigue in air and in steam at 1200°C

Fatigue Stress (MPa)	Fatigue Environment	Retained Strength (MPa)	Retained modulus (GPa)	Failure Strain (%)
Prior fatigue at 0.1 Hz				
170	Air	194	51.7	0.38
Prior fatigue at 1.0 Hz				
100[a]	Air	194	56.6	0.44
125[a]	Air	199	54.9	0.44
150[a]	Air	199	43.4	0.53
170[a]	Air	192	40.7	0.51
100[a]	Steam	174	47.6	0.40
125[a]	Steam	168	52.0	0.43
Prior fatigue at 10 Hz				
150	Steam	184	55.4	0.48

[a] Data from Ruggles-Wrenn et al [43].

Retained strength and stiffness of the specimens, which achieved a run-out, are summarized in Table II. It is seen that all specimens tested in air exhibited no loss of tensile strength, irrespective of the fatigue stress level or the loading frequency. However, considerable stiffness loss (18-33%) was observed. Full retention of tensile strength indicates that no damage occurred to the fibers. The reduction in stiffness is most likely due to additional matrix cracking. Conversely, prior fatigue in steam caused reduction in both strength and stiffness. In steam, retained strength decreased by as much as 12% and stiffness decreased by as much as 20%. Strength and stiffness degradation increases with decreasing frequency of prior fatigue. As the loading frequency in the 170 MPa tests decreased from 10 to 1.0 Hz, retained strength decreased

Effects of Frequency on Fatigue Behavior of an Oxide-Oxide Ceramic Composite at 1200°C

by 12% and stiffness decreased by 20%. The discrepancy between the retained modulus of a runout specimen and the decrease in hysteresis modulus observed during fatigue testing most likely stems from different methods used to determine the retained and hysteresis moduli.

Prior work [43] revealed the degrading effect of steam on fatigue performance of the N720/A composite at 1200°C. Mehrman et al [44] demonstrated that superposition of a hold time at maximum stress onto a fatigue cycle degraded fatigue life of N720/A at 1200°C. The presence of steam amplified the deleterious effect of the hold time. Results of the current study support the earlier observations. Degrading effects of steam on fatigue performance are evident. Furthermore, a marked influence of the loading frequency on fatigue life in steam suggests that a time-dependent damage process akin to that associated with creep affects the cyclic life. A certain minimum time under stress is required to activate the said damage process on the first cycle. As the fatigue stress level increases, less time under stress would be required for damage activation on the first cycle and damage growth on subsequent cycles. A lower loading frequency provides for longer time under stress per cycle, accelerating damage growth and degradation of fatigue life. Consider results of the 150 MPa tests in steam. As the loading frequency decreases by an order of magnitude, so does the time to failure.

Composite Microstructure

Optical micrographs of fracture surfaces obtained in fatigue tests conducted at 1200°C in steam are shown in Figs. 8 and 9. The fracture planes of all specimens are not well defined. The 0° fiber tows break over a wide range of axial locations, in general spanning the entire width of the specimen. The fibers in the 0° tows in each cloth layer exhibit random failure producing fiber pullout. Fracture surfaces obtained at 0.1 Hz (Fig. 8) are dominated by coordinated failure of the fiber bundles with limited areas of fiber pull-out. It is also seen that the size of the damage zone decreases with increasing fatigue stress level. While a damage zone of ≈10 mm in length was produced in the 100 MPa test, the damage zone obtained in the 150 MPa test is only ~4 mm long. Note that the specimen with the largest damage zone also had the longest fatigue life.

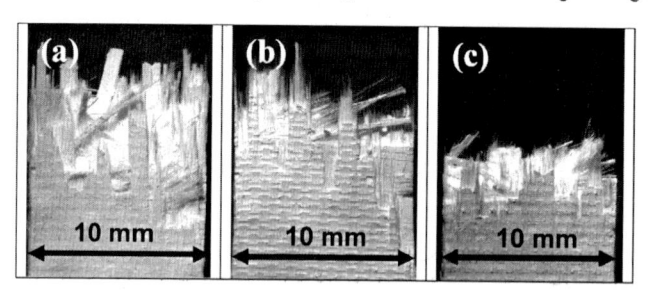

Fig. 8. Fracture surfaces of the N720/A specimens tested in fatigue at 1200°C in steam at 0.1 Hz with maximum stress of: (a) 100 MPa, (b) 125 MPa, and (c) 150 MPa.

As demonstrated above, loading frequency has a profound effect on the fatigue life. The influence of the loading frequency on the fracture surface topography is illustrated in Fig. 9 (a), (b), and (c), which display fracture surfaces obtained in 170 MPa fatigue tests conducted at 0.1, 1.0, and 10 Hz, respectively. The damage zone produced at 0.1 Hz is small, on the order of 2

Effects of Frequency on Fatigue Behavior of an Oxide-Oxide Ceramic Composite at 1200°C

mm. Near-planar failure dominates, but small areas of brushy failure are visible. A longer damage zone and a somewhat higher level of fiber pullout are seen at 1.0 Hz. A dramatically different fracture surface topography is obtained at 10 Hz. The length of the damage zone has increased to ~13 mm. Fiber pullout is considerable, as is the variation in pull-out length. Fiber pullout length as well as the damage zone size increase with increasing frequency. Once again, the greatest amount of fiber pullout and the largest damage zone are produced by the specimens with the longest fatigue life. Specimens tested in this effort produced damage zones ranging from ~ 1 to 14 mm in length. It is noteworthy that specimens that exhibited longer life invariably produced longer damage zone and more fiber pullout.

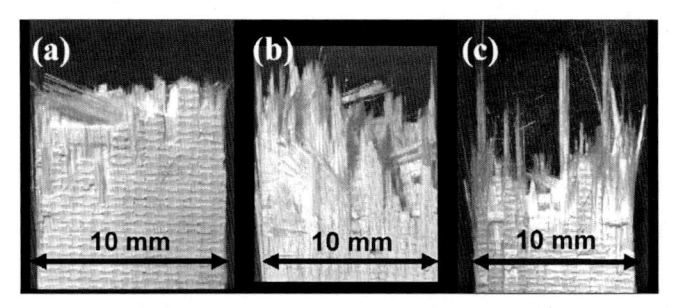

Fig. 9. Fracture surfaces of the N720/A specimens tested in fatigue with maximum stress of 170 MPa at 1200°C in steam at: (a) 0.1 Hz, (b) 1.0 Hz, from Ruggles-Wrenn et al [43], and (c) 10 Hz.

The SEM micrographs in Fig. 10 show features of the composite microstructure observed in the current research. All fracture surfaces contain areas of fiber pullout as well as regions of flatter, more coordinated fracture with little or no fiber pullout. An overall view of a typical fracture surface is presented in Fig. 10(a). Figure 10(b) shows pullout of individual fibers. Note that the locations of the fiber failure within an individual tow, and consequently the lengths of fiber pullout exhibit a broad distribution. A region of coordinated fracture is seen above the matrix-rich area in Fig. 10(c). Note that most of the fibers fracture on different planes, indicating that a single crack front did not cause this fracture topography. Evidence of the strong fiber-matrix bond is seen in Fig. 10(d), showing matrix material still bonded to the pulled out fiber.

While both the regions of fiber pullout and the regions of nearly planar fracture are present in all fracture surfaces, the balance of these two types of fracture topography within a given fracture surface is influenced by fatigue stress level and loading frequency. Effect of the fatigue stress is illustrated in Fig. 11 (a)-(c). The fracture surface produced in the 125 MPa (Fig. 11(a)) test is dominated by regions of fiber pullout where individual fibers are clearly discernable. While the overall appearance of the fracture surface is brushy, isolated areas of coordinated failure are also present. A rectangular region of nearly planar fracture surrounded by individual pulled out fibers can be seen in the middle of the micrograph. The fracture surface produced in the 150 MPa test (Fig. 11(b)) still exhibits individual fiber pullout, however, large areas of planar fracture are clearly visible. Finally, the fracture surface obtained in the 170 MPa test (Fig. 11(c)) is dominated by regions of coordinated planar fracture. Influence of the loading frequency is demonstrated in Fig. 11 (c) and (d), displaying fracture surfaces obtained in the 170 MPa tests conducted at 0.1 and 10 Hz, respectively. The difference between the two fracture

Effects of Frequency on Fatigue Behavior of an Oxide-Oxide Ceramic Composite at 1200°C

surfaces is striking. Regions of planar failure dominate the fracture surface produced at 0.1 Hz, while extensive fiber pullout is prevalent at 10 Hz.

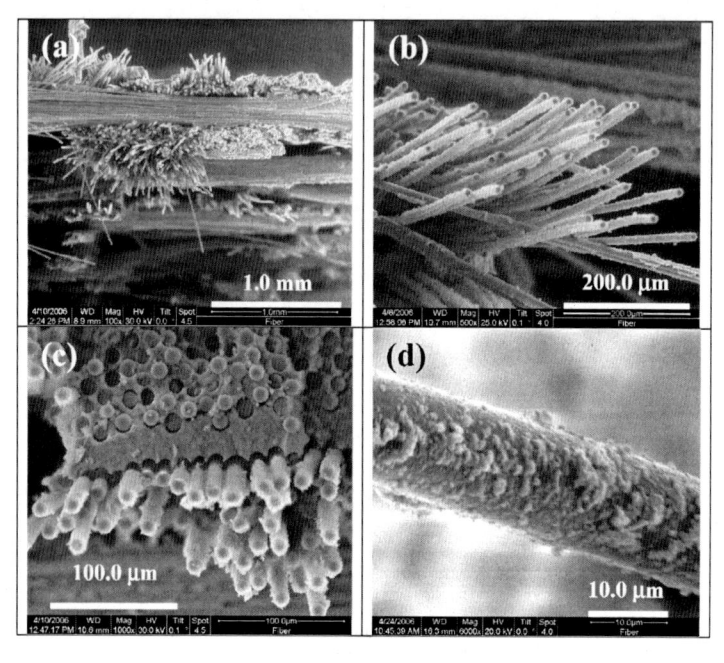

Fig.10. Fracture surfaces of N720/A specimens tested in fatigue at 1200°C in steam at 0.1 Hz: (a) overall view showing fiber pullout and regions of coordinated fracture, (b) individual fiber pullout, (c) fiber pullout below and region of coordinated failure above the matrix-rich area, (d) matrix particles attached to the pulled out fiber.

Predominantly planar fracture surface is indicative of a fast failure process where crack fronts propagate rapidly through both matrix and fibers. Note that the specimen subjected to fatigue with the maximum stress of 170 MPa at 0.1 Hz (Fig. 11(c)) survived only 12 cycles (~0.03 h). In contrast to coordinated planar fracture, a brushy fracture surface typically accompanies longer times to failure. Recall that the specimen subjected to the 170 MPa fatigue tests at 10 Hz (Fig. 11(d)) survived 11387 cycles (~0.32 h). The higher levels of fiber pull-out are consistent with larger failure strains, which are also observed in the cases of longer cyclic life. The fracture surfaces obtained at the loading frequency of 0.1 Hz can be readily distinguished from those produced at 10 Hz. In addition, the fracture surface appearance can be correlated with the failure time.

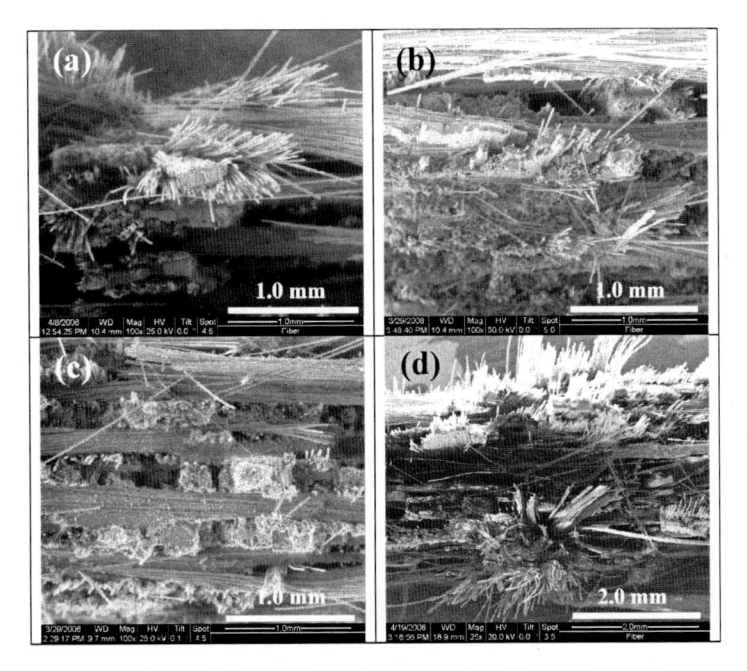

Fig. 11. Fracture surfaces of the N720/A specimens tested in fatigue at 1200°C in steam: (a) at 0.1 Hz and 125 MPa, (b) at 0.1 Hz and 150 MPa, (c) at 0.1 Hz and 170 MPa and (d) at 10 Hz and 170 MPa.

CONCLUDING REMARKS

Effect of loading frequency on fatigue behavior of the N720/A continuous fiber ceramic composite was investigated at 1200°C in laboratory air and in steam. Fatigue stress levels ranged from 100 to 170 MPa in air, and from 75 to 170 MPa in steam. Composite microstructure, as well as damage and failure mechanisms were explored.

In laboratory air, fatigue life appears to be independent of the loading frequency. In air, the fatigue limit (based on a run-out condition of 10^5 cycles) of 170 MPa (88% UTS at 1200°C) was achieved at both 0.1 and 1.0 Hz. The material retains 100% of its tensile strength. However, considerable stiffness loss (~30%) is observed.

Presence of steam causes noticeable degradation in fatigue performance for all loading frequencies investigated. In steam, loading frequency has a profound effect on fatigue performance. Fatigue limit, number of cycles to failure, and time to failure all decrease as the loading frequency decreases. A strong influence of the loading frequency on fatigue performance in steam suggests that a similar time-dependent failure mechanism may be operating under both cyclic and static loading at 1200 °C in steam. It is believed that stress corrosion of the N720 fibers is the mechanism behind the degraded mechanical performance in steam. It is likely that subcritical crack growth in the fiber is caused by a chemical interaction of water molecules with mechanically strained Si-O bonds at the crack tip, with the rate of chemical reaction increasing exponentially with applied stress.

Appearance of the fracture surfaces obtained in steam changes with the loading frequency. The amount of fiber pullout as well as the damage zone size increase with increasing frequency. Planar fracture dominates fracture surfaces obtained at 0.1 Hz. At 10 Hz, extensive fiber pullout is prevalent.

REFERENCES

[1] H. Ohnabe, S. Masaki, M. Onozuka, K. Miyahara, T. Sasa, "Potential Application of Ceramic Matrix Composites to Aero-Engine Components," *Composites: Part A*, **30**, 489-496 (1999).

[2] L. P. Zawada, J. Staehler, S. Steel, "Consequence of Intermittent Exposure to Moisture and Salt Fog on the High-Temperature Fatigue Durability of Several Ceramic-Matrix Composites," *J. Am. Ceram. Soc.*, **86**(8), 1282-1291 (2003).

[3] S. Schmidt, S. Beyer, H. Knabe, H. Immich, R. Meistring, A. Gessler, "Advanced Ceramic Matrix Composite Materials for Current and Future Propulsion Technology Applications," *Acta Astronautica*, **55**, 409-420 (2004).

[4] K. M. Prewo, J. A. Batt. The Oxidative Stability of Carbon Fibre Reinforced Glass-Matrix Composites," *J. Mater. Sci.*, **23**, 523-527 (1988).

[5] T. Mah, N. L. Hecht, D. E. McCullum, J. R. Hoenigman, H. M. Kim, A. P. Katz, H. A. Lipsitt, "Thermal Stability of SiC Fibres (Nicalon)," *J. Mater. Sci.*, **19**, 1191-1201 (1984).

[6] K. L. More, P. F. Tortorelli, M. K. Ferber, J. R. Keiser, "Observations of Accelerated Silicon Carbide Recession by Oxidation at High Water-Vapor Pressures," *J. Am. Ceram. Soc.*, **83**(1), 11-213 (2000).

[7] M. K. Ferber, H. T. Lin, J. R. Keiser, "Oxidation Behavior of Non-Oxide Ceramics in a High-Pressure, High-Temperature Steam Environment," *Mechanical, Thermal, and Environmental Testing and Performance of Ceramic Composites and Components*. M. G. Jenkins, E. Lara-Curzio, and S. T. Gonczy, editors. ASTM STP 1392, 210-215 (2000).

[8] E. J. Opila, "Variation of the Oxidation Rate of Silicon Carbide with Water Vapor Pressure," *J. Am. Ceram. Soc.*, **82**(3), 625-636 (1999).

[9] E. E. Hermes, R. J. Kerans, "Degradation of Non-Oxide Reinforcement and Oxide Matrix Composites," *Mat. Res. Soc., Symposium Proceedings*, **125**, 73-78 (1988).

[10] A. Szweda, M. L. Millard, M. G. Harrison, *Fiber-Reinforced Ceramic-Matrix Composite Member and Method for Making*, U. S. Pat. No. 5 601 674, (1997).

[11] S. M. Sim, R. J. Kerans, "Slurry Infiltration and 3-D Woven Composites," *Ceram. Eng. Sci. Proc.*, **13**(9-10), 632-641 (1992).

[12] E. H. Moore, T. Mah, and K. A. Keller, "3D Composite Fabrication Through Matrix Slurry Pressure Infiltration," *Ceram. Eng. Sci. Proc.*, **15**(4), 113-120 (1994).

[13] M. H. Lewis, M. G. Cain, P. Doleman, A. G. Razzell, J. Gent, "Development of Interfaces in Oxide and Silicate Matrix Composites," *High-Temperature Ceramic–Matrix Composites II: Manufacturing and Materials Development*, A. G. Evans, and R. G. Naslain, editors, American Ceramic Society, 41–52 (1995).

[14] F. F. Lange, W. C. Tu, A. G. Evans, "Processing of Damage-Tolerant, Oxidation-Resistant Ceramic Matrix Composites by a Precursor Infiltration and Pyrolysis Method," *Mater. Sci. Eng. A*, **A195**, 145–150 (1995).

[15] R. Lunderberg, L. Eckerbom, "Design and Processing of All-Oxide Composites," *High-Temperature Ceramic–Matrix Composites II: Manufacturing and Materials Development*, A. G. Evans, and R. G. Naslain, editors, American Ceramic Society, 95–104 (1995).

[16] E. Mouchon, P. Colomban, "Oxide Ceramic Matrix/Oxide Fiber Woven Fabric Composites Exhibiting Dissipative Fracture Behavior," *Composites*, **26**, 175–182 (1995).

[17] P. E. D. Morgan and D. B. Marshall, "Ceramic Composites of Monazite and Alumina," *J. Am. Ceram. Soc.*, **78**(6), 1553–1563 (1995).

[18] W. C. Tu, F. F. Lange, A. G. Evans, "Concept for a Damage-Tolerant Ceramic Composite with Strong Interfaces," *J. Am. Ceram. Soc.*, **79**(2), 417–424 (1996).

[19] R. J. Kerans, R. S. Hay, N. J. Pagano, T. A. Parthasarathy, "The Role of the Fiber-Matrix Interface in Ceramic Composites," *Am. Ceram. Soc. Bull.*, **68**(2), 429-442 (1989).

[20] A. G. Evans, F. W. Zok, "Review: the Physics and Mechanics of Fiber-Reinforced Brittle Matrix Composites," *J. Mater. Science*, **29**, 3857-3896 (1994).

[21] R. J. Kerans, T. A. Parthasarathy, "Crack Deflection in Ceramic Composites and Fiber Coating Design Criteria," *Composites: Part A*, **30**, 521-524 (1999).

[22] R. J. Kerans, R. S. Hay, T. A. Parthasarathy, M. K. Cinibulk, "Interface Design for Oxidation-Resistant Ceramic Composites," *J. Am. Ceram. Soc.*, **85**(11), 2599- 2632 (2002).

[23] C. G. Levi, J. Y. Yang, B. J. Dalgleish, F. W. Zok, A. G. Evans, "Processing and Performance of an All-Oxide Ceramic Composite," *J. Am. Ceram. Soc.*, **81**, 2077-2086 (1998).

[24] A. G. Hegedus, *Ceramic Bodies of Controlled Porosity and Process for Making Same*, U. S. Pat. No. 5 0177 522, May 21, (1991).

[25] T. J. Dunyak, D. R. Chang, M. L. Millard, "Thermal Aging Effects on Oxide/Oxide Ceramic-Matrix Composites," *Proceedings of 17th Conference on Metal Matrix, Carbon, and Ceramic Matrix Composites. NASA Conference Publication 3235, Part 2*, 675-90 (1993)

[26] L. P. Zawada, S. S. Lee, "Mechanical Behavior of CMCs for Flaps and Seals," *ARPA Ceramic Technology Insertion Program (DARPA)*, W. S. Coblenz WS, editor. Annapolis MD, 267-322 (1994).

[27] L. P. Zawada, S. S. Lee, "Evaluation of the Fatigue Performance of Five CMCs for Aerospace Applications," *Proceedings of the Sixth International Fatigue Congress*, 1669-1674 (1996).

[28] T. J. Lu, "Crack Branching in All-Oxide Ceramic Composites," *J. Am. Ceram. Soc.*, **79**(1), 266-274 (1996).

[29] F. W. Zok, C. G. Levi, "Mechanical Properties of Porous-Matrix Ceramic Composites," *Adv. Eng. Mater.*, **3**(1-2), 15-23 (2001).

[30] C. Q. Rousseau, "Monotonic and Cyclic Behavior of Silicon Carbide/Calcium Aluminosilicate Ceramic Composite," *Thermal and Mechanical Behavior of Metal Matrix and Ceramic Matrix Composites*. J. Kennedy J, H. Moeller and W. Johnson, editors. ASTM STP 1080, 136-51 (1990).

[31] J. Holmes, "Influence of Stress Ratio on the Elevated-Temperature Fatigue of a Silicon Carbide Fiber-Reinforced Silicon Nitride Composite," *J. Am. Ceram. Soc.*, **74**(7), 1639-45 (1991).

[32] J. C. McNulty, F. W. Zok, "Low-Cycle Fatigue of Nicalon™-Fiber Reinforced Ceramic Composites," *Comp. Sci. Tech.*, **59**, 1597-1607 (1999).

[33] E. Lara-Curzio, M. Ferber, R. Boisvert, A. Szweda, "The High Temperature Tensile Fatigue Behavior of a Polymer-Derived Ceramic Matrix Composite," *Ceram. Eng. Sci. Proc.*, **16**, 341-9 (1995).

[34]F. Heredia, J. McNulty, F. Zok, A. G. Evans, "Oxidation Embrittlement Probe for Ceramic Matrix Composites," *J. Am. Ceram. Soc.*, **78**(8), 2097-2100 (1995).

[35]P. Reynaud, "Cyclic Fatigue of Ceramic-Matrix Composites at Ambient and Elevated Temperatures," *Comp. Sci. Tech.*, **56**(7), 809-14 (1996).

[36]M. Mizuno, Y. Nagano, Y. Sakaida, Y. Kagawa, M. Watanabe, 'Cyclic-Fatigue Behavior of SiC/SiC Composites at Room and High Temperatures," *J. Am. Ceram. Soc.*, **79**(12), 3065-77 (1996).

[37]S. F. Shuler, J. W. Holmes, X. Wu, D. Roach, "Influence of Loading Frequency on the Room-Temperature Fatigue of a Carbon-Fiber/SiC-Matrix Composite," *J. Am. Ceram. Soc.*, **76**(9), 2327-36 (1993) .

[38]J. M. Staehler, S. Mall, L. P. Zawada, "Frequency Dependence of High-Cycle Fatigue Behavior of CVI C/SiC at Room Temperature," *Comp. Sci. Tech.*, **63**, 2121-31 (2003).

[39]J. W. Holmes, X. Wu, B. F. Sorensen, "Frequency Dependence of Fatigue Life and Internal Heating of a Fiber-Reinforced/Ceramic-Matrix Composite," *J. Am. Ceram. Soc.*, **77**(12), 3284-86 (1994).

[40]N. Chawla, Y. K. Tur, J. W. Holmes, J. R. Barber, A. Szweda, "High-Frequency Fatigue Behavior of Woven-Fiber-Fabric-Reinforced Polymer-Derived Ceramic-Matrix Composites," *J. Am. Ceram. Soc.*, **81**(5), 1221-30 (1998).

[41]E. Vanswijgenhoven, J. W. Holmes, M. Wevers, A. Szweda A, "The Influence of Loading Frequency on High-Temperature Fatigue Behavior of a Nicalon- Fabric-Reinforced Polymer-Derived Ceramic-Matrix Composite," *Scripta Mater.*, **38**(12): 1781-88 (1998).

[42]L. P. Zawada, R. S. Hay, S. S. Lee, J. Staehler, "Characterization and High-Temperature Mechanical Behavior of an Oxide/Oxide Composite," *J. Am. Ceram. Soc.*, **86**(6), 981-90 (2003).

[43]M. B. Ruggles-Wrenn, S. Mall, C. A. Eber, L. B. Harlan, "Effects of Steam Environment on High-Temperature Mechanical Behavior of Nextel^{TM}720/Alumina (N720/A) Continuous Fiber Ceramic Composite," *Composites: Part A*, **37**(11), 2029-40 (2006).

[44]J. M. Mehrman, M. B. Ruggles-Wrenn, S. S. Baek, "Influence of Hold Times on the Elevated-Temperature Fatigue Behavior of an Oxide-Oxide Ceramic Composite in Air and in Steam Environment," *Comp. Sci. Tech.*, **67**, 1425-1438 (2007).

[45]R. A. Jurf, S. C. Butner, "Advances in Oxide-Oxide CMC," *Proceedings of the 44th ASME Gas Turbine and Aeroengine Congress and Exhibition. June 7-10*, (1999).

[46]COI Ceramics, Unpublished Data.

[47]B. F. Sorensen, J. W. Holmes, E. L. Vanswijgenhoven, "Does a True Fatigue Limit Exist for Continuous Fiber-Reinforced Ceramic Matrix Composites?" *J. Am. Ceram. Soc.*, **85**(2), 359-365 (2002).

POST CREEP/DWELL FATIGUE TESTING OF MI SIC/SIC COMPOSITES

Ojard, G[2]., Calomino, A[3]., Morscher, G[4]., Gowayed, Y[5]., Santhosh, U[6]., Ahmad, J[5]., Miller, R[2]., and John, R[1].

[1] Air Force Research Laboratory, AFRL/MLLMN, Wright-Patterson AFB, OH
[2] Pratt & Whitney, East Hartford, CT
[3] NASA-Glenn Research Center, Cleveland, OH
[4] Ohio Aerospace Institute, Cleveland, OH
[5] Auburn University, Auburn, AL
[6] Research Applications, Inc., San Diego, CA

ABSTRACT:
As increased interest is seen in high performance ceramic matrix composites, residual capability of the material after exposure to the environment needs to be measured. Residual tensile testing was carried out on a series of previously tested creep and dwell fatigue samples obtained from several lots of melt infiltrated SiC/SiC composite panels. Most of the time-dependent testing was done at 1204°C but tests were also done at other temperatures of interest at a wide range of stresses and durations. It was found that the average strain to failure for the samples was 0.5% for the as-received tensile tests, during failure of creep tests, or as a summation of creep strain and strain of residual tensile tests.

INTRODUCTION
Interest in Ceramic Matrix Composites (CMCs) is high since this class of material has properties that are relatively constant with increasing temperatures up to near its maximum use temperature [1]. This is shown in the interest of CMCs for extended high temperature uses such as combustor liners and turbine vanes[2,3]. These applications take benefit of the high temperature capability of the material with the added benefit of weight reduction (density) as well as durability improvement.

As part of the effort to demonstrate and verify material performance, long term testing at temperatures is typically undertaken. This can take the forms of creep, dwell fatigue or fatigue testing [4,5]. What is typically not considered as part of the characterization is the retained properties (or residual properties) of the material after these long-term tests. The retained properties can give insight into material degradation with time (exposure) as well as how the material may be expected to perform near the end of application being considered.

In this paper, a series of creep and dwell fatigue tests were conducted at various conditions of temperature, stress and time. All testing was done in air. Since several of the tests were stopped at select points without failure being seen, residual testing was done. For the vast majority of the exposures done, room temperature tensile tests were performed to note retained properties. There were a few cases where the retained properties were tested at elevated temperatures. The resulting tensile properties were then reviewed against the exposure conditions. The resulting capabilities were seen to fit a distinct trend of retained strain capability that could be useful in looking at material performance with exposure.

PROCEDURE

Material Description
For this testing, the composite system interrogated was a Melt Infiltrated In-Situ BN SiC/SiC composite (MI SiC/SiC). This system has the most characterization done to date and is being considered for high temperature applications [6, 1, 2]. The MI SiC/SiC system has a stochiometric SiC (Sylramic™) fiber in a multiphase matrix of SiC deposited by chemical vapor deposition followed by slurry casting of SiC particulates with a final melt infiltration of Si metal. The specific MI SiC/SiC tested for this effort had 36% volume fraction fibers using a 5 HS weave at 20 EPI. The fibers are 10 μm diameter and there are 800 fibers per tow. This material system was developed by NASA-GRC and is sometimes referred to as the 01/01 material [7]. A cross section of this material is shown in Figure 1.

Si doped BN coat

Sylramic™ fiber

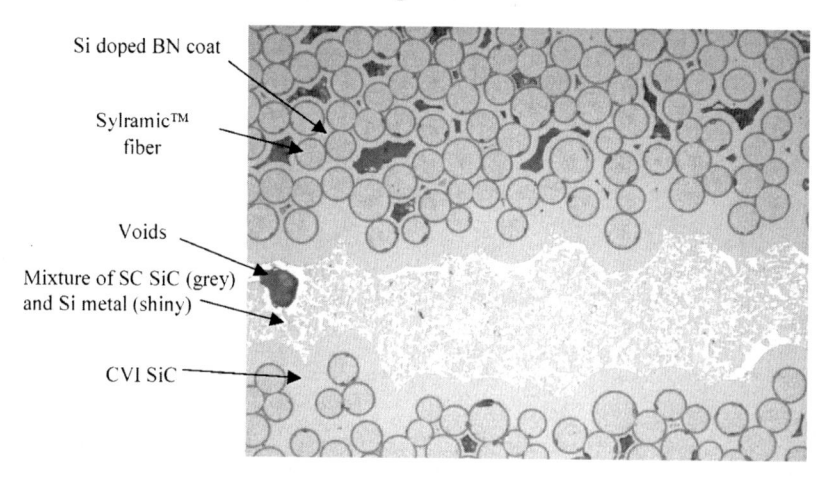

Voids

Mixture of SC SiC (grey) and Si metal (shiny)

CVI SiC

Figure 1. Cross section of Melt Infiltrated In-Situ BN SiC/SiC composite

Creep and Dwell Fatigue Testing (Exposure)
Creep testing is a test to determine the strain-time dependence of a material under a constant load. This test is also used to determine the long-term behavior of the material under combination of load and temperature. For ceramic matrix composites, when this testing is done in air, there is an added complexity of environmental exposure. Creep testing has three regimes, primary (or initial), secondary (steady state) and tertiary creep. The creep rate for the material is typically measured during the secondary (steady state) regime. For this series of testing, side extensometry was used to either measure the creep strain (the initial strain during loading was not recorded) or the total strain during the test.

The dwell fatigue testing in this case was an interrupted creep test. The load was taken off and placed back on every 2 hours. This type of dwell fatigue has been used extensively for previous material characterization efforts [2]. The R ratio for these tests was 0.1. Since the test was cycled every two hours, total strain was recorded for the test as this would allow the modulus to be determined as a function of time. The modulus work will not be discussed here.

Post Creep/Dwell Fatigue Testing of MI SiC/SiC Composites

A total of 47 creep and dwell fatigue tests were performed (19 creep tests and 28 dwell fatigue). The temperatures used were 815°C and 1204°C. The max applied stress levels for this testing were 110, 166, 193, 221, 235 and 248 MPa. The exposure duration ranged from 0.5 to 2036 hours.

Of the testing described above, 32 samples were tensile tested at either 24°C, 815°C or 1204°C. Tensile testing was done per ASTM C1358 and C1359. Strain was recorded during these tests using side extensometry. Samples that were not creep or dwell fatigue exposed were also tensile tested to note the baseline material performance.

RESULTS

Creep and Dwell Fatigue Testing

The range of stresses and times that the material was exposed for under creep and dwell fatigue conditions are shown in Figure 2. As can be seen, failures only occurred during creep testing when the material was exposed for high stresses or long times.

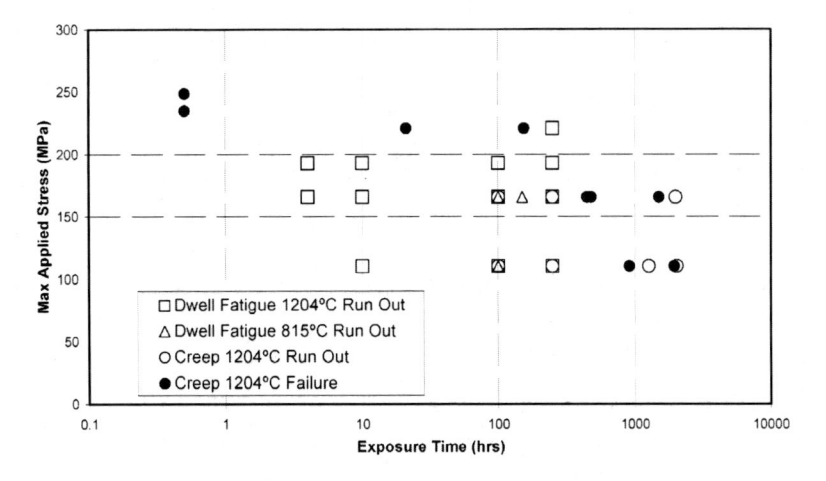

Figure 2. Creep and Dwell Fatigue Testing Summary

Baseline Tensile Testing (No Exposure)

Tensile testing was done at 24°C, 815°C and 1204°C. The results of this testing is shown in Figure 3. The curves generated at 24°C and 825°C are almost identical. (The 24°C test was a cyclic tensile test where the sample was initially loaded to 300 MPa and then unloaded and reloaded to failure.) The tensile curve generated at 1204°C shows some strength and strain degradation. These results are in line with past experience on this material system.

Post Creep/Dwell Fatigue Testing of MI SiC/SiC Composites

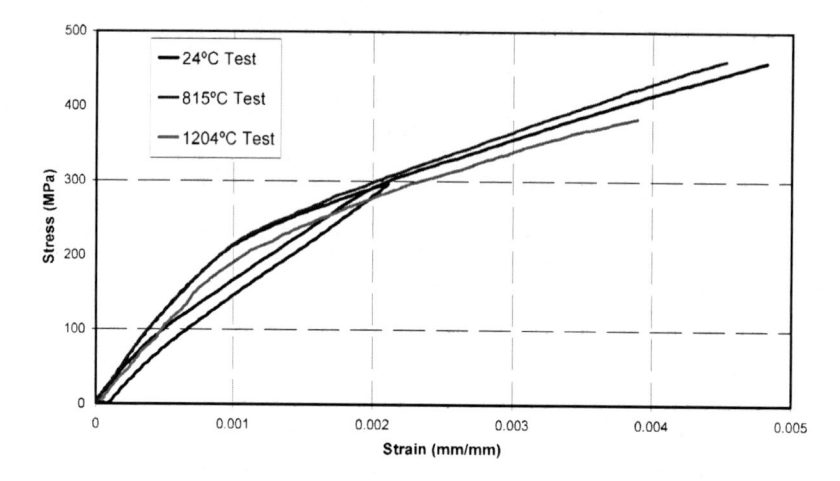

Figure 3. Baseline stress-strain curves at 24°C, 815°C and 1204°C

Tensile Testing (Post Exposure)

Most of the residual tensile testing was done at 24°C. A few tests were done at 815°C and 1204°C. Select residual stress-strain curves are shown in Figure 4. These initial curves looking only at exposure at 110 MPa show that there can be a significant drop in the retained strength and strain to failure.

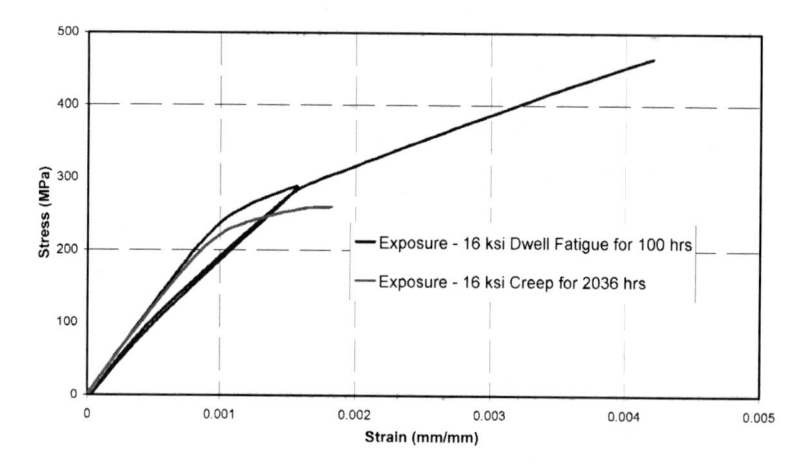

Figure 4. Post-exposure stress-strain curves for samples tested at 110 MPa (dwell fatigue or creep)

One of the most interesting properties is the elastic modulus that can be determined from the stress-strain curve. A decrease in modulus can indicate the presence of cracks formed during testing [8]. An increase in the modulus can indicate that the fiber interface coating could be attacked and that the sliding is inhibited [9]. An initial review of the modulus for all the samples tested at 1204°C and stresses of 110, 166 and 193 MPa is shown in Figure 5. As can be seen, the values are close to the baseline material capability with no trend with temperature.

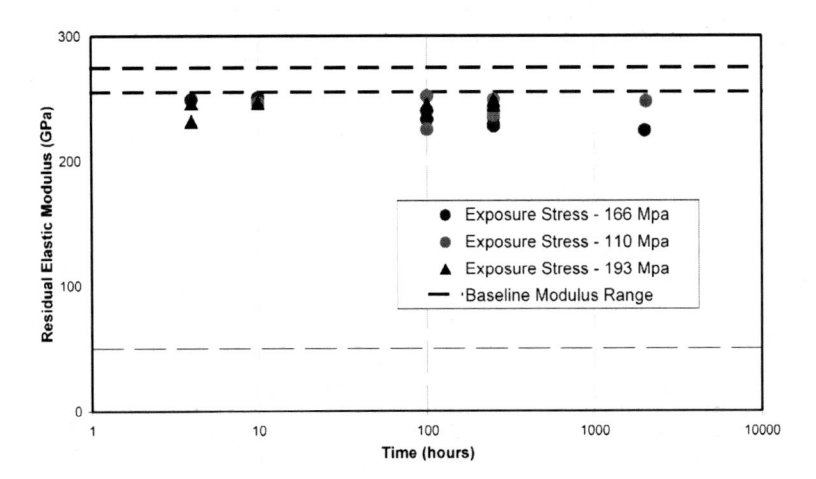

Figure 5. Post-exposure Modulus for samples after dwell fatigue or creep testing

The residual strength of the material can also be plotted against exposure time. This is done for the testing at 1204°C and is shown in Figure 6. As can be seen, there is no clear trend with exposure on the remaining residual strength capability. Only when exposures approach 2,000 hours is there a significant reduction in material capability that can be clearly seen.

The strain to failure can also be viewed as a function of exposure time. This is done for the testing at 1204°C and is shown in Figure 7. Again, no clear differentiation of the samples after exposure is seen. Only significant decreases are noted at long exposure times.

Additionally, all the results shown in Figures 5, 6 and 7 were done with the exposure temperature held constant. Additional limited exposure testing was done at 815°C. There were also limited residual testing done at 815°C and 1204°C. There is no distinct benefit for showing this data on these plots due to the lack of clear differentiation of exposure durations.

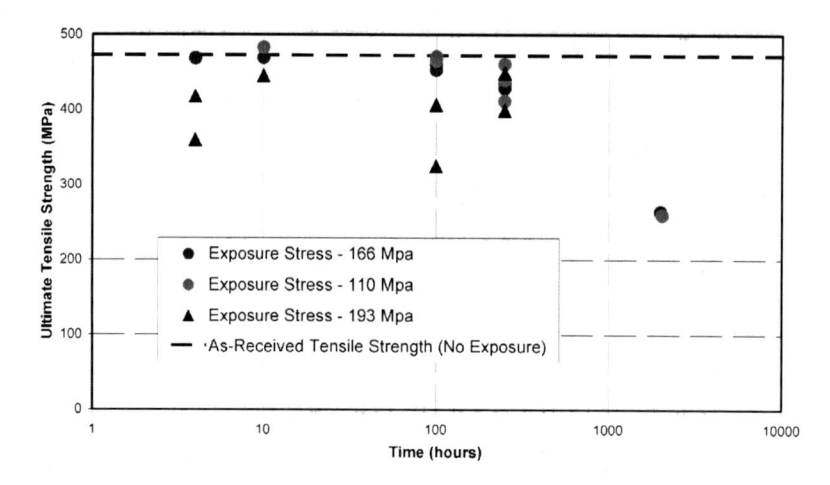

Post Creep/Dwell Fatigue Testing of MI SiC/SiC Composites

Figure 6. Post-exposure tensile strength for samples after dwell fatigue or creep testing

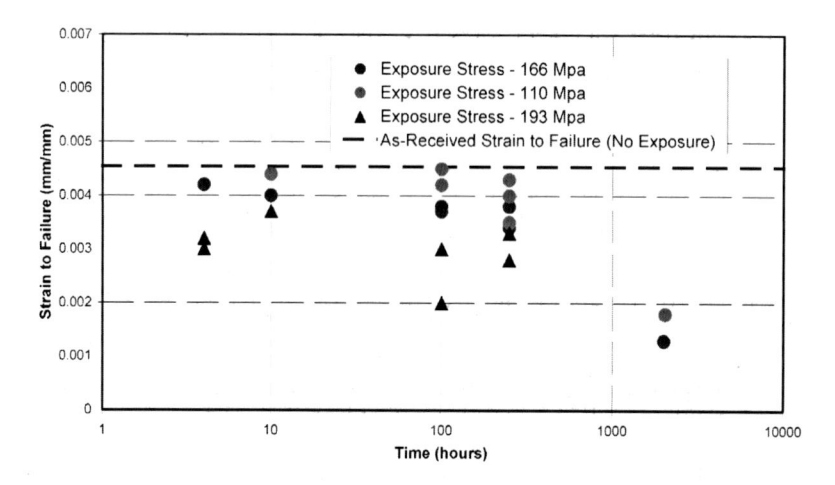

Figure 7. Post-exposure strain to failure for samples after dwell fatigue or creep testing

DISCUSSION

There is a need to clearly see the effects of exposure on CMC materials. All of the plots shown do not differentiate the exposures done. While there is interest that the material can maintain load at various temperatures and durations, this does not state clearly what the material is still capable of after exposure. The results as presented previously are academic at best.

Since strain was measured during the dwell fatigue and creep testing, the evolved strain could give insight into the exposure that the CMC saw during testing. The data shown in Figures 5-7 are re-plotted based on this proposed exposure parameter. Figure 8 clearly shows a stronger differentiation of the data than that seen previously. In Figure 8, there appears to be a line that the data clusters around. Additionally, this type of parameter can now be extended to the full range of exposures and residual tests that were done. These two ideas are shown in Figure 9.

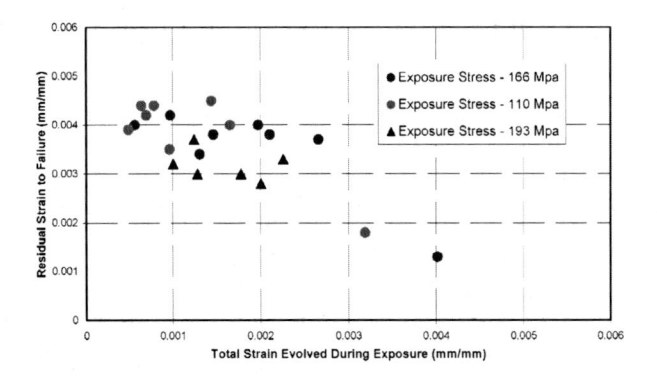

Figure 8. Data from Figures 5-7 re-plotted against the evolved strain during exposure

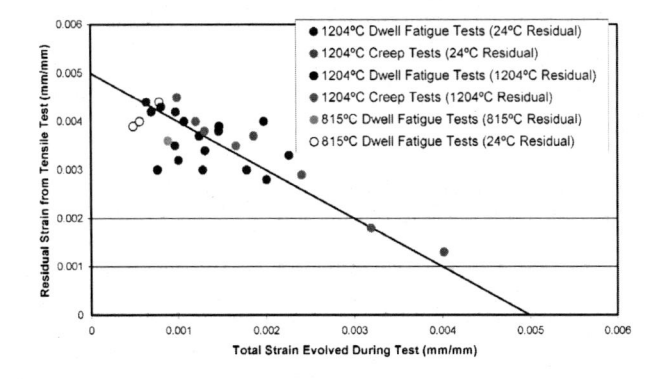

Figure 9. All exposure data shown with the 0.005 strain line the data clusters around

Post Creep/Dwell Fatigue Testing of MI SiC/SiC Composites

The proposed line at 0.005 mm/mm strain was not arbitrary. The tensile tests done on the material on average fail at a strain near 0.005 mm/mm. In addition, the creep tests that failed under loads of 110 and 166 MPa also failed at a strain near 0.005 mm/mm. This is shown in Figure 10. There is a strong conclusion that the strain capability of the fiber (and hence the material) is 0.005 mm/mm. This is supported by multiple tests that were done for this experimental effort. There is no temperature effect seen for the temperature range studied (24°C, 815°C and 1204°C). This is confirmed by the tensile tests done at different temperatures (see Figure 3) as there was only a slight drop with temperature on the strain to failure capability of the material.

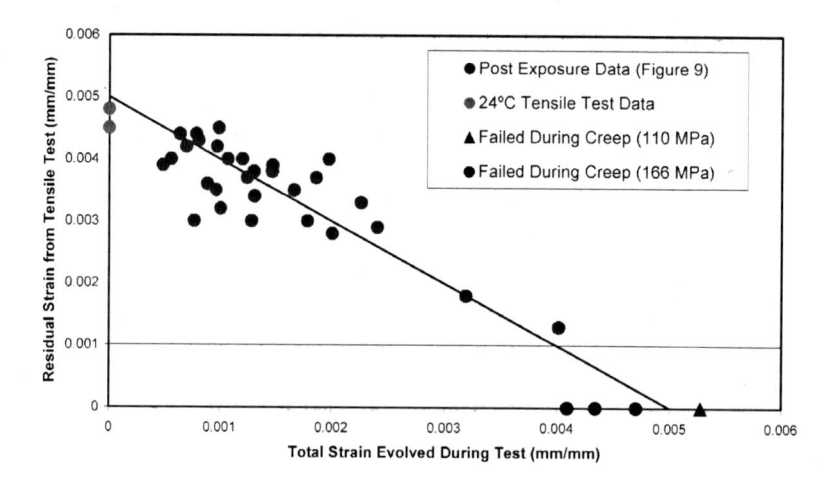

Figure 10. Figure 9 with tensile and creep failures plotted over the data

CONCLUSION

The use of evolved strain has been demonstrated as a damage parameter that can be used for a wide range of exposures for the MI SiC/SiC CMC system. If there is a means to predict the strain evolved during an application, the residual capability at any given point could be predicted since the sum of the evolved strain and the residual strain capability sum to 0.005 mm/mm, approximately. This was proven for the temperatures and stresses studied.

Additionally, that the data was so well behaved indicates that this material does not suffer from intermediate temperature embrittlement. Several CMC systems are known to have such phenomena where at intermediate temperatures, there is aggressive oxidative attack where the fiber interface coating is lost since a protective oxide scale does not form. If such an attack were to occur, the correlation seen against the strain line of 0.005 mm would not be observed. The data would be expected to fall off of this line and show a hook at lower strains and not extend as seen in Figures 9 and 10. The fully dense nature of the MI SiC/SiC microstructure is protecting the fiber interface.

ACKNOWLEDGMENTS

The Materials & Manufacturing Directorate, Air Force Research Laboratory under contract F33615-01-C-5234 and contract F33615-03-D-2354-D04 sponsored portions of this work

REFERENCES
[1] Wedell, James K. and Ahluwalia, K.S., "Development of CVI SiC/SiC CFCCs for Industrial Applications" 39th International SAMPE Symposium April 11- 14, 1994, Anaheim California Volume 2 pg. 2326.
[2] Brewer, D., Ojard, G. and Gibler, M., "Ceramic Matrix Composite Combustor Liner Rig Test", ASME Turbo Expo 2000, Munich, Germany, May 8-11, 2000, ASME Paper 2000-GT-0670.
[3] Calomino, A., and Verrilli, M., "Ceramic Matrix Composite Vane Sub-element Fabrication", ASME Turbo Expo 2004, Vienna, Austria, June 14-17, 2004, ASME Paper 2004-53974.
[4] Ojard, G., Gowayed, Y., Chen, J., Santhosh, U., Ahmad J., Miller, R., and John, R., "Time-Dependent Response of MI SiC/SiC Composites Part 1: Standard Samples" to be published in Ceramic Engineering and Science Proceedings, 2007.
[5] Y. Gowayed, G. Ojard, J. Chen, R. Miller, U. Santhosh, J. Ahmad and John, R., "Time-Dependent Response of MI SiC/SiC Composites Part 2: Samples with Holes", to be published in Ceramic Engineering and Science Proceedings, 2007.
[6] Bhatt, R.T., McCue, T.R., and DiCarlo, J.A., "Thermal Stability of Melt Infiltrated SiC/SiC Composites", Ceramic Engineering and Science Proceedings, Vol. 24, Issue 4, 2003, pg. 295.
[7] Hurwitz, F.I., Calomino, A.M., McCue, T.R., and Morscher, G.N., "C-Coupon Studies of SiC/SiC Composites Part II: Microstructrual Characterization", Ceramic Engineering and Science Proceedings, Vol. 23, Issue 3, 2002, pg 387.
[8] Darzens, S., Chermant, J-L., Vicens, J. and Sangleboeuf, J.C., "Damage Creep Mechanisms of SiCf-SiBC Composites", Ceramic Engineering and Science Proceedings, 23(3) 2002, pp. 395.
[9] Sun, E.Y., Lin, H-T., and Brennan, J.J., "Intermediate-Temperature Environmental Effects on Boron Nitride-Coated Silicon Carbide-Fibre-Reinforced Glass-Ceramic Composites", J.Am.Ceram.Soc. Vol. 80, no. 3, pp. 609-614. 1997

TIME-DEPENDENT RESPONSE OF MI SIC/SIC COMPOSITES PART 1: STANDARD SAMPLES

Ojard, G[2]., Gowayed, Y[3]., Chen, J[3]., Santhosh, U[4]., Ahmad, J[4]., Miller, R[2]., and John, R[1].

[1] Air Force Research Laboratory, AFRL/MLLMN, Wright-Patterson AFB, OH
[2] Pratt & Whitney, East Hartford, CT
[3] Auburn University, Auburn, AL
[4] Research Applications, Inc., San Diego, CA

ABSTRACT:
With the increased interest in using high performance ceramic matrix composites for advanced applications, long-term property behavior is of interest. In this work, time-dependent response of MI SiC/SiC composites (01/01 material) was experimentally evaluated under creep and dwell fatigue loading. A series of standard samples were tested at 815°C and 1204°C at various stress levels and multiple durations. All specimens showed primary and steady state creep responses. There were also some samples that showed tertiary creep response. Environmental degradation was empirically related to material response at different stress levels. Micrographic images of failed specimens revealed the existence of cavitations that were possibly caused by the creep strain at high stress areas.

INTRODUCTION

As Ceramic Matrix Composites (CMCs) are being considered for long duration applications, testing is needed to understand the material behavior under conditions of sustained load and temperature. Examples of long term applications can be found in ground base turbines for power generation where CMCs are being considered for combustor liners, turbine vanes and shroud applications [1,2]. These applications can see design times of up to 30,000 hours. Such long term applications are working to leverage the high temperature material capability while taking advantage of the weight reduction, reduced cooling and durability improvements that CMCs can provide over typical metals used in such applications.

As part of an effort to look into the long-term behavior of a CMC, a series of creep and dwell fatigue tests were undertaken. During the tests, strain was measured. The testing was done at 1204°C with some testing done at 815°C. The results of this testing will be shown and discussed.

PROCEDURE

Material Description
For this testing, the composite system interrogated was a Melt Infiltrated In-Situ BN SiC/SiC composite (MI SiC/SiC). This system has the most characterization done to date and is being considered for high temperature applications [3]. The MI SiC/SiC system has a stochiometric SiC (Sylramic™) fiber in a multiphase matrix of SiC deposited by chemical vapor deposition followed by slurry casting of SiC particulates with a final melt infiltration of Si metal. The specific MI SiC/SiC tested for this effort had 36% volume fraction fibers using a 5 HS weave at 20 EPI. The fibers are 10 μm diameter and there are 800 fibers per tow. This material system was developed by NASA-GRC and is sometimes referred to as the 01/01 material [4]. A cross section of this material is shown in Figure 1.

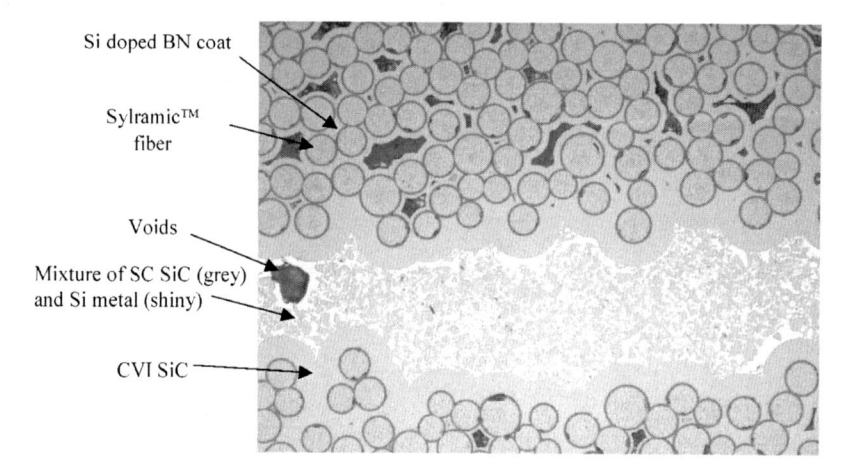

Figure 1. Cross section of Melt Infiltrated In-Situ BN SiC/SiC composite

Creep Testing

All creep testing was done in air using a SiC furnace. Temperature was controlled by thermocouples placed on the sample or by optical pyrometry. Load was applied using a dead weight using a lever arm. Strain was recorded using a 1" extensometer. Some of the testing allowed total strain to be recorded while other tests could only capture the creep strain. Testing was done at either 1204°C or 815°C.

Dwell Fatigue Testing

The dwell fatigue cycle for this effort was a creep type but the load was cycled off and on every 2 hours (R ratio of 0.1). All the dwell fatigue testing was done in air using a SiC furnace. Temperature was controlled by thermocouples placed on the sample. Load was applied using a dead weight with a lever arm. The lever arm is controlled by a cam system so that the load could be cycled off in a controlled manner every 2 hours. Strain was recorded using a 1" extensometer. One extensometer was used during the entire test so that the load and unload cycle was recorded and the modulus value could be calculated for every cycle. Testing was done at either 1204°C or 815°C.

Stress Level for Testing:

The vast majority of testing was done at stress levels of 110.4, 165.6 and 193.2 MPa. These values were picked based on the tensile curves at 1204°C (See Figure 2.) The stress of 110.4 MPa is still in the elastic linear region of the material. The stresses of 165.6 and 193.2 MPa are in the knee of the stress-strain curve where some in-elastic strain is occurring indicating that the matrix is cracking allowing for some environmental attack into the material. Additionally, some limited testing was done at stress levels of 220.8 MPa and higher.

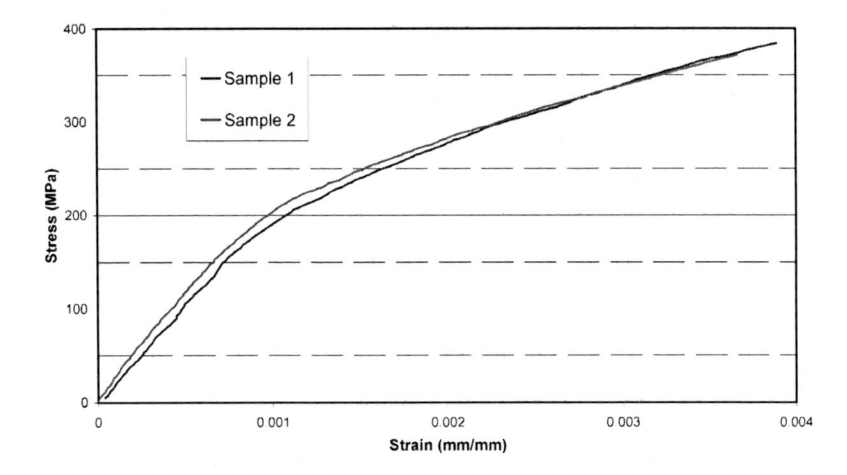

Figure 2. Stress-strain curves at 1204°C for the MI SiC/SiC System being studied

RESULTS

Creep and Dwell Fatigue Testing Summary
 The range of stresses and times that the material was exposed for under creep and dwell fatigue conditions are shown in Figure 3 where the range of stresses and exposure times are shown (without and with failure noted). All of this testing was done at 1204°C. As can be seen, failures only occurred during creep testing when the material was exposed to high stresses or for long times. There was limited testing at 815°C for durations of less than 250 hours and there were no sample failures.

Long Term Creep Testing Analysis:
 As can be seen in Figure 3, there were only 8 samples tested for relatively long durations (>250 hours) at stress levels of 110.4 and 165.6 MPa. The resulting creep curves for this effort are shown in Figure 4. The curves for both the 110.4 and 165.6 MPa stress levels show both a primary and a steady state creep region. Some of the samples show tertiary creep and those samples that did show tertiary creep failed. The steady state creep rate was determined for this testing after a review of the instantaneous creep versus time (see Figure 5). As can be seen in Figure 5 for these series of tests, the steady state creep region does not occur until near 250-300 hours. Hence, the steady state creep rate was determined from 250 hours on. At 110.4 MPa, the creep rate was found to be 3.4×10^{-10} while at 165.6 Mpa, the creep rate was found to be 6.3×10^{-10}. The creep rates for these conditions do not appear to be highly stress related since the exponents of the creep rates are the same and there is not a large difference in the magnitudes. In addition, this data is based on limited testing (2 samples for each condition) as shown in Figure 4.

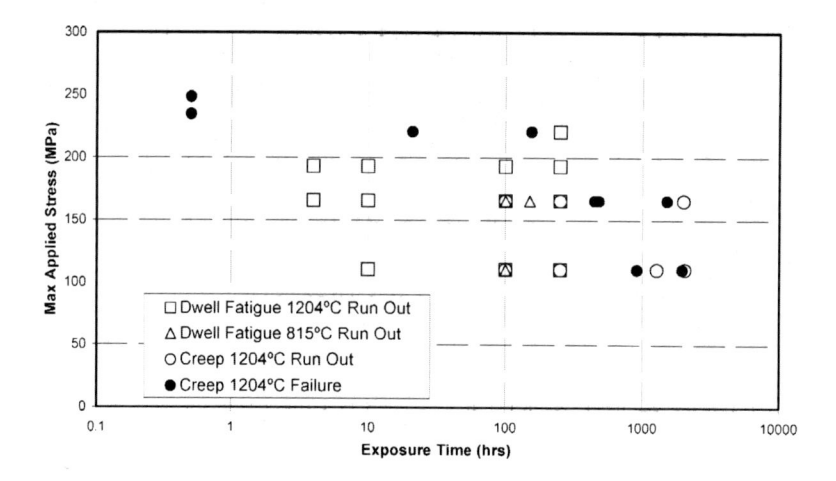

Time-Dependent Response of MI SiC/SiC Composites Part I: Standard Samples

Figure 3. Creep and Dwell Fatigue Testing Summary

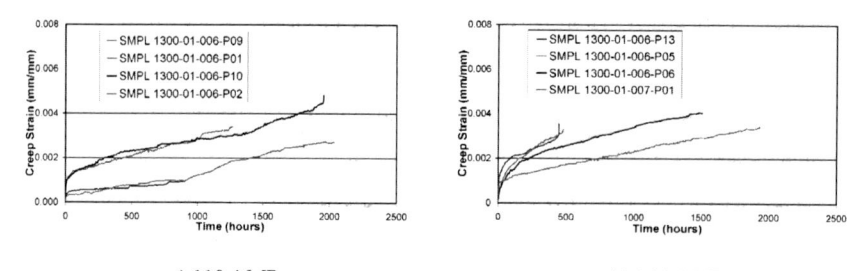

a) 110.4 MPa
b) 165.6 MPa
Figure 4. Long-term Creep Curves

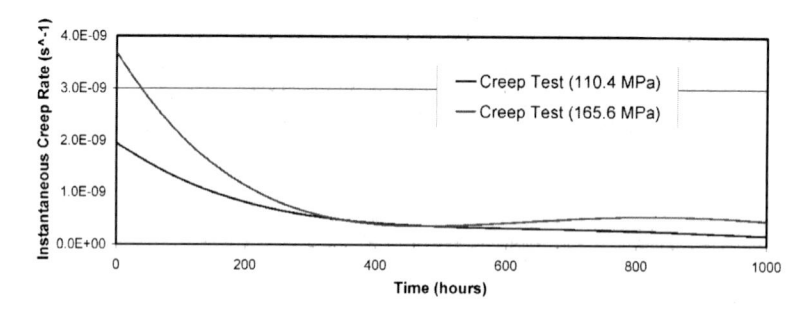

Figure 5. Instantaneous creep rate for tests done at 110.4 and 165.6 MPa

Time-Dependent Response of MI SiC/SiC Composites Part I: Standard Samples

Short Term Creep Testing Analysis:
A few samples were tested in creep for durations less than 250 hours. Due to the shorter duration, an additional stress levels were added over the work shown in Figure 4. The additional stresses were 220.8 MPa, 234.6 MPa and 248.4 MPa. The two highest stress levels were very short in duration (<0.5 hours) and data collection during those tests proved difficult. Some of the longer lasting tests are shown in Figure 6 showing the resulting creep curves at 1204°C. The curves shown for 110.4 and 165.6 MPa are consistent with the curves shown in Figure 4 in that the same strain evolution is seen (both samples were stopped at 250 hours). The strain evolution at 220.8 MPa is significantly greater and consistent with the fact that the sample failed at 153 hours. For the test at 220.8 MPa, it is not clear that a steady state region was ever achieved.

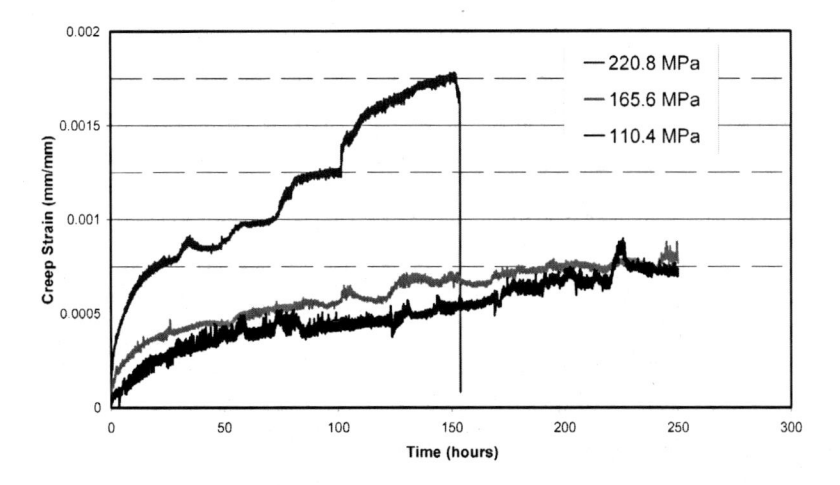

Figure 6. Short term creep curves at multiple stress levels

Dwell Fatigue Testing Analysis – Strain Evolution:
Dwell fatigue testing was done at stress levels of 110.4, 165.6, 193.2 and 220.8 MPa and no testing exceeded 250 hours. As noted earlier, the dwell duration was 2 hours before the load was cycled off and on. During the tests, total strain was recorded. For the series of tests that went to 250 hours, the total strain history is shown in Figure 7. The total strain evolved increases with increasing stress level consistent with most of the creep data shown previously. For this series of tests, the 110.4 MPa data shows less strain evolution than the creep testing effort.

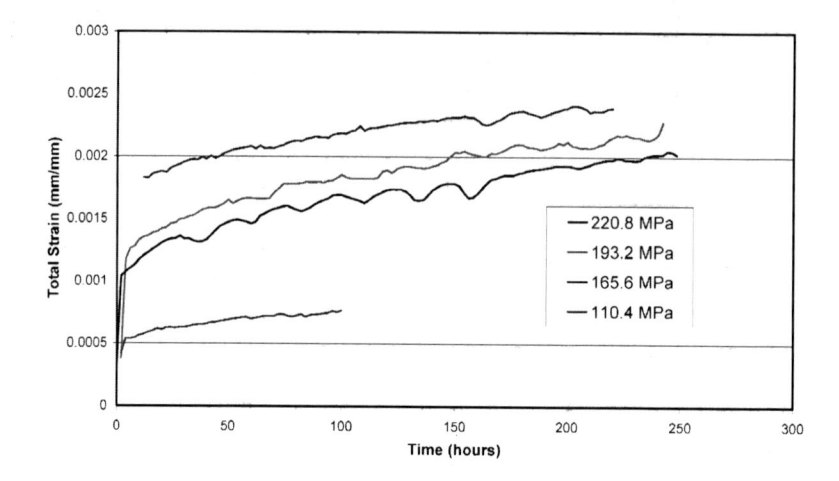

Time-Dependent Response of MI SiC/SiC Composites Part I: Standard Samples

Figure 7. Strain-time history for dwell fatigue samples at various stress levels

Dwell Fatigue Testing Analysis – Modulus Evolution:
During the dwell fatigue testing, stress-strain was recorded for load-unload cycles occurred and this allowed the modulus to be determined at the start of each cycle. The change in modulus was used as an indication of damage [5]. The modulus values versus time is shown in Figure 8. As can be seen, there is early and consistent damage evolution in the high stress test results (220.8 MPa) that is not seen in the other stress levels sine they show relatively constant modulus during the test.

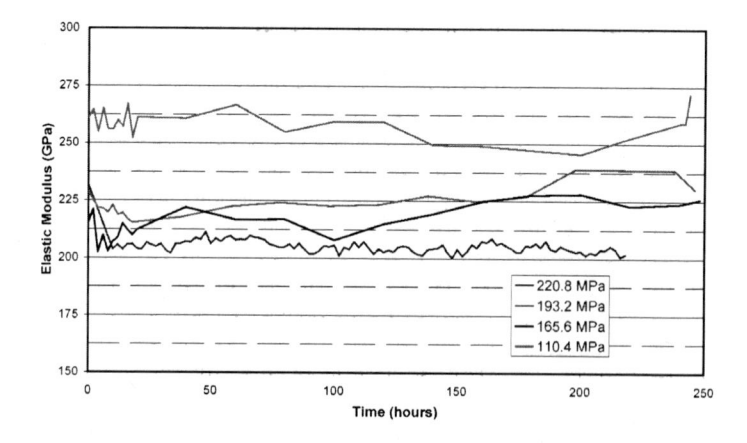

Figure 8. Modulus-time history for dwell fatigue samples at various stress levels

DISCUSSION

The work done to date shows that the MI SiC/SiC CMC is capable of sustaining loads at temperatures up to 1204°C for long durations. As shown in Figure 3, the material did not show failures in a region of stress and time of 193.2 MPa by 250 hours. Short test durations were seen once stresses of 220.8 MPa were applied where cracking would be occurring allowing environmental attack of the fiber-interface allowing embrittlement and resulting in short term tests (note that this testing is above the proportional limit of the material as seen in Figure 2). In addition, failures at lower stresses were only seen at 450 hours or greater. At stresses of 110.4 and 165.6 MPa, durations as long as 2,000 hours were achieved. (At these times, the material still shows residual tensile capability [5].) Also, the work shows that very low creep rates are seen in this material and that the overall strain evolution is low indicating that this material will remain relatively dimensionally stable in any long-term application (see Figure 4).

As was shown in Figures 4 and 5, the material has a relatively long primary creep region. This indicates that the damage needed to evolve in the material is time dependent and does not occur instantaneously. This is consistent, for the most part, with the modulus evolution shown in Figure 8. At the lower stress levels, the modulus values measured are remaining relatively constant showing that there is no damage evolution in the material. This was not the case for the test at 220.8 MPa where the modulus dropped after the first few cycles. This would be indicative of matrix cracking in the material at the high stress level seen (when compared to Figure 2, the stress level is above the proportional limit of the material). This is consistent with work shown by other investigators [5,7].

In order to investigate the creep behavior, some samples were sectioned and polished both near and away from the fracture location. In particular, a sample tested at 165.6 MPa and 1204°C that showed tertiary creep was sectioned (See Figure 4b.). Cross section of this material is shown in Figure 9. This gives indication that during testing voids are forming in the material under tensile load. In addition, additional cross section work was done even closer to the fracture face as shown in Figure 10 for this sample. At the lower magnification, even additional void formation is seen. Significant voids are seen when sectioned at the failure face. There is even indications that the voids are linking up to form cracks.

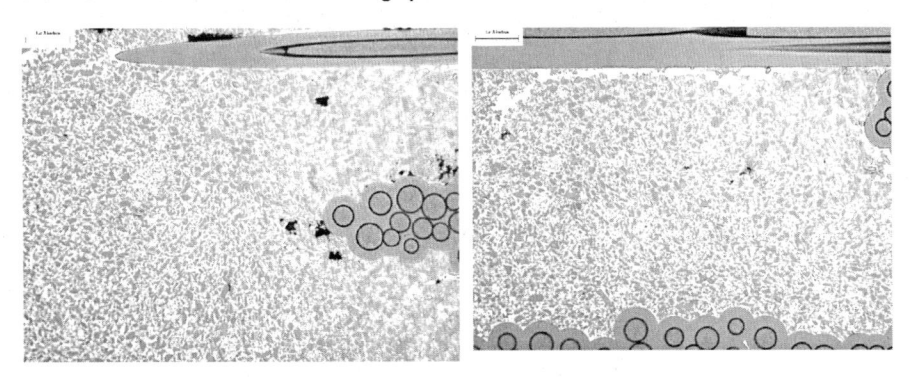

a) away from failure surface   b) as manufactured

Figure 9. Optical cross section of sample tested at 165.6 MPa and 1204°C in creep

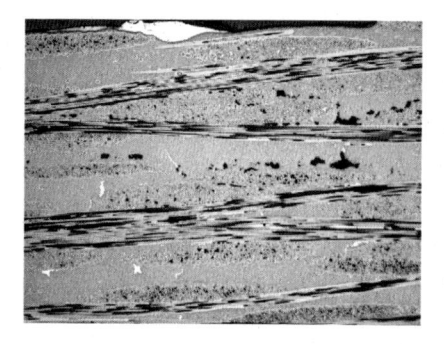

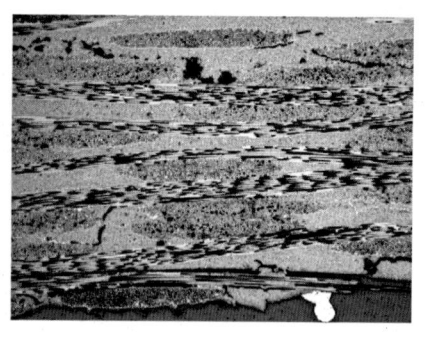

a) away from failure face   b) at failure face

Figure 10. Additional microstructure work from sample showed in Figure 9

This effort is consistent with work done on Si/SiC where creep testing showed that voids formed at the boundaries of the Si with the SiC particulates [8-10]. As part of that work, the authors showed that the creep rate in tension was 20 times that shown in compression. As part of the confirmation of the creep testing done here, two tests were done at 165.6 MPa and 1204°C in compression. These tests were only run for a short duration (10 hours), but no primary creep was seen. This supports the position that cavitation/void formation is occurring for the MI SiC/SiC composite at the Si-SiC particulate boundary.

CONCLUSIONS

The work has demonstrated and documented the long-term behavior of the MI SiC/SiC system. The material shows a very low creep rate and a large primary creep region. This is consistent with expectations of this CMC system. The work has shown that at the high stress of 220.8 MPa, there is clear damage measurement as seen in the modulus work. In addition, it was shown that the main damage evolution in this material was creep in the Silicon metal forming voids both in the Silicon as well as between the Silicon/SiC particulates. This is consistent with other investigators who looked at Si/SiC particulate material (same creep rates within temperature bounds and percent material) [9] and consistent with the fact that SiC typically shows creep at temperatures of 1600°C and higher [11].

This work shows that the material is well suited for applications at the test temperatures explored here. Additionally, work should be done on the effects of holes and that is reported in the companion paper to this paper [12]. Applications that consider even longer durations than tested here should consider additional testing.

ACKNOWLEDGMENTS

The Materials & Manufacturing Directorate, Air Force Research Laboratory under contract F33615-01-C-5234 and contract F33615-03-D-2354-D04 sponsored portions of this work

REFERENCES:

[1]Brewer, D., Ojard, G. and Gibler, M., "Ceramic Matrix Composite Combustor Liner Rig Test", ASME Turbo Expo 2000, Munich, Germany, May 8-11, 2000, ASME Paper 2000-GT-0670.
[2]Calomino, A., and Verrilli, M., "Ceramic Matrix Composite Vane Sub-element Fabrication", ASME Turbo Expo 2004, Vienna, Austria, June 14-17, 2004, ASME Paper 2004-53974.
[3]Bhatt, R.T., McCue, T.R., and DiCarlo, J.A., "Thermal Stability of Melt Infiltrated SiC/SiC Composites", Ceramic Engineering and Science Proceedings, Vol. 24, Issue 4, 2003, pg. 295.
[4]Hurwitz, F.I., Calomino, A.M., McCue, T.R., and Morscher, G.N., "C-Coupon Studies of SiC/SiC Composites Part II: Microstructrual Characterization", Ceramic Engineering and Science Proceedings, Vol. 23, Issue 3, 2002, pg 387.
[5]Darzens, S., Chermant, J-L., Vicens, J. and Sangleboeuf, J.C., "Damage Creep Mechanisms of SiCf-SiBC Composites", Ceramic Engineering and Science Proceedings, 23(3) 2002, pp. 395.
[6]Ojard, G., Calomino, A., Morscher, G., Gowayed, Y., Santhosh, U., Ahmad J., Miller, R. and John, R., "Post Creep/Dwell Fatigue Testing of MI SiC/SiC Composites", to be published in Ceramic Engineering and Science Proceedings, 2007.
[7]Bouillon, E.P., Ojard, G.C., Habarou, G., Spriet, P.C., Arnold T., Feindel, D.T., Logan, C., Rogers, K., Doppes, G., Miller, R., Grabowski, Z. and Stetson, D.P., "Engine Test Experience and Characterization Of Self Sealing Ceramic Matrix Composites For Nozzle Applications in Gas Turbine Engines", ASME Turbo Expo 2003, Atlanta, Georgia, June 16-19, 2003, ASME Paper GT2003-38967.
[8]T. Chuang and S. Wiederhorn, "Damage enhanced creep in siliconized silicon carbide: mechanics of deformation", J. Am. Ceramic Society, 71, [7] 595-601, 1988.
[9]S. Wiederhorn, D. Roberts, T. Chuang, and L. Chuck, "Damage enhanced creep in siliconized silicon carbide: Phenomenology", J. Am. Ceramic Society, 71, [7] 602-608, 1988.
[10]S. Widerhorn, B. Hockey and J. French, "Mechanisms of deformation of silicon nitride and silicon carbide at high temperature", J. of Eurepean Ceramic Society, 19, 2273-2284, 1999.
[11]Lane, J.E., Carter, Jr., C.H., and Davies, R.F., "Kinetics and Mechanisms of High-Temperature Creep in Silicon Carbide: III, Sintered α-Silicon Carbine", J. Am. Ceramic Society, Vol. 71 [4], 281-295, 1998.
[12]Y. Gowayed, G. Ojard, J. Chen, R. Miller, U. Santhosh, J. Ahmad and John, R., "Time-Dependent Response of MI SiC/SiC Composites Part 2: Samples with Holes", to be published in Ceramic Engineering and Science Proceedings, 2007.

TIME-DEPENDENT RESPONSE OF MI SIC/SIC COMPOSITES PART 2: SAMPLES WITH HOLES

Y. Gowayed[3], G. Ojard[2], J. Chen[3], R. Miller[2], U. Santhosh[4], J. Ahmad[4], and R. John[1]

[1] Air Force Research Laboratory, AFRL/MLLMN, Wright-Patterson AFB, OH
[2] Pratt & Whitney, East Hartford, CT
[3] Auburn University, Auburn, AL
[4] Research Applications, Inc., San Diego, CA

ABSTRACT
Time-dependent response of samples with holes manufactured from MI SiC/SiC composites (01/01 material) was experimentally evaluated under creep and dwell fatigue loading. Forty specimens with a central hole with sizes 2.286, and 4.572 mm were tested. All specimens showed primary and steady state creep responses. Environmental degradation was empirically related to material response at different stress levels utilizing information from samples without holes. ANSYS Finite Element Analysis was used to map the creep strain in the vicinity of the hole utilizing the empirical correlation.

INTRODUCTION
Testing is conducted to understand the behavior of Ceramic Matrix Composites (CMCs) under conditions of sustained load at high temperatures. Examples of long term environments would include ground base turbines for power generation where CMCs are being considered for combustor liners, turbine vanes and shroud applications. These applications can see design times of up to 30,000 hours. Such long term applications are working to leverage the high temperature material capability while taking advantage of the weight reduction, reduced cooling and durability improvements that CMCs can provide over typical metals.

As part of an effort to look into the long-term behavior of a CMC, a series of creep and dwell fatigue tests were undertaken for specimens with and without a central hole at 1204 °C and various stress levels. This paper reports data and analysis of specimens with holes. Data on specimens without holes can be found in part 1 of this series of papers.

EXPERIMENTAL PROGRAM
Materials and Manufacturing
The material chosen for the study was the Melt Infiltrated SiC/SiC CMC system, which was initially developed under the Enabling Propulsion Materials Program (EPM) and is still under further refinement at NASA-Glenn Research Center (GRC). This material system has been systematically studied at various development periods and the most promising was the 01/01 Melt Infiltrated iBN SiC/SiC (01/01 is indicative of the month and year that development was frozen). There is a wide set of data from NASA for this system as well as a broad historic database from the material development. This allowed a testing system to be put into place to look for key development properties which would be needed from a modeling effort and would hence leverage existing data generated by NASA-GRC.

The Sylramic® fiber was fabricated by DuPont as a 10 μm diameter stochiometric SiC fiber and bundled into tows of 800 fibers each. The sizing applied was polyvinyl alcohol (PVA). For this study, the four lots of fibers, which were used, were wound on 19 different spools. The tow spools were then woven into a 5HS balanced weave at 20 EPI. An in-situ Boron Nitride (iBN) treatment was performed on the weave (at NASA-GRC), which created a fine layer of BN on every fiber. The fabric was then laid in graphite tooling to correspond to the final part design (flat plates for this experimental program). All the panels were manufactured from a symmetric cross ply laminate using a total of 8 plies. The graphite tooling has holes to allow the CVI deposition to occur. At this stage, another BN coat layer was applied. This BN coating was doped with Si to provide better environmental protection of the interface. This was followed by SiC vapor deposition around the tows. Typically, densification is done to about 30% open porosity. SiC particulates are then slurry cast into the material followed by melt infiltration of a Si alloy to arrive at a nearly full density material. The material at this time has less than 2% open porosity. Through this process, 15 panels were fabricated in 3 lots of material.

After fabrication, all the panels were interrogated by pulse echo ultrasound (10 MHz) and film X-ray. There was no indication of any delamination and no large scale porosity was noted in the panels. In addition, each panel had two tensile bars extracted for witness testing at room temperature. All samples tested failed above a 0.3% strain to failure requirement. Hence, all panels were accepted into the testing effort.

Creep testing at 1204 °C
Creep testing is a test to determine the strain-time dependence of a material under a constant load. This test is also used to determine the long-term behavior of the material under combination of load and temperature. For ceramic matrix composites, when this testing is done in air, there is an added complexity of environmental exposure. Creep testing typically has three regimes, primary (or initial), secondary (steady state) and tertiary creep. The creep rate for the material is typically measured during the secondary (steady state) regime.

A total of 12 creep tests were done on straight-sided tensile bars with two different central hole sizes (2.286 and 4.572 mm) at 1204 °C. An extensometer was mounted on the specimen side to measure the strain with a gage length of 25.4 mm. The stress-time data for this experiment is summarized in Figure 1. There were no failures in this testing effort. During the creep testing, the creep strain was recorded. Typical creep strain-time data for some of the longer duration samples are shown in Figures 2 and 3 for 2.286 and 4.572 mm holes, respectively.

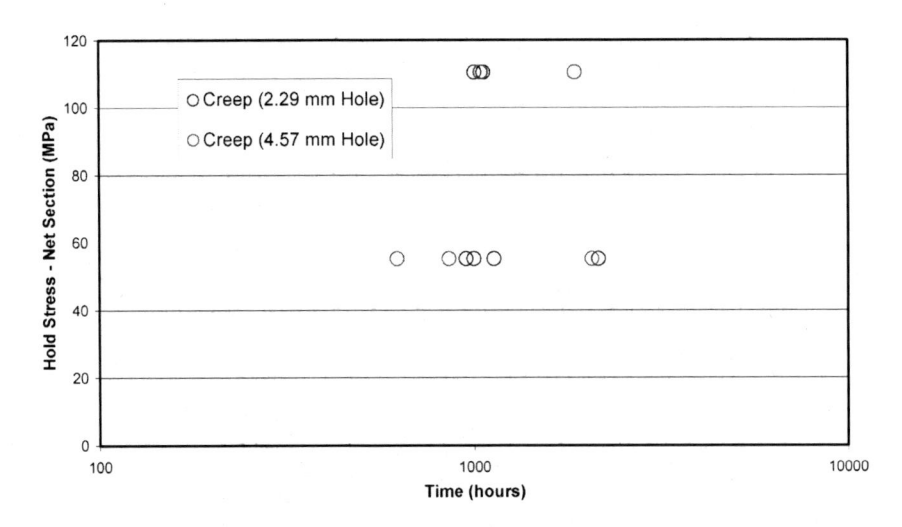

Figure 1: Net Section Stress-Time data for creep testing of samples with holes

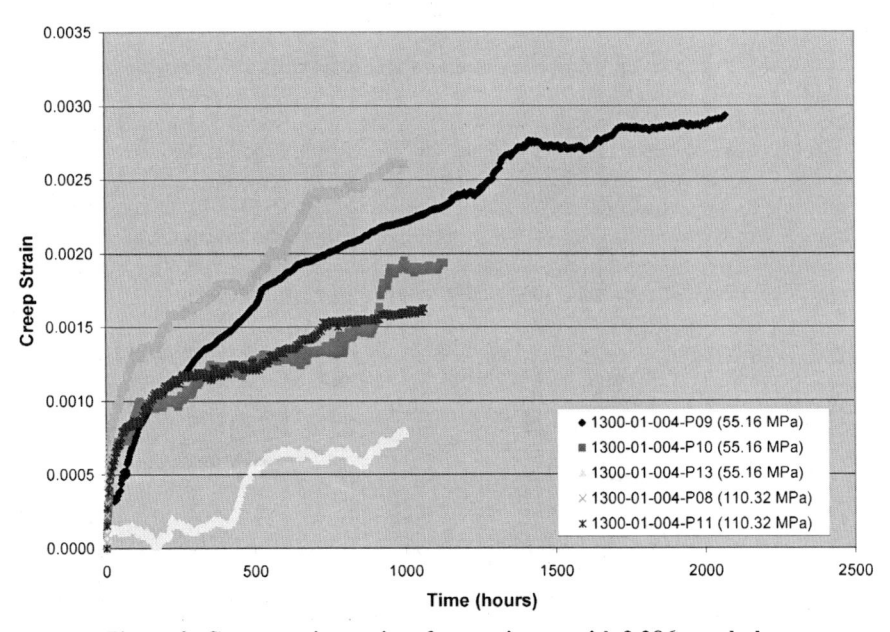

Figure 2: Creep strain vs. time for specimens with 2.286 mm hole

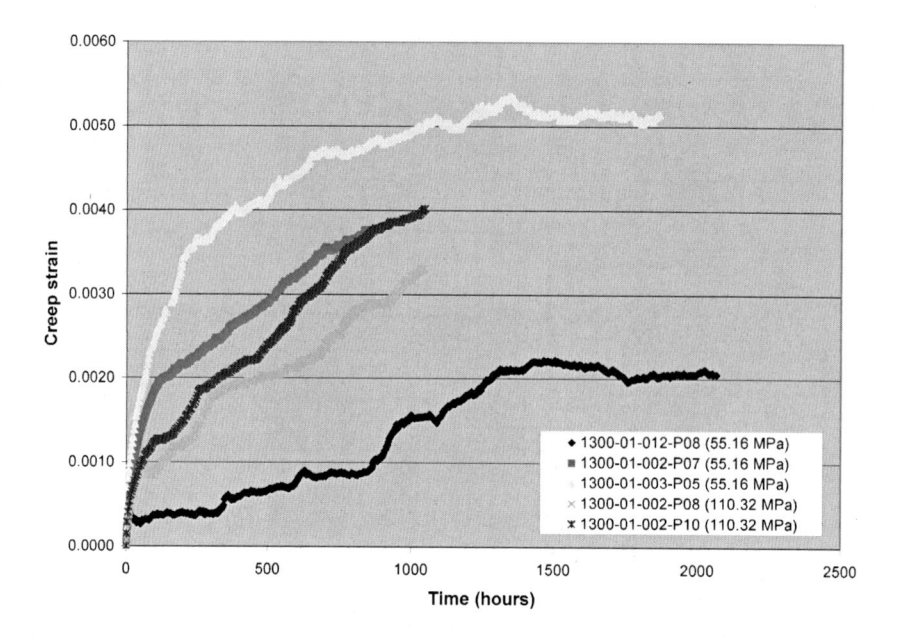

Figure 3: creep strain vs. time for specimens with 4.572 mm hole

The average slopes of the steady state region of the 2.286 and 4.572 mm holes and no hole specimens are listed in Table 1 along with the standard deviation values. It can be seen that the standard deviation value is high as compared to the average value for some of the experiments, especially the 2.286 mm hole loaded at 55.16 MPa and the 4.572 mm hole at 55.16 MPa. The impact of the increase in the hole radius can be seen from the data of the 4.572 mm hole especially at the high stress level, but the effect of the existence of the 2.286 mm hole is not evident at both stress levels.

Table 1: Average and standard deviation of the slope (/sec) of the steady state region

Net section stress	Average	Standard deviation
No hole		
55.16 MPa	---	---
110.32 MPa	3.4xE-10	1.3E-10
2.286 mm *hole*		
55.16 MPa	2.62E-10	2.05E-10
110.32 MPa	3.16E-10	1.79E-10
4.572 mm *hole*		
55.16MPa	3.59E-10	3.19E-10
110.32 MPa	7.46E-10	2.02E-10

Dwell Fatigue at 1204 °C
Dwell fatigue testing, sometimes referred to as "low cycle fatigue tests", is the superposition of a stress hold at the peak stress of a fatigue cycle. Dwell fatigue tests were conducted where the dwell time was 2 hours with a 1 minute load and unload at the beginning and end of each cycle. Twenty eight samples were tested at 55.16, 110.32, 165.48 and 193.06 MPa net-section stress levels. Data acquired during these tests are typically large in size and include load-reload tests at different time intervals. A Matlab code was developed to extract this data and allow characterization of the change in strain with time. Figure 4 shows the Matlab code output for the change of total strain with time for samples with 2.286 mm hole.

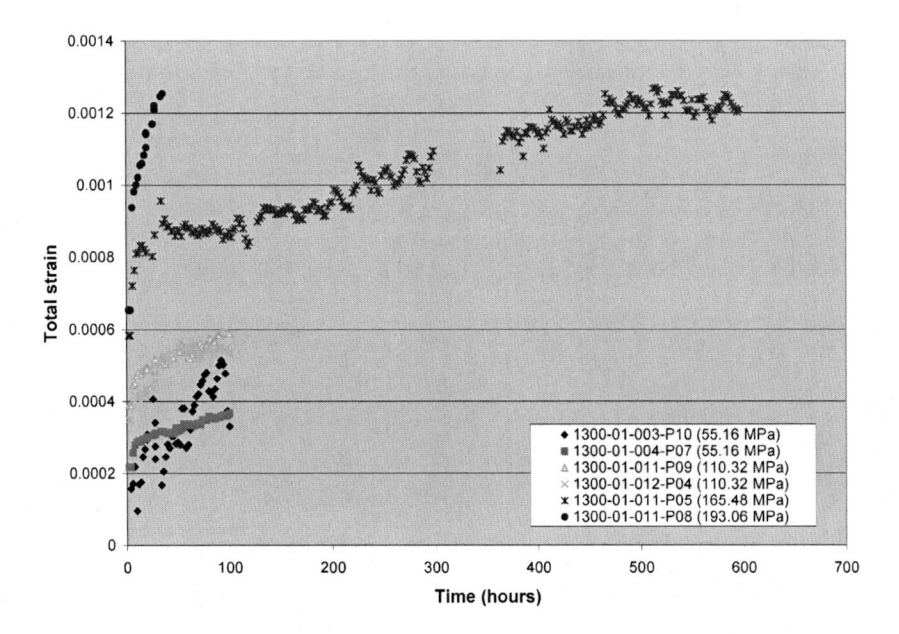

Figure 4: Change in the value of total strain for specimens with 2.286 mm hole with time at 1204 °C

From Figure 4, it can be seen that the value of the total strain is consistently higher at higher stress levels. The primary creep region was much smaller in size than that observed in creep tests (refer to paper 1 in this series).

Creep analysis using ANSYS for samples with a central hole

A 3 parameter curve fitting equation $strain\_creep = A\sigma^n t^p + \varepsilon_o$ (with A=2.51e-23, n=2.3, p=0.30, σ = applied remote stress (Pa), t = time (hours), and ε_o = elastic strain) was used to fit the data for the standard dwell fatigue samples. These 3 parameters were chosen to provide the best fit possible for the data of the standard samples without holes cat all stress levels. This equation was used to map the strain for specimens with holes. The existence of the hole causes stress concentrations that alter the strain values in the area close to the hole. In order to develop an understanding of the impact of the existence of the hole on the strain values, ANSYS FEA software was used to model the behavior of the material as an orthotropic material. Element type PLANE 182 was used to generate the 3D model and element type SOLID185 was used to conduct analysis using the mesh shown in Figure 5. The orthotropic creep behavior was represented by the combination of Hill's anisotropy model and the implicit Creep Model 6 in ANSYS. The Hill option's material behavior is described by six constants that define the creep ratios in different directions [1, 2].

The change of creep strain with time at four different locations around the hole shown in Figure 6 was reported by ANSYS for undamaged material. It can be seen that the local creep at the edge of the hole is large and may cause local damage. Figure 7 shows the comparison between the measured data and model calculations for 55.16 and 110.32 MPa loaded specimens measured at point 1 with a reasonable level of accuracy.

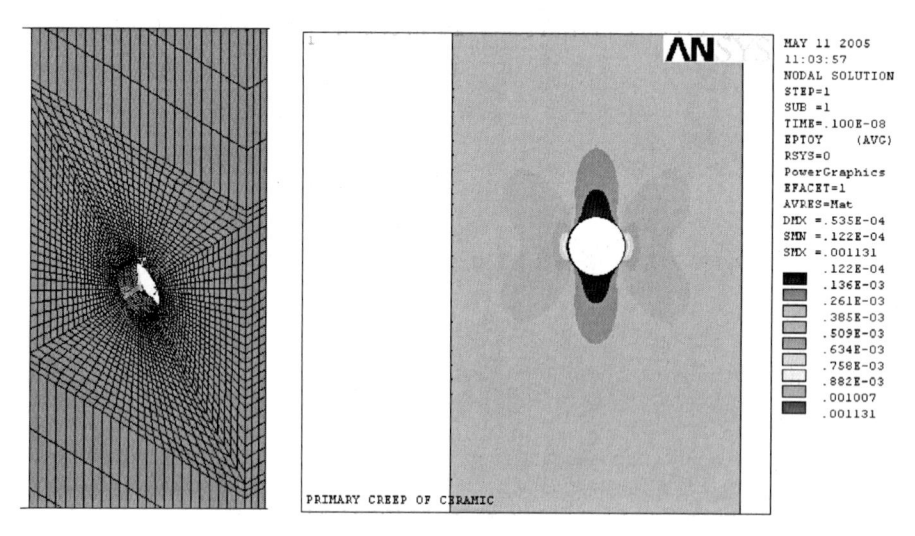

Figure 5: ANSYS creep-strain distribution around the hole at time t=0

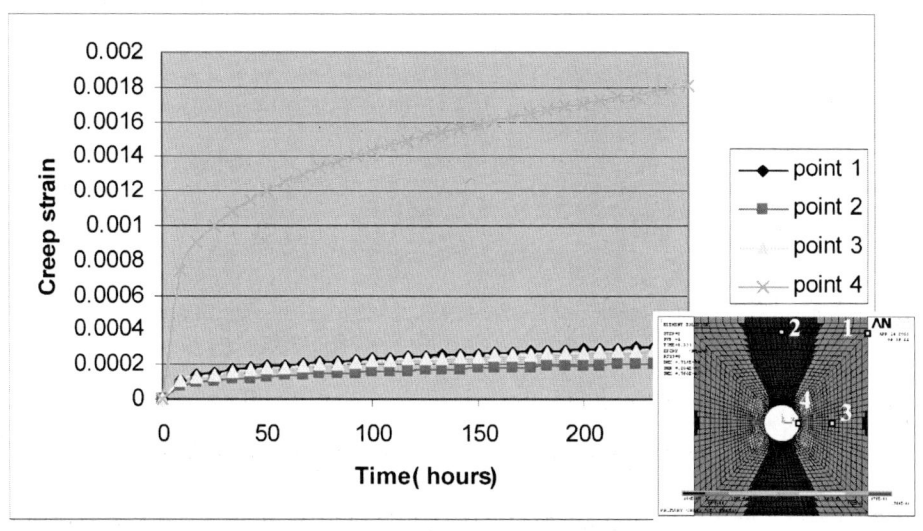

Figure 6: ANSYS creep-strain distribution at t=8.33 h and location of monitoring points

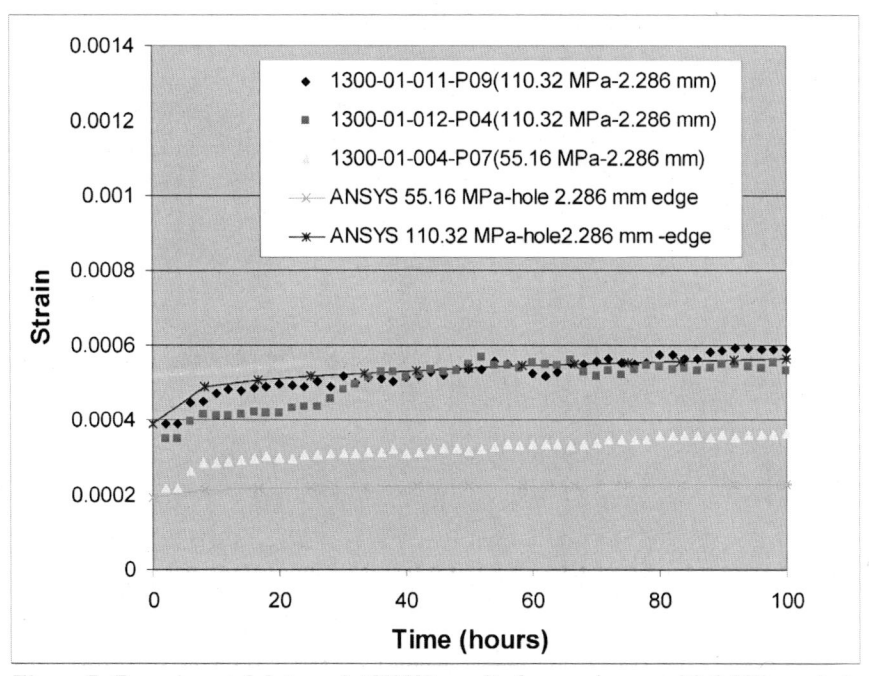

Figure 7: Experimental data and ANSYS results for specimens with 2.286 mm hole

CONCLUSIONS

Experimental evaluation of time-dependent response of SiC/SiC composites was conducted utilizing creep and dwell fatigue experiments. Samples had central holes with sizes of 2.286 and 4.572 mm forming 20% of the width of the specimen. Data from dwell fatigue experiments were more consistent than those in creep experiments. Dwell fatigue data fit for standard samples without holes were used to model creep strain around holes utilizing ANSYS. The model showed a 9 fold increase in creep strain values in the vicinity of the hole when compared to at the sample edge.

ACKNOWLEDGMENTS

The Materials & Manufacturing Directorate, Air Force Research Laboratory under contract F33615-01-C-5234 and contract F33615-03-D-2354-D04 sponsored portions of this work

REFERENCES

1. R. Hill, "A theory of yielding and plastic flow of anisotropic metals," Proc. Roy. Soc. A 193, 281, 1948
2. R. Hill, "The Mathematical Theory of Plasticity," Oxford University Press, Oxford, 1950.

EFFECTS OF ENVIRONMENT ON CREEP BEHAVIOR OF AN OXIDE-OXIDE CERAMIC COMPOSITE WITH ±45° FIBER ORIENTATION AT 1200 °C

G. T. Siegert, M. B. Ruggles-Wrenn[*]
Department of Aeronautics and Astronautics
Air Force Institute of Technology
Wright-Patterson Air Force Base, Ohio 45433-7765

S. S. Baek
Agency for Defense Development
Daejeon, Korea

ABSTRACT

The tensile creep behavior of an oxide-oxide continuous fiber ceramic composite (CFCC) with ±45° fiber orientation was investigated at 1200 °C in laboratory air, in steam and in argon. The composite consists of a porous alumina matrix reinforced with laminated, woven mullite/alumina (Nextel™720) fibers, has no interface between the fiber and matrix, and relies on the porous matrix for flaw tolerance. The tensile stress-strain behavior was investigated and the tensile properties measured at 1200 °C. The elastic modulus was 45 GPa, UTS was 55 MPa, and failure strain was 0.27%. Tensile creep behavior was examined for creep stresses in the 15-45 MPa range. Primary and secondary creep regimes were observed in all tests. Tertiary creep was observed at 45 MPa in air and at stress levels ≥ 40 MPa, in steam and argon environments. For creep stress levels ≤35 MPa, creep run-out (set to 100 h) was achieved in all test environments. At creep stresses > 35 MPa, creep performance was best in laboratory air and worst in argon. The presence of either steam or argon accelerated creep rates and significantly reduced creep life. Composite microstructure, as well as damage and failure mechanisms were investigated. Matrix degradation appears to be the cause of early failures in argon.

INTRODUCTION

Advances in power generation systems for aircraft engines, land-based turbines, rockets, and, most recently, hypersonic missiles and flight vehicles have raised the demand for structural materials that have superior long-term mechanical properties and retained properties under high temperature, high pressure, and varying environmental factors, such as moisture [1]. Typical components include combustors, nozzles and thermal insulation. Ceramic-matrix composites (CMCs), capable of maintaining excellent strength and fracture toughness at high temperatures are prime candidate materials for such applications. Additionally, the lower densities of CMCs and their higher use temperatures, together with a reduced need for cooling air, allow for improved high-temperature performance when compared to conventional nickel-based superalloys [2]. Advanced reusable space launch vehicles will likely incorporate fiber-reinforced CMCs in critical propulsion components. In these applications, CMCs will be subjected to mechanical loading in complex environments. For example, a typical service environment for a

---

[*] Corresponding author
The views expressed are those of the authors and do not reflect the official policy or position of the United States Air Force, Department of Defense or the U. S. Government.

reusable rocket engine turbopump rotor includes hydrogen, oxygen and steam, at pressures > 200 atm. Furthermore, concurrent efforts in optimization of the CMCs and in design of the combustion chamber are expected to accelerate the insertion of the CMCs into aerospace turbine engine applications, such as combustor walls [3-5]. Because these applications require exposure to oxidizing environments, the thermodynamic stability and oxidation resistance of CMCs are vital issues.

The main advantage of CMCs over monolithic ceramics is their superior toughness, tolerance to the presence of cracks and defects, and non-catastrophic mode of failure. It is widely accepted that in order to avoid brittle fracture behavior in CMCs and improve the damage tolerance, a weak fiber/matrix interface is needed, which serves to deflect matrix cracks and to allow subsequent fiber pullout [6-9]. Historically, following the development of SiC fibers, fiber coatings such as C or BN have been employed to promote the desired composite behavior. However, the non-oxide fiber/non-oxide matrix composites generally show poor oxidation resistance [10, 11], particularly at intermediate temperatures (~800 °C). These systems are susceptible to embrittlement due to oxygen entering through the matrix cracks and then reacting with the interphase and the fibers [12-15]. The degradation, which involves oxidation of fibers and fiber coatings, is typically accelerated by the presence of moisture [16-22]. Using oxide fiber/ non-oxide matrix or non-oxide fiber/oxide matrix composites generally does not substantially improve the high-temperature oxidation resistance [23]. The need for environmentally stable composites motivated the development of CMCs based on environmentally stable oxide constituents [24-32].

More recently it has been demonstrated that similar crack-deflecting behavior can also be achieved by means of a finely distributed porosity in the matrix instead of a separate interface between matrix and fibers [33]. This microstructural design philosophy implicitly accepts the strong fiber/matrix interface. It builds on the experience with porous interlayers as crack deflection paths [34, 35] and extends the concept to utilize a porous matrix as a surrogate. The concept has been successfully demonstrated for oxide-oxide composites [24, 28, 32, 36-40]. Resulting oxide/oxide CMCs exhibit damage tolerance combined with inherent oxidation resistance. However, due to the strong bonding between the fiber and matrix, a minimum matrix porosity is needed for this concept to work [41]. An extensive review of the mechanisms and mechanical properties of porous-matrix CMCs is given in [42].

In many potential applications oxide-oxide CMCs will be subject to multiaxial states of stress. The woven CMC materials developed for use in aerospace engine components are typically made from 0°/90° fiber architectures. However, the highest loads in structural components are not always applied along the direction of the reinforcing fibers. As a result, the components could experience stresses approaching the off-axis tensile and creep strengths. The objective of this effort is to investigate the off-axis tensile and creep behaviors of an oxide-oxide CMC consisting of a porous alumina matrix reinforced with the Nextel™720 fibers. Several previous studies examined high temperature mechanical behavior of this composite in the 0°/90° fiber orientation [2, 43-45]. This study investigates tensile and creep behavior of the Nextel™720/alumina (N720/A) composite in the ±45° orientation at 1200 °C in oxidizing (air and steam) and inert (argon) environments. Creep tests were conducted in air, steam and argon environments at stress levels ranging from 15 to 45 MPa. Results reveal that test environment has a noticeable effect on creep life. The composite microstructure, as well as damage and failure mechanisms are discussed.

MATERIAL AND EXPERIMENTAL ARRANGEMENTS

The material studied was Nextel™720/Alumina (N720/A), a commercially available oxide-oxide ceramic composite (COI Ceramics, San Diego, CA), consisting of a porous alumina matrix reinforced with Nextel™720 fibers. The composite was supplied in a form of 2.8 mm thick plates, comprised of 12 0°/90° woven layers, with a density of ~2.77 g/cm^3 and a fiber volume of approximately 45%. Matrix porosity was ~24%. The fiber fabric was infiltrated with the matrix in a sol-gel process. The laminate was dried with a "vacuum bag" technique under low pressure and low temperature, then pressureless sintered [46]. No coating was applied to the fibers. The damage tolerance of the N720/A composite is enabled by a porous matrix. Representative micrograph of the untested material is presented in Fig. 1, which shows 0° and 90° fiber tows as well as numerous matrix cracks. In the case of the as-processed material, most are shrinkage cracks formed during processing rather than matrix cracks generated during loading. Porous nature of the matrix is seen in Fig. 1 (b).

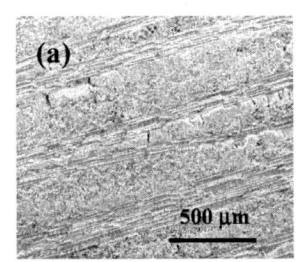

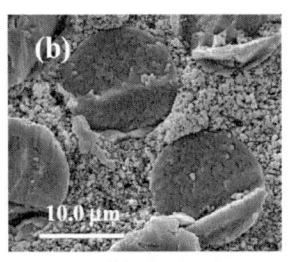

Fig. 1. As-received material: (a) overview, (b) porous nature of the matrix is evident.

A servocontrolled MTS mechanical testing machine equipped with hydraulic water-cooled collet grips, a compact two-zone resistance-heated furnace, and two temperature controllers was used in all tests. An MTS TestStar digital controller was employed for input signal generation and data acquisition. Strain measurement was accomplished with an MTS high-temperature air-cooled uniaxial extensometer. For elevated temperature testing, thermocouples were bonded to the specimens using alumina cement (Zircar) to calibrate the furnace on a periodic basis. The furnace controller (using a non-contacting thermocouple exposed to the ambient environment near the test specimen) was adjusted to determine the power setting needed to achieve the desired temperature of the test specimen. The determined power setting was then used in actual tests. The power setting for testing in steam was determined by placing the specimen instrumented with thermocouples in steam environment and repeating the furnace calibration procedure. Thermocouples were not bonded to the test specimens after the furnace was calibrated. Tests in steam environment employed an alumina susceptor (tube with end caps), which fits inside the furnace. The specimen gage section is located inside the susceptor, with the ends of the specimen passing through slots in the susceptor. Steam is introduced into the susceptor (through a feeding tube) in a continuous stream with a slightly positive pressure, expelling the dry air and creating 100% steam environment inside the susceptor. An alumina susceptor was also used in tests conducted in argon environment. In this case high purity argon was introduced into the susceptor creating an inert gas environment around the test section of the specimen. Fracture surfaces of failed specimens were examined using SEM (FEI Quanta 200 HV) as well as an optical microscope (Zeiss Discovery V12). The SEM specimens were carbon coated.

All tests were performed at 1200 °C. In all tests, a specimen was heated to test temperature in 25 min, and held at temperature for additional 15 min prior to testing. Dog bone shaped specimens of 152 mm total length with a 10-mm-wide gage section were used in all tests. Tensile tests were performed in stroke control with a constant displacement rate of 0.05 mm/s in laboratory air. Creep-rupture tests were conducted in load control in accordance with the procedure in ASTM standard C 1337 in laboratory air, steam and argon environments. In all creep tests the specimens were loaded to the creep stress level at the stress rate of 15 MPa/s. Creep run-out was defined as 100 h at a given creep stress. In each test, stress-strain data were recorded during the loading to the creep stress level and the actual creep period. Thus both total strain and creep strain could be calculated and examined. To determine the retained tensile strength and modulus, specimens that achieved run-out were subjected tensile test to failure at 1200 °C. It is worthy of note that in all tests reported below, the failure occurred within the gage section of the extensometer.

RESULTS AND DISCUSSION
Monotonic Tension

Tensile stress-strain behavior at 1200 °C is typified in Fig. 2. The stress-strain curves obtained for the 0°/90° fiber orientation are nearly linear to failure. Material exhibits typical fiber-dominated composite behavior. The average ultimate tensile strength (UTS) was 190 MPa, elastic modulus, 76 GPa, and failure strain, 0.38 %. These results agree well with the data reported earlier [44, 47]. In the case of the ±45° orientation, the nonlinear stress-strain behavior sets in at fairly low stresses (~15 MPa). As the stress exceeds 50 MPa, appreciable inelastic strains develop rapidly. The specimen achieves a strain of 0.3% at the maximum load. Once the UTS is reached, the softening commences. These observations are consistent with the results reported earlier for the porous-matrix ceramic composites [48, 49]. The elastic modulus (46 GPa) and UTS (55 MPa) obtained for the ±45° orientation are considerably lower than the corresponding values for the 0°/90° specimens.

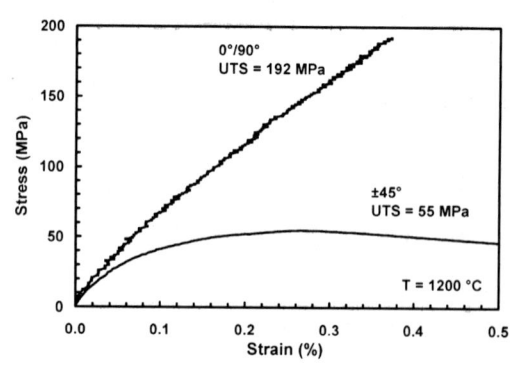

Fig. 2. Tensile stress-strain curves for N720/A ceramic composite at 1200 °C. Data for the 0°/90° fiber orientation from Ruggles-Wrenn et al [44] is also shown.

Creep-Rupture

Results of the creep-rupture tests for N720/A composite with ±45° fiber orientation are summarized in Table I, where creep strain accumulation and rupture time are shown for each creep stress level and test environment. Creep curves obtained in air, steam and argon are shown in Figs.3, 4 and 5, respectively. The time scale in Figs. 3(b), 4(b), and 5(b) is reduced to clearly show the creep curves produced at 45 MPa.

Table I. Summary of creep-rupture results for the N720/A ceramic composite with ±45° fiber orientation at 1200 °C in laboratory air, steam and argon environments

Environment	Creep Stress (MPa)	Creep Strain (%)	Time to Rupture (s)
Air	15	3.38	360,000*
Air	35	13.3	360,000*
Air	40	13.1	85,379
Air	45	1.48	119
Steam	15	4.80	360,000*
Steam	35	16.5	360,000*
Steam	40	4.92	2,616
Steam	45	0.65	58
Argon	15	6.67	360000*
Argon	35	20.5	360000*
Argon	40	3.58	64
Argon	45	4.12	22

* Run-out

Creep curves produced in all tests at 15 and 35 MPa exhibit primary and secondary creep regimes, but no tertiary creep. Transition from primary to secondary creep occurs late in creep life, primary creep persists during the first 40-50 h of the creep test. Note that creep run-out of 100 h was achieved in all 15 and 35 MPa tests, regardless of environment. While the test environment appears to have little influence on the appearance of the creep curves obtained at 15 and 35 MPa, it has a noticeable effect on the strain accumulated during 100 h of creep. For a given creep stress, the largest creep strains were accumulated in argon, followed by those accumulated in steam and in air. In contrast, all creep curves obtained at 45 MPa show primary, secondary and tertiary creep. At 45 MPa, transition from primary to secondary creep occurs almost immediately, and secondary creep transitions to tertiary creep during the first third of the creep life. At the intermediate stress level of 40 MPa, only primary and secondary creep are observed in air and in steam, but all three creep regimes are seen in argon.

Minimum creep rate was reached in all tests. Creep rate as a function of applied stress is presented in Fig. 6, where results for N720/A composite with 0°/90° fiber orientation from prior work [44] are included for comparison. It is seen that in air the secondary creep rate of the ±45° orientation can be as high as 10^6 times that of the 0°/90° orientation. In steam, the ±45° creep rate can be as high as 10^5 times the 0°/90° rate. This result is hardly surprising, considering that the creep rupture of the 0°/90° orientation is likely dominated by creep rupture of the Nextel 720 fibers. It is recognized that Nextel[TM]720 fiber has the best creep performance of any commercially available polycrystalline oxide fiber. The superior high-temperature creep

Effects of Environment on Creep Behavior of an Oxide-Oxide Ceramic Composite

performance of the Nextel™720 fibers results from the high content of mullite, which has a much better creep resistance than alumina [50]. Conversely the creep rupture of the ±45° orientation is largely dominated by an exceptionally weak porous alumina matrix.

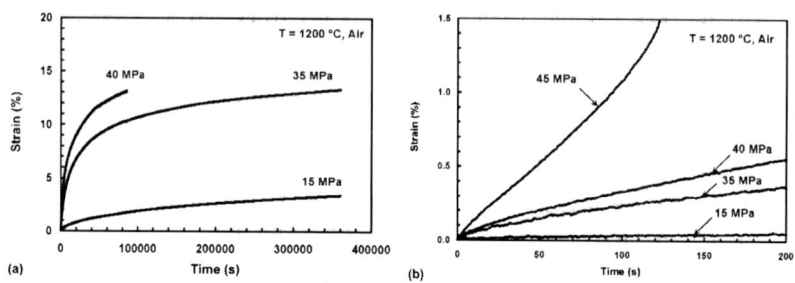

Fig. 3. Creep curves for N720/A composite with ±45° fiber orientation at 1200 °C in air: (a) time scale chosen to show creep strains accumulated at 15 and 35 MPa and (b) time scale reduced to show the creep curve obtained at 45 MPa.

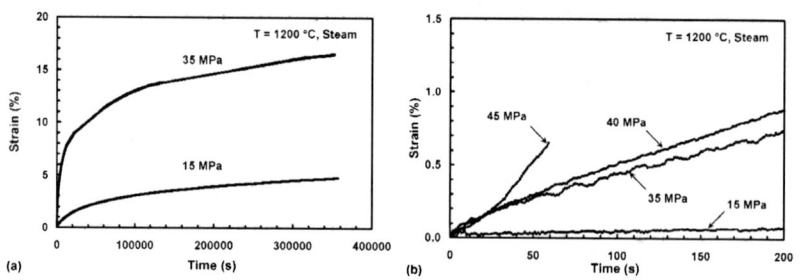

Fig. 4. Creep curves for N720/A composite with ±45° fiber orientation at 1200 °C in steam: (a) time scale chosen to show creep strains accumulated at 15 and 35 MPa and (b) time scale reduced to show creep the curve obtained at 45 MPa.

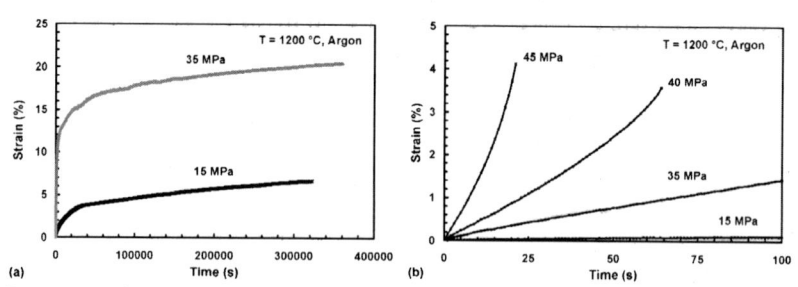

Fig. 5. Creep curves for N720/A composite with ±45° fiber orientation at 1200 °C in argon: (a) time scale chosen to show creep strains accumulated at 15 and 35 MPa and (b) time scale reduced to show creep curves obtained at 40 and 45 MPa.

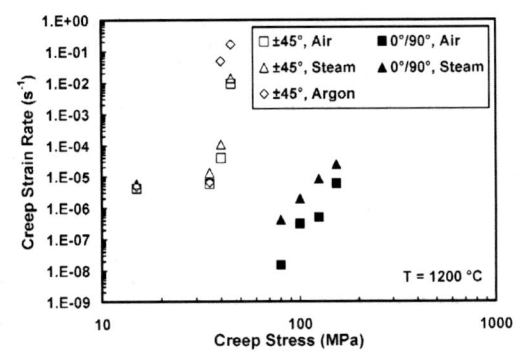

Effects of Environment on Creep Behavior of an Oxide-Oxide Ceramic Composite

Fig. 6. Minimum creep rate as a function of applied stress for N720/A ceramic composite with ±45° fiber orientation at 1200°C in laboratory air, steam and argon. Data for the 0°/90° fiber orientation from Ruggles-Wrenn et al [44] are also shown.

For both fiber orientations, the minimum creep rates increase with increasing applied stress. In the case of the 0°/90° orientation, the secondary creep rate increases by two orders of magnitude as the creep stress increases from 80 to 154 MPa. For a given creep stress, creep rate in steam is approximately an order of magnitude higher than that in air. In the case of the ±45° fiber orientation, for stresses < 40 MPa creep rate is relatively unaffected by environment. Creep rates obtained in all tests at 15 and 35 MPa are $\leq 10^{-5}$ s^{-1}. As the creep stress increases to 45 MPa, the creep rate in air increases by ~ 3 orders of magnitude. Note that the creep rates obtained in steam remain close to those obtained in air, while the creep rates obtained in argon accelerate dramatically. The creep rate in argon is at least two orders of magnitude higher than the rates obtained in air and in steam at 40 MPa, and one order of magnitude higher at 45 MPa.

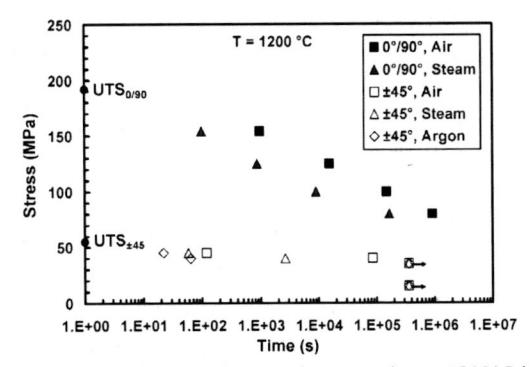

Fig.7. Creep stress vs time to rupture for N720/A ceramic composites at 1200°C in laboratory air, steam and argon. Data for 0°/90° fiber orientation from Ruggles-Wrenn et al [44].

Stress-rupture behavior is summarized in Fig. 7, where results for N720/A composite with 0°/90° fiber orientation from prior work [44] are included for comparison. As expected, creep life decreases with increasing applied stress for both fiber orientations. In the case of the 0°/90° orientation, the presence of steam dramatically reduced creep lifetimes. The reduction in creep life due to steam was at least 90% for applied stress levels ≥100 MPa, and 82% for the applied stress of 80 MPa. Because the creep performance of the 0°/90° orientation is dominated by the fibers, fiber degradation is a likely source of the composite degradation. Recent studies [51, 52] suggest that the loss of mullite from the fiber may be the mechanism behind the degraded creep performance in steam. Alternatively, poor creep resistance in steam may be due to a stress-corrosion mechanism. In this case, crack growth in the fiber is caused by a chemical interaction of water molecules with mechanically strained Si-O bonds at the crack tip with the rate of chemical reaction increasing exponentially with applied stress [53-61]. In the case of the ±45° orientation, environment has little effect on the creep lifetimes (up to 100 h) for applied stresses ≤ 35 MPa. For stresses ≥ 40 MPa, creep lifetimes can be reduced by as much as an order of magnitude in the presence of steam. An even greater reduction in creep life is seen in the presence of argon. Further experiments would be required to understand the cause of such drastic degradation of the creep performance in argon.

Retained strength and modulus of the specimens that achieved a run-out are summarized in Table II. Tensile stress-strain curves obtained for the specimens subjected to prior creep at 15 MPa are presented in Fig. 8 together with the tensile stress-strain curve for the as-processed material. Prior creep in air and in steam appears to have increased tensile strength and stiffness. Furthermore, prior creep considerably reduced the composite's capacity for inelastic straining. The pre-crept specimens produced higher proportional limits and much lower failure strains than the as-processed material. This indicates that additional matrix sintering may be taking place which causes strengthening of the matrix.

Table II. Retained properties of the N720/A specimens with ±45° fiber orientation subjected to prior creep at 1200 °C in laboratory air, steam and argon environments

Environment	Creep Stress (MPa)	Retained Strength (MPa)	Retained Modulus (GPa)	Strain at Failure (%)
Air	15	61.1	53.6	0.17
Air	35	60.0	45.3	0.16
Steam	15	67.0	52.1	0.16
Steam	35	58.4	47.3	0.13
Argon	15	58.3	47.7	0.14
Argon	35	59.4	46.1	0.10

Effects of Environment on Creep Behavior of an Oxide-Oxide Ceramic Composite

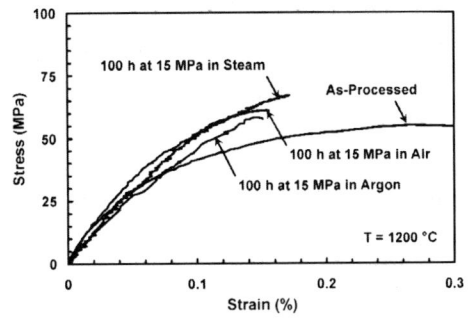

Fig.8. Effects of prior creep at 1200°C in laboratory air, steam and argon environments on tensile stress-strain behavior of N720/A ceramic composite with ±45° fiber orientation.

Composite Microstructure

Fracture surfaces of the N720/A specimens with ±45° fiber orientation tested in creep at 45 are shown in Figs. 9 and 10. Figures 11 and 12 show the fracture surfaces produced in tensile tests on specimens subjected to prior creep at 35 MPa.

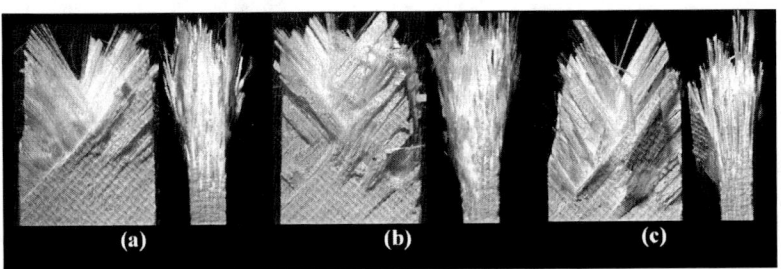

Fig. 9. Fracture surfaces obtained in creep tests at 45 MPa conducted at 1200 °C in: (a) air, (b) steam and (c) argon.

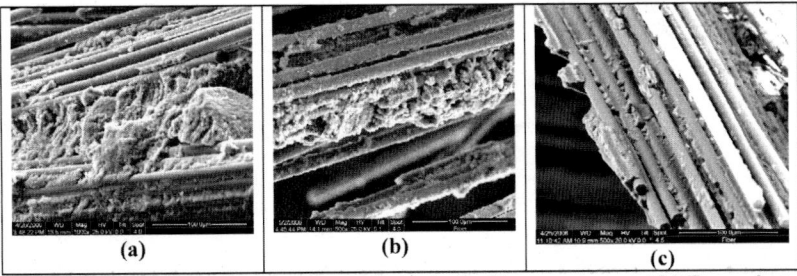

Fig. 10. SEM micrographs of the fracture surfaces obtained in creep tests at 45 MPa conducted at 1200 °C in: (a) air, (b) steam and (c) argon.

By contrast, the fracture surfaces obtained in creep tests at 35 MPa reveal that the failure mechanism in this case includes extensive fiber fracture. Only the specimen pre-crept in air exhibits the V-shaped fracture typically seen in the porous matrix composites with ±45° orientation. Specimens subjected to prior creep in steam and in argon fractured along the planes nearly orthogonal to the loading direction. Fracture surfaces in Fig. 11 (b) and (c) suggest that specimens pre-crept in steam and in argon failed catastrophically at the maximum load, with the majority of the fibers breaking in the process. Such failure process is typical for dense-matrix CMCs. As seen in Fig. 12, the amount of matrix material remaining bonded to the fiber surfaces is greater than that in the specimens subjected to creep at 45 MPa. The difference is particularly pronounced for specimens tested in steam.

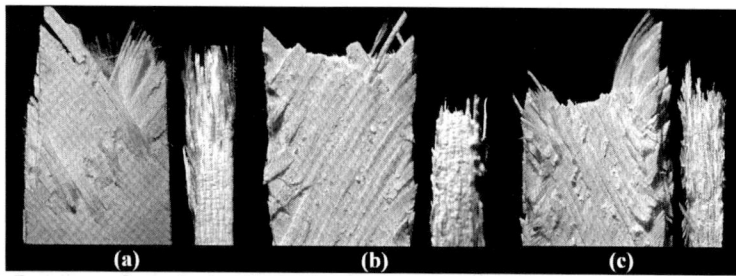

Fig. 11. Fracture surfaces obtained in tensile tests on specimens subjected to prior creep at 35 MPa at 1200°C in: (a) air, (b) steam and (c) argon.

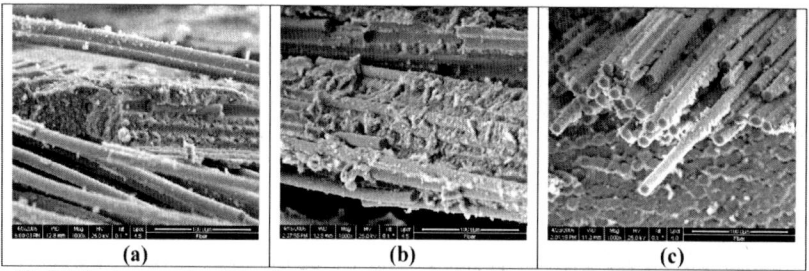

Fig. 12. SEM micrographs of the fracture surfaces obtained in tensile tests on specimens subjected to prior creep at 35 MPa at 1200°C in: (a) air, (b) steam and (c) argon.

Recent studies [62-64] investigated effects of thermal aging on the physical and mechanical properties of composites consisting of Nextel™720 fibers and a porous matrix of mullite and alumina. For a composite with a pure alumina matrix, a porosity reduction of ~6% was observed after a 10-min exposure at 1200 °C [63, 64]. For a composite with a mullite/alumina matrix, strengthening of the matrix and the fiber-matrix interfaces was observed following aging at 1200 °C [62]. Additional sintering of the matrix during the aging treatments was considered to be associated predominantly with Al_2O_3. It is likely that additional sintering of the matrix occurred during 100-h creep tests at 35 MPa conducted in steam and in argon. The resultant strengthening of the matrix is manifested in the retained properties of the composite.

Results in Table II show that after prior creep at 35 MPa, the modulus and the tensile strength increase, and the inelastic straining capabilities of the composite decrease. The strengthening is also manifested in the change in the failure mechanism. The failure of the pre-crept composite is dominated by fiber fracture, while the as-processed material fails predominantly through matrix damage and interplay delamination. The specimens pre-crept in air also exhibit an increase in the modulus and strength after 100 h of creep. However, in this case the fracture surface does not suggest a change in the failure mechanism. It is possible that the sintering of the matrix is accelerated in the presence of steam. It is also somewhat surprising that strengthening of the matrix occurs in argon environment, where no oxygen is present.

CONCLUDING REMARKS

The tensile stress-strain behavior of the N720/A composite with ±45° fiber orientation was investigated and the tensile properties measured at 1200 °C. The stress-strain behavior departs from linearity at a low stress of ~15 MPa. Once the UTS = 55 MPa is reached, the softening commences. Considerable inelastic strains develop at stresses ≈ 50 MPa.

The creep-rupture behavior of the N720/A composite with ±45° fiber orientation was characterized for stress levels ranging from 15 to 45 MPa at 1200°C in laboratory air, steam and argon environments. For the stress levels ≤ 35 MPa the material exhibits primary and secondary creep regimes. At 45 MPa, primary, secondary and tertiary creep regimes are observed. For a given applied stress creep strain accumulation is highest in argon, followed by that in steam and in air. Creep strain rates range from 4.2×10^{-6} to 9.4×10^{-3} s^{-1} in air, from 5.8×10^{-6} to 1.4×10^{-2} s^{-1} in steam, and from 4.9×10^{-6} to 1.6×10^{-1} s^{-1} in argon. For creep stress levels ≤ 35 MPa creep rates are less than 10^{-5} s^{-1}. At 45 MPa, creep rates in all environments increase by at least three orders of magnitude. At 45 MPa, creep rate in argon is about one order of magnitude higher than the rates produced in air and in steam.

Creep run-out of 100 h was achieved at applied stress levels ≤ 35 MPa. The run-out specimens exhibited an increase in stiffness and in strength. Prior creep significantly diminished composite's capability for inelastic straining. For applied stresses ≥ 40 MPa, the presence of steam and especially the presence of argon drastically reduced creep lifetimes.

In tests of less than 100-h duration, the failure occurs primarily through matrix damage and interplay delamination, with minimal fiber fracture. For test durations > 100 h, the failure mechanism is dominated by fiber fracture. It is possible that the matrix undergoes additional sintering during the long-term tests. Additional sintering and consequently strengthening of the matrix may be behind the change in failure mechanisms.

REFERENCES

[1]F. Zok, "Developments in Oxide Fiber Composites," *J. Am. Ceram. Soc.*, **89**(11), 3309-3324 (2006).

[2]L. P. Zawada, J. Staehler, S. Steel, "Consequence of Intermittent Exposure to Moisture and Salt Fog on the High-Temperature Fatigue Durability of Several Ceramic-Matrix Composites," *J. Am. Ceram. Soc.*, **86**(8), 1282-1291 (2003).

[3]M. Parlier, M. H. Ritti, "State of the Art and Perspectives for Oxide/Oxide Composites," *Aerospace Sci. Technol.*, **7**, 211-221 (2003).

[4]M. A. Mattoni, J. Y. Yang, C. G. Levi, F. W. Zok, L. P. Zawada, "Effects of Combustor Rig Exposure on a Porous-Matrix Oxide Composite," *Int. J. Applied Cer Technol.*, **2**(2), 133-140 (2005).

[5]T. A. Parthasarathy, L. P. Zawada, R. John, M. K. Cinibulk, J. Zelina, "Evaluation of Oxide-Oxide Composites in a Novel Combustor Wall Application," *Int. J. Applied. Cer. Technol.*, **2**(2), 122-132 (2005).

[6]R. J. Kerans, R. S. Hay, N. J. Pagano, T. A. Parthasarathy, "The Role of the Fiber-Matrix Interface in Ceramic Composites," *Am. Ceram. Soc. Bull.*, **68**(2), 429-442 (1989).

[7]A. G. Evans, F. W. Zok, "Review: the Physics and Mechanics of Fiber-Reinforced Brittle Matrix Composites," *J. Mater. Sci.*, **29**:3857-3896 (1994).

[8]R. J. Kerans, T. A. Parthasarathy, "Crack Deflection in Ceramic Composites and Fiber Coating Design Criteria," *Composites: Part A*, **30**, 521-524 (1999).

[9]R. J. Kerans, R. S. Hay, T. A. Parthasarathy, M. K. Cinibulk, "Interface Design for Oxidation-Resistant Ceramic Composites," *J. Am. Ceram. Soc.*, **85**(11), 2599- 2632 (2002).

[10]K. M. Prewo, J. A. Batt. The Oxidative Stability of Carbon Fibre Reinforced Glass-Matrix Composites," *J. Mater. Sci.*, **23**, 523-527 (1988).

[11]T. Mah, N. L. Hecht, D. E. McCullum, J. R. Hoenigman, H. M. Kim, A. P. Katz, H. A. Lipsitt, "Thermal Stability of SiC Fibres (Nicalon)," *J. Mater. Sci.*, **19**, 1191-1201 (1984).

[12]J. J. Brennan. *Fiber Reinforced Ceramic Composites; Ch. 8*. Masdayazni KC ed. Noyes, New York, 1990.

[13]F. Heredia, J. McNulty, F. Zok, A. G. Evans, "Oxidation Embrittlement Probe for Ceramic Matrix Composites," *J. Am. Ceram. Soc.*, **78**(8), 2097-2100 (1995).

[14]R. S. Nutt, "Environmental Effects on High-Temperature Mechanical Behavior of Ceramic Matrix Composites," *High-Temperature Mechanical Behavior of Ceramic Composites*. S. V. Nair, and K. Jakus, editors. Butterworth-Heineman, Boston, MA, (1995).

[15]A. G. Evans, F. W. Zok, R. M. McMeeking, Z. Z. Du, "Models of High-Temperature Environmentally-Assisted Embrittlement in Ceramic Matrix Composites," *J. Am. Ceram. Soc.*, **79**, 2345-52 (1996).

[16]K. L. More, P. F. Tortorelli, M. K. Ferber, J. R. Keiser, "Observations of Accelerated Silicon Carbide Recession by Oxidation at High Water-Vapor Pressures," *J. Am. Ceram. Soc.*, **83**(1), 11-213 (2000).

[17]K. L. More, P. F. Tortorelli, M. K. Ferber, L. R. Walker, J. R. Keiser, W. D. Brentnall, N. Miralya, J. B. Price, "Exposure of Ceramic and Ceramic-Matrix Composites in Simulated and Actual Combustor Environments," *Proceedings of International Gas Turbine and Aerospace Congress*, Paper No. 99-GT-292 (1999).

[18]M. K. Ferber, H. T. Lin, J. R. Keiser, "Oxidation Behavior of Non-Oxide Ceramics in a High-Pressure, High-Temperature Steam Environment," *Mechanical, Thermal, and*

Environmental Testing and Performance of Ceramic Composites and Components. M. G. Jenkins, E. Lara-Curzio, and S. T. Gonczy, editors. ASTM STP 1392, 210-215 (2000).

[19]J. A. Haynes, M. J. Lance, K. M. Cooley, M. K. Ferber, R. A. Lowden, D. P. Stinton, "CVD Mullite Coatings in High-Temperature, High-Pressure Air-H_2O," *J. Am. Ceram. Soc.*, **83**(3), 657-659 (2000).

[20]E. J. Opila, R. E. Hann Jr., "Paralinear Oxidation of SiC in Water Vapor," *J. Am. Ceram. Soc.*, **80**(1), 197-205 (1997).

[21]E. J. Opila, "Oxidation Kinetics of Chemically Vapor Deposited Silicon Carbide in Wet Oxygen," *J. Am. Ceram. Soc.*, **77**(3), 730-736 (1994).

[22]E. J. Opila, "Variation of the Oxidation Rate of Silicon Carbide with Water Vapor Pressure," *J. Am. Ceram. Soc.*, **82**(3), 625-636 (1999).

[23]E. E. Hermes, R. J. Kerans, "Degradation of Non-Oxide Reinforcement and Oxide Matrix Composites," *Mat. Res. Soc., Symposium Proceedings*, **125**, 73-78 (1988).

[24]A. Szweda, M. L. Millard, M. G. Harrison, *Fiber-Reinforced Ceramic-Matrix Composite Member and Method for Making*, U. S. Pat. No. 5 601 674, (1997).

[25]S. M. Sim, R. J. Kerans, "Slurry Infiltration and 3-D Woven Composites," *Ceram. Eng. Sci. Proc.*, **13**(9-10), 632-641 (1992).

[26]E. H. Moore, T. Mah, and K. A. Keller, "3D Composite Fabrication Through Matrix Slurry Pressure Infiltration," *Ceram. Eng. Sci. Proc.*, **15**(4), 113-120 (1994).

[27]M. H. Lewis, M. G. Cain, P. Doleman, A. G. Razzell, J. Gent, "Development of Interfaces in Oxide and Silicate Matrix Composites," *High-Temperature Ceramic–Matrix Composites II: Manufacturing and Materials Development*, A. G. Evans, and R. G. Naslain, editors, American Ceramic Society, 41–52 (1995).

[28]F. F. Lange, W. C. Tu, A. G. Evans, "Processing of Damage-Tolerant, Oxidation-Resistant Ceramic Matrix Composites by a Precursor Infiltration and Pyrolysis Method," *Mater. Sci. Eng. A*, **A195**, 145–150 (1995).

[29]R. Lunderberg, L. Eckerbom, "Design and Processing of All-Oxide Composites," *High-Temperature Ceramic–Matrix Composites II: Manufacturing and Materials Development*, A. G. Evans, and R. G. Naslain, editors, American Ceramic Society, 95–104 (1995).

[30]E. Mouchon, P. Colomban, "Oxide Ceramic Matrix/Oxide Fiber Woven Fabric Composites Exhibiting Dissipative Fracture Behavior," *Composites*, **26**, 175–182 (1995).

[31]P. E. D. Morgan and D. B. Marshall, "Ceramic Composites of Monazite and Alumina," *J. Am. Ceram. Soc.*, **78**(6), 1553–1563 (1995).

[32]W. C. Tu, F. F. Lange, A. G. Evans, "Concept for a Damage-Tolerant Ceramic Composite with Strong Interfaces," *J. Am. Ceram. Soc.*, **79**(2), 417–424 (1996).

[33]C. G. Levi, J. Y. Yang, B. J. Dalgleish, F. W. Zok, A. G. Evans, "Processing and Performance of an All-Oxide Ceramic Composite," *J. Am. Ceram. Soc.*, **81**, 2077-2086 (1998).

[34]J. B. Davis, J. P. A. Lofvander, A. G. Evans, "Fiber Coating Concepts for Brittle Matrix Composites," *J. Am. Ceram. Soc.*, **76**(5), 1249–57 (1993).

[35]T. J. Mackin, J. Y. Yang, C. G. Levi, A. G. Evans, "Environmentally Compatible Double Coating Concepts for Sapphire Fiber Reinforced γ-TiAl," *Mater. Sci. Eng.*, **A161**, 285–93 (1993).

[36]A. G. Hegedus, *Ceramic Bodies of Controlled Porosity and Process for Making Same*, U. S. Pat. No. 5 0177 522, May 21, (1991).

[37]T. J. Dunyak, D. R. Chang, M. L. Millard, "Thermal Aging Effects on Oxide/Oxide Ceramic-Matrix Composites," *Proceedings of 17th Conference on Metal Matrix, Carbon, and Ceramic Matrix Composites. NASA Conference Publication 3235, Part 2,* 675-90 (1993).

[38]L. P. Zawada, S. S. Lee, "Mechanical Behavior of CMCs for Flaps and Seals," *ARPA Ceramic Technology Insertion Program (DARPA), W. S. Coblenz WS, editor*. Annapolis MD, 267-322 (1994).

[39]L. P. Zawada, S. S. Lee, "Evaluation of the Fatigue Performance of Five CMCs for Aerospace Applications," *Proceedings of the Sixth International Fatigue Congress*, 1669-1674 (1996).

[40]T. J. Lu, "Crack Branching in All-Oxide Ceramic Composites," *J. Am. Ceram. Soc.*, **79**(1), 266-274 (1996).

[41]Mattoni MA, Yang JY, Levi CG, Zok FW. Effects of Matrix Porosity on the Mechanical Properties of a Porous Matrix, All-Oxide Ceramic Composite. J Am Ceram Soc 2003; 84(11):2594-2602.

[42]F. W. Zok, C. G. Levi, "Mechanical Properties of Porous-Matrix Ceramic Composites," *Adv. Eng. Mater.*, **3**(1-2), 15-23 (2001).

[43]S. Steel, L. P. Zawada, S. Mall, "Fatigue Behavior of a Nextel 720/Alumina (N720/A) Composite at Room and Elevated Temperature," *Ceram. Eng. Sci. Proc.*, **22**(3), 695-702, (2001).

[44]M. B. Ruggles-Wrenn, S. Mall, C. A. Eber, L. B. Harlan, "Effects of Steam Environment on High-Temperature Mechanical Behavior of Nextel™720/Alumina (N720/A) Continuous Fiber Ceramic Composite," *Composites: Part A*, **37**(11), 2029-40 (2006).

[45]J. M. Mehrman, M. B. Ruggles-Wrenn, S. S. Baek, "Influence of Hold Times on the Elevated-Temperature Fatigue Behavior of an Oxide-Oxide Ceramic Composite in Air and in Steam Environment," *Comp. Sci. Tech.*, in press.

[46]R. A. Jurf, S. C. Butner, "Advances in Oxide-Oxide CMC," *Proceedings of the 44th ASME Gas Turbine and Aeroengine Congress and Exhibition. June 7-10,* (1999).

[47]COI Ceramics, Unpublished Data.

[48]L. P. Zawada, R. S. Hay, S. S. Lee, J. Staehler, "Characterization and High-Temperature Mechanical Behavior of an Oxide/Oxide Composite," *J. Am. Ceram. Soc.*, **86**(6), 981-90 (2003).

[49]J. A. Heathcote, X. Y. Gong, J. Y. Yang, U. Ramamurty, F. W. Zok, "In-Plane Mechanical Properties of an All-Oxide Ceramic Composite", *J. Am. Ceram. Soc.*, **82**(10), 2721-30 (1999).

[50]D. M. Wilson, L. R. Visser, "High Performance Oxide Fibers for Metal and Ceramic Composites," *Composites: Part A*, **32**, 1143-1153 (2001).

[51]S. Wannaparhun, S.Seal, "A Combined Spectroscopic and Thermodynamic Investigation of Nextel-720/Alumina Ceramic Matrix Composite in Air and Water Vapor at 1100°C," *J. Am. Ceram. Soc.*, **86**(9), 1628-30 (2003).

[52]C. X. Campbell, E. V. Carelli, K. L. More, P. Varghese, S. Seal, V. H. Desai, "Effect of High-Temperature Water Vapor Exposure on Nextel 720 in an Alumina-Matrix CMC," *Siemens Westinghouse Power Corporation Technical Document TP-02076*.

[53]R. J. Charles and W. B. Hillig WB, "The Kinetics of glass Failure by Stress Corrosion, *Symposium on Mechanical Strength of Glass and Ways of Improving It*. Florence, Italy. September 25-29 (1961). Union Scientifique Continentale du Verre, Charleroi, Belgium, 511-27 (1962).

[54]R. J. Charles and W. B. Hillig WB, "Surfaces, Stress-Dependent Surface Reactions, and Strength," *High-Strength Materials*. V. F. Zackey, editor. John Wiley & Sons, Inc., New York, 682-705 (1965).

[55]S. M. Wiederhorn, "Influence of Water Vapor on Crack Propagation in Soda-Lime Glass," *J. Am. Ceram. Soc.*, **50**(8), 407-14 (1967).

[56]S. M. Wiederhorn, L. H. Bolz, "Stress Corrosion and Static Fatigue of Glass," *J. Am. Ceram. Soc.*, **53**(10), 543-48 (1970).

[57]S. M. Wiederhorn, "A Chemical Interpretation of Static Fatigue," *J. Am. Ceram. Soc.*, **55**(2):81-85 (1972).

[58]S. M. Wiederhorn, S. W. Freiman, E. R. Fuller, C. J. Simmons, "Effects of Water and Other Dielectrics on Crack Growth," *J. Matl. Sci.*, **17**,3460-78 (1982).

[59]M T. A. Michalske, S. W. Freiman, "A Molecular Mechanism for Stress Corrosion in Vitreous Silica," *J. Am. Ceram. Soc.*, **66**(4):284-288 (1983).

[60]T. A. Michalske, B. C. Bunker, "Slow Fracture Model Based on Strained Silicate Structures. J Appl. Phys., **56**(10), 2686-93 (1984).

[61]T. A. Michalske, B. C. Bunker, "A Chemical Kinetics Model for Glass Fracture," *J. Am. Ceram. Soc.*, **76**(10), 2613-18 (1993).

[62]E. A. V. Carelli, H. Fujita, J. Y. Yang, F. W. Zok, "Effects of Thermal Aging on the Mechanical Properties of a Porous-Matrix Ceramic Composite," *J. Am. Ceram. Soc.*, **85**(3), 595-602 (2002).

[63]Fujita H, Jefferson G, McMeeking RM, Zok FW. Mullite/Alumina Mixtures for Use as Porous Matrices in Oxide Fiber Composites," *J. Am. Ceram. Soc.*, **87**(2), 261-67 (2004).

[64]H. Fujita, C. G. Levi, F. W. Zok, G. Jefferson, "Controlling Mechanical Properties of Porous Mullite/Alumina Mixtures via Precursor-Derived Alumina," *J. Am. Ceram. Soc.*, **88**(2), 367-75 (2005).

ASSESSMENTS OF LIFE LIMITING BEHAVIOR IN INTERLAMINAR SHEAR FOR Hi-Nic SiC/SiC CERAMIC MATRIX COMPOSITE AT ELEVATED TEMPERATURE

Sung R. Choi,[†] Robert W. Kowalik, Donald J. Alexander
Naval Air Systems Command
Patuxent River, MD 20670

Narottam P. Bansal
National Aeronautics & Space Administration
Glenn Research Center
Cleveland, OH 44135

ABSTRACT
Assessments of life limiting behavior of a gas-turbine grade, melt-infiltrated Hi-Nicalon™ SiC continuous fiber-reinforced SiC ceramic matrix composite (CMC) were made in interlaminar shear using both stress rupture and constant stress-rate testing at 1316 °C in air. The composite exhibited appreciable life limiting behavior with a life susceptibility parameter of n_s=22-24, estimated based on a proposed phenomenological power-law life prediction model. The phenomenological life model was in good agreement in prediction between the stress rupture and the constants stress-rate data, validating its appropriateness in describing the life limiting phenomenon of the CMC coupons subjected to interlaminar shear. The governing mechanism(s) associated with failure in interlaminar shear appeared to remain unchanged, independent of the type of loading configurations. Constant shear stress-rate testing could be a possible means of life prediction test methodology for CMCs in interlaminar shear at elevated temperatures when relatively short lifetimes are expected.

INTRODUCTION
Fiber-reinforced ceramic matrix composites (CMCs) have shown much improved resistance to fracture and increased damage tolerance in in-plane direction, as compared with the monolithic ceramics. However, inherent material/processing defects, voids, and/or cracks in the matrix-rich or fiber-matrix interface regions can still cause delamination under interlaminar normal or shear stress, resulting in loss of stiffness or in some cases structural failure. Interlaminar tensile and shear strength behaviors of CMCs have been characterized in view of their unique interfacial architectures and importance in structural applications [1-4]. It has been oserved that many 2-D woven CMCs exhibited poor interlaminar properties with interlminar shear strength of 30-50 MPa and interlaminar tensile strength of 10-20 MPa [5].

Most of efforts regarding the assessments of life limiting properties of CMCs have been made for in-plane direction. Few studies have been done on the issue of life limiting of CMCs in *interlaminar shear* at elevated temperatures. In a previous study [6], the life limiting properties

---
[†] Corresponding author; email address: sung.choi1@navy.mil

of a cross-plied glass ceramic composite (Hi-Nicalon™ SiC fiber-reinforced barium strontium aluminosilicate matrix composite, SiC/BSAS) were evaluated in shear at 1100 °C in air by using double-notch shear test specimens subjected to constant shear stress-rate loading. The composite exhibited shear strength degradation with decreasing stress rates, analogous to the slow crack growth (SCG) process occurring in tension for many advanced ceramics at elevated temperatures. The life limiting behavior in shear was modeled using a power-law SCG process of a crack located at fiber-matrix interfaces. This model has been applied to other CMCs such as SiC/SiCs, SiC/MAS (magnesium aluminosilicate), and C/SiC [7].

This paper, as an extension of the previous work [6,7], describes life limiting behavior of a commercial, gas-turbine grade, melt infiltration (MI) Hi-Nicalon™ SiC fiber-reinforced SiC ceramic matrix composite (designated Hi-Nic SiC/SiC) in interlaminar shear. Double notch shear (DNS) test specimens were tested at 1316 °C in air under stress rupture loading. Life limiting behavior of the Hi-Nic SiC/SiC composite was analyzed using a power-law type of phenomenological life model proposed previously [6,7]. Additional experimental data, determined in constant shear stress-rate testing at 1316 °C in air, were also used to further validate the proposed model.

EXPERIMENTAL PROCEDURES

Material

A 2-D woven Hi-Nicalon™ SiC fiber-reinforced SiC ceramic matrix composite (Hi-Nic SiC/SiC), fabricated by GE Power System Composites (Newark, DL; vintage '02), was used in this study. Detailed descriptions of the composite and its processing can be found elsewhere [8]. Briefly, Hi-Nic SiC fibers, produced in tow, were woven into 2-D 5 harness-satin cloth. The cloth preforms of the composite were cut into 200 mm x 150 mm, 8 ply-stacked, and chemically vapor infiltrated (CVI) with a thin BN-based interface coating followed by SiC matrix overcoating. Remaining matrix porosity was filled with SiC particulates and then with molten silicon at 1400 °C, a process termed slurry casting and melt infiltration (MI). The MI SiC/SiC composite was composed of about 39 vol% fibers. The nominal dimensions of the composite panels fabricated were about 200 mm by 150 mm with a thickness of about 2.0 mm. Typical microstructure is shown in Figure 1.

Stress Rupture Testing

Stress rupture testing for the Hi-Nic SiC/SiC composite was conducted in interlaminar shear at 1316 °C in air. The double-notch-shear (DNS) test specimens were machined from the composite panels. Test specimens were 12.7 mm wide (W) and 30 mm long (L). The thickness of test specimens corresponded to the nominal thickness (≈2 mm) of the composite. Two notches, 0.3 mm wide (h) and 6 mm (L_n) away from each other, were made into each test specimen such that the two notches were extended to the middle of the specimen so that shear failure occurred on the plane between the notch tips. Schematics of a DNS test specimen and the test setup are shown in Figure 2. Test fixtures were all made of α-SiC. A specially designed, tubular type of anti-buckling guides was also used. A total of 22 test specimens were tested over a total of five different levels of applied shear stresses ranging from 8 to 17.8 MPa. Test specimen configurations and test procedure were followed in accordance with ASTM C 1425 [9]. All testing was performed using an electromechanical test frame (Model 8562, Instron, Canton, MA). Time to failure of each specimen tested was determined from the data acquisition system.

Life Limiting Behavior in Interlaminar Shear for Hi-Nic SiC/SiC Ceramic Matrix Composite

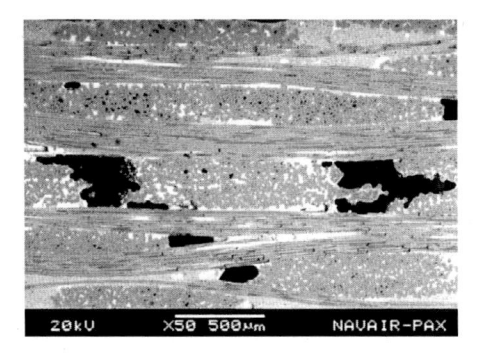

Figure 1. Microstructure of Hi-Nic SiC/Sic composite used in this study.

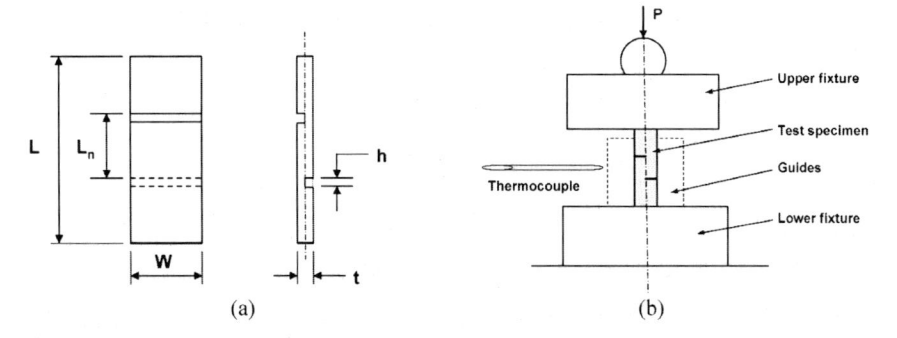

Figure 2. (a) Configurations of double notch shear (DNS) test specimen and (b) a schematic showing a test setup used in this work

Interlaminar shear stress, i.e., the average nominal shear stress, was calculated using the following relation

$$\tau = \frac{P}{WL_n} \quad (1)$$

where τ is the applied shear stress, P is the applied load (in compression), and W and L_n are the specimen width and the distance between the two notches, respectively (see Figure 2).

Constant Shear Stress-Rate Testing
　　Additional interlaminar shear testing was also conducted for the composite using constant shear stress-rate testing. Each test specimen was subjected to a given applied shear stress-rate

until it failed. A total of three different shear stress rates ranging from 5 to 0.005 MPa/s were employed with a total of five specimens tested at each applied shear stress rate. Test fixtures, test temperature, test specimen configuration, and test frame were the same as those used in stress rupture testing. This type of testing, in which strength is determined as a function of stress rate, is often called 'constant stress-rate' or 'dynamic fatigue' testing when applied to glasses and monolithic ceramics to determine their slow crack growth (SCG) behavior in tension or flexure [10,11]. The purpose of this supplementary testing was to determine life limiting behavior in constant stress-rate loading and to compare it with that in stress rupture, with which the phenomenological life prediction model can be validated. Applied shear stress rate ($\dot{\tau}$) was calculated using the relation

$$\dot{\tau} = \frac{\dot{P}}{WL_n} \qquad (2)$$

where $\dot{P}$ is the applied load rate (in compression), which can be applied directly to test specimens via a test frame in load control.

RESULTS AND DISCUSSION

Stress Rupture

All specimens tested in stress rupture at 1316 °C failed in typical shear mode along their respective interlaminar shear planes. The results of stress rupture testing are presented in Figure 3, where time to failure is plotted as a function of applied shear stress. The data clearly show an evidence of life limiting behavior, where time to failure decreased with increasing applied shear stress rate. The solid line represents the best-fit based on the log (*time to failure*) vs. log (*applied interlaminar shear stress*) relation, which will be discussed later. A relatively large scatter in time to failure is noted, similar to the feature shown in many CMCs and advanced monolithic ceramics subjected to stress rupture in tension or flexure at ambient or elevated temperatures.

Fracture surfaces of specimens tested showed discoloration due to oxidation, which appeared to be greater for lower stress-tested specimens than for higher stress-tested specimens. Specimens tested at lower stresses showed typically smooth and clean fracture surfaces with little broken fibers and debris; whereas, specimens tested at higher stresses revealed increased damage with fiber breakage, as shown in Figure 4. This implies that interface-associated failure through slow crack growth or damage accumulation would be more dominant at the lower stress regime than at the higher stress regime.

Although not presented here, it was observed from fracture surfaces that there were some 'blind' regions where no matrix or silicon was filled between plies or tows, resulting in significant gaps or voids or pores in the composite. This certainly contributes to decreased interlaminar shear or tensile properties.

Constant Stress-Rate Test

Without exception, all specimens tested in constant stress-rate loading failed in interlaminar shear. The results of constant stress-rate testing are presented in Figure 5, where interlaminar shear strength was plotted as a function of applied shear stress rate in a log-log scheme. The solid line represents the best fit. Despite some scatter in the data, the overall interlaminar shear strength decreased with decreasing applied shear stress rates. This phenomenon of strength degradation with decreasing test rate, often called slow crack growth or dynamic fatigue when referred to monolithic brittle materials in tension or in flexure [10,11], is an evidence of slow crack growth or damage accumulation occurring at the fiber-matrix interfaces along a respective shear plane under loading. This type of life limiting behavior, associated with strength degradation in interlaminar shear, was also observed for other CMCs including SiC/SiCs, SiC/MAS, C/SiC, and SiC/BSAS at elevated temperatures [6,7]. Based on the results of Figures 3 and 5, it can be stated that life limiting behavior of the composite occurred in interlaminar shear, either in constant loading (stress rupture) or in time varying loading (constant stress rate).

Unlike those tested in stress rupture, the specimens tested in constant stress rates did not show any significant difference in discoloration of fracture surfaces between high and low stress-rate tested specimens. This was due to a relatively short test time (≤ 1 hr) in constant stress-rate testing. The fracture surfaces of specimens tested at lower rates was somewhat smoother with less damage in fibers and matrix than those of specimens tested at higher test rates, as seen in Figure 6.

Assessment of Life Limiting Parameters in Stress Rupture
A phenomenological slow crack growth (SCG) model proposed previously [6,7] will be applied to the stress rupture data determined in this study for the Hi-Nic SiC/SiC composite. The proposed SCG model in mode II is similar in expression to the power-law relation in mode I loading and takes the following empirical formulation

$$v_s = \frac{da}{dt} = \alpha_s (K_{II} / K_{IIc})^{n_s} \qquad (3)$$

where v_s, a, t, K_{II}, and K_{IIc} are crack-growth rate in interlaminar shear, crack size, time, mode II stress intensity factor, and mode II fracture toughness, respectively. α_s and n_s are life limiting (or SCG) parameters in interlaminar shear.

The generalized expression of K_{II} along the crack front of a penny- or half-penny shaped crack subjected to shear loading either on crack planes or on remote material body is [12]

$$K_{II} = Y_s \tau a^{1/2} f(\theta, \varphi) \qquad (4)$$

where Y_s is a crack geometry factor related to a function of $f(\theta,\varphi)$ with the angles θ and φ being related to load and a particular point of the crack front. Using Eqs. (3) and (4) together with some mathematical manipulations, one can obtain time to failure (t_f) as a function of applied shear stress, as done for brittle materials in mode I loading [13]:

$$t_f = D_s [\tau]^{-n_s} \qquad (5)$$

where

$$D_s = B_s [\tau_i]^{n_s - 2} \qquad (6)$$

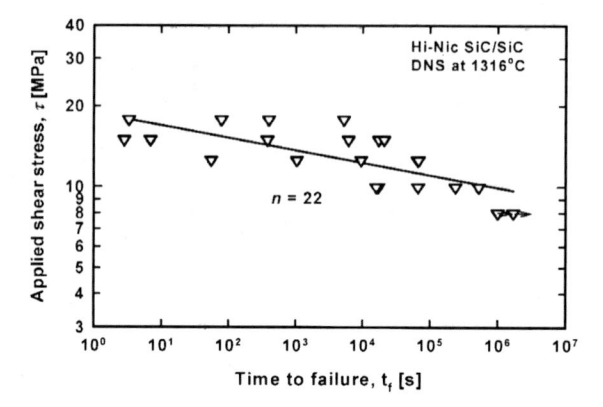

Figure 3. Results of stress rupture testing for Hi-Nic SiC/SiC composite in interlaminar shear at 1316 °C in air. The solid line represents the best fit based on Eq. (7).

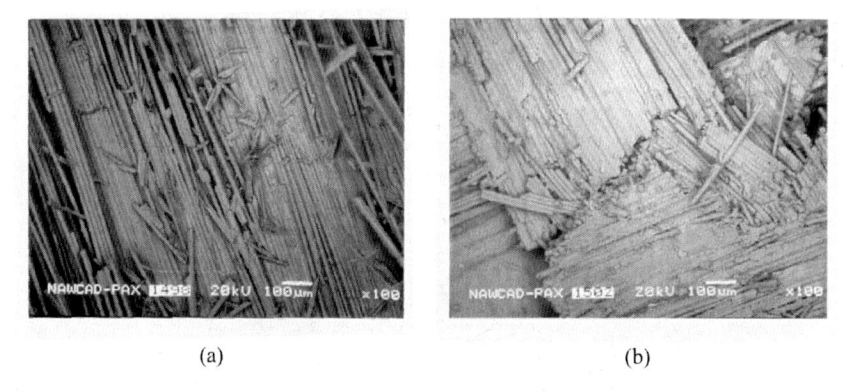

Figure 4. Fracture surfaces of specimens subjected to stress rupture in interlaminar shear tested at 1316 °C in air for Hi-Nic SiC/SiC composite: (a) a high stress of 17.8 MPa (time to failure=3 s) and (b) a low stress of 10 MPa (time to failure=524,000 s).

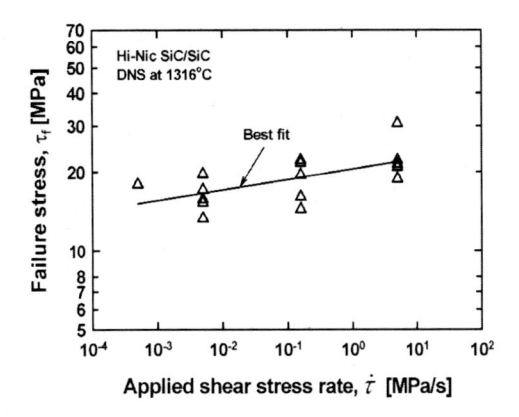

Figure 5. Results of failure stress (interlaminar shear strength) as a function of applied shear stress rate determined in constant stress-rate testing at 1316 °C in air for Hi-Nic SiC/SiC composite. The solid line represents the best fit based on log τ_f vs. log $\dot{\tau}$.

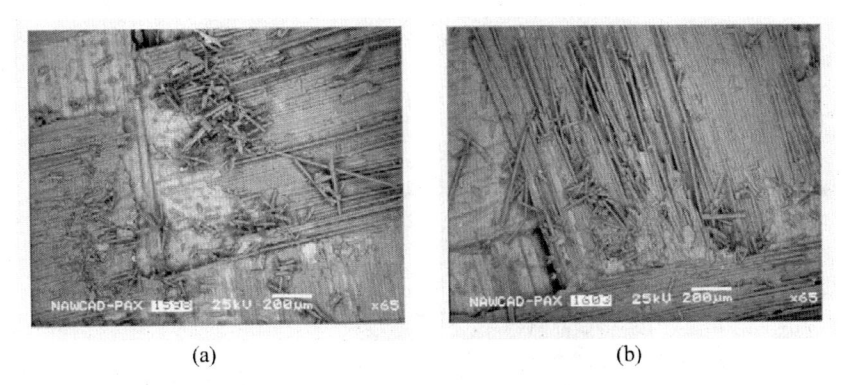

Figure 6. Fracture surfaces of specimens subjected to constant stress-rate testing in interlaminar shear tested at 1316 °C in air for Hi-Nic SiC/SiC composite: (a) a high stress rate of 5 MPa/s (shear strength=22 MPa) and (b) a low stress rate of 0.005 MPa/s (shear strength=13 MPa).

where $B_s = 2K_{IIc}[\alpha_s[Y_s f(\theta,\varphi)]^2(n_s-2)]$ and τ_i is the inert shear strength. The geometry function may be simplified as $f(\theta,\varphi) = 1$ in the case of double-notch shear loading for an infinite material body‡. Equation (5) can be expressed in a more convenient form by taking logarithms of both sides

$$\log t_f = -n_s \log \tau + \log D_s \qquad (7)$$

which is identical in form to the case in mode I loading used in monolithic ceramics [13]. Life limiting parameters n_s and D_s in interlaminar shear can be determined based on Eq. (7), respectively, from the slope and the intercept of a linear regression analysis of the log (*individual time to failure with unit of second*) vs. log (*individual applied interlaminar shear stress with units of MPa*) data, Figure 3, as also suggested in ASTM C 1576 [13]. Use of the data in Figure 3 and the linear regression, life limiting parameters of the composite were found to be

$$n_s = 21.5 \text{ and } \log D_s = 27.5$$

with the coefficient of correlation of curve fit of $r_{coef} = 0.7303$. The best fit was indicated as a solid line in the figure. As seen from the figure, statistically good agreement exists between the model and the data, although the data scatter in time to failure was in a few orders of magnitude, which is typical of most brittle materials in stress rupture or in cyclic fatigue. It has been generally categorized for brittle materials under mode I that life limiting susceptibility is significant for $n<30$, intermediate for $30<n<70$, and insignificant for $n>70$. Hence, in terms of this categorization, the Hi-Nic SiC/SiC composite with $n_s=22$ can be said to exhibit *significant* life limiting in interlaminar shear at 1316 °C in air.

Verification
The proposed crack growth formulation, Eq. (3), indicates that for a given material/environmental condition, crack velocity depends on K_{II} so that in principle the life limiting parameters can be determined in any loading configuration which is either static, cyclic, or any time varying. Therefore, it should be possible to make a life prediction from one loading configuration to another provided that the same failure mechanism is operative. In this section, the life limiting parameters that were determined in stress rupture will be used to predict the strength degradation behavior in constant stress-rate loading and to validate the proposed crack growth model. Using Eqs. (5) and (6) with some mathematical stipulations, a relationship between interlaminar shear strength (τ_f) and applied shear stress rate ($\dot{\tau}$) can be derived as follows:

$$\tau_f = D_d \, [\dot{\tau}]^{\frac{1}{n_s+1}} \qquad (8)$$

where

---

‡ The geometry factor $Y_s f(\theta,\varphi)$ can change as a crack grows through a SCG process if the crack becomes finite relative to the material body (test coupon). This change, however, occurs typically close to an instability for a material exhibiting a life limiting parameter $n \geq 20$, resulting in little change in the value of time to failure (or strength in constant stress-rate loading) [14]. Hence, the approach using $f(\theta,\varphi)=1$ with a constant Y_s in this work is reasonable throughout the lives of test coupons.

$$D_d = [D_s (n_s + 1)]^{\frac{1}{n_s+1}} \qquad (9)$$

Equation (8) shows that interlaminar shear strength is a function of applied shear stress rate for a given material/temperature/environment (expressed typically in a log-log scheme). Therefore, a strength prediction requires only the life prediction parameters n_s and D_s that are estimated from stress rupture data.

The resulting prediction of interlaminar shear strength, based on Eq. (8) with the estimated parameters n_s and D_s, is presented as a solid line (in log-log scheme) in Figure 7. Despite some scatters in shear strength, the prediction was in good agreement with the best fit of the experimental data. The discrepancy in shear strength between the prediction and the data was only 6 to 9 %. Particularly, the life limiting parameter of n_s=22 estimated from stress rupture was in excellent agreement with n_s=24 that was evaluated from the constant stress-rate data by a regression analysis of $\log(\tau_f)$-vs.-$\log(\dot{\tau})$ in Eq. (8). The parameter D_d was also in good agreement with $\log D_d$=27.5 and 31.4, estimated from the stress rupture and the constant stress-rate data, respectively. These results indicated that the *overall* governing failure mechanism of the composite subjected to interlaminar shear remained almost unchanged, regardless of loading configurations, and that the failure mechanism could be described by the power-law type of crack growth formulation, Eq. 3. Statistically, the prediction made above represents a failure probability of approximately 50 %. Of course, different levels of failure probability in shear strength can be made if reliable Weibull strength data are available.

Although exploration of detailed failure mechanism(s) is beyond the scope of this work, a more detailed microscopic failure analysis regarding matrix/fiber interaction, matrix cracking, localized SCG, creep, and environmental effects [15-17], is still needed. It should be kept in mind that the phenomenological model proposed here may incorporate other operative models such as viscous sliding, void nucleation, and coalescence, etc., which can be all covered under a generic term of delayed failure, slow crack growth, fatigue, creep, or damage initiation/accumulation. An important note from the results of this work is that since life limiting of CMCs occurs not only in tension but in interlaminar shear, modification or development of any integrated reliability/life prediction code should take into account a phenomenon of time-dependent interlaminar shear failure at elevated temperatures.

CONCLUSIONS

Life limiting behavior of a Hi-Nic SiC/SiC composite was assessed in interlaminar shear using stress rupture testing at 1316 °C in air. The composite exhibited a significant life limiting susceptibility with n_s=22 that was evaluated based on a proposed phenomenological crack growth law. Good agreement was found in interlaminar shear strength between the prediction from the stress rupture data and the actual constant stress-rate data. The results indicated that governing failure mechanism(s) was almost unchanged, regardless of loading configurations, either in stress rupture or in constant stress rate loading. Constant stress-rate testing could be possible to use as a means of life prediction test methodology for CMCs in interlaminar shear when short lives of components are anticipated.

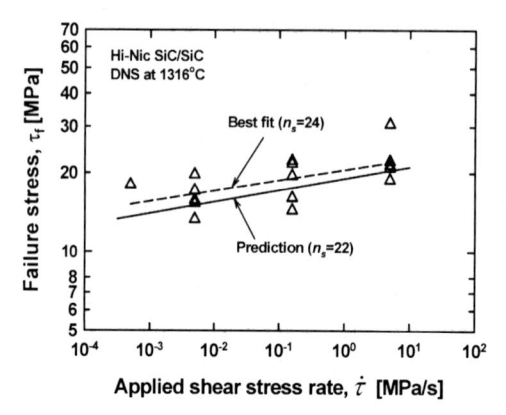

Figure 7. Comparison in interlaminar shear strength between the predicted and the actual data for Hi-Nic SiC/SiC composite in interlaminar shear tested at 1316 °C in air.

Acknowledgements

This work was supported by the Office of Naval Research (ONR) and the NASA Ultra-Efficient Engine Technology (UEET) Project. Mechanical testing by R. Pawlik of NASA Glenn is acknowledged.

REFERENCES
1. P. Brondsted, F. E. Heredia, and A. G. Evans, "In-Plane Shear Properties of 2-D Ceramic Composites," *J. Am. Ceram. Soc.*, **77**[10] 2569-2574 (1994).
2. E. Lara-Curzio and M. K. Ferber, "Shear Strength of Continuous Fiber Ceramic Composites," ASTM STP 1309, p. 31, American Society for Testing & Material, West Conshohocken, PA (1997).
3. N. J. J. Fang and T. W. Chou, "Characterization of Interlaminar Shear Strength of Ceramic Matrix Composites," *J. Am. Ceram. Soc.*, **76**[10] 2539-2548 (1993).
4. Ö. Ünal and N. P. Bansal, "In-Plane and Interlaminar Shear Strength of a Unidirectional Hi-Nicalon Fiber-Reinforced Celsian Matrix Composite," *Ceramics International*, **28** 527-540 (2002).
5. S. R. Choi and N. P. Bansal, "Interlaminar Tension/Shear Properties and Stress Rupture in Shear of Various Continuous Fiber-reinforced Ceramic Matrix Composites," *Advances in Ceramic Matrix Composites XI*, Edited by N. P. Bansal, J. P. Singh, and W. M. Kriven, The American Ceramic Society, Westerville, Ohio; *Ceramic Transactions*, **175** 119-134 (2006).
6. S. R. Choi and N. P. Bansal, "Shear Strength as a Function of Test Rate for SiC_f/BSAS Ceramic Matrix Composite at Elevated Temperature," *J. Am. Ceram. Soc.*, **87**[10] 1912-1918 (2004).

7. S. R. Choi, N. P. Bansal, A. M. Calomino, and M. J. Verrilli, "Shear Strength Behavior of Ceramic Matrix Composites at Elevated Temperatures," *Advances in Ceramic Matrix Composites X*, Edited by J. P. Singh, N. P. Bansal, and W. M. Kriven, The American Ceramic Society, Westerville, Ohio; *Ceramic Transactions*, **165** 131-145 (2005).
8. D. Brewer, "HSR/EPM Combustor Materials Development Program," *Mat. Sci. Eng.*, **A261** 284-291 (1999).
9. ASTM C 1425, "Test Method for Interlaminar Shear Strength of 1-D and 2-D Continuous Fiber-Reinforced Advanced Ceramics at Elevated Temperatures," *Annual Book of ASTM Standards*, Vol.15.01, American Society for Testing & Materials, West Conshohocken, PA (2006).
10. ASTM C 1368, "Standard Test Method for Determination of Slow Crack Growth Parameters of Advanced Ceramics by Constant Stress-Rate Flexural Testing at Ambient Temperature," *Annual Book of ASTM Standards*, Vol. 15.01, American Society for Testing and Materials, West Conshohocken, PA (2006).
11. ASTM C 1465, "Standard Test Method for Determination of Slow Crack Growth Parameters of Advanced Ceramics by Constant Stress-Rate Flexural Testing at Elevated Temperatures," *Annual Book of ASTM Standards*, Vol. 15.01, American Society for Testing and Materials, West Conshohocken, PA (2006).
12. H. Tada, P.C. Paris, and G. R. Irwin, The Stress Analysis of Cracks Handbook, p. 418, ASME, NY (2000).
13. ASTM C 1576 "Standard Test Method for Determination of Slow Crack Growth Parameters of Advanced Ceramics by Constant Stress Flexural Testing (Stress Rupture) at Ambient Temperature," *Annual Book of ASTM Standards*, Vol. 15.01, American Society for Testing and Materials, West Conshohocken, PA (2006).
14. S. R. Choi and J. P. Gyekenyesi, "Slow Crack Growth Analysis of Brittle Materials with Finite Thickness Subjected to Constant Stress-Rate Testing," *J. Mater. Sci.*, **34** 3875-3882 (1999).
15. C. A. Lewinsohn, C. H. Henager and R. H. Jones, "Environmentally Induced Time-Dependent Failure Mechanism in CFCCS at Elevated Temperatures," *Ceram. Eng. Sic. Proc.*, **19**[4] 11-18 (1998).
16. C. H. Henager and R. H. Jones, "Subcritical Crack Growth in CVI Silicon Carbide Reinforced with Nicalon Fibers: Experiment and Model," *J. Am. Ceram. Soc.*, **77**[9] 2381-94 (1994).
17. S. M. Spearing, F. W. Zok and A. G. Evans, "Stress Corrosion Cracking in a Unidirectional Ceramic-Matrix Composite," *J. Am. Ceram. Soc.*, **77**[2] 562-70 (1994).

ARCHITECTURAL DESIGN OF PREFORMS AND THEIR EFFECTS ON MECHANICAL PROPERTY OF HIGH TEMPERATURE COMPOSITES

Jae Yeol Lee, Tae Jin Kang
School of Materials Science & Engineering, Seoul National University
Shinlim-Dong, Kwanak-Ku
Seoul, South Korea, 151-742

Joon-Hyung Byun
Composite Materials Lab, Korea Institute of Machinery & Materials
66 Sangnam-dong,
Changwon, South Korea, 641-831

ABSTRACT
 To study the mechanical properties of high temperature composites, three types of needle punched performs with three different punching densities were prepared and their respective carbon/carbon composites were produced. In order to determine the proper level of needle punching, process parameters and material parameters have been considered. Optimal needle punching was designed to have the most uniform interval of needle trace and the least overlaps of needle sites. Mechanical properties such as tensile strength and flexural strength showed close relationship with the needle punching density, and they showed large differences between the x (in-plane) and z (through- thickness) directions. However, compressive strength had little correlations with the needle punching density and the direction. OxiPAN fibers showed better mechanical properties than carbon fibers because they suffered less damage by needle punching, and fabrics made of filament fibers showed higher mechanical properties than those made of spun or staple fibers.

INTRODUCTION
 High temperature composites such as SiC/SiC, C/SiC or C/C composites have been extensively used in a number of aeronautics and space propulsion applications[1,2]. Proper design and manufacturing of preforms is critical to composite manufacturing and optimum thermostructural performance because preforms are the structural backbone of a composite[3,4,5]. Among many techniques for preform fabrication, the needle punching process attracts much interest because it provides a uniform fibrous structure and can be easily applied to large parts with cost-effectiveness. In spite of its wide application, inter-relationships between the preform characteristics and the composites performance have not been fully explored. This is mainly due to that the control of fiber content in the thickness direction is not as feasible as 3D performs by the weaving or braiding technique. In this work, mechanical properties of needle punched performs and their carbon/carbon composites were investigated and compared with various needle punching densities, raw materials, and textile structures.

DESIGN OF NEEDLE PUNCHING PROCESS
 In the needle punching process, determination of the appropriate level of needle punching is not simple. Excessive punching results in considerable damage to the fabric layers, but insufficient punching results in minimal bridging between the fabric layers, which increases the likelihood of perform delamination. To choose the proper level of needle punching and to

increase the needle punching efficiency, two kinds of parameters should be taken into account. The one is process parameters related with the needle punching machine and its process, another is the material parameters associated with the properties of the fabric layers.

Process parameters

Process parameters are related with the needle punching machine and its process as depicted in Figure 1. The needle punching density, D, which is the number of punching repetitions in the unit area, is expressed in the following equation.

$$D = \frac{fN}{VW} \tag{1}$$

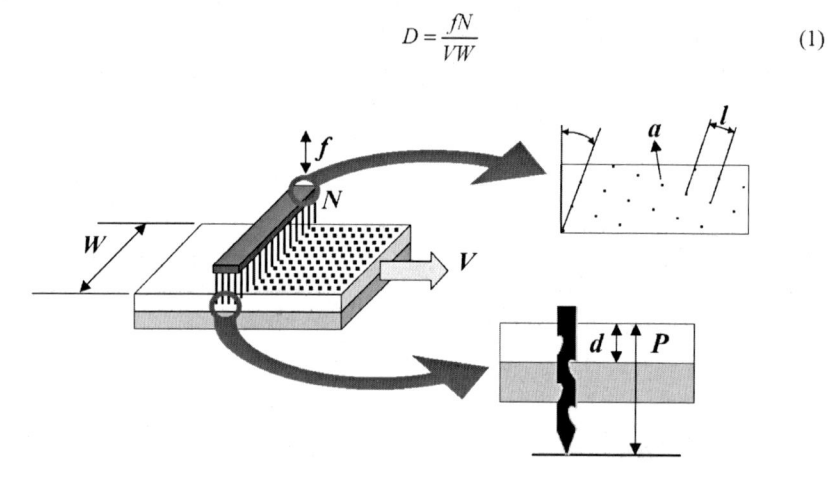

Figure 1. Schematic diagram of needle punching process and the processing parameters (f: stroke frequency, N: number of needles, W: width of punched material, V: feeding speed, θ: angle of needle arrangement, a: sectional area of a needle, l: distance between needles, P: penetration depth, d: thickness of a unit layer).

The ratio, P/d, is the number of unit layers that are penetrated by the needle. While the effect of needle punching is increased with the increase of the ratio, it is not directly proportional to the ratio. As a result, the effective density of needle punching for each unit layer is obtained by taking account of P/d in the equation, and the value of constant K is determined as 0.43, empirically.

$$D = \frac{fN}{VW} \times \left(\frac{P}{d} - 1\right)^K = \frac{fN}{VW} \times \left(\frac{P-d}{d}\right)^K \tag{2}$$

The arrangement of needles on the needle board can be another process parameter. The arrangement of needles is in square lattice, and they are rotated with the angle of θ to avoid the interference between needles (Fig. 1).

Preforms and their Effects on Mechanical Property of High Temperature Composites

Material parameters
Since, the needle punching process is accomplished by the encounter of the needle and the fiber, the properties of the fabric layer play an important role. In the case of woven fabric layers, there are four material parameters, which are the weaving densities of warp and weft yarns, and the yarn effective diameters. The area occupied by the fiber in the woven fabric can be determined with these parameters. In the case of non-woven fabrics, it is composed of arc-shaped short fibers. They can be characterized with four parameters: the curvature radius, the center angle, the thickness, and the number of fibers in unit area. The area occupied by the fiber in the non-woven fabric can be also determined with these parameters.

Distribution of punching pattern
For higher efficiency of needle punching, it is desirable to have evenly distributed needle punching with less overlapping. Distribution of punching pattern is dependent on the arrangement of needles and their progress. Figure 2 shows the arrangement of needles and their traces after repeated punching. For the evenly distributed needle punching pattern, the interval of each trace, Δx and Δy, should be consistent.

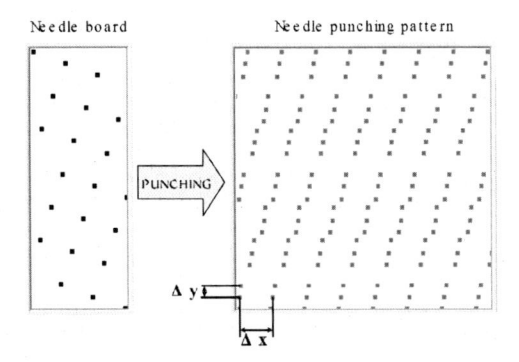

Figure 2. Arrangement of needles and resultant needle punching pattern.

While the vertical interval, Δy, depends only upon the arrangement of needles, the horizontal interval, Δx, depends upon the feeding speed, the occupied area by fibers, and the overlapping among the needle traces, as well as the arrangement of needles. Uniform horizontal intervals can be accomplished by the constant feeding speed, but the avoidance of overlapping is not simple. Figure 3 is the schematic diagram to show the magnified trace of needles and their overlapping. A needle with square 4x4 cells makes traces in the punching direction, and only the traces on the occupied area by fibers are valid. The traces of a second needle then produces two kinds of needle punched area because of the overlapping: the one is 1st punched area and the other is 2nd punched area. A 3rd, or more punched area, can be also made by the repeated overlapping. Needle punching efficiency can be evaluated with the total amount of needle punched area. Effective needle punching is accomplished with large punched area and small overlapped area.

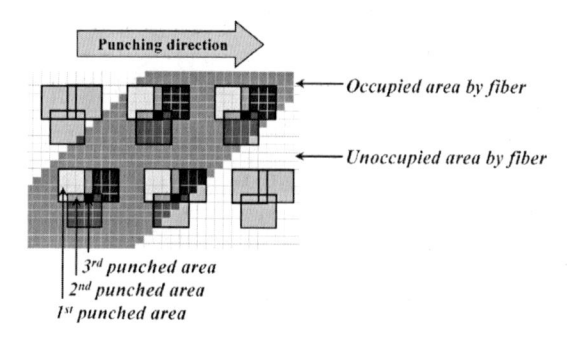

Figure 3. Schematic diagram of repeated needle traces.

Based on the above-mentioned concept, a needle punching process was designed to have the most uniform interval of needle trace and the least overlapping punched area.

PREPARATION OF CARBON/CARBON COMPOSITE

Figure 4 summarizes the preparation procedures of the fabrics for the needle punching process. They are composed of various textile structures and raw materials. The '3D structures' in bottom line denote the completed needle punched preforms of three dimensional structures. Abbreviations of 'O/P', 'C/P' and 'C/F' are used to refer to preforms with different structures. Table 1 summarizes the level of needle punching in terms of the needle punching density and the needle punched area. After preforms were fabricated by needle punching, they were carbonized to convert OxiPAN fibers into carbon fibers. Carbonization was carried out at 1300 °C under the flow of nitrogen gas. After the carbonization process, carbon matrix was formed in the preform to make carbon/carbon composites. The method of thermal gradient chemical vapor infiltration was adopted for the formation of carbon matrix. Propane gas was used as carbon source and the furnace temperature was 1000 °C.

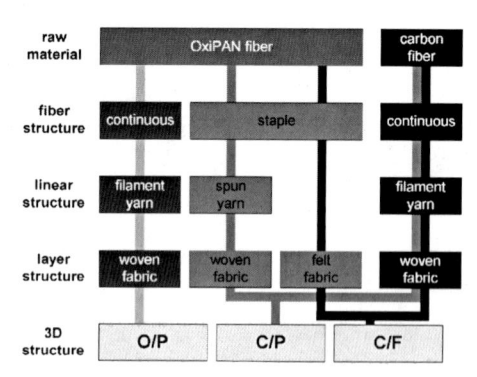

Figure 4. Types of needle punched performs with various materials and fiber/textile structures.

Preforms and their Effects on Mechanical Property of High Temperature Composites

Table I. Level of needle punching

Preform Code		C/F			C/P			O/P		
		A	B	C	A	B	C	A	B	C
Needle Punching Density (/cm^2)		1273	875	444	1692	1163	591	1462	1005	510
Needle Punched Area (%)	1st	32.2	26.4	16.4	37.2	36.0	26.9	35.7	30.9	20.5
	2nd	8.5	4.8	1.6	20.1	13.5	5.2	12.7	7.6	2.6
	3rd	1.4	0.5	0.1	7.1	3.3	0.7	2.9	1.2	0.2
	More	0.2	0.0	0.0	2.2	0.7	0.1	0.5	0.1	0.0
	Total	42.3	31.7	18.1	66.6	53.5	32.9	51.8	39.8	23.3

MECHANICAL PROPERTIES OF NEEDLE PUNCHED PREFORMS

Figure 5 shows the tensile strength of needle punched performs after carbonization. It was measured using universal testing machine at the cross-head speed of 1 mm/min according to ASTM D 638M. The specimens had rectangular cross-section of 5x4 mm and the gauge length was 10mm. With the increase of needle punching density, tensile strength in x-direction deceases due to the fiber breakage and that in z-direction increases because of the rearrangement of fibers. In the case of x-direction, C/F specimen exhibits the lowest tensile strength because of the weak felt fabrics. But O/P specimen shows very high tensile strength because it is composed of only filament fibers. In the case of z-direction, tensile strength is proportional to needle punching density. The level of tensile strength of C/F, C/P and O/P specimens has relation with the easiness of fiber rearrangement by needle punching. Tensile strength in z-direction shows approximately 10 % of that in x-direction. It is the limit of needle punching preforms, but it is remarkably improved in carbon/carbon composite.

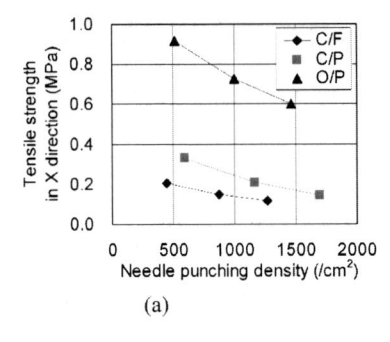

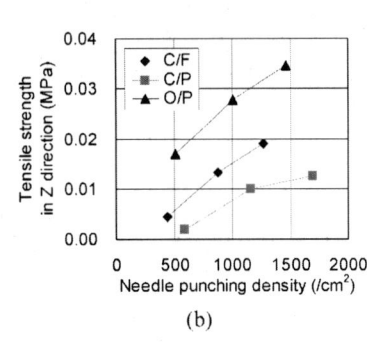

Figure 5. Tensile strength of needle punched preforms: (a) in x-direction; (b) in z-direction.

MECHANICAL PROPERTIES OF CARBON/CARBON COMPOSITES

In Figure 6, tensile strength of carbon/carbon composites in x-direction decreases and that in z-direction increases as the needle punching density increases. This is due to the fiber breakage and rearrangement by needle punching, respectively. The results of tensile strength of carbon/carbon composites are consistent with those of needle punched preforms. In both directions of x and z, tensile strength shows remarkable improvement as compared with that of the preform state. And owing to the existence of carbon matrix, differences of tensile strength in x- and z-direction are smaller than those of preforms. Tensile strength of carbon/carbon composite was measured using universal testing machine at the cross-head speed of 1 mm/min according to ASTM D 638M. The specimens had rectangular cross-section of 5x4 mm and the gauge length was 10mm.

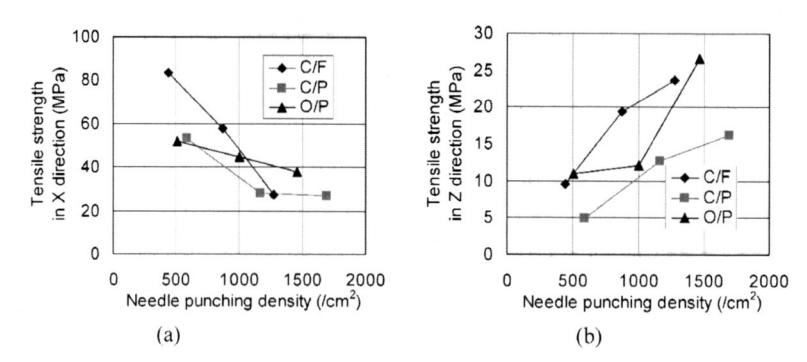

Figure 6. Tensile strength of carbon/carbon composites: (a) in x-direction; (b) in z-direction.

Results of compressive strength of composites are shown in Figure 7. It was measured using universal testing machine at the cross-head speed of 2 mm/min according to ASTM C 1358. The specimens had rectangular cross-section of 12.7x4 mm and the gauge length was 20 mm. It is difficult to find the effect of the needle punching density on the compressive strength, and the differences between the x- and z-directions are insignificant. It suggests that the carbon matrix plays a dominant role under compressive load in composite made of needle punched preform.

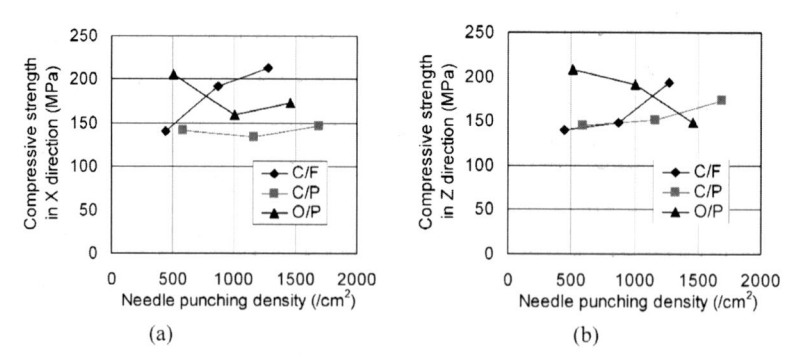

Figure 7. Compressive strength of carbon/carbon composites: (a) in x-direction; (b) in z-direction.

Flexural strength of carbon/carbon composite was measured using universal testing machine at the cross-head speed of 1 mm/min according to ASTM D 790M. The specimens had rectangular cross-section of 25x3 mm and the support span was 48 mm. With the increase of needle punching density, the flexural strength of composites decreases in x-direction, while it increases in z-direction as shown in Figure 8. It is consistent with the results of tensile properties by fiber breakage and rearrangement.In the case of x-direction, it is notable that C/F and C/P specimens exhibit very high flexural strength at the low level of needle punching. As the needle punching continues, the flexural strength decreases very rapidly due to the damage of carbon fabrics during the needle punching. On the other hand, the strength of O/P specimen composed of only OxiPAN fabrics reduces a little. Figure 9 shows the broken carbon fibers in C/F and C/P specimens. In the case of filament OxiPAN fabrics in O/P specimen, however, entangled or pulled fibers by needle punching are easily observed. As a result, O/P specimen is more durable at the high level of needle punching than C/F and C/P specimens. In case of z-direction, C/P and O/P specimens show similar trend with the tensile strength. However, C/F specimen shows better results than the preform state due to the contribution of carbon matrix.

CONCLUSION

Tensile strength of needle punched preforms showed close relationships with the needle punching density. With the increase of the needle punching density, tensile strength in x-direction decreased, while that in z-direction increased. Tensile strength in z-direction was under 20% of that in x-direction, which revealed large difference in x- and z-direction of needle punched preform. Carbon/carbon composites were fabricated by thermal gradient chemical vapor infiltration. Tensile strength of the composite showed similar result with that of the preform. Tensile strength in z-direction was about 10% to 80% of that in x-direction. Although the tensile strength showed a large difference between x- and z-directions, it was remarkably improved than the preform state. In the case of compressive strength of composites, there was little effect of the level of needle punching and the direction of x and z because the carbon matrix was more dominant than the fiber in transferring compressive load. The ratio of compressive strength in z- to x-directions was about 80% to 120%. Flexural strength of carbon/carbon composites showed similar results with those of tensile strength, but the effect of needle punching was more notable. The ratio of flexural strength in z- to x-direction was 10% to 40% at the low level of needle punching similarly to the case of tensile strength. At the high level of needle punching, however, the ratio increased to 70% to 150%.

Specimens made only of OxiPAN fibers showed better results than those made of the hybrid materials because they are less vulnerable in the needle punching. Among the three types of fabric structure made of OxiPAN fibers, woven fabrics of continuous fibers showed better mechanical properties than those of staple fibers.

REFERENCES

[1] K. Upadhya, "High Performance High Temperature Materials for Rocket Engines and Space Environment," *Proceedings of the Conference on Processing Fabrication and Application of Advanced Composites, ASM International*, 1-9 (1993).
[2] D. L. Schmidt, K. E. Davidson, and L. S. Theibert, "Carbon-Carbon Composites (CCC): a Historical Perspective," *41st International SAMPE Symposium*, 24-28 (1996).

[3]R. F. Gibson, *Principles of Composite Material Mechanics*, New York, McGraw-Hill, Inc., 1-20 (1994).
[4]L. M. Manocha and O. P. Bahl, "Influence of Carbon Fiber Type and Weave Pattern on the Development of 2D Carbon-Carbon Composites," *Carbon*, **26**, 13-21 (1988).
[5]R. Luo, C. Yang, and J. Cheng, "Effect of Preform Architecture on the Mechanical Properties of 2D C/C Composites Prepared Using Rapid Directional Diffused CVI Processes", *Carbon*, **40**, 2221-2228 (2002).

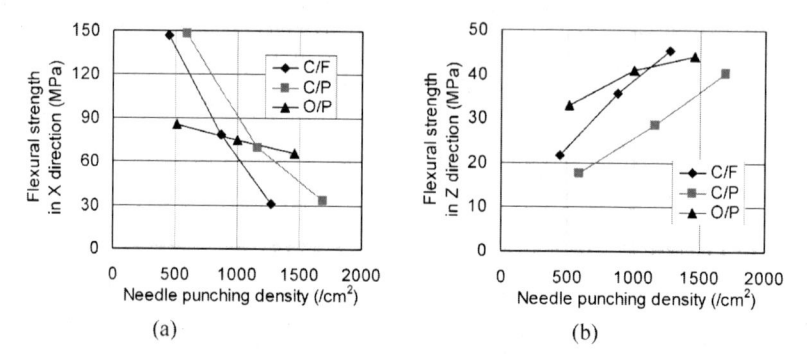

Figure 8. Flexural strength of carbon/carbon composites: (a) in x-direction; (b) in z-direction.

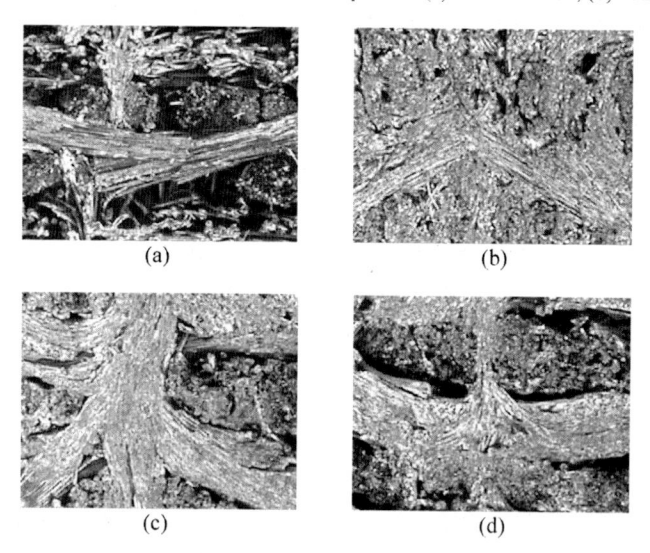

Figure 9. Fiber deformation by needle punching: (a) broken fiber in carbon fabric of a C/F specimen; (b) broken fiber in carbon fabric of a C/P specimen; (c) & (d) entangled fiber and pulled fiber, respectively, in filament OxiPAN fabric of an O/P specimen.

DESIGN FACTOR USING A SiC/SiC COMPOSITES FOR CORE COMPONENT OF GAS COOLED FAST REACTOR. 2: THERMAL STRESS

Jae-Kwang Lee and Masayuki Naganuma
Reactor Core and Fuel Design Group, O-arai Research and Development Center,
Japan Atomic Energy Agency
Naritacho 4002, O-arai.machi, Ibaraki, Japan, 311-1393

ABSTRACT

Thermal stress in SiC/SiC composite cladding for a sealed pin-type core of the gas-cooled fast reactor was estimated based on a preliminary model of thermal conductivity degradation in high temperature neutron irradiation environment and the typical mechanical properties of advanced SiC/SiC composites. Temperatures (T) of 1000, 750, and 500°C and linear heat rates (q') of 200, 400, and 200 W/cm were assumed at the top, middle, and bottom of the core, respectively. The clad wall thickness (t_c) was varied in a range 0.5 to 5 mm. In the range studied in this work, the largest thermal stress occurred in the middle of core, about 180 MPa, at the condition of $T = 750°C$, $q' = 400$ W/cm, and $t_c = 5$ mm. The results will be utilized for a core designs such as fuel cladding tube specifications and linear heat rate.

INTRODUCTION

In recent years, concerns about global warming, fossil fuels exhaustion, and growing demand for energy are imposing renewed attention on nuclear power. As the next generation nuclear power plants, Variations of advanced reactor concepts have been proposed and studied in the Generation IV International Forum. Five of the six proposed reactor concepts are based on the a fast spectrum in combination with a closed fuel cycle[1,2]. The Gas-cooled Fast Reactor (GFR) is the most advanced concept among them. For an outstanding attractiveness and cost competitiveness, core components of GFR will be placed in a demanding environment of high gas pressure, high operating temperature, and high fast neutron flux. Hence, high temperature resistant structural material will be required to enable a design concept of GFR. Currently, ceramics, particularly silicon carbide (SiC), are considered the only viable materials for the core component of GFR. SiC has a good heat resistance and low activation properties[3-6], and recent reports about remarkable advances in toughness improvement by reinforcing with SiC fiber encourages the application of fiber-reinforced SiC/SiC composites[7-9].

Application of a new material to core components must be first studied from the viewpoint of engineering viability. Particularly, thermal conductivity degradation of SiC-based materials anticipated in the high temperature and neutron irradiation environment is among the primary concern [10-12]. The ultimate goal of this work is to provide an evaluation regarding the applicability of SiC/SiC composites for core components and to recommend appropriate core configuration options. In this work, it was attempted to estimate the thermal stress in SiC/SiC core component considering the thermal conductivity degradation in a prototypical reactor environment. First, recent studies on GFR & SiC/SiC composites were surveyed and a simple stress analysis model of cladding tube using SiC/SiC composites was developed. Subsequently, thermal conductivity of unirradiated and irradiated SiC/SiC composites were estimated from the survey results. Thermal stress was estimated for normal and transient conditions. Specifically, the effects of the wall thickness of cladding tubes and the linear heat rate were investigated.

Table 1 GFR Core Design & Performance Parameters[13]

Core	Thermal Power (Electric Output)	2400 MWt (1124 MWe)
	Height/Equivalent Diameter	120/504cm
	Power Density	84W/cm³
	Maximum linear heat rate (in/out)	371/345W/cm
Sealed Pin	Material	Si,Zr,
	Outside diameter	9.9mm
	Wall thickness	0.45mm

REFERENCE GFR CORE DESIGN

Previously, GFR using coated fuel particle has been mainly studied. However, recently it appeared that the fuel pin type GFR core has advantages such as the convenience in reprocessing of spent fuel. Therefore, the core conditions assumed in this study is based on a design study of fuel pin type helium-cooled Fast Breeder Reactor (FBR)[13]. The core parameters are characterized by 850°C reactor outlet temperature, 371 W/cm linear heat rate, and 9.9 mm outer diameter for the sealed fuel pin. Core design and performance parameters were shown in Table 1.

REFERENCE SiC/SiC COMPOSITES

Irradiation behavior of SiC based materials
According to the recent investigations, thermal conductivity behavior of SiC based material depends primarily on irradiation temperature under 1000°C. However, at the temperature above 1000°C, thermal conductivity behavior of SiC based material depends on neutron fluence, and it is not linear as shown in Fig. 1 [10, 11, 14, 15]. Mechanical property of SiC/SiC composites SiC/SiC composite by NITE process (Nano-Infiltration and Transient Eutectic phase) showed a enough manufacturability for core component like a tube & drum products [8, 7, 16].

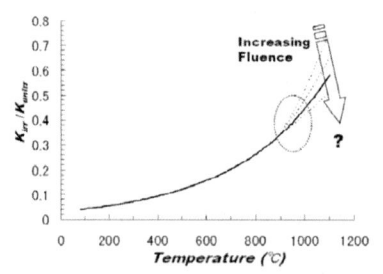

Fig. 1 Thermal conductivity behavior caused by temperature increasing & neutron irradiation.

Table 2 Properties of NITE SiC/SiC[16]

Key SiC/SiC properties	NITE (lab grade)
Density	2800–3000 kg/m³
Porosity	3–6%
Young's modulus	190–220 GPa
Thermal expansion coefficient	3.3–4.7 × 10⁻⁶ K⁻¹ (20–1000°C)
Thermal conductivity through thickness	17–29 W/m-K (20°C)ᵃ 15–20 W/m-K (1000°C)ᵃ
Maximum allowable combined stress	~150 MPaᵇ

ᵃ Unirradiated.
ᵇ 2/3 of tensile proportional limit stress.

Table 3 Estimate conditions of assumed model

Item	Simbol	Unit	
Power density	q'	W/cm	200-500
Outer diameter	D_{out}	mm	11-20
Inner diameter	D_{in}	mm	10
Thickness of cladding tube	t_c	mm	0.5-5
Mean radius of cladding tube	R_m	mm	5.25-7.5
Temperature	T	°C	RT-1000
Thermal expansion	α_{th}	1/K	4×10^{-6}
Young's modulus	E	GPa	200
Poisson's ratio	μ	-	0.18
Thermal conductivity of irradiated SiC/SiC composite	k_{irr}	W/m-K	6.2 (RT) 15.8 (1000°C)

Hence, the properties of Table 2 are referred to as properties of assumed cladding tube.

Assumed model

Thermal stress of cladding wall was estimated with a condition of linear heat rate up to 400W/cm, temperature up to 1000°C as normal state condition. Inner diameter of cladding tube was fixed in 10mm, and then outer diameter was changed by wall thicknesses. Estimation model and parameters are shown in Table 3 and Fig. 2.

Fig. 2 Schematic diagram of cladding pin using SiC/SiC composites.

METHOD & RESULTS OF ESTIMATION

Thermal conductivity degradation

The thermal conductivity of SiC/SiC composites, under neutron irradiation environments, was estimated from the unirradiated thermal conductivity (K_{unirr}) and the relation with thermal defect resistance (K_{rd}). Here, thermal defect resistance term is the material response to irradiation in the absence of grain boundary alteration or other factors such as internal cracking [17]. The temperature dependence of K_{unirr} was approximated by the linear fitting with the value at RT and 1000°C in Table 2. The irradiated thermal conductivity (K_{irr}) is given as follows[17];

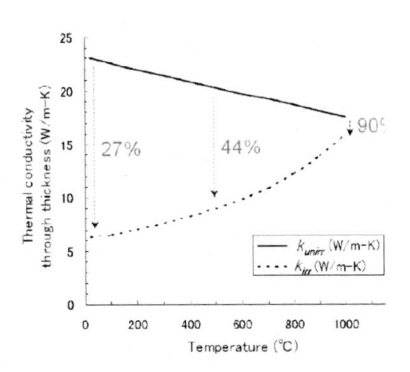

• $1/K_{irr} = 1/K_{unirr} + 1/K_{rd}$

Fig. 3 Thermal conductivity alteration of SiC/SiC composites by temperatur

According to the recent study, K_{rd} exhibits an approximately linear dependence on the irradiation temperature up to about 1000°C and the data could be fitted to the equation below [18];

• $1/K_{rd} = 0.119 - 1.13 \times 10^{-4} T$ (°C)

The result in Fig. 3 indicates that K_{unirr} decreases but K_{irr} increases with the increasing temperature. This result reflects a dynamic recovery of radiation defects in SiC based material irradiated at high temperatures. In this way, the alteration of thermal conductivity of SiC/SiC composites in the GFR environment was simplified. Specific neutron fluence level was not considered, because the thermal conductivity of irradiated SiC is considered to be saturated at a relatively low dose below 1000°C.

Thermal stress
To estimate thermal stress in a cladding tube wall caused by temperature drop (ΔT) and thermal conductivity (K) alteration, the following important assumptions are settled:
• Heat transfers linearly across the cladding tube wall.
• Heat generation within SiC/SiC wall is negligible.
• Density reduction of SiC/SiC composites by swelling is negligible.
• Properties of SiC/SiC composites are quasi-isotropic.
Temperature drop across the cladding tube wall is given as follows:

$$\Delta T = \frac{q'}{2\pi R_m} \frac{t_c}{K} K$$

Thermal stress is analytically derived by the following equation based on mechanical engineering design theory.

$$\sigma_\theta = \sigma_z = \frac{\alpha_{th} E \cdot \Delta T}{2(1-\mu)}$$

Properties of thermal expansion and irradiation creep of SiC/SiC composite are not significant [8,19]. Hence, in the range of RT-1000°C, constant values of thermal expansion coefficient (α_{th}) and Young's modulus (E) in Table 3 were used. Temperatures were 1000, 750, and 500°C and linear heat rates were 200, 400, and 200 W/cm, respectively, at the top, middle and bottom area of Fig. 2. Furthermore, cladding wall thickness was varied from 0.5mm up to 5mm.

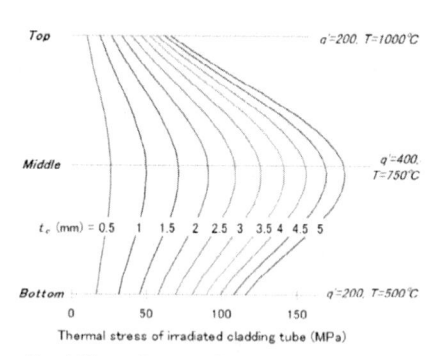

Fig. 4 Thermal stress of assumed cladding tube by changing of q' and t_c

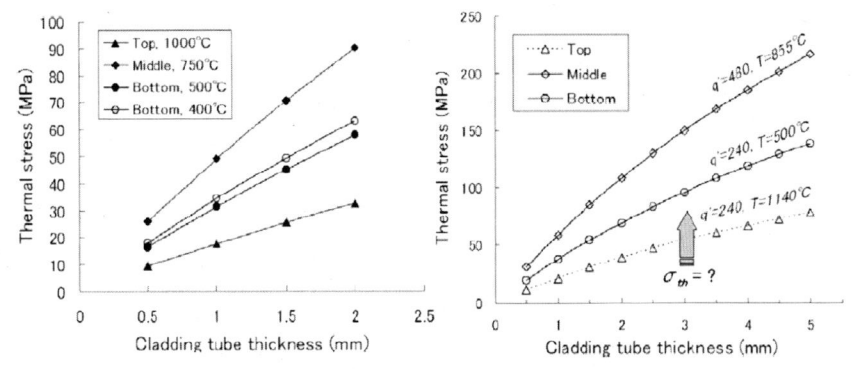

Fig. 5 Thermal stress of assumed cladding tube by changing of t_c - normal condition

Fig. 6 Thermal stress of assumed cladding tube by changing of t_c - transient condition

DISCUSSION

Estimation of the thermal stress distribution provides a significant design criterion for core specification, such as a maximum linear heat rate and cladding dimensions. Thus, the relation between the linear heat rate and the tube wall thickness to thermal stress was evaluated. As shown in Fig. 4, the largest thermal stress of about 180 MPa occurred in approximately the middle of core at the condition of $T = 750°C$, $q' = 400$ W/cm, and $t_c = 5$ mm. The thermal conductivity of the bottom of core should be paid attention to, because it is relatively high and variable with the design condition. As shown in Fig. 5, thermal stress in the condition of 400°C bottom area temperature was higher than that of 500°C. Thus, thermal stress in the bottom area varies significantly depending on the design condition such as the coolant inlet temperature or linear heat rate caused by reflector disposition.

Thermal stress in irradiated cladding tubes in transient conditions was also estimated with 20% increased linear heat rate is shown in Fig. 6. The thermal stress of top area is highly uncertain and may be higher than shown in Fig. 6, because it was estimated with the thermal conductivity at 1000°C. In the parameter range studied in this work, thermal stress at the top of cladding tubes in a transient condition is most uncertain. Hence, the thermal stress should be estimated carefully based on the properties of SiC irradiated at over 1000°C, which are not currently available. However, designing GFR cladding tubes with sufficient safety margins may be possible using realistic mechanical strength of reference SiC/SiC composites in Table 2. Thinner cladding wall thickness is recommended, because of the significantly lower strength of the fiber reinforce composites along the radial direction compared to along the axial or hoop directions.

SUMMARY

Thermal stress in the SiC/SiC fuel cladding tubes for a pin-type GFR core configuration was estimated based on the presently available design parameters and the properties data for neutron irradiated SiC/SiC. The results indicated that:
• The largest thermal stress occurs at approximately the middle of the core, where the linear heat rate is highest but the temperature is lower than the top,
• Thermal stress at the bottom of the core is uncertain; it will be affected by various design parameters including the coolant inlet temperature, and
• Thermal stress at the top of the core is currently unpredictable, especially for transient conditions; it will be strongly affected by various design conditions.

As a future work, more detailed and precise definitions of the reactor design parameters in both normal and transient conditions are required. Moreover, more complete database and model-based predictive capabilities for properties of SiC/SiC composites irradiated at above 1000°C are required.

REFERENCES

[1] B. Frois, "Advances in Nuclear Energy", *Nuclear Physics A*, **752**, 611-622 (2005).

[2] S. David, "Future Scenarios for Fission Based Reactors", *Nuclear Physics A*, **751**, 429-441 (2005).

[3] Pavel Hejzlar, etc., "Gas cooled fast reactor for generation IV service", *Progress in Nuclear Energy*, **47**, No. 1-4, 271-282 (2005).

[4] L.K. Mansur, etc., "Materials needs for fusion, Generation IV fission reactors and spallation neutron sources - similarities and differences", *Journal of Nuclear Materials*, **329-333**, 166-172 (2004).

[5] A.R. Raffray, etc., "Design and material issues for high performance SiCf/SiC-based fusion power cores", *Fusion Engineering and Design*, **55**, 55-95 (2001).

[6] D.B. Marshall, J.B. Davis, "Ceramics for future power generation technology: fiber reinforced oxide composites", *Current Opinion in Solid State and Materials Science*, **5**, 283-289 (2001).

[7] Akira Kohyama, "Advanced SiC/SiC Composite Materials for Fourth Generation Gas Cooled Fast Reactors", *Key Engineering Materials*, **287**, 16-21 (2005).

[8] B. Riccardi, etc., "Issues and advances in SiCf/SiC composites development for fusion reactors", *Journal of Nuclear Materials*, **329-333**, 56-65 (2004).

[9] A. Hasegawa, etc., "Critical issues and current status of SiC/SiC composites for fusion", *Journal of Nuclear Materials*, **283-287**, 128-137 (2000).

[10] D.J. Senor, etc., "Defect structure and evolution in silicon carbide irradiated to 1 dpa-SiC at 1100°C", *Journal of Nuclear Materials*, **317**, 145-159 (2003).

[11] L.L. Snead, etc., "Status of silicon carbide composites for fusion", *Journal of Nuclear Materials*, **233-237**, 26-36 (1996).

[12]Hanchen Huang, Nasr Ghoniem, "A swelling model for stoichiometric SiC at temperatures below 1000°C under neutron irradiation", *Journal of Nuclear Materials*, **250**, 192-199 (1997).

[13]JNC Report, "Feasibility Study on Commercialization of Fast Breeder Reactor Cycle Systems Interim Report of Phase II", *JNC TN9400 2004-035*, 1189-1190 (2004).

[14]G.W. Hollenberg, etc., "The effect of irradiation on the stability and properties of monolithic silicon carbide and SiCf/SiC composites up to 25dpa", *Journal of Nuclear Materials*, **219**, 70-86 (1995).

[15]Tadashi Maruyama, Masaaki Harayama, "Relationship between dimensional changes and the thermal conductivity of neutron-irradiated SiC", *Journal of Nuclear Materials*, **329-333**, 1022-1028 (2004).

[16]Y. Katoh, etc., "SiC/SiC composites through transient eutectic-phase route for fusion applications", *Journal of Nuclear Materials*, **329-333**, Part 1, 587-591 (2004).

[17]L.L. Snead, "Limits on irradiation-induced thermal conductivity and electrical resistivity in silicon carbide materials", *Journal of Nuclear Materials*, **329-333**, 524-529 (2004).

[18]Y. Katoh, etc., "Property tailorability for advanced CVI silicon carbide composites for fusion", *Fusion Engineering and Design*, **81**, Issues 8-14, 937-944 (2006).

[19]C.A. Lewinsohn, etc., "Irradiation-enhanced creep in SiC: data summary and planned experiments", *Journal of Nuclear Materials*, **253**, 36-46 (1998).

DEVELOPMENT OF NOVEL FABRICATION PROCESS FOR HIGHLY-DENSE & POROUS SiC/SiC COMPOSITES WITH EXCELLNT MECHANICAL PROPERTIES

Kazuya Shimoda
Graduate School of Energy Science, Kyoto University
Gokasho Uji, Kyoto 611-0011, Japan

Joon-Soon Park, Tatsuya Hinoki and Akira Kohyama
Institute of Advanced Energy, Kyoto University
Gokasho Uji, Kyoto 611-0011, Japan

ABSTRACT

Unidirectional continuous carbon coated SiC fiber-reinforced SiC matrix composites are prepared by nano-infiltration and transient eutectic phase (NITE) process using SiC "nano"-slurry infiltration technique and pressure sintering, and the effects of fiber volume fraction of sintering temperature evolutions on the microstructure and mechanical properties were characterized. Densification of the composites with standard fiber volume fraction (appropriately 30vol%) was developed even at lower sintering temperature and then saturated at 3rd stage of matrix densification. Hence densification of the composites with high volume fraction (above 50vol%) became restricted because the fibers retarded the infiltration of SiC nano-particles at lower temperature. With the increase of sintering temperature, density of the composites dramatically increased and simultaneously, the interaction between fiber and matrix was strengthened. SEM observation on the fracture surface revealed that fiber pull-out length is accordingly changed with sintering temperature as well as fiber volume fraction, resulting in the variation of tensile fracture behavior. SiC/SiC composites of various fracture types were successfully developed by NITE process as follows: (1) High ductility (highly dense & porous) composites and (2) High strength (highly dense) composites.

INTRODUCTION

Advanced nuclear energy systems, such as gas cooled fast reactor (GFR), very high temperature reactor (VHTR) and fusion reactor are potential candidates for sustainable energy systems in the future. In order to realize these attractive energy systems, structural materials must be responsible to keep their performance under very severe environment including high-temperature, high energy neutron bombardment and surrounding coolants and fuels. Today a major thrust is by the development of fiber-reinforced ceramic matrix composites (CMCs) in general and silicon carbide fiber-reinforced silicon carbide matrix (SiC/SiC) composites in particular. Because of fiber-reinforcement, SiC/SiC composites are more damage tolerant to mechanical and thermal loading (thermal shock) and have the capability for larger components than their SiC monolithic form. Also in comparison to the best high-temperature metallic alloys, SiC/SiC composites are lower density and thermal expansion, and have the potential for displaying excellent high-temperature thermo-mechanical properties under high energy neutron bombardment. However, realization of these reactor will be strongly depend on optimization of SiC/SiC composites microstructure, particularly in regard to the materials and processes used for the fiber, interphase and matrix constituents [1-3].

The objective of this paper is to provide the results of recent activities in our group at Kyoto University, aimed at developing advanced SiC/SiC composites with various attractiveness by the innovate process, called nano-infiltration and transient eutectic phase (NITE) process [4, 5]. Fig.1 shows two concepts on typical tensile fracture behavior for continuous fiber-reinforced CMCs [6, 7]. A primary need in their as-producted condition is to display as high a proportional limit stress (PLS) as possible. The high PLS value will allow the materials to carry high combinations of mechanical and thermal tensile stress without cracking, and be considered as the useful index of material/component design base on elastic mechanical behavior to a high stress. And thus, the composites, as shown in Fig. 1(a), are potentially considered as suitable for fusion blankets, heat exchangers and turbine disks. However, during service time, unexpectedly higher stresses may arise that can locally crack the matrix, causing immediate material failure if the composites don't display a large ductile fracture behavior with a high ultimate tensile strength (UTS). The composites, as shown in Fig. 1(b), are potentially considered as suitable for combustor liners, thermal protection tiles and after burner flaps.

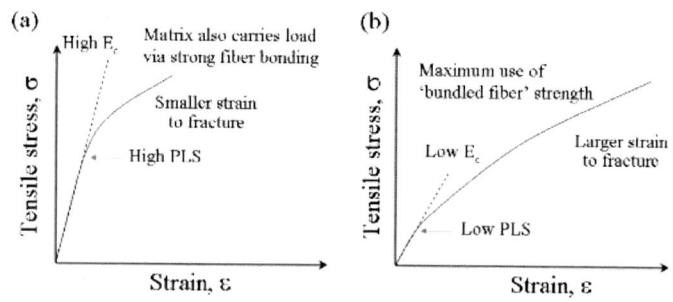

Fig. 1 Concept on tensile fracture behavior for continuous fiber-reinforced CMCs:
(a) High strength type and (b) High ductility type

EXPERIMENTAL PROCEDURE
NITE process for SiC/SiC composites
Pyrolytic carbon (PyC) coated Tyranno^{TM}-SA grade-3 polycrystalline SiC fibers (Ube Industrials Ltd., Japan) were employed as the reinforcement to fabricate SiC/SiC composites. The thickness of the carbon coating is about 0.5μm through chemical vapor deposition (CVD) process. Typical properties of the fiber are listed in Table. 1. It has reported that Tyranno^{TM}-SA fibers have high tensile strength and modulus shows no degradation in strength or change in composition on heating to 1900°C in an inert atmosphere and in air at 1000°C [8]. Schematic illustration of the fabrication process of the SiC/SiC composites is shown in Fig. 2. In this study, unidirectional (UD) PyC coated fibers were impregnated in "nano"-slurry, which mixed SiC nano-sized powder with 30 nm average grain size and processing additives, to form prepreg sheets. Those prepreg sheets were cut to 40 mm x 40 mm. Those prepared prepreg sheets were

Table. 1 Properties of Tyranno^{TM}-SA fiber (grade-3)

SiC Fiber	Atomic ratio (C/Si)	Diameter (μm)	Density (Mg/m³)	Filaments /yarn	Tensile strength (GPa)	Elastic modulus (GPa)
Tyranno^{TM}-SA	1.08	7.5	3.10	1600	2.51	409

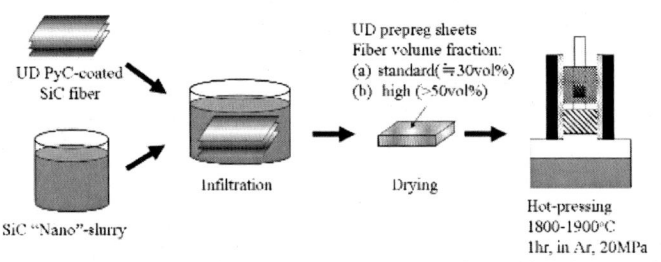

Fig. 2 Schematic illustration of the fabrication process of the SiC/SiC composites.

stacked unidirectionally in a graphite die, and then hot pressed at 1800-1900°C for 1 hour in Ar atmosphere under a pressure of 20MPa with two kinds of fiber volume fraction (V_f). One is standard fiber volume fraction, appropriately 30vol%. Another is high fiber volume fraction above 50vol%.

Characterization of fabricated SiC/SiC composites
The bulk density of fabricated composites was measured by the Archimedes' method, using distilled water as the immersion medium. Theoretical density of fabricated composites was calculated using the rule of mixtures, which consist of SiC matrix with processing additives and carbon coated Tyranno[TM]-SA fiber. Fabricated composites were subsequently cut into 3.0 x 1.5 x 40 mm in size for monotonic tensile test following the general guidelines of *ASTM C-1275*. On both edges of the tensile bars, the aluminum tabs were affixed to each side using a kind of standard Araldite binder. The gauge length was designated to be 20 mm. Tensile strain was recorded from the extensometer fixed on both sides of the gauge areas. Both the polished cross-section and the tensile fracture surface were observed by field-emission electron microscopy (FE-SEM).

RESULTS AND DISSCUSSIONS
Densification process
Fig. 3 shows the relative density of sintering temperature evolution. For monolithic SiC, the densification was dramatically promoted at the temperature of 1750-1800°C probably due to the liquid phase formation and saturated above 1800°C [9]. The corresponding composites with

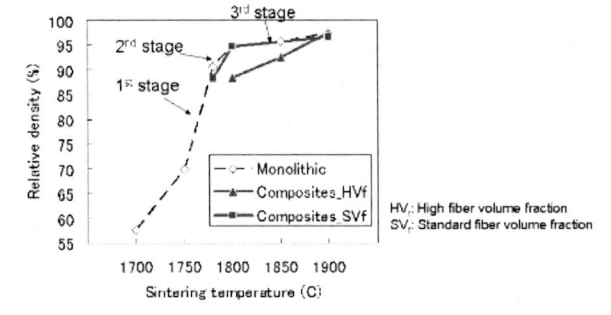

Fig. 3 Bulk relative density of the monolithic SiC and SiC/SiC composites by NITE process.

standard fiber volume fraction demonstrated similar characteristics although the introduction of fibers might retard the densification. One the other hand, the composites with high fiber volume fraction showed relatively lower density at lower temperature (1800°C). The increasing temperature could effectively enhance the densification. At 1900°C, almost 98% of the theoretical density could be obtained. SEM observation on polished cross-section shown in Fig. 4 indicated that no large pores could be identified in the inter-bundle regions. Pores were mainly distributed in the intra-bundle regions. The composites with high fiber volume fraction fabricated at 1800°C maintained 6.2% porosity, so that the formation of porous intra-bundle matrix could be identified, as shown in Fig. 4 (b). With increasing temperature, densification was dramatically promoted. At 1900°C, inter-bundle pores were strongly eliminated without induced-processing damage to fiber and interface, as shown in Fig. 4(c).

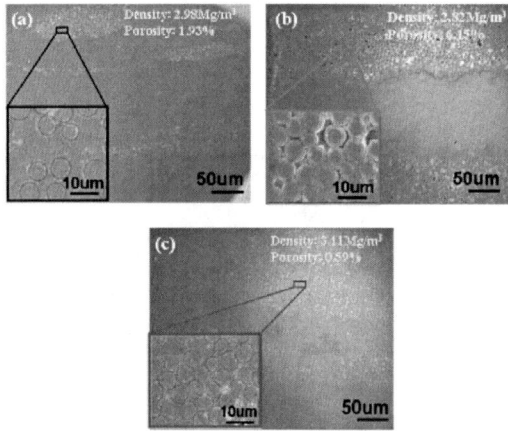

Fig. 4 Polished cross-section surface of the SiC/SiC composites: (a) fabricated at 1800°C with standard-V_f, (b) fabricated at 1800°C with high-V_f and (c) fabricated at 1900°C with high-V_f.

Tensile properties and fracture behavior

In Fig. 5, the stress-strain curves of typical SiC/SiC composites on two concepts of fracture behavior during monotonic tensile tests are presented. High ductility type composites in Fig. 1(b) were fabricated at lower temperature (1800°C). In addition, the composites were categorized in two types from the difference on the densification in the inter-fiber bundle regions, as shown in Fig. 5 (b) and (c). The dense composites with standard fiber volume fraction exhibited very large ductile fracture region after PLS with relatively high UTS (~380MPa). The porous composites with high volume fraction displayed lower strength with relatively large ductile fracture region after PLS. Fracture surface of those composites with standard and high volume fraction are shown in Fig. 6(a) and (b), respectively. At lower temperature, less densified matrix might aid the formation of a relatively weak PyC coating/matrix interface. This feature reflects on the strain-stress curves, which exhibit a wide ductile domain after PLS with relatively longer fiber full-out [10, 11]. High strength type composites in Fig. 1(a) were fabricated at higher temperature (1900°C) with high fiber volume fraction. This composites displayed very high PLS (~360MPa), UTS (~410MPa) and elastic modulus (~350GPa). The failure of the composites occurred soon after the PLS and far from the matrix crack saturation. This behavior inhibited load transfer from matrix to fibers and restricted fiber pull-out, as shown in Fig. 6(c). This shows that the enhanced matrix densification through a sufficient infiltration of SiC nano-power via transient eutectic phase is beneficial at least in terms of first fracture properties and significant damage is not occurring to the fibers and interface even at the harshest process conditions employed. However, even in the higher temperature fabricated composites, cracks deflected along PyC coating/matrix interface, including the interaction between PyC coating and matrix was strengthened.

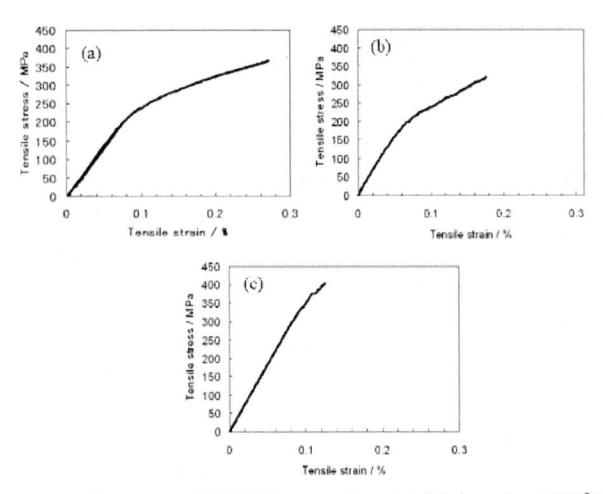

Fig. 5 Typical stress-strain curves of SiC/SiC composites: (a) fabricated at 1800°C with standard-V_f, (b) fabricated at 1800°C with high-V_f and (c) fabricated at 1900°C with high-V_f.

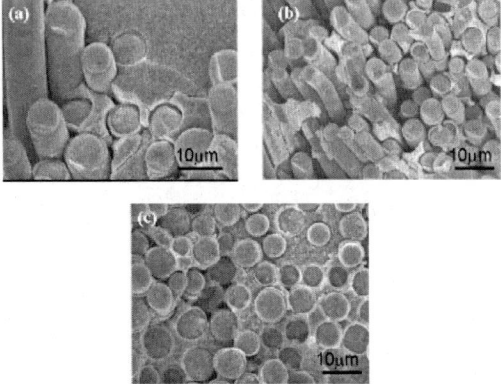

Fig. 6 Fracture surface of tensile-tested SiC/SiC composites: (a) fabricated at 1800°C with standard-V_f, (b) fabricated at 1800°C with high-V_f and (c) fabricated at 1900°C with high-V_f.

SUMMARY
Densification and tensile properties of SiC/SiC composites by nano-infiltration and transient eutectic phase (NITE) process were highly dependent on fiber volume fraction and sintering temperature. In the composites with standard fiber volume fraction, densification was achieved even at lower sintering temperature and then saturated (over 95% of theoretical density) at the 3rd stage of matrix (monolithic) densification. In the composites with high volume fraction, densification process became restricted because the fibers retarded the infiltration of SiC nano-particles at lower sintering temperature. With the increase of sintering temperature, density of the composites dramatically increased and simultaneously, the interaction between PyC coating and matrix was strengthened. Hence, significant damage of fibers and coating was not evidenced from SEM observation and tensile properties. SiC/SiC composites, through NITE process, of various fracture types were successfully developed by the tailoring of appropriate fabrication temperature to fiber volume fraction as follows: (1) High ductility (highly dense & porous) composites and (2) High strength (highly dense) composites.

REFERENCES

[1] "Ceramic Matrix Composites -Microstructure, Properties and Applications-," edited by I. M. Low. Woodhead Publishing, Abington, England, 2006.

[2] B. Riccardi, L. Giancarli, A. Hasegawa, Y. Katoh, A. Kohyama, R.H. Jones and L.L. Snead, "Issues and advances in SiC$_f$/SiC composites development for fusion reactors," *Journal of Nuclear Materials.*, **329–333**, 56–65 (2004).

[3] R. Naslain, "Design, Preparation and Properties of Non-Oxide CMCs for Application in Engines and Nuclear Reactors: An Overview," *Composites Science and Technology.*, **64**, 155-170 (2004).

[4] A. Kohyama, S. M. Dong and Y. Katoh, "Development of SiC/SiC Composites by Nano-Infiltration and Transient Eutectic (NITE) Process," *Ceramic Engineering and Science Proceedings.*, 311-318 (2000).

[5] S. M. Dong, Y. Katoh and A. Kohyama, "Processing Optimization and Mechanical Evaluation of Hot Pressed 2D Tyranno-SA/SiC Composites," *Journal of the European Ceramic Society.*, **23**, 1223-1231 (2003).

[6] J. A. DiCarlo, "Microstructural Optimization of High Temperature SiC/SiC Composites," Proceedings of the 5th International Conference on High Temperature Ceramic Matrix Composites (HTCMC5), p. 187-192, The American Ceramics Society, Ohio (2004).

[7] Y. Katoh, A. Kohyama and T. Nozawa, "SiC/SiC Composites through Transient Eutectic-phase Route for Fusion Applications," 11th International Conference on Fusion Reactor Materials (ICFRM-11) Presented at, Kyoto, Japan (2003).

[8] T. Ishikawa, Y. Kohtoku, K. Kumagawa, T. Yamamura and T. Nagasawa, "High-Strength Alkali-Resistant Sintered SiC fiber Stable to 2200°C," *Nature.*, **391**, 773-775 (1998).

[9] K. Shimoda, J.S. Park, T. Hinoki and A. Kohyama, "Densification Mechanism and Microstructural Evolution of SiC Matrix in NITE Process", *Ceramic Engineering and Science Proceedings.*, **27** [5], 19-27 (2006).

[10] F. Rebillat, J. Lamon, R. Naslain, E. L. Curzio, M. K. Feber and T. M. Besmann, "Properties of Multilayered Interphases in SiC/SiC Chemical-Vapor-Infiltrated Composites with 'Weak' and 'Strong' Interfaces," *Journal of the American Ceramics Society.*, **81** [9], 2315-2326 (1998).

[11] F. Rebillat, J. Lamon and A. Guette, "The Concept of a Strong Interface Applied to SiC/SiC Composites with a BN Interphase," *Acta Materialia.*, **48**, 4609-46018 (2000).

EFFECTS OF INTERFACE LAYER AND MATRIX MICROSTRUCTURE ON THE TENSILE PROPERTIES OF UNIDIRECTIONAL SiC/SiC COMPOSITES

Masaki KOTANI, Toshio OGASAWARA, Hiroshi HATTA and Takashi ISHIKAWA
Japan Aerospace Exploration Agency (JAXA)
6-13-1 Ohsawa, Mitaka, Tokyo, 181-0015 Japan

ABSTRACT
Tensile properties of unidirectional SiC fiber reinforced SiC matrix composites that were fabricated by polymer impregnation and pyrolysis method were systematically studied. In order to know the preferable conditions around the fibers, the specimens of various conditions of interface layer thickness and matrix microstructure were evaluated. The microstructure of the matrix was controlled by the kind of precursor polymer, the addition rate of filler particle and the number of the densification processing. The evaluation was conducted through tensile test, fiber push-out test and microscopy. The carbon layer of several hundred thickness was mostly enough to develop good fiber-sliding mechanism. The densification of the matrix was quite effective in tensile strength by limiting the location of fiber breakage. The precursor polymer and the filler addition mainly changed the morphology of the matrix and the interfacial property respectively. Those favorable conditions were found.

INTRODUCTION
Silicon carbide fiber reinforced silicon carbide matrix composite (SiC/SiC composite) is expected to be used as the structural material of airframe and engine of future aerospace transportation system, because of its superior mechanical performances at high temperature[1-3]. In ceramics matrix composites (CMCs) entirely consisted of essentially brittle materials such as SiC/SiC composite, a fiber/matrix interface bears an important role in many mechanical properties[4]. According to Prof. W. A. Curtin's theory[5], a matrix also has an important role on tensile strength as well as the coefficient of elasticity and proportional limit. In a fracturing process, it receives a distributed load from the broken fibers due to the frictional resistance. This provides the material with pseudo-ductility and consequently enhance the performance of CMC. A microstructure to yield many long pull-out fibers on fracture surface is not sufficient, but the microstructure in which the load carrying-capability of fiber and the probability of fiber breakage are well balanced and those efficiently contribute to bear a load is preferable. In the aspect of fabrication process, the matrix of CMC is generally difficult to densify sufficientlly and its microstructure much varies depending on process conditions[6-9]. So that, CMCs need to be considered as "process defect tolerant material". The important matter is to prepare the best condition that enables fibers to maximize their potential performances in limited fabrication techniques. For this purpose, the effects of interface layer and matrix on tensile strength and their relationship need to be sytematically understood.
    In this work, the effects of the fiber/matrix interface and matrix of the SiC/SiC composites on the tensile properties were experimentally investigated. Tests were conducted for the bundle composite, that was composed of a fiber bundle densified with a matrix, because of its simplicity of structure and the applicability of its data to higher order reinforcement structures. The samples were prepared by polymer impregnation and pyrolysis (PIP) method, forcucsing on its wide applicability and controllablility. The microstructure of the samples was systematically controlled

by the deposition time of carbon interface layer, the precursor polymer of the matrix, filler blending rate and the number of densification processing.

EXPERIMENTAL PROCEDURE

Bundle composites that are fiber bundle with interface layer and matrix in various conditions were prepared for the evaluation.

Tyranno-ZMI SiC fiber (Ube Industries LTD., Japan) was used for the reinforcement. As the precursor for the matrix, allylhydridepolycarbosilane (AHPCS, Starfire Systems Inc., USA) and polycarbosilane (PCS, Nippon Carbon Co., Ltd., Japan) were adopted. For the filler material of the matrix, SiC fine powder, ultra-fine grade of β-random^{TM}, (Ibiden Co., Ltd., Japan) was used. Its average particle size was 270 nm.

Fabrication procedures for the composite samples were as follows; 1) fiber bundle was fixed to the carbon fixture, 2) C layer was formed by chemical vapor infiltration (CVI) method on the fibers, SiC layer about 100nm thickness was subsequently formed on the layer as its barrier coating, 3) the precursor polymer or its powdery slurry was impregnated into the bundle in vacuum, 4) the prepreg bundle was heated in argon atmosphere, 5) the sample was subjected to multiple impregnation of the polymer and heating. The procedure until 4) is called first densification processing (N=1) in this article. For the source gas of C layer and SiC layer, methane and methyl trichlorosilane were used respectively. PCS was used as a hexane solution. Through these processes, linear composite sample of more than 150mm length with a diameter around 1mm was obtained.

The samples were evaluated by tensile test in the fiber direction, fiber/matrix interface shear strength test by monofilament push-out method, and the observations for the polished section and the fractured section using optical microscope (OM) and scanning electron microscope (SEM). The tensile test was conducted under the conditions of 90 mm of gage length, 0.5 mm/min of displacement rate. 5 samples were tested for each fabrication condition. The samples fractured at more than 5 mm distance from the grip part were defined as "5 mm condition clearing sample". The push-out test were conducted using a Berkovich-shape indenter with flat bottom at the loading rate of 20 mN/s. The disc specimens of about 100 μm thickness with both sides polished were prepared for the test. At least 20 effective data points were obtained for each sample. The interfacial shear strength was calculated from a load at the flat area of a load-displacement curve being divided by the area of an interface area that contributed to the slide. The position of fiber breakage was manually scanned using OM. The pull-out fiber length of the fractured surface was manually measured using SEM micrograph. All fiber that could be observed for a bundle were counted.

RESULTS AND DISCUSSIONS

No significant difference was recognized in either sample as a result of comparison between the average values of the maximum load obtained by the tensile test for the "5 mm condition clearing samples" and "all samples". Since there was no significant effect to the tensile characteristics due to the position of the breaking point and the dispersion within the same sample, it can be concluded that the analyses were properly conducted. The average values of the "5 mm condition clearing samples" are shown as follows.

Effects of Interface Layer and Matrix Microstructure on Tensile Properties

Thickness of interface layer
The maximum tensile load of each carbon layer thickness sample is shown in Figure 1. Testing of the samples without carbon layer was not possible because they were so weak that they fractured during preparation. Mechanical properties were greatly improved by adding a carbon layer of only 50 nm. Then the maximum load improved as the thickness was increased. In consideration of the interfacial properties, these behaviors are quite reasonable . Although there have been done many studies for the range of some 100 nm to 1000 nm in the past[10], no clear difference was recognized in this work. Rather, mechanical properties have even improved by increasing the carbon layer thickness up to the range of micron order.

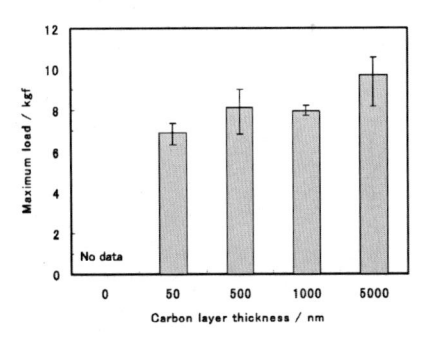

Fig. 1. Maximum loads of the composites of various carbon layer thicknesses.

Figure 2 exhibits the distribution of the pull-out fiber length obtained for the samples having the carbon layer of 50nm, 1000 nm and 5000 nm thickness. The average lengths of each sample are also noted near their lines. It was clearly seen that the distribution of the pull-out fiber length was shifted to higher value as the thickness of the interface layer increased. In case of the sample of 500 nm, only a few fibers were beyond 100 μm. Consequently the average length became higher from 12 μm at 50 nm to 116 μm at 5000 nm.

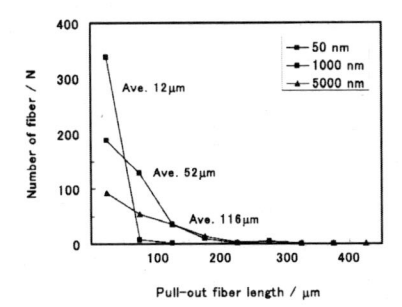

Fig. 2. Distributions of pull-out fiber length for the composites of the carbon layer thickness of 50 nm, 1000 nm and 5000 nm.

Figure 3 shows the interfacial shear strength of the composites of the carbon layer thickness of 1000 nm and 5000 nm. Significant difference of the strength was not identified. It was considered that the effect of a carbon layer as a interfacial sliding layer was almost saturated below 1000 nm.

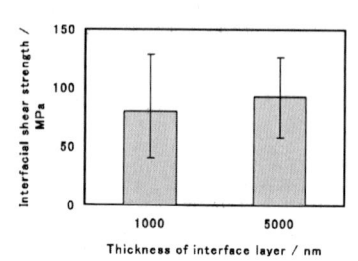

Fig. 3. Interfacial shear strengths of the composites of the carbon layer thickness of 1000 nm and 5000 nm.

Concerning the samples of 1000 nm and 5000 nm, longer pull-out fibers were formed for the sample of 5000 nm than that of 1000 nm, while the both had almost same interfacial properties. The reason for those tensile strengths could be explained by the total resisting force to break which was obtained by multipling the interfacial sliding area and its shear strength. Structural reliability and/or uniformity of micron order thick interface would contribute to increasing the length of pull-out fiber.

Number of densification processing

The maximum tensile loads of the samples after respective times of densification processing is shown in Figure 4. The maximum tensile load improved by applying PIP processing 1 through 8 times, in comparison with the fiber bundle only with interface treatment (n=0). It was confirmed that the densification degree of a sample after 8 times PIP processing almost reached saturation in the previous work[11]. Accordingly, the tensile strength of a bundle with coating layer was around half of that of a composite of highly densified matrix.

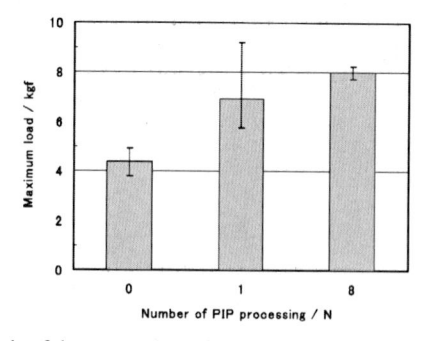

Fig. 4. Maximum loads of the composites of various numbers of densification processing.

Figure 5 exhibits the distribution of fiber breakage location in the gage area. For the sample of n=0, each fiber was mechanically almost independent, so that the fiber was broken randomly at its weakest points through the entire gauge length. On the other hand, fiber breakage area was limited in some ten milimeter and a milimeter for the sample of n=1 and n=8, respectively. It is thought that as the matrix surrounding the fiber was increasingly densified, matrix break in fiber direction decreased and breaking at a weakest point of each fiber was suppressed so that the fracture was limited to the neighborhood of the main crack point, resulting in the increase of the maximum load. As a issue of tensile property, the importance of the densification of a matrix was confirmed.

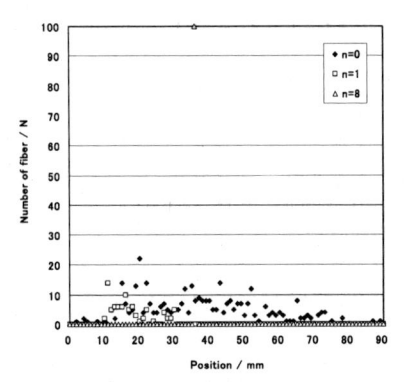

Fig. 5. Distributions of fiber breakage location in the entire gauge length for the bundles after various number of densification processing.

Precursor polymer

The maximum tensile load of the composites fabricated using AHPCS and PCS is exhibited in Figure 6. The composite obtained from PCS showed higher maximum load than the one from AHPCS. It is known that AHPCS has better weight yield, which contributes to the higher densification for the same number of PIP processing as compared with PCS[11]. Contrary to the degree of densification, higher maximum load was obtained for PCS-derived composite rather than AHPCS-derived one. Although an effectiveness of matrix densification on tensile strength was confirmed above, it was not directly shown in this case. There should be another critical factor in tensile property.

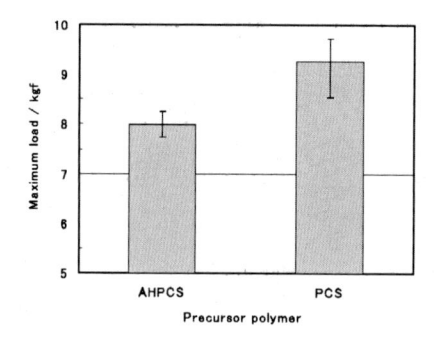

Fig. 6. Maximum loads of the composites obtained using AHPCS and PCS.

Figure 7 shows the length distribution of pull-out fiber for the samples produced using AHPCS and PCS. Compared with the sample of AHPCS, much longer pull-out fibers was observed in the sample of PCS for the same thickness of interface layer. The average length of PCS sample was 116 μm which was almost same value as that shown in the AHPCS sample with the carbon layer thickness of 5000 nm.

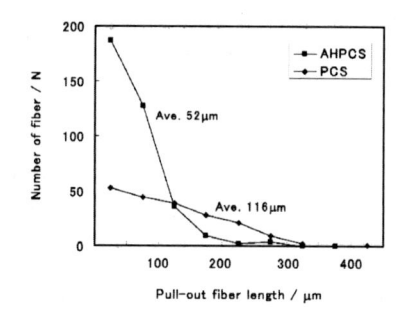

Fig. 7. Distributions of pull-out fiber length for the composites fabricated using AHPCS and PCS.

The interfacial shear strength of both composites were exhibited in Figure 8. The values obtained were almost the same to each other. Although the interfacial property was almost the same, the length distribution of pull-out fiber was apparently different. This relationship of the results was similar to that shown for the samples of the carbon layer thickness of 1000 nm and 5000 nm. As expected at that section, there would be some morphological and/or compositional advantage in improving structural reliability and/or uniformity for PCS-derived matrix rather than AHPCS-derived one. This matter needs to be more closely examined.

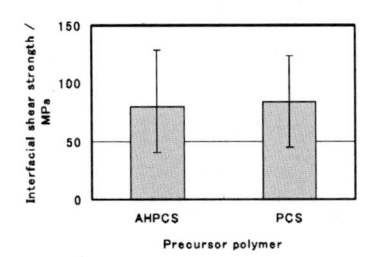

Effects of Interface Layer and Matrix Microstructure on Tensile Properties

Fig. 8. Interfacial shear strength of the composites fabricated using AHPCS and PCS.

Filler blending rate
The maximum tensile load of the composites to which SiC powder was added at each admixture rate in the first PIP processing is shown in Figure 9. In comparison with the samples without filler, maximum load was improved by mixing up to 30 wt.%. However, there approved to be no significant effect on tensile strength by increasing the mixture rate from 30 wt.% to 60 wt.%.

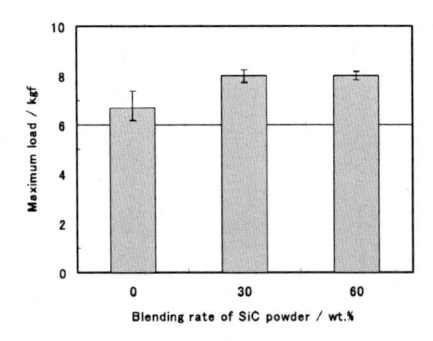

Fig. 9. Maximum loads of the composites fabricated at various filler blending rates.

The distribution of pull-out fiber length of the composites fabricated at various filler blending rate was exhibited in Fig. 10. The length of pull-out fiber increased along with the blending rate. The behavior between 30 wt.% and 60 wt.% was inconsistent with the result of tensile strength.

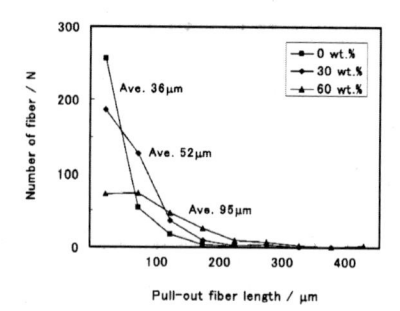

Effects of Interface Layer and Matrix Microstructure on Tensile Properties

Fig. 10. Distributions of pull-out fiber length of the composites fabricated at various filler blending rates.

Figure 11 shows the interfacial shear strength of the composites fabricated at the filler blending rates of 30 wt.% and 60 wt.%. Compared with the interfacial shear strength obtained for 30 wt.%, that obtained for 60 wt.% apparently decreased. It was found that an interfacial property of a composite could be affected by the distribution of fine filler particle in a matrix. Therefore, the result of tensile strength for the samples of 30 wt.% and 60 wt.% can be explained by a fracture-resistant factor cinsisted of an sliding area of interface and an interfacial sliding resistance.

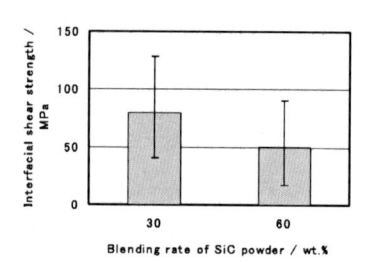

Fig. 11. Interfacial shear strength of the composites of the filler blending rates of 30 wt.% and 60 wt.%

SUMMARY

The tensile property of polymer-derived unidirectional SiC/SiC composite were experimentally evaluated. The following findings are obtained.
1) For an interface, it was approved that some hundred nanometer thickness of carbon layer provided sufficient interfacial function for tensile strength. More improvement was achieved by increasing the thickness up to a micron order.
2) For a matrix, it was confirmed that sufficient densification by repeating the PIP processing and adding filler materials was effective for the improvement of tensile strength.
3) Effectiveness of morphological and/or compositional control of a matrix to improve structural reliability and/or uniformity of an interface was implied.

4) Average pull-out fiber length of the composite that showed high tensile strength was found to be around 100 μm.

REFERENCES
[1] D. Brewer, "HSR/EPM Combustor Materials Development Program", *Mater. Sci. Eng.*, **A261**, 284--291 (1999).
[2] R. Jones, A. Szweda and D. Petrak, "Polymer Derived Ceramic Matrix Composites", *Composite*, **A30**, 569-575 (1999).
[3] K. Sato, A. Tezuka, O. Funayama, T. Isoda, Y. Terada, S. Kato and M. Iwata, "Fabrication and Pressure Testing of a Gas-turbine Compone t Manufactured by a Preceramic-polymer-impregnation Method", *Comp. Sci. Tech.*, **59**, 853-859 (1999).
[4] A. G. Evans and F. W. Zok, "Review: The Physics and Mechanics of Fiber-reinforced Brittle Matrix Composites", *J. Mater. Sci.*, **29**, 3857-3896 (1994).
[5] W. A. Curtin, "Theory of Mechanical Properties of Ceramic-Matrix Composites", *J. Am. Ceram. Soc.*, **74**, 2837-45 (1991).
[6] H. Yoshida, N. Miyata, M. Sagawa, S. Ishikawa, K. Naito, N. Enomoto and C. Yamagishi, "Preparation of Unidirectionally Reinforced Carbon-SiC Composite by Repeated Infiltration of Polycarbosilane", *J. Ceram. Soc. Japan*, **100**, 454-458 (1992).
[7] K. Nakano, A. Kamiya, H. Ogawa and Y. Nishino, "Fabrication and Mechanical Properties of Carbon Fiber Reinforced Silicon Carbide Composites", *J. Ceram. Soc. Japan*, **100**, 472-475 (1992).
[8] T. Tanaka, N. Tamari, I. Kondoh and M. Iwasa, "Fabrication and Evaluation of 3-dimensional Tyranno Fiber Reinforced SiC Composites by Repeated Infiltration of Polycarbosilane", *J. Ceram. Soc. Japan*, **103**, 1-5 (1995).
[9] M. Kotani, T. Inoue, A. Kohyama, K. Okamura and Y. Katoh, "Consolidation of Polymer-derived SiC Matrix Composites: Processing and Microstructure", *Comp. Sci. Tech*, **62**, 2179-2188 (2002).
[10] R. Naslain, "The Concept of Layered Interfases in SiC/SiC", *Ceramic Transactions*, **58**, High-Temperature Ceramic-Matrix Composite II: Manufacturing and Materials Development, 23-39 (1995).
[11] M. Kotani, Y. Katoh, A. Kohyama and M. Narisawa, "Fabrication and Oxidation-Resistance Property of Allylhydridopolycarbosilane-Derived SiC/SiC Composites", *J. Ceram. Soc. Japan*, **111**, 300-307 (2003).

TENSILE PROPERTIES OF ADVANCED SiC/SiC COMPOSITES FOR NUCLEAR CONTROL ROD APPLICATIONS

Takashi Nozawa, Edgar Lara-Curzio, Yutai Katoh
Materials Science and Technology Division, Oak Ridge National Laboratory
P.O. Box 2008, Oak Ridge, TN 37831, USA

Robert J. Shinavski
Hyper-Therm High-Temperature Composites, Inc.
18411 Gothard Street, Unit B, Huntington Beach, CA 92648, USA

ABSTRACT
Recent progress toward development of irradiation resistant SiC/SiC composites has shown great promise for their application in nuclear systems. In this paper, we report the results of a study on the evaluation of the mechanical properties of a nuclear-grade SiC/SiC composite, which is being considered as a candidate control rod material for a very-high temperature gas-cooled reactor. Specifically, this study evaluates the anisotropy in tensile properties of the composites to provide a basis of the practical component design. The materials were satin-woven or biaxially braided Hi-Nicalon™ Type-S fiber reinforced chemical-vapor-infiltrated (CVI) SiC matrix composites with multilayered interphase. Results indicate excellent axial and off-axial tensile fracture behaviors for the satin-woven composites. In contrast, the braided composites failed at unexpectedly lower stresses. The primary cause for this difference was the varied in-plane shear properties, on which off-axial tensile properties significantly depend. Superior in-plane shear properties for the satin-woven composites were achieved by increasing the volume fraction of transverse fibers normal to the fracture plane. Considering the failure modes depend on the off-axis angle, the anisotropy of proportional limit tensile stress and fracture strength were preliminary evaluated by a simple stress criterion model. It is worth noting that specimen size effect on axial and off-axial tensile properties seems very minor for nuclear-grade SiC/SiC composites with rigid CVI-SiC matrix.

INTRODUCTION
Silicon carbide (SiC) ceramics and composites are candidate materials for advanced fission and nuclear fusion applications due to several characteristics such as excellent stability of strength and thermal properties at elevated temperatures, chemical inertness, low radiation-induced activation and low after-heat. Additionally, a crystalline, high-purity, and stoichiometric SiC form such as chemically-vapor-deposited (CVD) SiC provides superior irradiation resistance, e.g., good strength retention [1], improved fracture toughness [2, 3], and moderate thermal transport properties under irradiation [4]. It is now recognized that nuclear-grade SiC/SiC composites reinforced by highly-crystalline and near-stoichiometric "Generation III" SiC fibers, i.e., Hi-Nicalon™ Type-S or Tyranno™-SA, with the high-quality SiC matrix, are similarly radiation-damage resistant [5–7]. Due to the recent advances of composite fabrication techniques and optimization of the functional fiber/matrix interface, remarkable improvement of the composite performance and reliability under neutron irradiation has been achieved [8].

The current Next Generation Nuclear Power (NGNP) projects a high temperature gas cooled reactor will likely operate at temperatures above which conventional or advanced alloys can be employed. For this reason, composite materials are being investigated. The U.S. NGNP

Tensile Properties of Advanced SiC/SiC Composites for Nuclear Control Rod Applications

composite R&D program focuses on 1) confirmative feasibility issues such as fabrication of desired shapes and sizes, fundamentals on mechanical and thermal properties, and irradiation effect study, 2) lifetime issues such as time-dependent fracture and irradiation creep, and 3) test standards and design codes development for composites. In this program, biaxially-braided composite tubes as a model component for the practical control rod are being evaluated.

One of the important design aspects for the use of tubular composites is fiber architecture to ensure the best optimized margins against stresses generated by mechanical loading or thermal expansion in the axial and hoop directions, i.e., anisotropy. Multi-axial fiber reinforcement is often applied to mitigate the issue of composite anisotropic properties. Historically, studies on anisotropic tensile fracture behaviors of conventional SiC/SiC composites have been conducted [9, 10]. For advanced SiC/SiC composites with rigid matrix such as CVI-SiC, understanding the mechanical contribution from the matrix as well as the fiber should be essential since the rigid SiC matrix enables to transfer load as large as the reinforcing fiber can.

The objective of this study is to characterize tensile properties of nuclear-grade SiC/SiC composites. Understanding anisotropy in tensile properties is emphasized to provide a basis for the practical component design. This study is also a part of an effort to develop design codes for the aforementioned components by establishing correlations between the mechanical behavior exhibited by test specimens obtained from plate composites and that of tubular components with the same fiber architecture and constituents.

Table I. NGNP-Grade SiC/SiC Composite Materials.

Material	ID	Loading Axis	Density (g/cm^3)	Fiber Contents	Porosity
Biaxial braid [±55°]	NG1 / Braid-55	Axial (x)	2.83	~0.3	0.08
	NG1 / Braid-35	Transverse (y)	2.87	~0.3	0.09
Biaxial braid [±53°]	NG2 / Braid-53	Axial (x)	2.62	~0.3	0.16
Satin-weave [0°/90°]	NG3 / S/W-0/90	Axial (x)	2.69	~0.4	0.14
	NG3 / S/W-45	±45° off-axis	2.61	~0.4	0.16

EXPERIMENTAL
Materials

SiC/SiC composite flat-plates with different fiber architectures were fabricated by Hyper-Therm High-Temperature Composites, Inc. (Huntington Beach, CA) and their key characteristics are summarized in Table I. The reinforcement fiber was Hi-Nicalon™ Type-S and the matrix was CVI-SiC. The fiber bundles were biaxially braided for NG1 and NG2, or satin-woven (S/W) for NG3. A triaxial braided architecture would be preferred as the degree of off-fiber axis loading of the composites is reduced and eliminated for longitudinal loads. When we originally chose a biaxial braid, the selection was made based upon the fact that we knew that biaxial braids result in higher fiber volume fractions than triaxial braids, and that this difference was through to be the dominant effect. The off-axis angle rotated from the axial loading direction was ±55° for NG1 and ±53° for NG2. Note that the fabrics of the NG1 and NG2 braid composites were prepared by opening a preform of the tubular composites. A multilayered interface composed of a sequence of pyrolytic carbon (PyC) and CVI-SiC was formed on the fiber. The carbon adjacent to the fiber has a thickness of ~150 nm, while the other 4 surrounding layers have a thickness of ~20 nm. The thickness of the SiC inserts between carbon interlayers was ~100 nm. Typical images of Braid-55 are shown in Figure 1. In the figures, the axial in-plane direction, the transverse in-plane direction, and through-thickness direction are defined as X, Y, and Z, respectively. The densities of NG1 and NG2 braid composites were ~2.9 g/cm^3 and ~2.7 g/cm^3, respectively. The porosity, V_p, of the NG1 braid was quite low, ~9%, while the NG2 composites possess the porosity of ~16%. The density and porosity of the NG3 S/W composites were ~2.7 g/cm^3 and ~16%, respectively. The total fiber volume fraction of 30~40% was designed for each material. The braid composites were targeted for a slightly lower fiber volume fraction so as to correspond to the tube architecture.

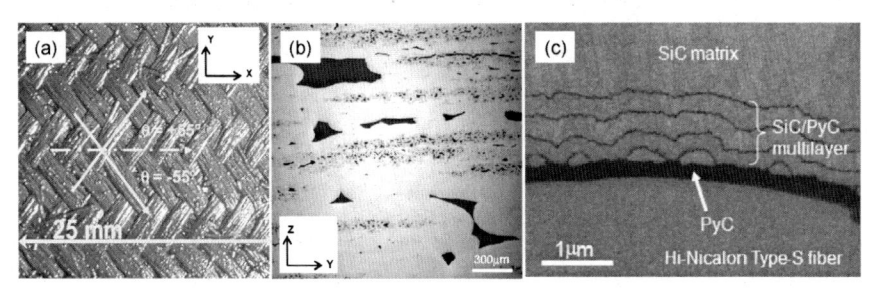

Figure 1. Typical microstructural images of Braid-55 SiC/SiC composites. (a) as-received composite surface (X-Y plane), (b) cross-section (Y-Z plane) and (c) fiber/matrix interface.

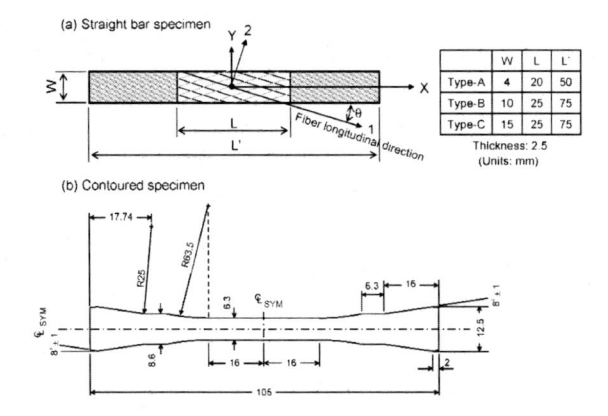

Figure 2. Schematic illustrations of tensile specimens: (a) face-loaded straight bar and (b) edge-loaded contoured specimens.

Mechanical Tests
Room-temperature tensile tests were performed following general guidelines of ASTM C 1275-00. Two types of tensile specimens: a face-loaded straight bar specimen and an edge-loaded contoured specimen were used (Figure 2). To investigate the effect of specimen size on tensile properties, the gauge width was varied from 4.0 to 15.0 mm. Sub-sets of tensile tests were conducted accompanied with unloading/reloading cycles to evaluate the hysteresis response

during damage accumulation for NG3 S/W composites. Tensile strain was measured using a pair of strain gauges bonded in the middle gauge section of the specimen. The length of the stain gauges was 5 mm. For off-axis tensile tests, 90°-axis (transverse) and 45°-axis strains as well as 0°-axis (axial) strain were measured to determine the in-plane shear strain using 3 mm-long strain gauges. A crosshead displacement rate of 0.5 mm/min was applied. Fracture surfaces were observed by the scanning electron microscopy (SEM).

In advance of tensile tests, dynamic Young's modulus was measured by the sonic resonance method according to ASTM C 1259-01 only for straight bar specimens.

RESULTS

Figure 3 exhibits typical tensile stress-stain curves of the SiC/SiC composites tested and key tensile properties are summarized in Table II. In the table, the tensile Young's modulus of the composites, E, was defined as a tangential modulus from the initial linearity. The proportional limit tensile stress (PLS) was defined by 5% deviation from linearity as described in ASTM C 1275-00. The tensile strength was defined as a fracture stress. Experimental errors indicated were deviation from the maximum or the minimum due to the scarce of valid tests, while scatter of dynamic Young's modulus means ± one standard deviation. For Young's modulus measurement, the sonic resonance data is statistically more reliable. Note that the fracture strain does not strictly correspond to the actual strain since the strain measurement area does not always cover the fracture section. The fracture strain and related strain values during non-linear damage accumulation segment are therefore used only for implication in this study.

Nuclear-grade S/W composites (NG3) show superior tensile fracture behaviors coupled with graceful unloading/reloading hysteresis curves. The tensile strength of S/W-0/90 was quite high over 300 MPa. Due primary to the anisotropy issue, the S/W-45 failed at a lower fracture stress (~210 MPa) than obtained in the axial tests. For both S/W-0/90 and S/W-45, the considerably high PLS (~130 MPa) was obtained. It is worth noting that such a high PLS was never achieved for conventional composites that possess low fracture toughness matrix such as polymer-derived amorphous-based SiC. Meanwhile, NG1 and NG2 braid composites show comparably low tensile strength (~110 MPa) and PLS (~80 MPa) with no correlation with the loading axis. The major difference between NG1 and NG2 composites is the magnitude of fracture strain. The fracture strain for NG2 braid composites was larger (~0.51%) than for NG1. Most of NG1 composites failed by first matrix cracking at the proportional limit (0.03~0.16%) with limited secondary strain accumulations. Of particularly important is that no major size effect seems anticipated in the specimen size range of this study for PLS and tensile strength data regardless of the loading axis preferences.

Young's modulus data is slightly questionable. The large experimental scatter for Braid-55 masks the actual size effect. Comparing the data for NG3 S/W composites, there seems minor size effect on Young's modulus but further investigation is required to conclude this due to the limited size range discussed in this study.

Figure 3 shows typical tensile fracture surfaces. The figure does not list fracture surfaces of Braid-35 due to the similarity with those of Braid-55. The first matrix cracking behavior seems dominate a fracture of Braid-35 and Braid-55, the fracture surface of which shows small amounts of very short fiber pullouts embedded in rich SiC matrix, causing a brittle-like fracture. In contrast, the fracture planes of Braid-53 and S/W-45 were parallel to the longitudinal fiber direction. Major cracks propagated within the fiber bundles for Braid-53. It is obvious that there exist fiber pullouts on the fracture surface. Specifically brush-like fibers for S/W-45 indicate

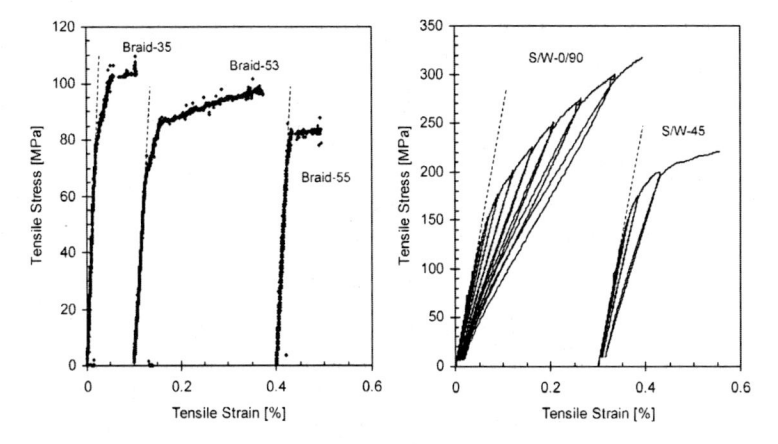

Figure 3. Tensile stress-strain curves of nuclear-grade SiC/SiC composites.

Table II. Tensile Properties of Nuclear-Grade SiC/SiC Composites.

Material	Spec. type / Gauge width / number of valid tests	Dynamic Young's modulus (GPa)	Tensile Young's modulus (GPa)	PLS (MPa)	Tensile strength (MPa)	Fracture strain (%)
Braid-35	Straight bar / 10.0 mm / 3	314 ±1	369 +49/-66	68 +12/-6	105 +2/-1	0.16 +0.09/-0.05
Braid-55	Straight bar / 4.0 mm / 3	338 ±6	302 +101/-73	86 +16/-18	87 +15/-19	0.03 +0.02/-0.01
	Contoured / 6.3 mm / 4	-	237 +16/-13	79 +8/-8	82 +8/-9	0.04 +0.01/-0.01
	Straight bar / 10.0 mm / 4	290 ±19	315 +62/-49	75 +3/-9	78 +6/-7	0.10 +0.15/-0.07
	Straight bar / 15.0 mm / 2	282 ±0	395 +9/-9	82 +4/-4	91 +1/-1	0.04 +0.01/-0.01
Braid-53	Straight bar / 10.0 mm / 5	218 ±12	174 +88/-55	62 +8/-7	108 +11/-11	0.51 +0.10/-0.24
S/W-0/90	Straight bar / 4.0 mm / 9	276 ±9	277 +7/-11	111 +21/-19	346 +67/-50	0.34 +0.05/-0.05
	Contoured / 6.3 mm / 4	-	256 +38/-26	114 +7/-10	348 +29/-43	0.62 +0.07/-0.12
S/W-45	Straight bar / 4.0 mm / 5	232 ±11	248 +12/-32	136 +10/-9	206 +15/-32	0.33 +0.23/-0.24
	Contoured / 6.3 mm / 4	-	232 +18/-32	102 +12/-14	206 +7/-4	0.44 +0.06/-0.07

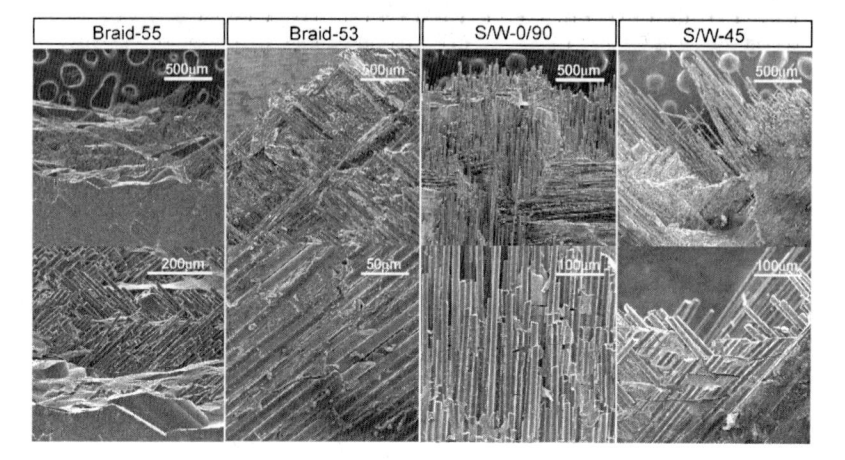

Figure 4. Typical fracture surface images of nuclear-grade SiC/SiC composites.

progressive debonding of transverse fibers normal to the fracture plane. These facts promise the in-plane shear as a primary fracture mode in off-axis tension. Meanwhile, the fracture surface of S/W-0/90 was very fibrous coupled with significant transverse cracks in the 0°-bundles. Similarly, very limited amounts of transverse cracks were observed for S/W-45. However, they were very minor and most of them were localized near the cross sections of the 45°-bundles.

Figure 5 shows results of hysteresis analysis for S/W-0/90 and S/W-45. In Fig. 5, the damage parameter, D, was defined as:

$$D = \frac{E}{E^*} - 1 \qquad (1)$$

where E^* is the unloading Young's modulus of each unloading curve. The loop width was the maximum differential strain of a hysteresis curve. The inelastic strain index, L, was a parameter defined by the inverse tangent modulus, I_p, the peak stress, σ_p, and the debond stress, σ_i [11]:

$$I_p = 4(\sigma_p - \sigma_i)L + \frac{1}{E^*} \qquad (2)$$

The inelastic strain index was also expressed as another form [11]:

$$L = \frac{b_2(1 - a_1 f_b)^2}{4 f_b^2 \tau E_m} \cdot \frac{R}{d} \qquad (3)$$

where a_1 and b_2 are coefficients given by Hutchinson and Jensen [12], f_b is the fiber volume fraction in the 0°-ply, E_m is the matrix modulus, τ is the interfacial friction stress, R is the fiber radius, and d is the mean length of matrix crack spacing.

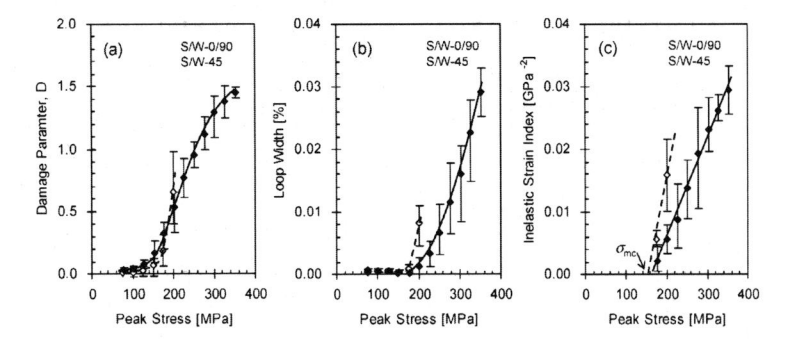

Figure 5. Results of hysteresis analysis: (a) damage parameter, (b) loop width and (c) inelastic strain index.

The damage parameter was first induced beyond the proportional limit and rapidly increased when the applied stress exceeded ~165 MPa for both S/W-0/90 and S/W-45. In contrast, both loop width and inelastic strain index were initiated at the stress of ~165 MPa and they increased monotonically with increasing applied stress. For a [0°/90°] fiber configuration, it is well-known that the damage accumulation first occurs by cracking in the 90°-bundles and then the 90°-bundle cracks extend to form transverse cracks, i.e., matrix cracking (in some papers this is referred as tunnel cracking), in the 0°-bundles [13, 14]. The stress to initiate matrix cracking, σ_{mc}, can be estimated as a stress at $L = 0$, i.e., $\sigma_{mc} = 165$ MPa. In Fig. 4, apparently the matrix cracks were not saturated. Thus an actual mean matrix cracking space length cannot be defined in this study. However, assuming a mean matrix crack spacing as a minimum matrix cracking space of ~25 μm observed in Fig. 3, an interfacial friction stress of ~100 MPa can be estimated for S/W-0/90. This seems reasonable within the same order obtained by the single fiber push-out test for unidirectional (UD) Hi-Nicalon™ Type-S/CVI-SiC composites [15]. Consequently, the high interfacial strength definitely contributes the high matrix cracking stress of S/W-0/90.

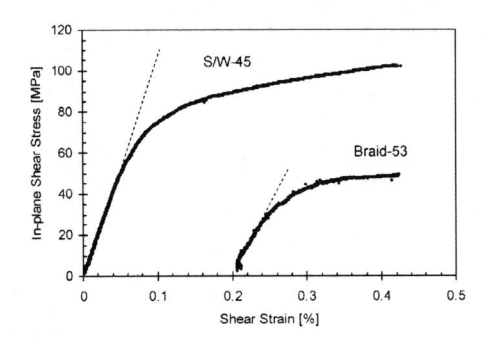

Figure 6. In-plane shear stress vs. shear strain.

Presently, the similar hysteresis analysis is not always guaranteed for the off-axis tensile data due to the coexistence of multiple fracture modes: tension, in-plane shear and detachment. If such an analysis is allowed, the results suggest the initiation of damage accumulation >150 MPa. Further investigation is required to discuss the detailed damage accumulation process for off-axis tensile specimens.

Off-axis tensile tests provide in-plane shear fracture behavior by simple stress conversion [16]. Figure 6 shows in-plane shear stress vs. shear strain curves for S/W-45 and Braid-53 (no measurements for Braid-35 and Braid-55). The non-linear segment followed by the initial linearity indicates failure with mixed modes: in-plane shear and bridging by fiber pullouts of transverse fibers. In Fig. 5, the lower proportional limit shear stress (~30 MPa) and shear failure stress (~55 MPa) for Braid-53 were obvious, while the proportional limit shear stress and the shear strength for S/W-45 were ~50 MPa and ~100 MPa, respectively. The in-plane shear properties were slightly improved for braid composites and considerably improved for S/W composites, compared with a proportional limit shear stress of ~15 MPa and a shear failure strength of ~50 MPa for early generation poor stiffness SiC matrix composites fabricated by the polymer impregnation and pyrolysis (PIP) process [17].

DISCUSSION
Test Validity

The Young's modulus data for Braid-55 is questionable due to large scatter. The possible explanation for the uncertainty is the limited surface area covered by the strain gauge. Strain gauge measurement depends on the surface condition of the composites. Due to the presence of pocket area between the weaving cross-sections (Fig. 1), in which SiC matrix is primarily filled for dense NG1 or inter-bundle pores for NG2, this effect is indispensable for Braid composites. Even though using an average strain data of two measurements, experimental scatter can be inevitable. By contrast, the tightly woven S/W composites gave consistent Young's modulus data. The uncertainty is also due to the statistically small number of tests. From this aspect, the sonic resonance data seems comparably reliable. As aforementioned, the fracture strain is used for implication because of uncertainty of the strain measurement beyond PLS.

With the fracture within the gauge section for all tensile specimens without twisting or bending, the stress data such as PLS and tensile strength are valid for any size of specimens.

Specimen Size Effect

The effect of specimen size and geometry on axial and off-axial tensile properties for PIP SiC/SiC composites was reported by the authors [17]. The key conclusions of this study are 1) very minor length, width and thickness effects on axial tensile properties if the fiber volume fraction in the loading direction in a unit structure is unchanged by specimen size, 2) the size dependency of off-axis tensile properties due to the size-relevant change of fracture modes, and 3) very minor effect of specimen geometry. For instance, very narrow specimens show much lower in-plane shear properties due to the size effect. A probable explanation for no major systematic size effect observed for CVI SiC/SiC composites of this study is improved in-plane shear properties. The in-plane shear strength and detachment strength of the low-stiffness PIP-SiC matrix composites discussed in the size effect study [17] are inherently low. The high load transferability at the fiber/matrix interface for the CVI-SiC/SiC composites can achieve superior in-plane shear properties. Besides, comparably dense matrix can reduce open pores as a crack initiation site. The similar trend was reported for the dense NITE (nano-infiltration transient eutectic phase sintered) composites [18].

In Ref. 17, the size effect of in-plane shear strength is also reported. The in-plane shear strength obtained by the off-axis tension depends significantly on specimen width, while the in-plane shear strength by Iosipescu shear test was unchanged. Very minor size effect of the off-axial tensile properties may support constant in-plane shear strength regardless of specimen size.

It is speculated that minor size effect on Young's modulus is also anticipated but presently no conclusion can be drawn due to uncertainty of the data.

Anisotropy of Tensile Properties

Continuous fiber reinforcement of the composites provides higher reliability in fracture behaviors as compared with those of brittle ceramics, however, simultaneously imposes anisotropy on material properties. In Figure 7, the anisotropy of PLS and tensile strength was summarized. For comparison, tensile data of "Generation III" SiC/SiC composites in literature were also included. Both PLS and tensile strength decrease with increasing the off-axis tensile loading angle.

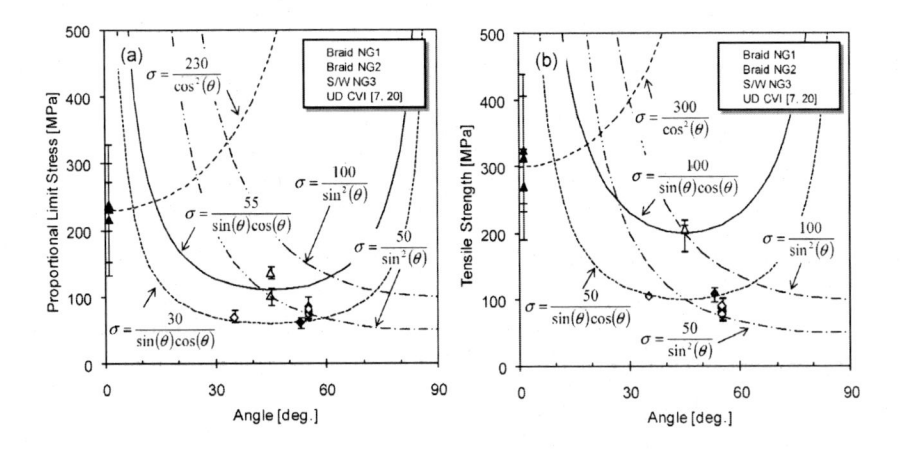

Figure 7. Anisotropy of (a) PLS and (b) tensile strength. Estimates of PLS and tensile strength by stress criterion models for the unique and multiple failure mode cases are also plotted. Materials discussed are Hi-Nicalon Type-S/CVI-SiC composites with V_f = 0.3~0.4.

Key crack initiation mechanisms in axial and off-axial tension are supposed to be 1) transverse cracking perpendicular to the fiber longitudinal direction, i.e., matrix cracking, 2) in-plane shear cracking within the fiber bundles along the longitudinal fibers, and 3) interlaminar shear cracking along the longitudinal fibers by fiber detachment. Theoretically the applied stress can be separated into three stress elements: tensile stress in the longitudinal fiber direction, σ_{11} (=$\sigma_x \cos^2\theta$), tensile stress in the transverse fiber direction, σ_{22} (=$\sigma_x \sin^2\theta$), and in-plane shear stress on the plane parallel to the longitudinal fiber direction, τ_{12} (=$\sigma_x \sin\theta \cos\theta$). Then the strictest criterion dominates the ultimate failure of the composites. In general, the multiple failure modes should operate to cause the composites' failure but, for simplicity, this study assumes a unique failure mode. Maki, et al. [20] also proposed the same criterions for the first cracking event during off-axis tension for UD composites. A critical stress to initiate matrix cracking, i.e., transverse cracking in 0°-bundles, can be defined as σ_{11} = 230 MPa from the tensile result of UD composites. Also a critical in-plane shear stress to induce first matrix cracking is equivalent to a proportional limit shear stress obtained from Fig. 6 (τ_{12} = 30 MPa for NG1 and NG2 braid composites vs. τ_{12} = 55 MPa for NG3 S/W composites). Only a critical detachment stress is unknown. In Fig. 7(a), the detachment stresses of 50 and 100 MPa are simulated to consider the effect of detachment stress.

From Fig. 7(a), it is speculated that the high PLS for S/W composites was due primarily to the high proportional limit in-plane shear stress to initiate a parallel crack in the fiber bundles. Slight improvement of in-plane shear strength can significantly increase the PLS. Note that the large scatter of PLS data at θ = 0 is primarily as a consequence of scattered fiber volume fraction. The varied failure mode may also influence on the experimental error. Another explanation is the detachment dominant failure initiation mechanism but this might be unlikely. Fracture images suggest that the detachment is not a primary mechanism in the off-axis angle range of concern. Typically the detachment failure shows a V-shaped fracture plane with intact fibers. Fracture image of S/W-45 exhibits blush-like fracture surface with a V-shape fracture plane but there were many broken fibers by shear and fiber pullouts normal to the fracture plane. Although further careful investigation is mandatory, the high detachment stress would be anticipated from the high PLS (~100 MPa) for 2D composites.

Once a crack is initiated, composites' damage accumulates progressively. For [0°/90°] composites, micro-cracks first initiate in the 90°-bundles and transverse cracks in the 0°-bundles becomes dominant with increasing applied stress. Many of the 90°-bundle cracks grow to penetrate into the 0°-bundles to form transverse cracks at higher stresses [13]. Since damages in composites rapidly increase beyond the materials proportional limit, the transverse cracking for S/W-0/90 generally occurred at lower stress (~165 MPa from Fig. 4) than that of UD composites (~230 MPa). In contrast, it is speculated that the secondary damage for S/W-45 was caused by the combined effect of in-plane shear and bridging with fiber sliding at the debonded fiber/matrix interface.

Similar to the PLS case, a fracture mechanism anisotropy map for tensile strength can be defined in Figure 7(b). The critical in-plane shear fracture strengths of 50 MPa for both braid composites, and 100 MPa for NG3 S/W composites, were derived in Fig. 6. The detachment failure strength is assumed to be identical with a detachment initiation stress. This is a reasonable assumption since the trans-thickness tensile specimen generally fails with a very short non-linear damage accumulation stage [20]. Similar to the PLS case, large scatter of tensile strength for UD composites (θ = 0) was due to scatter of the axial fiber volume fraction in the cross-section. It is well-known that the composite fracture strength depends significantly on the axial fiber volume fraction [21]. According to the preliminary mapping evaluation, the co-operation of multiple failure modes is speculated for S/W-45. In contrast, the failure mode of braid composites appears only in-plane shear, although the failure behavior beyond the proportional limit was quite different in each braid composite as discussed.

In-Plane Shear Properties
As discussed previously, off-axis tensile properties depend significantly on in-plane shear

properties. The higher proportional limit shear stress (~55 MPa) and in-plane shear fracture strength (~100 MPa) for S/W-45 resulted in the superior off-axis tensile properties. Generally, the pore distribution would affect the in-plane shear data. The braid composites preferably possess pocket pores between fiber bundles however the effect of pocket pores should be minor comparing between NG1 (V_p = 0.08) and NG2 (V_p = 0.16). The probable explanation for such difference in in-plane shear data is the contribution from continuous fibers aligned perpendicular to the shear fracture plane. Denk et al. [22] specified the effect of the volume fraction of transverse fibers on in-plane shear strength by Iosipescu shear testing of carbon/carbon composites. A high fiber volume fraction in a unit structure was achieved due probably to the tightly-woven architecture for S/W-45. Therefore, it is reasonable to conclude that S/W-45 with a slightly higher fiber volume fraction of ~40% exhibits the higher in-plane shear properties. Additionally, the high interfacial friction (~100 MPa) for nuclear-grade SiC/SiC composites can allow significant load transfer via the fiber/matrix interface beyond first in-plane shear cracking.

CONCLUSIONS

This study aims to evaluate tensile properties for nuclear-grade SiC/SiC composites and to provide a basis for the design of tubular components for control rod applications. Anisotropy of tensile properties was specifically emphasized. This study is also a part of an effort to develop test standards and design codes by establishing correlations between the mechanical behavior of plate composites and that of tubular components. For these purposes, Hi-Nicalon™ Type-S fiber reinforced CVI-SiC matrix composites with multilayered interphase were evaluated.

A nuclear-grade satin-woven composite (NG3) exhibited excellent axial and off-axis tensile behavior with a graceful unloading/reloading hysteresis response, although the braided composites (NG1 and NG2) failed at unexpectedly lower stresses. The possible cause for such a difference was the magnitude of in-plane shear properties. For NG3 satin-woven composites, the higher in-plane shear properties, which have been achieved by increasing the volume fraction of transverse fibers normal to the fracture plane, enable higher off-axis tensile properties. Applying a simple stress criterion model, the anisotropy of proportional limit tensile stress and fracture strength was preliminary evaluated. It is worth noting that very minor size effect on axial and off-axis tensile properties seems anticipated for the nuclear-grade SiC/SiC composites with rigid CVI-SiC matrix.

ACKNOWLEDGEMENTS

The authors would like to thank Dr. Snead and Dr. Kondo for reviewing the manuscript. This work was sponsored by the U.S. Department of Energy, Office of Nuclear Energy Science and Technology under contract DE-AC05-00OR22725 with Oak Ridge National Laboratory, managed by UT-Battelle, LLC.

REFERENCES

[1] Y. Katoh and L.L. Snead, Mechanical Properties of Cubic Silicon Carbide after Neutron Irradiation at Elevated Temperatures, *J. ASTM Int.*, **2**, JAI12377 (2005).
[2] S. Nogami, A. Hasegawa and L.L. Snead, Indentation Fracture Toughness of Neutron Irradiated Silicon Carbide, *J. Nucl. Mater.*, **307-311**, 1163-67 (2002).
[3] Y. Katoh, private communication.
[4] L.L. Snead, Limits on Irradiation-Induced Thermal Conductivity and Electrical Resistivity in Silicon Carbide Materials, *J. Nucl. Mater.*, **329-333**, 524-29 (2004).

[5]L. L. Snead, Y. Katoh, A. Kohyama, J. L. Bailey, N. L. Vaughn and R. A. Lowden, Evaluation of Neutron Irradiated Near-Stoichiometric Silicon Carbide Fiber Composites, *J. Nucl. Mater.*, **283-287**, 551-55 (2000).

[6]J.B.J. Hegeman, J.G. van der Laan, M. van Kranenburg, M. Jong, D. d'hulst and P. ten Pierick, Mechanical and Thermal Properties of SiCf/SiC Composites Irradiated with Neutrons at High Temperatures, *Fusion Engng. Des.*, **75-79**, 789-93 (2005).

[7]Y. Katoh, T. Nozawa, L.L. Snead and T. Hinoki, Effect of Neutron Irradiation on Tensile Properties of Unidirectional Silicon Carbide Composites, *J. Nucl. Mater.*, in press (2007).

[8]Y. Katoh, L.L. Snead, C.H. Henager, Jr., A. Hasegawa, A. Kohyama, B. Riccardi and H. Hegeman, Current Status and Critical Issues for Development of SiC Composites for Fusion Applications, *J. Nucl. Mater.*, in press (2007).

[9]C. Cady, F.E. Heredia and A.G. Evans, In-Plane Mechanical Properties of Several Ceramic-Matrix Composites, *J. Am. Ceram. Soc.*, **78**, 2065-78 (1995).

[10]T. Nozawa, Y. Katoh, A. Kohyama and E. Lara-Curzio, Effect of Specimen Size and Fiber Orientation on the Tensile Properties of SiC/SiC Composites, 25th Annual International Conference on Advanced Ceramics & Composites, 2001, Cocoa Beach, FL, USA.

[11]E. Vagaggini, J.-M. Domergue and A.G. Evans, Relationships between Hysteresis Measurements and the Constituent Properties of Ceramic Matrix Composites: I, Theory, *J. Am. Ceram. Soc.*, **78**, 2709-20 (1995).

[12]J.W. Hutchinson and H.M. Jensen, Models of Fiber Debonding and Pullout in Brittle Composites with Friction, *Mech. Mater.*, **9**, 139-63 (1990).

[13]Z.C. Xia, R.R. Carr and J.W. Hutchinson, Transverse Cracking in Fiber-Reinforced Brittle Matrix, Cross-Ply Laminates, *Acta Metal. Mater.*, **41**, 2365-76 (1993).

[14]Z.C. Xia and J.W. Hutchinson, Matrix Cracking of Cross-Ply Ceramic Composites, *Acta Metal. Mater.*, **42**, 1933-45 (1994).

[15]T. Nozawa, Y. Katoh and L.L. Snead, The Effects of Neutron Irradiation on Shear Properties of Monolayered PyC and Multilayered PyC/SiC Interfaces of SiC/SiC Composites, *J. Nucl. Mater.*, in press (2007).

[16]W.R. Broughton, Shear, *Mechanical Testing of Advanced Fibre Composites*, J.M. Hodgkinson, Eds., Woodhead Publishing Limited, Cambridge, England, 2000, pp. 100–23.

[17]T. Nozawa, Y. Katoh and A. Kohyama, Evaluation of Tensile Properties of SiC/SiC Composites with Miniaturized Specimens, *Mater. Trans.*, **46**, 543-51 (2005).

[18]T. Nozawa, Y. Katoh, A. Kohyama and E. Lara-Curzio, Specimen Size Effect on the Tensile and Shear Properties of the High-Crystalline and High-Dense SiC/SiC Composites, *Ceram. Eng. Sci. Proc.*, **B24**, 415-20 (2003).

[19]T. Nozawa, unpublished work.

[20]Y. Maki, T. Hinoki and A. Kohyama, Comprehensive Evaluation of the Mechanical Properties of Advanced SiC/SiC Composites at High Temperature, 29th International Conference of Advanced Ceramics and Composites, 2005, Cocoa Beach, FL, USA.

[21]W.A. Curtin, Theory of Mechanical Properties of Ceramic-Matrix Composites, *J. Am. Ceram. Soc.*, **74**, 2837-45 (1991).

[22]L. Denk, H. Hatta, A. Misawa and S. Somiya, Shear Fracture of C/C Composites with Variable Stacking Sequence, *Carbon*, **39**, 1505-13 (2001).

Particulate Reinforced and Laminated Composites

INFLUENCE OF THE ARCHITECTURE ON THE MECHANICAL PERFORMANCES OF ALUMINA-MULLITE AND ALUMINA-MULLITE-ZIRCONIA CERAMIC LAMINATES

Alessandra Costabile and Vincenzo M. Sglavo
DIMTI, University of Trento
Via Mesiano 77
38050 Trento (Italy)

ABSTRACT

Ceramic laminates with different architecture have been produced and characterized in this work. Thickness of single layers, composition and stacking sequence have been changed in order to modify the residual stress profile generated after sintering upon cooling and corresponding apparent fracture toughness. Laminates composed of alumina/mullite and alumina/zirconia layers have been considered. The intensity of the compressive stresses within the laminates is shown to scale to the final average strength. Moreover, if the laminate architecture and corresponding apparent fracture toughness curve allow the stable growth of surface defects, a minimum failure stress in agreement to theoretical value is shown even when large surface artificial flaws are introduced. In any case, the engineered laminates fail at loads larger than fracture stress estimated from the residual stress profile; such calculated strength can be therefore considered as the reliable mechanical resistance of the material.

INTRODUCTION

The main limitation to the use of ceramic materials in structural applications resides in their scarce mechanical reliability. The brittleness is directly related to the low value of fracture toughness and to the presence of flaws generated either during the production process or in service. The consequence is a scatter of strength data too large to allow safe design, unless statistical approaches embodying acceptable minimum failure risk are used. In order to overcome such problems many efforts have been made in the past. The fracture behaviour of ceramics has been improved by using the reinforcing action of grain anisotropy or second phases,[1] by the promotion of crack shielding effects by the phase-transformation or micro-cracking[1] and by introducing low-energy paths for crack propagation in porous[2] or within weak interlayers in laminates.[3-6] As an alternative, laminated structures characterized by the presence of thin layers in residual compression alternated to thicker layers in tension[7-12] have been shown to possess a minimum failure stress (threshold stress) or fracture toughness values as high as 17 MPa m$^{0.5}$. More recently, Sglavo and co-workers[13-15] have also described the possibility to improve the mechanical behaviour of ceramic laminates by introducing a residual stress profiles originated from differences in thermal expansion coefficient of the different constituting layers. If the development of residual stresses in ceramic multilayers is opportunely controlled, materials characterized by high fracture resistance and limited strength scatter can be designed and produced.

The aim of the present work is to investigate the relationship existing between different architectures and mechanical properties of engineered ceramic laminates produced by using alumina/mullite and alumina/zirconia composites layers. Such composites were labelled as AMy and AZy, where "A", "M", "Z", and "y" stay for alumina, mullite, zirconia and the volume percent content of mullite or zirconia, respectively.

Mechanical Performances of Alumina-Mullite and Alumina-Mullite-Zirconia Laminates

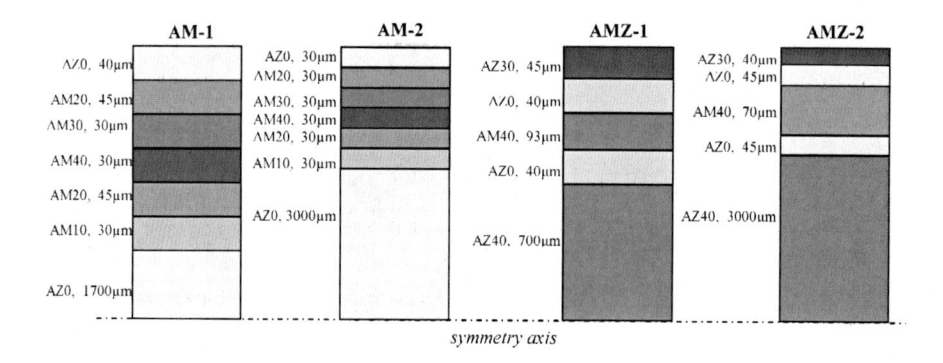

[Figure showing laminate architectures AM-1, AM-2, AMZ-1, AMZ-2 with labeled layers]

AM-1: AZ0, 40μm; AM20, 45μm; AM30, 30μm; AM40, 30μm; AM20, 45μm; AM10, 30μm; AZ0, 1700μm

AM-2: AZ0, 30μm; AM20, 30μm; AM30, 30μm; AM40, 30μm; AM20, 30μm; AM10, 30μm; AZ0, 3000μm

AMZ-1: AZ30, 45μm; AZ0, 40μm; AM40, 93μm; AZ0, 40μm; AZ40, 700μm

AMZ-2: AZ30, 40μm; AZ0, 45μm; AM40, 70μm; AZ0, 45μm; AZ40, 3000μm

symmetry axis

Figure 1. Architecture of AM-1, AM-2, AMZ-1 and AMZ-2 laminates. Layers thickness and composition are reported (dimensions are not in scale).

Four different laminate architectures were considered here. Figure 1 shows the composition, thickness of the layers and stacking order of the produced laminates labelled as AM-1, AM-2, AMZ-1 and AMZ-2. It is clear that all laminates possess a symmetrical architecture, which allows maintaining the plane geometry upon sintering, heating and cooling.[14] As a matter of fact, because of the different thermal expansion coefficient of the layers, different residual stress profiles are generated in the laminates. For example, in the AM-1 laminate, since AZ0 layers possess a thermal expansion coefficient higher than in AM10/AM20/AM40 layers,[13,14] one can easily predict that the AM10/AM20/AM40 sheets will be subjected to residual compressive stresses while the AZ0 layer will be in residual tension. More precisely, the difference between free deformation or free deformation rate of the single lamina with respect to the average value of the whole laminate accounts for the creation of residual stresses upon sintering and successive cooling. It is important here to differentiate between sintering stresses due to constrained sintering and elastic stresses related to thermal expansion coefficient mismatch. Stresses created during sintering are usually immediately removed because of the viscous-elastic behaviour of the system.[16-20] Conversely, macroscopic elastic stresses are generated upon successive cooling. With the exception of the edges, if thickness is much smaller than the other dimensions, each lamina can be considered to be in a biaxial stress state. In the common case where stresses are developed upon cooling from differences in thermal expansion coefficients only, the residual stress in layer i (Fig. 2) can be written as:[14,15]

$$\sigma_i = E_i^* (\overline{\alpha} - \alpha_i) \Delta T \qquad (1)$$

where α_i is the thermal expansion coefficient, $E_i^* = E_i/(1-v_i)$ (v_i = Poisson's ratio, E_i = Young modulus) $\Delta T = T_{SF} - T_{RT}$ (T_{SF} = stress free temperature = 1200°C,[9] T_{RT} = room temperature) and $\overline{\alpha}$ is the average thermal expansion coefficient of whole laminate equal to:

$$\overline{\alpha} = \frac{\sum_1^n E_i^* t_i \alpha_i}{\sum_1^n E_i^* t_i}. \qquad (2)$$

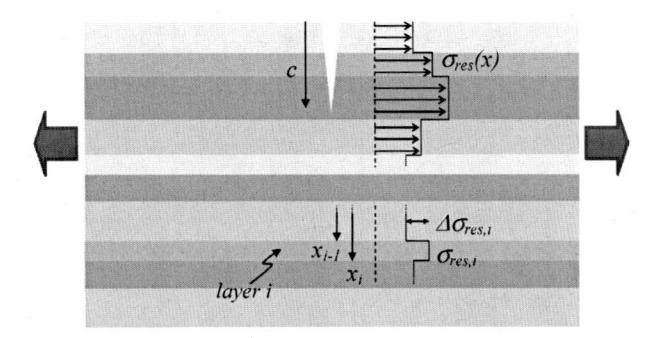

Figure 2. Model of the laminate considered for residual stresses and fracture toughness calculation.

As for the laminates of interest in this work, AM-1 and AM-2 composites are characterized by the same layers sequence and composition; different thicknesses were used to modify the residual stress profile and specifically to shift (in AM-2) the maximum compression towards the surface. The same considerations are valid if AMZ-1 and AMZ-2 laminates are compared. In these latter laminates the use of both AM and AZ layers allows the generation of higher intensity compressive residual stresses in comparison to AM-1/2 composites.[13-15]

EXPERIMENTAL PROCEDURE

Ceramic laminates were produced starting from green laminae produced by tape casting water-based slurries. Alpha-alumina (ALCOA, A-16SG, D_{50} = 0.4 µm) was considered as the fundamental starting material. High purity mullite (KCM Corp., KM101, D_{50} = 0.77 µm) and yttria (3% mol) stabilized zirconia (TOSOH, TZ-3YS, D_{50} = 0.4 µm) powders were chosen as second phases. The experimental procedure used to produce the green tapes is described in details in previous works.[13-15] Alumina/mullite (AM) and alumina/zirconia (AZ) composite layers were prepared. For the production of the laminates, green disks of nominal diameter equal to 20 mm were cut by a hollow punch from different green laminae, stacked together, thermo-compressed at 70°C under a pressure of 30 MPa for 15 min. All samples were sintered in air at 1600°C for 2 h.

The ceramic laminates were mechanically characterized by the piston-on-three-balls test.[16] The disks were supported by three balls (3.2 mm diameter), lying on a circle (25.4 mm diameter) 120° apart. The load was applied at the centre by a hardened steel cylinder (1.6 mm diameter). A Vickers indentation (using loads from 10 N to 200 N) was introduced in the center of the prospective face subjected to tension in order to evaluate the effect of large artificial flaws on failure stress.

RESULTS AND DISCUSSION

The biaxial bending strength (σ_f) measured on AM-1 and AM-2 samples is equal to 368 ± 44 MPa and 461 ± 56 MPa, respectively. The strength distribution for the two laminates is shown in Fig. 3. Fitting of the failure stresses allows the calculation of the Weibull modulus, which is equal to 15 and 22 for AM-1 and AM-2 laminates, respectively. Both laminates show therefore a quite high mechanical reliability, AM-2 ceramic composite being the best between the two materials.

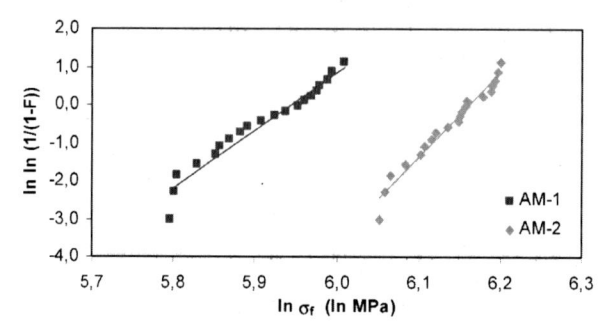

Figure 3. Weibull diagram for AM-1 and AM-2 engineered laminates. Straight lines correspond to linear fitting of experimental data.

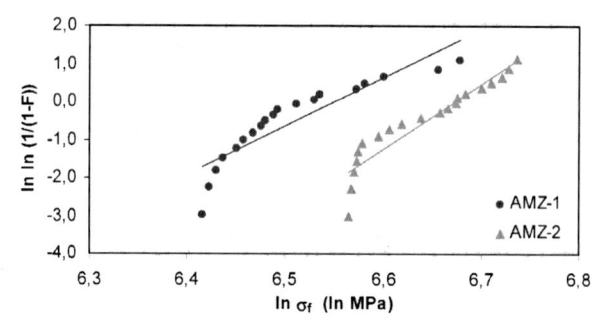

Figure 4. Weibull diagram for AMZ-1 and AMZ-2 engineered laminates. Straight lines correspond to linear fitting of experimental data.

The average bending strength measured on the AMZ-1 and AMZ-2 composites is equal to 673 ± 53 MPa and 768 ± 47 MPa, respectively. For the engineered AMZ-2 laminate Weibull modulus equal to 17 is calculated and this clearly points out the higher reliability of this engineered composite material with respect to the AMZ-1 laminate, characterized by the Weibull modulus equal to 13. It is interesting to observe that in spite of the limited number of samples considered for the mechanical strength characterization, the distributions for AMZ-1/2 laminates shown in Fig. 4 bend at low stress values, pointing out a minimum resistance values (*alias* a threshold stress[9]), equal to about 600 MPa and 700 MPa for AMZ-1 and AMZ-2, respectively. Similarly, from Fig. 3, though it is less evident, a minimum fracture resistance equal to ≈ 320 MPa and ≈ 410 MPa can be pointed out also for AM-1 and AM-2 laminates, respectively.

It is useful to analyze the obtained results on the basis of the laminates architectures and, therefore, of the residual stress profiles frozen in the composites. The residual stress profiles for the different architectures considered in this work calculated by Eq. (1) are shown in Figs 5 and 6. As expected the maximum compressive stress is deeper in AM-1 laminate than in AM-2. Similarly, the maximum residual compression (whose intensity is higher than in AM-1/2 composites) is closer to the surface in AMZ-2 laminate than in AMZ-1.

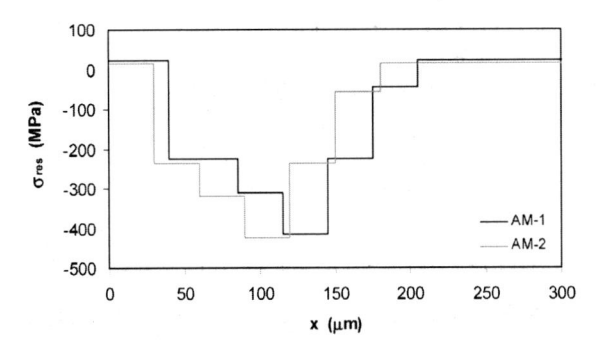

Figure 5. Calculated residual stress profiles of AM-1 and AM-2 engineered laminates (x = depth from the surface).

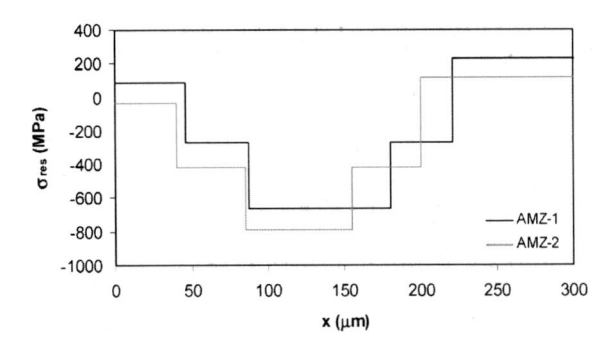

Figure 6. Calculated residual stress profile of AMZ-1 and AMZ-2 engineered laminates.

The residual stress can be considered as responsible for an increase of the apparent fracture toughness of the laminate.[13-15] If the simple model shown in Fig. 2 is considered, corresponding to a surface crack in a laminate ceramic, the apparent fracture toughness in layer i ($x_{i-1}<x<x_i$) can be defined as:[13-15]

$$T_i = K_{C,i} - \sum_{j=1}^{i}\left[2\psi\left(\frac{c}{\pi}\right)^{0.5} \Delta\sigma_{res,j} \left[\frac{\pi}{2} - \arcsin\left(\frac{x_{j-1}}{c}\right)\right]\right]. \qquad (3)$$

where $\psi \approx 1.12$, $K_{C,i}$ is the fracture toughness of layer i and c and $\Delta\sigma_{res,j}$ are described in Fig. 2. Here the approximation is made that the elastic modulus of the different layers is constant.[13-15] It is clear that the presence of an increasing apparent fracture toughness curve can improve the mechanical resistance and promote the stable growth of surface defects.[1,2]

Changes in the laminates architecture are reflected therefore also in the apparent fracture toughness curves as shown in Fig. 7. All laminates present an increasing apparent fracture toughness curve at least in the first three (AMZ-1/2) or four (AM-1/2) layers. A simple graphic constructions shown in Fig. 7 (where external applied stress intensity factors are represented as straight lines through the origin) allows to point out also that surface defects in a wide range of sizes can undergo stable growth before final failure. Only very small surface defects lead directly to catastrophic failure. The comparison of the stable growth interval with typical surface flaw size allows to point out that most of the defects in AMZ-1/2 engineered laminates can undergo stable growth before final failure; this effect can be considered as responsible for the presence of the minimum failure stress in the Weibull distribution (Fig. 4). The bimodal strength distribution can be therefore associated to surface defects undergoing stable growth in the low stress region and to natural flaws leading directly to catastrophic failure in the high stress region. It is exceptionally interesting to observe that the maximum strength that can be calculated from the graphic construction reported in Fig. 7 is equal to $\approx$ 600 MPa and $\approx$ 700 MPa for AMZ-1 and AMZ-2, respectively. Such vales well agree with the minimum failure stress values evidenced from Fig. 4.

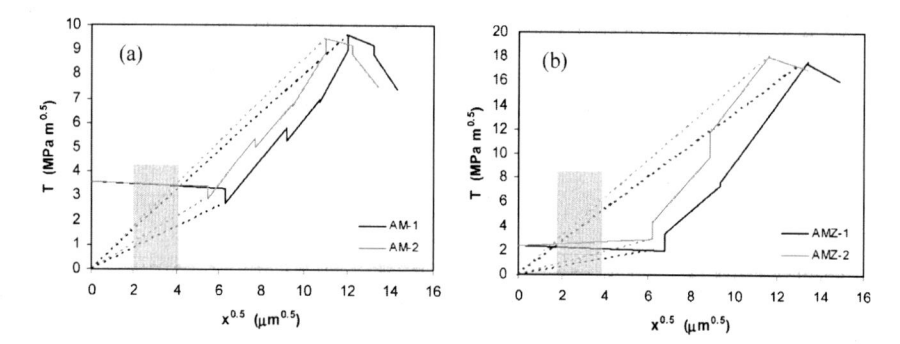

Figure 7. Apparent fracture toughness for AM-1 / AM-2 (a) and AMZ-1 / AMZ-2 (b) engineered laminates. The straight dashed lines are use to evaluate the stable growth interval and maximum theoretical strength. The grey bars represent the typical surface crack depth.

For AM-1 and AM-2 laminates, Fig. 7(a) shows that most typical surface defects fall outside of the stable growth interval. This explains the different trend of the strength distributions (Fig. 4). In addition, minimum failure stresses are therefore slightly larger that the theoretical strength, equal to $\approx$ 300 MPa and $\approx$ 400 MPa for AM-1 and AM-2 laminates, respectively.

The high mechanical reliability of the engineered laminates is confirmed also when large artificial surface defects are introduced. Figure 8 shows the strength of indented samples as a function of indentation load. It is clear that the failure stress of engineered laminates remains

Mechanical Performances of Alumina-Mullite and Alumina-Mullite-Zirconia Laminates

independent from initial flaw size, while the strength of monolithic laminates (made of pure alumina or AZ40 composite) decreases with the indentation load as typically observed in ceramic materials.[1] As expected from the chosen architecture, AMZ-2 laminate possesses the highest strength, slightly larger than 700 MPa, in good agreement with theoretical failure stress value and with data shown in Fig. 4. The same behaviour is also shown for AMZ-1 laminate. AM-2 and AM-1 composites show indentation strength values larger than theoretical ones though they are in good agreement with failure stresses reported in Fig. 4. This means that even large artificial flaws are not deep enough to undergo stable growth. In any case, one can declare that even when large defects are produced, the engineered laminates fail at loads larger than the designed fracture stress: this allows to state that it is safe to use such value as the trusted mechanical resistance of the material.

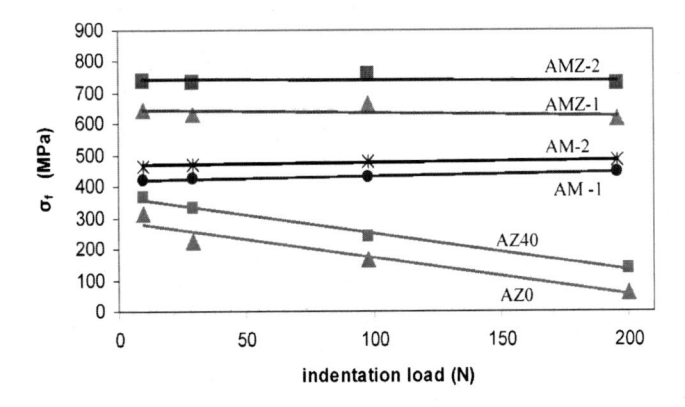

Figure 8. Failure stress as a function of the indentation load for engineered AM-1, AM-2, AMZ-1, AMZ-2 laminate and monolithic sample (AZ0 and AZ40).

CONCLUSIONS

The present work has shown that the laminate architecture deeply influences the mechanical performances of the material. Composition, thickness and stacking order of the layers are responsible for different residual stress profile generated upon cooling after sintering. The intensity of the maximum compression scales to the average failure stress. If the laminate architecture and corresponding apparent fracture toughness curve allow the stable growth of surface defects, a minimum failure stress can be measured even when large surface artificial flaws are introduced. Such threshold stress is also in agreement to the design value. In any case, the engineered laminates fail at loads larger than the designed fracture stress that can be considered as the reliable mechanical resistance of the material.

REFERENCES

[1] B. R. Lawn, "Fracture of brittle solids", Second Edition, *Cambridge University Press*, Cambridge, UK (1993).

[2] J. B. Davis, A. Kristoffersson, E. Carlstrom & W. J. Clegg, "Fabrication and crack deflection in ceramic laminates with porous interlayers", *J. Am. Ceram. Soc.*, **83** [10] 2369-74 (2000).

[3] W. J. Clegg, K. Kendall & McN. Alford, "A simple way to make tough ceramics", *Nature (London)*, **347**, 455-57 (1990).
[4] W. M. Kriven & D.-H- Kuo., "High-strength, flaw-tolerant, oxide ceramic composite", U.S. Pat. No. 5,948,516, September 7, 1999.
[5] R. E. Mistler, "Strengthening alumina substrates by incorporating grain growth inhibitor in surface and promoter in interior", U.S. Pat. No. 3,652,378, March 28, 1972.
[6] M. P. Harmer, H. M. Chan & G. A. Miller, "Unique opportunities for microstructural engineering with duplex and laminar ceramic composites", *J. Am. Ceram. Soc.*, **75** [7] 1715-28 (1992).
[7] C. J. Russo, M. P. Harmer, H. M. Chan & G. A. Miller, "Design of laminated ceramic composite for improved strength and toughness", *J. Am. Ceram. Soc.*, **75** [12] 3396-400 (1992).
[8] R. Latkshminarayanan, D. K. Shetty & R. A. Cutler, "Toughening of layered ceramic composites with residual surface compression", *J. Am. Ceram. Soc.*, **79** [1] 79-87 (1996).
[9] M. P. Rao, A. J. Sánchez-Herencia, G. E. Beltz, R. M. McMeeking, & F. F. Lange, "Laminar Ceramics That Exhibit a Threshold Strength", *Science*, **286**, 102-5 (1999).
[10] N. Orlovskaya, M. Lugovy, V. Subbotin, O. Radchenko, J. Adams, M. Chheda, J. Shih, J. Sankar, S. Yarmolenko, "Robust design and manufacturing of ceramic laminates with controlled thermal residual stresses for enhanced toughness", *J. Mater. Sci.*, **40**, 5483-5490 (2005).
[11] N. Orlovskaya, J. Kuebler, V. Subbotin, M. Lugovy, "Design of Si3N4-based ceramic laminates by the residual stresses", *J. Mater. Sci.*, **40**, 5443-5450 (2005).
[12] M. Lugovy, V. Slyunyayev, N. Orlovskaya, G. Blugan, J. Kuebler, M. Lewis, "Apparent fracture toughness of Si3N4-based laminates with residual compressive or tensile stresses in surface layers", *Acta Materialia*, **53**, 289-96 (2005).
[13] V. M. Sglavo, M. Paternoster and M.Bertoldi, "Tailored Residual Stresses in High Reliability Alumina-Mullite Ceramic Laminates", *J. Am. Ceram. Soc.*, **88** [10] 2826–2832, 2005.
[14] V. M. Sglavo and M. Bertoldi, "Design and Production of Ceramic Laminates with High Mechanical Reliability", *Composites: Part B*, **37**, 481-9, 2006.
[15] V. M. Sglavo and M. Bertoldi, "Design and Production of Ceramic Laminates with High Mechanical Resistance and Reliability", *Acta Mat.*, **54**, 4929-37, 2006.
[16] P. Z. Cai, D. J. Green, and G. L. Messing, "Constrained densification of alumina/zirconia hybrid laminates, I: experimental observations of processing defects", *J. Am. Ceram. Soc.*, **80** [8] 1929-39 (1997).
[17] P. Z. Cai, D. J. Green, and G. L. Messing, "Constrained densification of alumina/zirconia hybrid laminates, II: viscoelastic stress computation", *J. Am. Ceram. Soc.*, **80** [8] 1940-48 (1997).
[18] B. Kellett and F. F. Lange, "Stress induced by differential sintering in powder compacts", *J. Am. Ceram. Soc.*, **67** [5] 369-72 (1984).
[19]. K. Bordia and R. Raj, "Sintering behaviour of ceramic films constrained by a rigid substrate", *J. Am. Ceram. Soc.*, **68** [6] 287-92 (1985).
[20] V. . Sglavo, P. Z. Cai and D. J. Green, "Damage in Al_2O_3 Sintering Compacts under Very Low Tensile Stress", *J. Mater. Sci. Lett.*, **18** (1999) 895-900.
[16] D. K. Shetty, A.R. Rosenfield, P. McGuire, G.K. Bansal and W.H. Duckworth, "Biaxial Flexure Tests for Ceramics", *Ceram. Bull.*, **59**, Vol.12 (1980).

FABRICATION OF NOVEL ALUMINA COMPOSITES REINFORCED BY SiC NANO-PARTICLES AND MULTI-WALLED CARBON NANOTUBES

Kaleem Ahmad and Wei Pan
State Key Laboratory of New Ceramics and Fine Processing,
Department of Materials Science and Engineering,
Tsinghua University
Beijing, 100084,
P. R. China

ABSTRACT
 A novel microstructure design of three phase alumina composites reinforced by nanosize SiC particles and multiwalled carbon nanotubes (MWNTs) is proposed. Alumina composites reinforced by different MWNTs contents ranging from 5, 7 and 10 vol% along with concurrent reinforcement of low volume fractions of SiC nanoparticles ranging from 1, 2 and 3 vol% were fabricated by spark plasma sintering technique based on proposed microstructure design. The nanocomposites have shown significant improvements in fracture toughness and bending strength, while hardness remains almost stable. The enhancement in fracture toughness was observed by change in fracture mode of alumina from intergranular to transgranular due to presence of SiC nanosize particles at the intragranular positions and further improvement has also been observed due to presence of MWNTs at intergranular positions making intertwining network structure around alumina grains in the matrix. The overall improvements in mechanical properties were ascribed to the complementary effect of MWNTs and SiC in strengthening and toughening of alumina matrix. The three hallmarks of toughening mechanisms in ceramic fiber composites i.e. crack deflection at the MWNTs/matrix interface, crack bridging by MWNTs at the grain boundaries and MWNTs pullout at the fracture surface were observed at the nano scale in the composites.

INTRODUCTION
 Carbon nanotubes (CNTs) have excellent mechanical properties with elastic moduli of the order of ≈1 TPa for single walled carbon nanotubes and ≈950 GPa for multiwalled CNTs[1]. Their low density, high aspect ratio, and superior mechanical properties suggest that CNTs might be suitable as a novel fiber material for ceramic nanocomposites. Therefore, incorporating CNTs to improve mechanical properties of alumina especially fracture toughness has become an interesting area of research now a days.[2-5] Alumina is a low cost ceramic material having superior mechanical, thermal, chemical, and outstanding electrical properties. These properties make it attractive for use in a variety of applications ranging from armor to wear resistance products, electronic substrate to spark plug, insulations to magneto hydrodynamic power generators and coarse grinding grit to cutting tools.[6-8] The armor systems based on alumina ceramics can defeat high velocity projectiles with steel, lead and even tungsten carbide cores.[8] However, the present alumina ceramics have low performance limited by their toughness.[9-11] Many attempts have been made to enhance mechanical properties of alumina especially fracture toughness using CNTs but the results are disappointing and contradictory.[4, 5, 12] Moreover, in most of the cases if some improvement in fracture toughness of alumina was observed using CNTs that is at the expense of other properties such as bending strength or hardness.[3, 5] So significant improvements in mechanical properties specially fracture toughness without deteriorating other intrinsic properties

are yet to be demonstrated.[4, 5, 13] Not long ago, Zhan et al.[5] claimed three times improvements in fracture toughness over monolithic alumina with 10 volume (vol) % of SWNTs. However, the hardness of their composites decreases with increasing CNTs contents[5] and also with decreasing density[5]. On the contrary Wang et al.[4], claimed that CNTs do not improve fracture toughness of alumina and results of Zhan[5] were over estimated by measuring fracture toughness using indentation method.[13, 14] Furthermore, Wang and colleagues showed that no classical radial cracks were observed on polished surfaces of alumina/CNTs composite because CNTs allow shear deformation under indenter and crack length can no longer be used to measure the toughness, however he reported some increase in contact damage resistance.[4, 13] Recently Fan et al.[3] reported an increase of 80% in fracture toughness while no improvement in bending strength[3] were reported instead of some decrease. Most of the studies conducted recently were carried out using SWNTs[4, 5, 14] due to their perfect structure[5] as compared to MWNTs, on the contrary fractured end MWNTs (imperfect MWNTs) are found to be more effective to improve strength and toughness of the composites due to their enhanced load carrying capability.[15, 16] In fractured end MWNTs the pullout forces are controlled by mechanical interlocking of the end defect with outer walls or surrounding matrix that may provide more effective load transfer from outer to inner walls resulting enhanced strength and toughness.

On the other hand a lot of work[17-21] has also been done to improve the mechanical properties of alumina by addition of nano size SiC particles after new design concepts of ceramic nanocomposites proposed by Niiha[22] and consequently significant improvements in mechanical properties were reported by him. However, a lot of other subsequent studies showed similar but lower increase in bending strength accompanied by modest, if any increase in fracture toughness.[23-25] The purpose of the present study is to develop novel alumina composites reinforced by both SiC nano particles and MWNTs with improved mechanical properties based on a new three phase microstructure design. In this design alumina matrix is reinforced by nano size SiC particles and carbon nanotubes concurrently. The microstructure of the nanocomposite is constructed by dispersing second phase nanosize SiC particles within the matrix grains and on the grain boundaries present as intragranular and intergranular reinforcements while CNTs are present at grain boundaries as intergranular reinforcements as shown schematically in Figure 1. SiC has a lower thermal expansion coefficient than alumina, which means an embedded SiC particle in an alumina matrix is subjected to radial compression after sintering while immediate surrounding matrix is in radial compression but hoop tension as shown in Figure 1. The elastic energy associated with each SiC particle is insufficient to initiate and propagate microcracks, but a crack approaching tends to be deflected towards the particle and to be pinned by it, while CNTs at the grain boundaries act as crack stoppers (Figure 1.). The intragranular SiC particles change the fracture mode and strengthen the grain boundary consequently impede intergranular fracture mode of alumina. In addition, the ropes like intertwining network of MWNTs at the grain boundaries will bridge the crack propagation at the intergranular positions, consequently impede the fracture of grain and in this way improve the toughness of the composites. This novel design provides a redundant method to improve the strength and toughness of alumina matrix in addition to improvements in hardness. Low volume fractions of SiC were used to change the fracture mode of alumina from intergranular to transgranular,[18, 21] as one of the main toughening mechanism in alumina/SiC nanocomposites is change in fracture mode from intergranular to transgranular which implies a reinforcement of grain boundaries.[26]

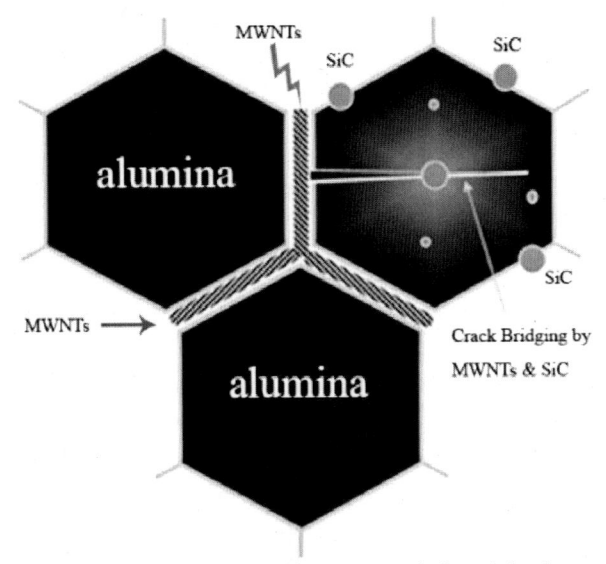

Figure 1. Schematic diagram of three phase microstructure design of alumina matrix reinforced by CNTs and SiC nano-particles

The next advantage of adding SiC is that classical radial cracks can be observed by indentation in contrast to Wang *et al.*[4] studies and these radial cracks help to find the direct evidence of toughening mechanisms in alumina matrix on nanoscale. The third benefit is that hardness remain almost stable which normally decreases with increases of CNTs contents in ceramics.[5,22] The SiC were used in 1, 2 and 3 vol % while, the MWNTs were used in 5, 7 and 10 vol%. In this way nine combinations of composites were prepared to optimize the best vol% fraction of SiC and CNTs. The starting powders in appropriate proportions were mixed using the process reported by Zhan *et al*. In brief, the starting material alumina (Chong Qing Tuo Yuan, China, purity 99.9%) was ultrasonically mixed with different volume percentages of MWNTs (Green Chemical Reaction Engineering and Technology, China) along with each (1, 2, 3) vol% of SiC (Shijia Zhuang High-tech Ceramics Material, Hebei Province P.R. China, purity 99.9%). After ultrasonic mixing the slurry was passed through 200mesh and then for further mixing, powder mixers were ball milled for 24h. After drying, powder mixtures were spark plasma sintered (SPS, Dr. Sinter 1050, Sumitomo Coal Mining Co., Japan) at 1550°C - 1600°C in vacuum under a uniaxial pressure of 50 MPa in a 20 mm inner diameter cylindrical graphite mold. It should be noted that CNTs have been proven stable under spark plasma sintering at 1700°C -2000°C.[27,28] So any damage or any change in structure of CNTs at 1550°C -1600 °C temperature is ruled out. For MWNTs, a value of 2.1 g/cm^3 was used for calculating the relative densities of the composites. The relative densities of sintered samples were measured by the Archimedes's method. The fracture toughness was measured by direct toughness measurement i.e. single edge notch beam (SENB) method containing pre-crack of length ≤1.5mm. SENB is far more reliable technique than indentation method[16] to measure the fracture toughness. The bending strength was measured by three point bending test. Hardness was measured by Vickers indentation method with a load of 49N for 15 sec and cracks on polished surfaces were produced to investigate the toughening mechanisms. X-ray diffraction analysis was carried out which showed no new phases were formed other than carbon, alumina and SiC.

RESULTS AND DISCUSSION
Densification
   The mechanical properties of 5, 7 and 10 vol% of MWNTs with different SiC contents (1, 2, 3 vol%) of alumina composites as a function of relative density and SiC vol% are shown in Figs 2, 3, and 4 respectively. The highest values of relative densities for 5 vol% MWNTs with 1 vol% SiC (Fig. 2) and for 7 vol% of MWNTs with 1vol% SiC (Fig. 3) of alumina composite were reached at 98% and 97% respectively while most of other composites have shown relative densities in the range of 95 to 96 % of theoretical one. In general for all the composites as the volume contents of SiC increases from 1 to 3 vol % or volume fraction of MWNTs increases from 5 to 7 and then 10 the relatively density decreases (Figures 2, 3, 4). This is due to the strong inhibiting effect of the nano-size SiC particles on the densification.[29] In addition increase of MWNTs contents from 5 to 7 and 10; also have some negative effect on densification. This behavior was also observed by Ning *et al.*[30] and explained that CNTs perhaps act as one kind of solid impurity to prevent the flowing of the matrix during sintering, so consequently inhibiting the densification.[30] In spite of special care to disperse MWNTs homogenously, some clustering might have negative effect on the densification of composites as the vol% of CNTs increases to 10.

Mechanical Properties
   The mechanical properties of 5 vol% MWNTs with 1, 2 and 3 vol % of SiC reinforced alumina composites are shown in Fig. 2. When we increase the SiC contents from 1 to 3 vol%, the density decreases from 98 to 96%, the fracture toughness decreases from a value ≈6.50 to ≈5.80 MPam$^{1/2}$, bending strength from ≈498.72 to ≈455.10 MPa and hardness from ≈16.50 to ≈15.79 GPa as shown in Fig. 2. These all properties show strong dependencies on density. The higher values of density shows good dispersion of MWNTs and as a result the mechanical properties are higher and this behavior is in accordance with the previous studies.[5,30] The highest improvement in bending strength reported so far for alumina composites is 10% with 1 wt% of MWNTs[12], while in our studies we are getting ≈45% improvement with 5vol% MWNTs and 1vol% SiC (Fig 2) alumina composites compared with monolithic alumina. Zhan *et al.* showed a drastic decrease in hardness with density, He reported 9.30 GPa for 10 vol% SWNTs/alumina composites at 95% relative density. In our studies at the same densification level for 10 vol% MWNTs, we have ≈14.30 GPa with just 1vol% addition of SiC (Fig 4.). The decrease in hardness is very low this is due to addition of SiC which improved hardness.
   For 7 vol% MWNTs with 1, 2 and 3 vol% of SiC, all the three samples show significant improvements in fracture toughness. The fracture toughness is in the range of 6 to 7 MPa m$^{1/2}$, bending strength is in the range of 450 to 500 MPa and hardness is in the range of 14 to 16 GPa as shown in Fig. 3. The highest value of fracture toughness obtained is ≈6.90 for one vol % SiC and 7 vol% MWNTs and there is an improvement of around 115% in fracture toughness over monolithic alumina. The decrease in hardness with decrease of density is also not very strong as shown by other researchers this might be due to presence of SiC nano particles.
   In case of 10 vol% MWNTs the improvements in bending strength and fracture toughness are lower and not smooth while hardness is almost stable as shown in Fig. 4. It seems

that as the vol % of MWNTs increased to 10, the level of clustering increased. The increase of clustering with the increase of CNTs vol % were also reported by other researcher.[2, 5, 30]

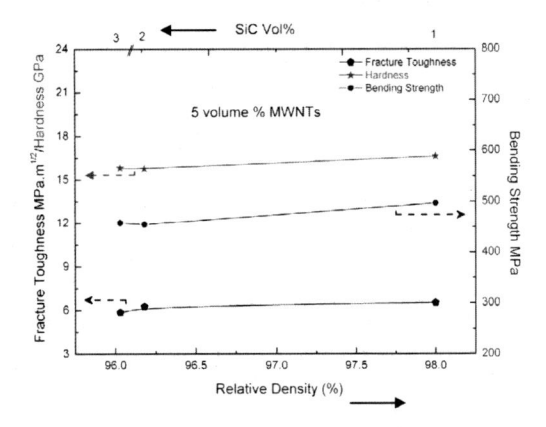

Figure 2. Mechanical Properties of alumina composites reinforced by MWNTs (5 vol%) and SiC nanoparticles (1, 2 and 3 vol %).

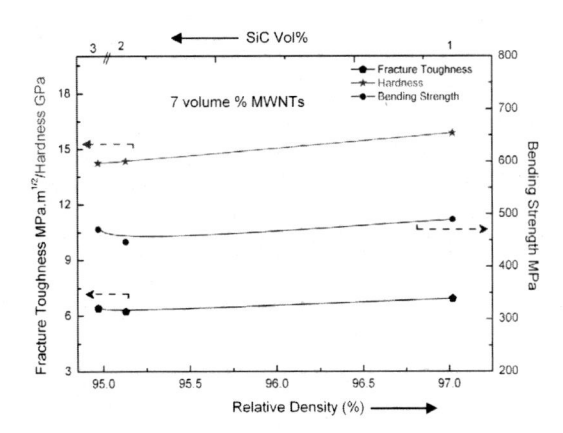

Figure 3. Mechanical Properties of alumina composites reinforced by MWNTs (7 vol%) and SiC nanoparticles (1, 2 and 3 vol %).

The agglomeration of MWNTs, caused low densification and improvements in the values of fracture toughness and bending strength are low.

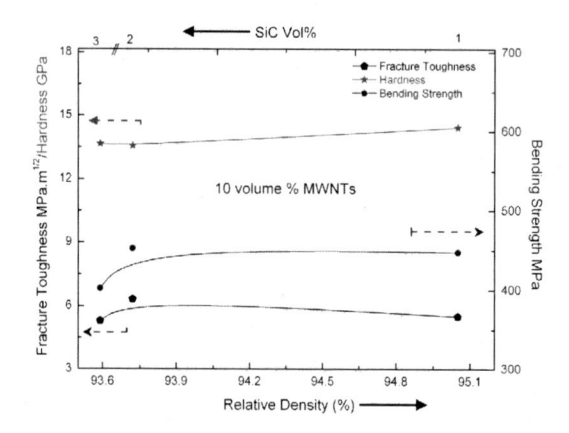

Figure 4. Mechanical Properties of alumina composites reinforced by MWNTs (10 vol%) and SiC nanoparticles (1, 2 and 3 vol %).

As we know that probability of agglomeration increases with MWNTs volume fraction and MWNTs hinder densification with the increase of their concentration. In case of agglomeration of MWNTs, the bonding is loose and agglomerates acts similar to pores having size of agglomerate, consequently these clusters causes degradation in mechanical properties. In our study the mechanical properties of 10 vol% MWNTs with 1, 2 and 3 vol% of SiC is somewhat lower. That may be due to presence of clustering which decrease the mechanical properties specially bending strength. Fan et al.[3] used 12 vol% of MWNTs/alumina composites and observed no increase in bending strength instead of some decrease due to agglomeration of CNTs.[3]

From the nine combinations of Al_2O_3-MWNTs-SiC composites the best volume fractions were found to be 5 and 7 vol% for MWNTs in combination with one or two vol% of SiC. These combinations give higher densification homogenous dispersion and consequently compare able better mechanical properties and almost no drastic decrease in hardness.

Microstructure

High resolution scanning electron micrographs of fractured surfaces show that MWNTs are uniformly distributed as shown in Figures 5(a) and 5(b). The fracture mode is mixture of intergranular and transgranular Figure 5(a). The mean grain size of the matrix is around 1μm. The ropes of MWNTs are strongly entangled with the alumina matrix and appear to envelop the alumina grains like a mesh Figure 5(b). The strong interface between alumina matrix and MWNTs is evident from the stamped imprints of MWNTs on the alumina grains Fig 5(b). The presence of network structure of MWNTs at intergranular positions of alumina matrix grains, good interfacial bonding between MWNTs and alumina, and grains size refinements, all of these phenomenon are responsible for consistent improvement in mechanical properties and are in good agreement with the others studies.[5, 31, 32] There are two factors which are complementing each other in toughening and strengthening of the alumina matrix, one is low volume fractions of SiC nano-particles that are strengthening the grain boundaries, consequently impeding the intergranular fracture mode and improving fracture toughness, harness and bending strength. The other is mesh like network structure of ropes of MWNTs around the grains of the alumina matrix which hinders both intergranular fracture mode due to strong interfacial bonding with alumina and as well as intragranular fracture mode due to network structure around the alumina grains (Figure 6.).

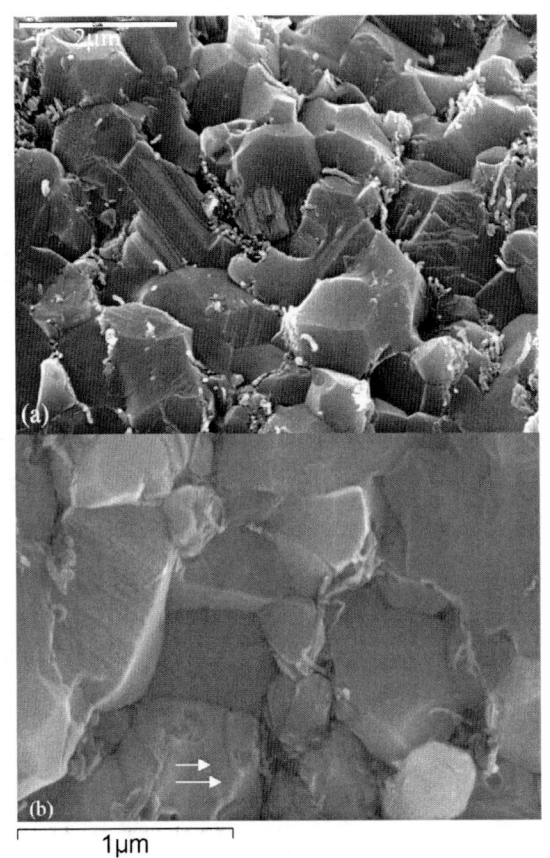

Figure 5. (a) Fracture surface of composite (5vol MWNTs-1vol SiC-94vol% alumina) showing inter/intra granular mix fracture mode (b) MWNTS uniformly distributed at the inter granular positions of alumina grains

The improvement in fracture toughness and no drastic decrease in hardness values show a positive effect of addition of low vol % of SiC and also help to observe the MWNTs toughening

Alumina Composites Reinforced by SiC Nanoparticles and Multi-Walled C Nanotubes

effects directly at the nanoscale. The one of the important feature in the fracture surface of composites is MWNTs pullout during fracture process as shown in Fig. 6. Since MWNTs were used instead of SWNTs to fabricate the composites so pull out of tubes is more obvious due to load transfer of alumina matrix to the outer shell of MWNTs. (Figure 6.)

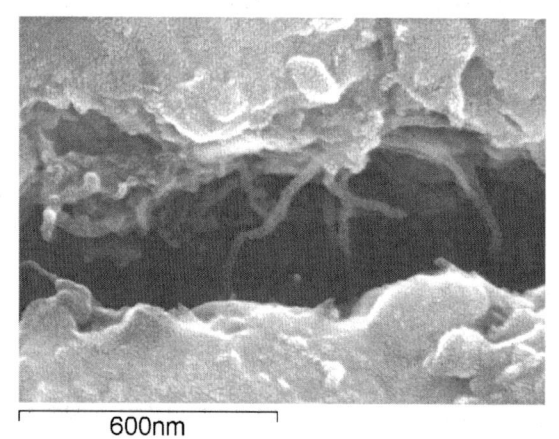

Figure 6. MWNTs pull out at perpendicular direction along the crack prorogation

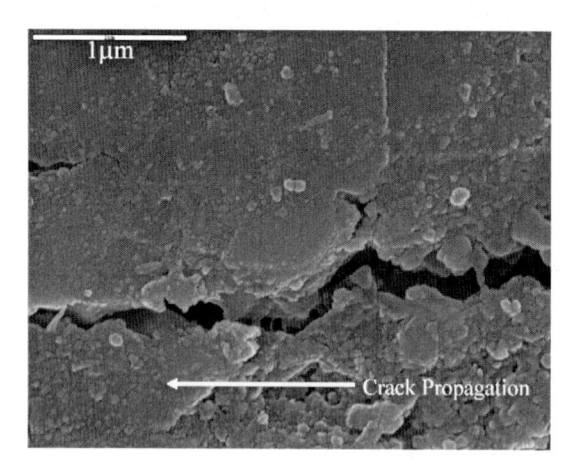

Figure 7. Crack bridging by MWNTs and crack opening size diminishing continuously along the arrow

Fig. 8. Crack deflection and crack bridging by MWNTs mesh on the surface of the composite

These pull outs show that MWNTs bear significant stress by sharing the portion of the load and in addition also toughen the matrix by crack bridging process as show in Figures 7, 8. Moreover crack deflection by MWNTs, in addition to crack bridging is also evident from Fig 8. The crack bridging of the MWNTs restrain the crack opening and reduces the driving force for crack propagation. The crack propagation energy is absorbed by all three standard approaches of toughening mechanisms observed in case of micron scale fiber reinforcement in addition to improvements by SiC nanoparticles. The circuitous path of crack propagation and crack opening size diminishing by MWNTs bridging effect show direct evidence of toughening and strengthening mechanism of alumina matrix. All the three standard toughening mechanisms in the present study were also reported by Xia[16] and colleagues on highly ordered parallel array of MWNTs on amorphous nanoporous alumina matrix[16] coating of 90µm. The direct evidence of toughness mechanisms clearly substantiate the improvements in mechanical properties based on novel three phase microstructure design of alumina composites without deteriorating other intrinsic properties.

CONCLUSION

In summary, nine alumina composites reinforced by different vol fractions of SiC (1, 2 and 3 vol %) in addition with different vol% of MWNTs (5, 7 and 10 vol%) were fabricated by spark plasma sintering based on novel three phase design of microstructure. Significant improvements in mechanical properties were observed due to change in fracture mode from intergranular to transgranular with the addition of SiC and further toughening is also observed due to presence of MWNTs at the grain boundaries. The small vol % of SiC has positive effect on the mechanical properties of alumina composites specially hardness and fracture toughness. The three hallmarks of toughening mechanism in standard ceramic matrix composites containing micron scale reinforcing fibers were also observed on nanoscale by MWNTs. Low vol% of CNTs and SiC can be used successfully to improve mechanical properties of alumina without compromising other intrinsic properties.

ACKNOWLEDGEMENTS
The author thanks to National Science Foundation of China (No. 50232020, 50572042), Chinese Scholarship Council, Higher Education Commission, PAEC Islamabad Pakistan

REFERENCES

[1] R. S. Ruoff, D. Qian, and W. K. Liu, Mechanical Properties of Carbon Nanotubes: Theoretical Predictions and Experimental Measurements, *C. R. Phys.*, **4**, 993-1008 (2003).
[2] S. I. Cha, K. T. Kim, K. H. Lee, C. B. Mo, and S. H. Hong, Strengthening and Toughening of Carbon Nanotube Reinforced Alumina Nanocomposite Fabricated by Molecular Level Mixing Process, *Scr. Mater.*, **53**, 793-97 (2005).
[3] J. P. Fan, D. Q. Zhao, M. S. Wu, Z. N. Xu, and J. Song, Preparation and Microstructure of Multi-Wall Carbon Nanotubes-Toughened Al2o3 Composite, *J. Am. Ceram. Soc.*, **89**, 750-53 (2006).
[4] X. T. Wang, N. P. Padture, and H. Tanaka, Contact-Damage-Resistant Ceramic/Single-Wall Carbon Nanotubes and Ceramic/Graphite Composites, *Nat. Mater.*, **3**, 539-44 (2004).
[5] G. D. Zhan, J. D. Kuntz, J. L. Wan, and A. K. Mukherjee, Single-Wall Carbon Nanotubes as Attractive Toughening Agents in Alumina-Based Nanocomposites, *Nat. Mater.*, **2**, 38-42 (2003).
[6] A. P. Goswami, S. Roy, and G. C. Das, Effect of Powder, Chemistry and Morphology on the Dielectric Properties of Liquid-Phase-Sintered Alumina, *Ceram. Int.*, **28**, 439-45 (2002).
[7] A. Krell, P. Blank, L. M. Berger, and V. Richter, Alumina Tools for Machining Chilled Cast Iron, Hardened Steel, *Am. Ceram. Soc. Bull.*, **78**, 65-73 (1999).
[8] E. Medvedovski, Alumina-Mullite Ceramics for Structural Applications, *Ceram. Int.*, **32**, 369-75 (2006).
[9] M. Cain, and R. Morrell, Nanostructured Ceramics: A Review of Their Potential, *Appl. Organomet. Chem.*, **15**, 321-30 (2001).
[10] N. Camuscu, Effect of Cutting Speed on the Performance of Al2o3 Based Ceramic Tools in Turning Nodular Cast Iron, *Mater. Des.*, **27**, 997-1006 (2006).
[11] C. H. Xu, C. Z. Huang, and X. Ai, Toughening and Strengthening of Advanced Ceramics with Rare Earth Additives, *Ceram. Int.*, **32**, 423-29 (2006).
[12] J. Sun, L. Gao, and X. H. Jin, Reinforcement of Alumina Matrix with Multi-Walled Carbon Nanotubes, *Ceram. Int.*, **31**, 893-96 (2005).
[13] B. W. Sheldon, and W. A. Curtin, Nanoceramic Composites: Tough to Test, *Nat. Mater.*, **3**, 505-06 (2004).
[14] J. P. Fan, D. M. Zhuang, D. Q. Zhao, G. Zhang, M. S. Wu, F. Wei, and Z. J. Fan, Toughening and Reinforcing Alumina Matrix Composite with Single-Wall Carbon Nanotubes, *Appl. Phys. Lett.*, **89**, 121910 (2006).
[15] Z. Xia, and W. A. Curtin, Pullout Forces and Friction in Multiwall Carbon Nanotubes, *Phys. Rev. B*, **69**, 233408 (2004).
[16] Z. Xia, L. Riester, W. A. Curtin, H. Li, B. W. Sheldon, J. Liang, B. Chang, and J. M. Xu, Direct Observation of Toughening Mechanisms in Carbon Nanotube Ceramic Matrix Composites, *Acta Mater.*, **52**, 931-44 (2004).
[17] B. Baron, C. S. Kumar, G. Le Gonidec, and S. Hampshire, Comparison of Different Alumina Powders for the Aqueous Processing and Pressureless Sintering of Al2o3-Sic Nanocomposites, *J. European Ceram. Soc.*, **22**, 1543-52 (2002).

[18] A. M. Cock, I. P. Shapiro, R. I. Todd, and S. G. Roberts, Effects of Yttrium on the Sintering and Microstructure of Alumina-Silicon Carbide "Nanocomposites", *J. Am. Ceram. Soc.*, **88**, 2354-61 (2005).

[19] Z. Z. Peng, S. Cai, Y. W. Wang, and H. Z. Wu, Pressureless Sintering the Al2o3/Sic Nanocomposites in Deoxidized Atmosphere, Trans Tech Publications Ltd, **280-283**, 1093-96 (2005)

[20] S. K. C. Pillai, B. Baron, M. J. Pomeroy, and S. Hampshire, Effect of Oxide Dopants on Densification, Microstructure and Mechanical Properties of Alumina-Silicon Carbide Nanocomposite Ceramics Prepared by Pressureless Sintering, *J. European Ceram. Soc.*, **24**, 3317-26 (2004).

[21] A. J. Winn, and R. I. Todd, Microstructural Requirements for Alumina-Sic Nanocomposites, *Br. Ceram. Trans.*, **98**, 219-24 (1999).

[22] J. Sun, L. Gao, M. Iwasa, T. Nakayama, and K. Niihara, Failure Investigation of Carbon Nanotube/3y-Tzp Nanocomposites, *Ceram. Int.*, **31**, 1131-34 (2005).

[23] L. Carroll, M. Sternitzke, and B. Derby, Silicon Carbide Particle Size Effects in Alumina-Based Nanocomposites, *Acta Mater.*, **44**, 4543-52 (1996).

[24] M. Sternitzke, B. Derby, and R. J. Brook, Alumina/Silicon Carbide Nanocomposites by Hybrid Polymer/Powder Processing: Microstructures and Mechanical Properties, *J. Am. Ceram. Soc.*, **81**, 41-48 (1998).

[25] H. Wu, S. G. Roberts, and B. Derby, Residual Stress and Subsurface Damage in Machined Alumina and Alumina/Silicon Carbide Nanocomposite Ceramics, *Acta Mater.*, **49**, 507-17 (2001).

[26] D. Sciti, J. Vicens, and A. Bellosi, Microstructure and Properties of Alumina-Sic Nanocomposites Prepared from Ultrafine Powders, *J. Mater. Sci.*, **37**, 3747-58 (2002).

[27] C. Qin, X. Shi, S. Q. Bai, L. D. Chen, and L. J. Wang, High Temperature Electrical and Thermal Properties of the Bulk Carbon Nanotube Prepared by Sps, *Mater. Sci. Eng. A-Struct. Mater. Prop. Microstruct. Process.*, **420**, 208-11 (2006).

[28] H. L. Zhang, J. F. Li, K. F. Yao, and L. D. Chen, Spark Plasma Sintering and Thermal Conductivity of Carbon Nanotube Bulk Materials, *J. Appl. Phys.*, **97**, 114310 (2005).

[29] C. E. Borsa, H. S. Ferreira, and R. Kiminami, Liquid Phase Sintering of Al2o3/Sic Nanocomposites, *J. European Ceram. Soc.*, **19**, 615-21 (1999).

[30] J. W. Ning, J. J. Zhang, Y. B. Pan, and J. K. Guo, Fabrication and Mechanical Properties of Sio2 Matrix Composites Reinforced by Carbon Nanotube, *Mater. Sci. Eng. A-Struct. Mater. Prop. Microstruct. Process.*, **357**, 392-96 (2003).

[31] G. D. Zhan, J. D. Kuntz, J. E. Garay, and A. K. Mukherjee, Electrical Properties of Nanoceramics Reinforced with Ropes of Single-Walled Carbon Nanotubes, *Appl. Phys. Lett.*, **83**, 1228-30 (2003).

[32] J. D. Kuntz, G. D. Zhan, and A. K. Mukherjee, Nanocriletalline-Matrix Ceramic Composites for Improved Fracture Toughness, *MRS Bull.*, **29**, 22-27 (2004).

EFFECT OF CARBON ADDITIONS AND B₄C PARTICLE SIZE ON THE MICROSTRUCTURE AND PROPERTIES OF B₄C - TiB₂ COMPOSITES

R.C. McCuiston[*] and J.C. LaSalvia
U.S. Army Research Laboratory
AMSRD-ARL-WM-MD
Aberdeen Proving Ground, MD 21005-5069

B. Moser
BAE Systems
Advanced Ceramics Division
991 Park Center Dr., Vista, CA 92083

ABSTRACT
The effects of carbon additions and mean B_4C particle size, on the microstructure and properties of TiB_2 particulate-reinforced (24 vol.%) B_4C-based composites, were investigated. Two different B_4C powder grades with mean particle sizes of 1 μm and 10 μm were used. The TiB_2 powder had a mean particle size of 5 μm. Excess carbon was created through the pyrolysis of sugar which was introduced into these powder blends. The C_{added}/B_4C ratios examined included 0, 0.021, and 0.035 and were based upon published data. The resulting powder blends (six variants) were hot-pressed near 2000°C to full density. Densities were determined using Archimedes' method, while elastic properties were determined using an ultrasound technique. Tiles were subsequently sectioned for microstructural and mechanical characterization. Knoop hardness was determined using both 2 kg and 4 kg loads, while fracture toughness was determined using the Chevron-notch technique. The microstructure of polished and fractured cross-sections was examined by electron microscopy. Results show that the fracture toughness values of all B_4C-TiB_2 composite variants are significantly higher than monolithic B_4C. Fractographic examination shows significant crack path deflection around the TiB_2 grains leading to the observed increase in fracture toughness. The fracture toughness of the composites produced using the 1 μm mean particle size powder were lower than those produced using the 10 μm mean particle size powder. The addition of carbon was found to reduce fracture toughness. The processing and properties of these composites will be presented and discussed.

INTRODUCTION
The armor community has long maintained an interest in low density ceramics exhibiting high hardness and fracture toughness. Low density provides for weight savings to the armor package. High hardness is required to overmatch the projectile causing projectile core fragmentation and high fracture toughness is desired for improved multi-hit capability, along with the added benefit of improved durability. Silicon carbide (SiC) is currently the ceramic of choice for armor applications across a wide spectrum of threats. However, by employing lower density ceramics, such as B_4C or B_4C based composites, the potential for weight saving across a fleet of vehicles is huge. That weight savings will also have benefits in terms of reduced vehicle wear and tear and fuel consumption as well as easier transportation logistics.

---

[*] Work performed with support by an appointment to the Research Participation Program at the U.S. ARL administered by the Oak Ridge Associated Universities through an interagency agreement between the U.S. Department of Energy and U.S. ARL.

Monolithic B$_4$C has a low density of 2.52 g/cc and a high hardness of 25+ GPa, but it has a very low fracture toughness of ~2.9 MPa√m, i.e. it is extremely brittle.[1] The low fracture toughness is likely one of the reasons B$_4$C has a reduced ballistic efficiency in particular armor configurations or against certain threats, when compared with an armor-grade hot pressed SiC which has a fracture toughness of ~5.1 MPa√m.[2] While conducting V50 experiments, Moynihan[3] et al. found a change in the fragmentation behavior of B$_4$C in the range of 20-23 GPa. The reason for change in fracture behavior was not explained. Sphere impact studies have shown that B$_4$C does not form a comminuted region below the impact site, as seen in SiC.[1,2,4] Figure 1 shows cross sections of B$_4$C and SiC that have been subjected to sphere impacts at ~ 300 m/s. The comminuted region or Mescall zone, seen in the SiC, can be thought of as effective or quasi-plastic mechanism that limits the propagation of damage more globally. No such zone is formed in B$_4$C and significant surface erosion and propagation of long cracks can occur as a result.

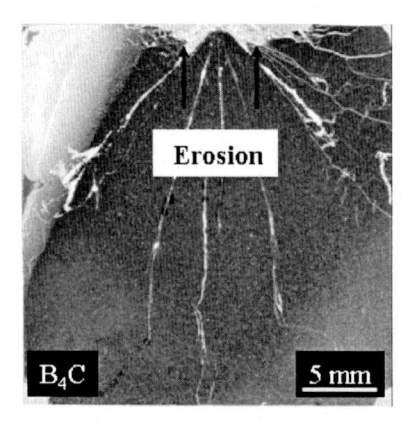

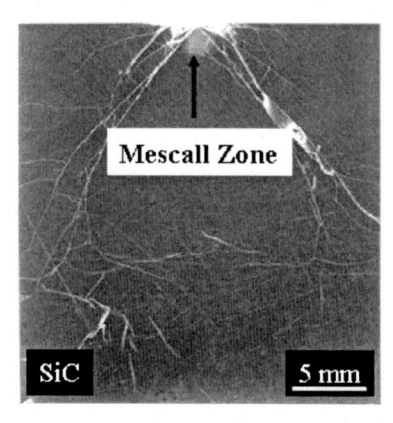

Figure 1. Comparison of the sub-surface damage in B$_4$C and SiC, that have both been sphere impacted at ~300 m/s. B$_4$C exhibits significant surface erosion while SiC shows a Mescall zone, associated with a quasi-plastic response.

It has recently been shown that B$_4$C has a pressure-induced localized-amorphization mechanism, which creates nanometer scale amorphous bands within the B$_4$C grains. This mechanism also appears to be activated within the pressure range of 20-23 GPa.[5] More recently shear localization has been observed in sphere impacted B$_4$C, which results in the creation of shear bands containing nanoscale comminuted B$_4$C.[6] The roles that pressure-induced localized-amorphization and shear localization play in the ballistic behavior of B$_4$C is still unknown.

The low fracture toughness of B$_4$C combined with a sub-optimal ballistic efficiency in particular situations, has prevented a more wide scale usage of B$_4$C. Increasing the fracture toughness of B$_4$C may not only provide improved multi-hit capability and durability, it may also provide for a competing fracture mechanism to localized-amorphization and shear localization.

A common approach to increasing fracture toughness in monolithic ceramics has been to add an amount of a secondary phase, typically another ceramic, thereby creating a composite.[7] The secondary phase alters the fracture behavior from almost entirely trans-granular in a monolithic ceramic like B$_4$C, to a combination of trans-granular and inter-granular fracture, thereby increasing the fracture toughness. The exact mechanism by which the propagating cracks are deflected can vary, with microcracking[8,9], residual stress and phase transformations being common. The choice of secondary phases that can be added to B_4C is somewhat limited when compared to oxide ceramics, due to B_4C's reactivity at high temperature. A ceramic that has been successfully added to B_4C as a secondary phase, is TiB_2.[9-19] The TiB_2 may be added directly as a powder[9,10,17-19] or as either an oxide[11,13-16] or carbide[12], either one of which will react in-situ to form TiB_2. It has been found that the addition of carbon is very important to the B_4C-TiB_2 system, as it segregates at the phase boundary as nanometer scale particles, which provides for easy crack propagation at the B_4C-TiB_2 interface.[9] Various researchers have shown that the addition of TiB_2 to B_4C can result in a composite with a high fracture toughness, flexural strength and hardness.[9-11] This work focused on the processing and characterization of a B_4C-TiB_2 composite with the long term goal of ballistic evaluation for comparison to monolithic B_4C.

EXPERIMENTAL PROCEDURE

Processing

Six different B_4C-TiB_2 composites were produced. All six variations contained 24 vol.% of TiB_2, which was calculated to give a B_4C-TB_2 composite with a theoretical density of 3.0 g/cm^3. The six variations were made with one of two different mean grain size B_4C powders, a 1 μm (ESK Tetrabor 3000F) and a 10 μm (Cercom PAD). One TiB_2 powder (H.C. Starck, Grade G), with a mean grain size of 5 μm, was used. Three different amounts of carbon were added, 0.0, 2.1 and 3.5 wt.%. The carbon was added as a weight percentage of the total B_4C content. These values were chosen based on Figure 7 in reference [9]. For a composite containing 20 vol.% TiB_2, the addition of ~2.1 wt.% carbon yielded a K_{IC} of ~ 4 MPa√m while the addition of ~3.5 wt.% carbon yielded a K_{IC} of ~ 5 MPa√m. The six variations produced in this research are shown below in Table I and will be referred to as V1 through V6.

Table I. Designations of the B_4C-TiB_2 (24 vol.%) composite variations.

C_{added}/B_4C (wt. %)	B$_4$C Powder	
	Cercom PAD	ESK Tetrabor
0	V1	V4
2.1	V2	V5
3.5	V3	V6

The six variations were batched from the as received powders and wet ball milled in polyethylene bottles using a reduced charge of Al_2O_3 milling media. The solvent used was water. Aqueous processing was used for environmental considerations, i.e. green processing. The carbon precursor was added as sugar, as it is soluble in water. The milled powder slurry was dried and granulated. The powder was hot pressed in a 150 mm diameter graphite die at a temperature near 2000°C, using approximately a 20 MPa load. The maximum temperature was held until zero displacement of the ram was attained, the hold time not to exceed 3 hours. A temperature hold was included near 600°C to pyrolize the sugar into carbon. An atmosphere of flowing argon was used throughout the run.

Effect of Carbon Additions and B$_4$C Particle Size on B$_4$C–TiB$_2$ Composites

Characterization
The hot pressed samples were machined into tiles of size 100 x 100 x 12 mm with a standard commercial finish. The Young's modulus, E, of the tiles was measured using the ultrasonic velocity technique following ASTM E 494-95. Several tiles were machined into size B bend bars for measurement of the four point flexure strength, S$_{(4,40)}$, following ASTM C 1161-02c (16 to 18 bars). Density of the bend bars was measured using the Archimedes' method in water. Several of the size B bend bars were machined with chevron notches, for measurement of fracture toughness, K$_{Ivb}$, following ASTM C 1421-01b (9-11 notched bars). Samples of a size 15 x 15 x 12 mm were cut from the tiles using a SiC grit blade. The samples were mounted in epoxy and polished to a final step of colloidal silica, following standard metallographic techniques. Knoop hardness, H$_K$, was measured using loads of 2 and 4 kg, following ASTM C 1326-03. The microstructural and chemical aspects of the polished and of fracture surfaces were examined using a FE-SEM (Hitachi, S4700), with an EDS attachment (EDAX, Genesis). Phase analysis was performed using XRD (Phillips) on bend bar surfaces, with Cu Kα radiation. The scan range was from 20°-80° using a step size of 0.02° with a dwell of 2 seconds per step.

RESULTS AND DISCUSSION

Microstructural Evaluation
The polished microstructures of V1 through V3 are shown in Figure 2 (a-c), respectively. In the images B$_4$C is visible as light gray and TiB$_2$ as dark gray. The TiB$_2$ distribution is uniform in some areas, but more processing work is required to optimize the distribution. The presence of graphitic carbon is noticeable and, as would be expected, appears to increase visibly with added carbon content. It is clear that there is a fourth and possibly a fifth phase present in V1-V3, as the images are punctuated by a bright white phase, that appears to have phase boundaries contained within it. EDS spectra collected on the bright white phase showed the presence of aluminum, oxygen and nitrogen, while another segregated phase contained silicon and carbon. XRD spectra collected on V1 and V2 detected B$_4$C and TiB$_2$ as the major phases. There was also detected a significant amount of an aluminum oxynitride phase, Al$_x$O$_y$N$_z$, and trace amounts of graphite, boron nitride (BN) and SiC. The XRD spectra of V3 showed the major phases of B$_4$C and TiB$_2$. There was also a significant amount of aluminum oxide (Al$_2$O$_3$) detected along with trace amounts of graphite and BN. The presence of an Al$_x$O$_y$N$_z$ or Al$_2$O$_3$ phases is not unexpected. The PAD B$_4$C powder used in V1-V3 does utilize an aluminum based sintering additive and in addition, the powders were ball milled using Al$_2$O$_3$ milling media. Judging from the SEM images and from the size of the XRD peak heights, there is a significant amount of Al$_x$O$_y$N$_z$ and/or Al$_2$O$_3$, which may have an affect of the mechanical properties.

The polished microstructures of V4 through V6 are shown in Figure 2 (d-f), respectively. Again, in the images, the B$_4$C is visible as light gray and the TiB$_2$ as dark gray. As with V1-V3, the TiB$_2$ distribution in V4-V6 is only uniform in select areas, indicating more processing work is required to optimize the TiB$_2$ distribution. The phase boundary of the TiB$_2$ grains appears scalloped due to the small grain size of the B$_4$C used. The only additional visible phase in V4-V6 is graphitic carbon. No Al$_x$O$_y$N$_z$ or Al$_2$O$_3$ is visible. XRD results on V4-V6 confirm the visual findings. The main phases detected were B$_4$C and TiB$_2$, with a trace amount of graphite. Less graphite appears to be present in V4-V6 when compared with V1-V3, likely due to it being consumed by a boron oxide phase. Due to the small grain size of the Tetrabor powder, it will naturally have a larger specific surface area than the PAD powder. As the boron oxide content is very likely surface area dependant, the Tetrabor powder will have more boron oxide content and will therefore consume more of the native as well as the added carbon, through a variation of the reaction $2B_2O_3 + 7C = B_4C + 6CO$.

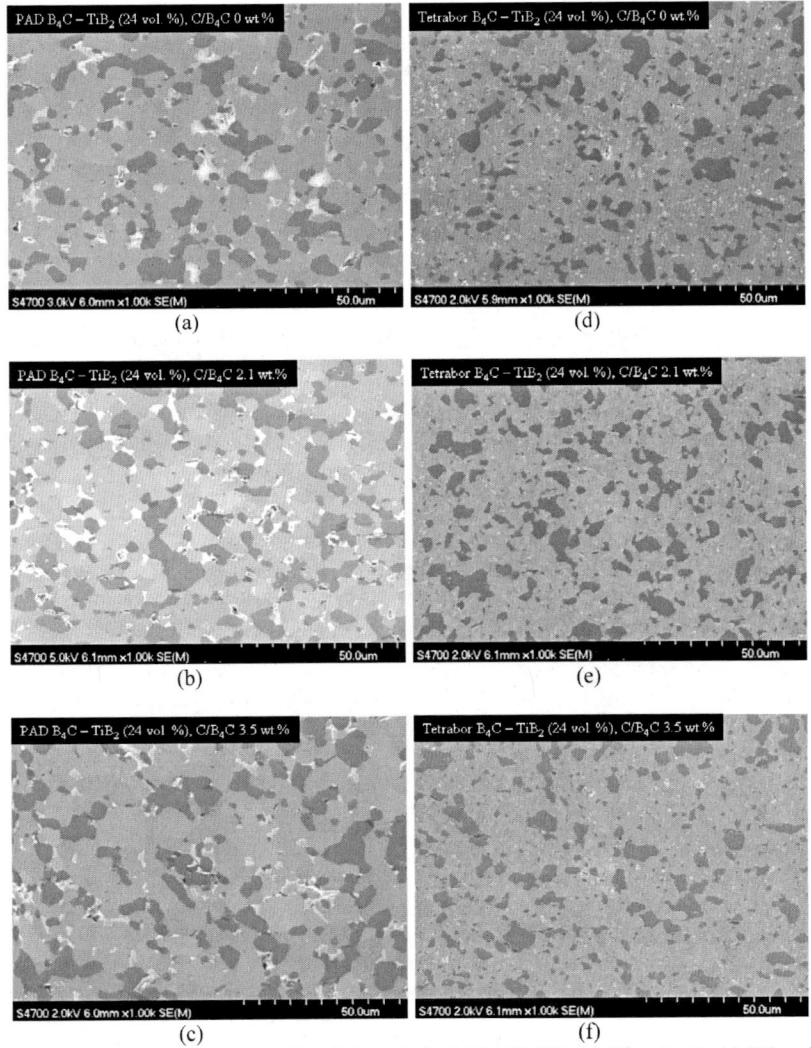

Figure 2. Polished cross sections of variations 1-6 (a) V1, (b) V2, (c) V3, (d) V4, (e) V5 and (f) V6. The dark gray phase is TiB$_2$, light gray phase is B$_4$C, the black phase is graphite and the white phase visible in V1-V3 is predominantly an Al$_x$O$_y$N$_z$ phase.

Effect of Carbon Additions and B$_4$C Particle Size on B$_4$C–TiB$_2$ Composites

Mechanical Properties

Table II lists the results for measurements of bulk density ρ, Young's modulus E, bend strength S, Knoop hardness H$_K$ and Chevron-notch fracture toughness K$_{Ivb}$ for the B$_4$C-TiB$_2$ composites V1-V6 and for monolithic PAD B$_4$C. The measured densities of V1-V6 are all near the designed density of 3.00 g/cm^3. The composites made with the Tetrabor powder, V4-V6, have a slightly lower density. This may be due to the smaller grain size of the Tetrabor powder being more prone to forming agglomerates. There is no observable trend of carbon additions influencing the density in the composites.

The Young's modulus of a B$_4$C-TiB$_2$ composite with 24 vol.% TiB$_2$ is ~ 470 GPa, calculated using the simple rule of mixtures. The composite modulus is slightly higher than that of monolithic B$_4$C, 450 GPa, because the modulus of monolithic TiB$_2$ is 550 GPa. The measured values of E for V1-V3, shown in Table II, are considerably lower than either the expected composite E of 470 GPa or the monolithic B$_4$C value of 450 GPa. A decrease in modulus of a monolithic material can typically be explained by the presence of cracks or porosity. There was no micro-cracking observed in the microstructures of V1-V3, shown in Figure 2 (a-c), and the measured densities would indicate that porosity is not entirely responsible. Porosity in B$_4$C can be considered to be comprised of both open porosity and excess graphitic carbon, because the modulus of graphite is only ~ 10 GPa. The addition of excess carbon did decrease the modulus, but the fact that V1 has such a low modulus, with no additional carbon, would indicate that the carbon alone was not entirely responsible for the decrease. The presence of the Al$_x$O$_y$N$_z$ and Al$_2$O$_3$ phases could also contribute to the decrease because the modulus of those phases is considerably less, ~ 250 - 300 GPa, and a significant amount is observed in the microstructure. It would therefore seem likely that the reduction in modulus is due to a combination of the excess graphitic carbon and the presence of the Al$_x$O$_y$N$_z$ and Al$_2$O$_3$ phases. The measured value of E for V4 is slightly higher than that of monolithic B$_4$C. The addition of carbon to V5 and V6 did decrease the modulus in a linear fashion back to that of monolithic boron carbide. As previously explained, the presence of excess carbon can suppress the modulus by acting as porosity.

The four point flexural strength measurements for V1-V6 and monolithic PAD B$_4$C are shown in Table II. The baseline flexural strength of PAD B$_4$C is ~ 450 MPa. The measured flexural strength of V1 is similar at ~ 430 MPa. The addition of only 2.1 wt.% C/B$_4$C decreased the flexural strength significantly to ~ 310 MPa. A further addition of carbon to 3.5 wt.% C/B$_4$C decreased the flexural strength only slightly more to ~ 296 MPa. Such a dramatic decrease in flexural strength, for the small carbon additions, would indicate, either that the carbon was improperly distributed and created large flaws, or that the presence of the Al$_x$O$_y$N$_z$ and Al$_2$O$_3$ phases had a large impact. The latter is unlikely due to the small grain size of the Al$_x$O$_y$N$_z$ and Al$_2$O$_3$ phases. As will be shown shortly, large porous regions within the composites V1-V3 were found at the fracture origin of several of the bend bar fracture surfaces that were examined. The flexural strengths of the composites V4-V6 are significantly higher than monolithic B$_4$C. The average flexural strength of V4 was measured at 550 MPa, a 100 MPa increase over monolithic B$_4$C. The large standard deviation though would indicate there are processing issues to be resolved. The significant difference between the flexural strengths of V1-V3 and V4-V6 may be due to either a difference in the carbon distribution or the grain size. A decrease in grain size does normally lead to an increase in flexural strength, but as fracture of B$_4$C is almost entirely trans-granular and insensitive to grain size, the decrease in grain size would more likely have resulted in a smaller average flaw size. The slight decrease in flexural strength from V4 to V6, with the addition of excess carbon, would indicate either that the carbon in V4-V6 was better distributed than in V1-V3, which seems unlikely as the same processing was used, or more likely that more of it was utilized in removing the surface boron oxide.

The Knoop hardness, measured using 2 and 4 kg, of composites V1-V6 and monolithic PAD B_4C are shown in Table II. An estimate of the hardness at 4 kg, for a B_4C-TiB_2 composite containing 24 vol.% TiB_2, was calculated to be ~ 18.3 GPa, using the simple rule of mixtures. The Knoop hardness of TiB_2 is lower than B_4C, so a slight decrease in hardness from monolithic B_4C is expected. There is minimal difference between the measured hardness values at 2 kg and 4 kg for composites V1-V6 because the plateau of the indentation size effect curve has likely been reached. The hardness values for composites V1-V3 are significantly lower than the calculated composite hardness of 18.3 GPa. This is likely due to the presence of the excess graphitic carbon as well as the $Al_xO_yN_z$ and Al_2O_3 phases, all of which have significantly lower Knoop hardness values than either B_4C or TiB_2. The hardness values for the composites V4-V6 are near the calculated value of 18.3 GPa, with the hardness value of V5 at 19GPa, actually higher than the estimated value or the measured monolithic B_4C value. The measured values are closer, than V1-V3 are, to the estimated hardness, likely due to the decreased amount of excess graphitic carbon and a lack of $Al_xO_yN_z$ and Al_2O_3 phases. The Knoop hardness of 18.9 GPa measured for monolithic PAD B_4C is not an absolute value though, given the difficulty of interpreting indents in B_4C at such large loads.

The measured values for Chevron notch fracture toughness of composites V1-V6 and monolithic PAD B_4C are shown in Table II. The baseline Chevron notch fracture toughness of monolithic B_4C was measured as 3.32 MPa$\sqrt{m}$. The Chevron notch fracture toughness measured for composite V1 is 5.70 MPa$\sqrt{m}$, a significant increase. Adding carbon reduced the Chevron notch fracture toughness of V2 to 5.42 MPa$\sqrt{m}$ and 5.51 MPa$\sqrt{m}$ for V3. Sigl et al. [9] showed that in properly processed B_4C-TiB_2 composites, the added carbon segregates to the B_4C-TiB_2 phase boundary and should increase the fracture toughness, up until a plateau in carbon additions is reached. As previously explained, the reason for adding 2.1 and 3.5 wt.% C/B_4C was to create composites with fracture toughness' of ~ 4 and 5 MPa$\sqrt{m}$, respectively. While high fracture toughness values were obtained, the addition of carbon has had the opposite intended effect. This is likely due to an improper distribution of carbon. The Chevron notch fracture toughness for composites V4-V6 were all measured as significantly greater than monolithic B_4C. The toughness of V4 was measured to be 5.19 MPa$\sqrt{m}$ and again the addition of carbon to V5 and V6 slightly decreased the fracture toughness, rather than increase it as postulated. The slight decrease in Chevron notch fracture toughness between V1-V3 and V4-V6 is likely again related to the presence of slightly more excess carbon and the $Al_xO_yN_z$ and Al_2O_3 phases in V1-V3. The fact that more graphitic carbon remains in V1-V3 means an increased probability of finding it at the B_4C-TiB_2 phase interface. The $Al_xO_yN_z$ and Al_2O_3 phases have also likely provided an increased number of weak phase interfaces, the same reason for adding TiB_2. Normally a reduction in grain size, as in V4-V6, can explain a decrease in fracture toughness, but as previously mentioned, B_4C fractures almost entirely in a trans-granular mode, so B_4C grain size likely had a minimal effect.

Effect of Carbon Additions and B₄C Particle Size on B₄C–TiB₂ Composites

Table II. Measured values of bulk density ρ, Young's modulus E, bend strength S, Knoop hardness H_K and Chevron-notch fracture toughness K_Ivb for composite variations 1-6 and PAD B₄C.

Variation	ρ (g/cm³)	E (GPa)	S_(4,40) (MPa)	H_K (GPa) H_K2	H_K4	K_Ivb (MPa √m)
V1, PAD 0.0% C_added	3.02±0.00	423±5	430.5±35.7	16.6±0.3	15.9±0.5	5.70±0.29
V2, PAD 2.1% C_added	2.99±0.02	412±5	312.4±25.4	16.6±0.8	15.2±0.6	5.42±0.28
V3, PAD 3.5% C_added	3.01±0.00	409±6	296.4±36.3	15.4±1.7	15.1±0.7	5.51±0.31
V4, Tetrabor 0.0% C_added	2.94±0.02	461±11	551.9±50.0	18.2±0.6	17.7±0.2	5.19±0.13
V5, Tetrabor 2.1% C_added	3.00±0.01	458±9	545.9±55.8	20.1±0.6	19.0±0.6	5.13±0.11
V6, Tetrabor 3.5% C_added	2.96±0.02	443±14	536.7±29.5	18.3±0.5	17.4±0.5	5.01±0.21
PAD B₄C	2.52±0.01	449±12	454.0±36.3	21.3±0.3	18.9±0.8	3.32±0.14

Fracture and Indentation Behavior

Figure 3 (a) and (b) shows representative SEM images of the fracture surface of composites V2 (PAD) and V5 (Tetrabor), respectively. The fracture surfaces were generated from the four point flexure test samples discussed previously. Observed on both fracture surfaces, are exposed TiB₂ grain facets and graphite, as well as cleavage steps of both large grains of B₄C and TiB₂. Such a mix of inter and trans-granular fracture would explain the increase in measured fracture toughness of composites V1-V6, over monolithic B₄C.

The SEM images seem in Figure 3 (c) and (d) are of fracture origins in bend bars of V2 (PAD) and V5 (Tetrabor), respectively. The fracture origin of the bend bars, in all cases examined, was either a porous region of grains or an agglomerate, composed of large TiB₂ grains. No fracture origins found resulting from machining damage were found. The porous regions appear to be the predominant failure origin type in V1-V3, while the agglomerates of large grained TiB₂ appear to be the predominant failure origin in V4-V6. Both the porous regions and TiB₂ agglomerates can be removed through processing improvements.

Figure 4 (a) and (b) show Knoop indents generate at 4 kg in monolithic PAD B₄C and the B₄C-TiB₂ composite V5, respectively. The image of the Knoop indent in PAD B₄C shows the common spalling problem associated with sharp indents at high loads in a brittle material, like B₄C. The spalling is an indication of the low fracture toughness of monolithic B₄C. Contrast this with the image of the Knoop indent in composite V5. This is no evidence of spallation in the B₄C-TiB₂ composite, due to the increase in fracture toughness. Figure 4 (c) is an image of the crack path found at the right tip of the Knoop indent in V5, Figure 4 (b). The grain structure of the B₄C is clearly visible. The crack path travels through the B₄C grains in a trans-granular mode and it is deflected around the TiB₂ grains at or very near to the TiB₂-B₄C interface.

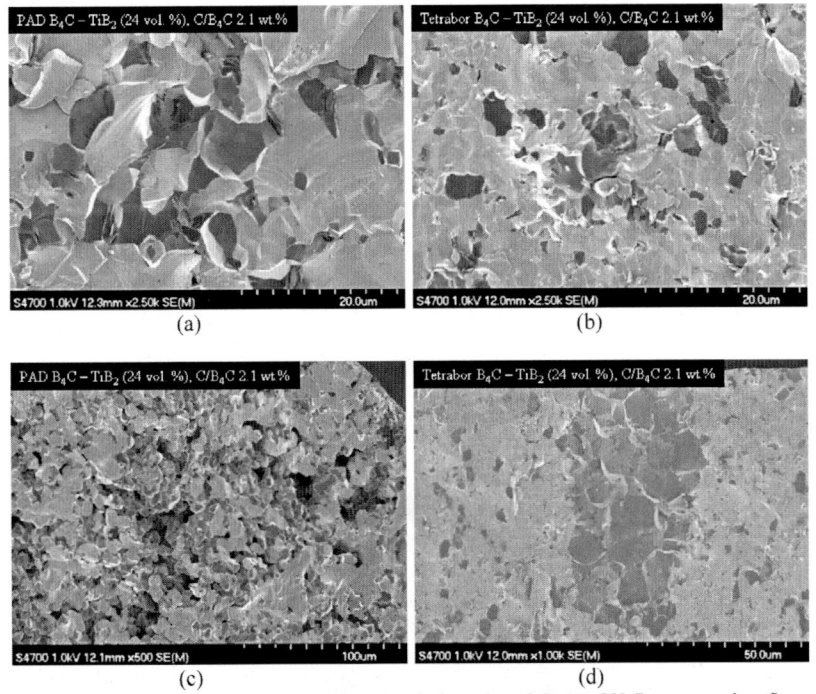

Figure 3. Fracture surface of bend bars from variations 2 and 5. (a,) V2 Porous region flaw, (b) V2 fracture surface, (c) V5 Large TiB$_2$ flaw, (d) V5 fracture surface.

Effect of Carbon Additions and B$_4$C Particle Size on B$_4$C–TiB$_2$ Composites

(a)

(b)

(c)

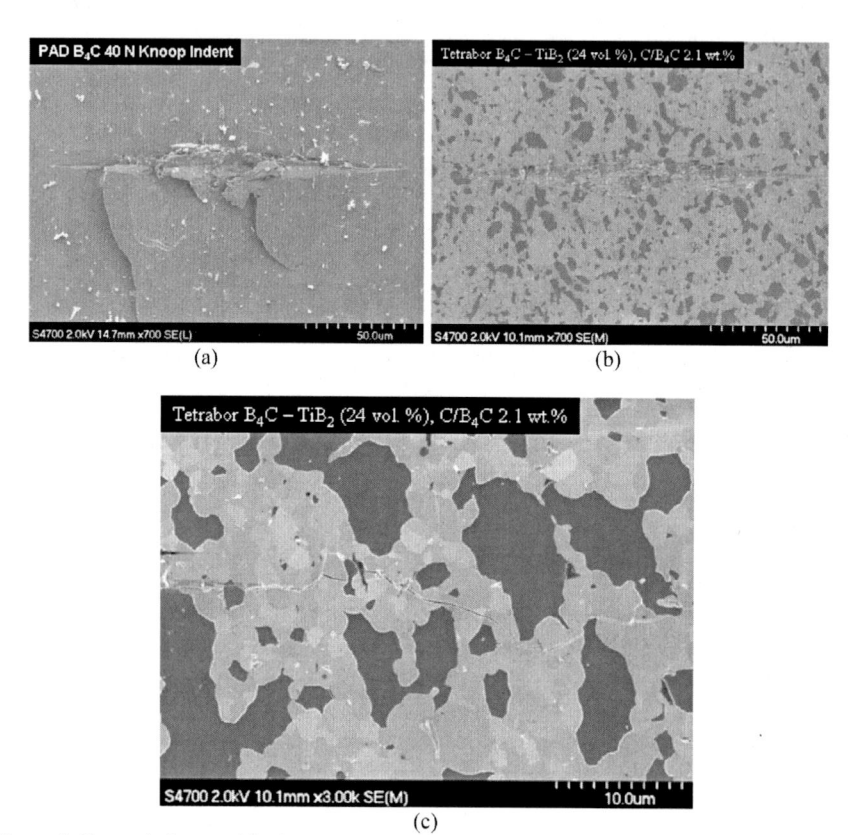

Figure 4. Knoop indents at 4 kg in (a) monolithic PAD B$_4$C and (b) the B$_4$C-TiB$_2$ composite V5. Notice that significant spalling occurs in the monolithic B$_4$C, while there is no spalling in the B$_4$C-TiB$_2$ composite. (c) Crack propagation at the tip of the Knoop indent shown in (b). The crack path is through the B$_4$C and around the TiB$_2$ and/or at the B$_4$C-TiB$_2$ interface.

SUMMARY AND CONCLUSIONS

The effects of B$_4$C grain size and carbon additions on the processing and mechanical properties of six variants of a B$_4$C-TiB$_2$ composite system were examined. The three composites fabricated using the smaller grain size B$_4$C Tetrabor powder, V4-V6, had higher measured values of Young's modulus, flexural strength and Knoop hardness, but lower fracture toughness values than the three composites fabricated using the larger grain size B$_4$C PAD powder, V1-V6. These differences in measured mechanical properties do not appear to a result of the grain size difference however. Rather, a combination of excess graphitic carbon, resulting from less surface oxide content and the presence of Al$_x$O$_y$N$_z$ and Al$_2$O$_3$ phases, appears to have impacted the mechanical properties of composites V1-V3. The effect of increasing carbon additions on composites V1-V6 was to reduce the Young's modulus, bend strength, Knoop hardness and fracture toughness. This result was contrary to the expected increase in fracture toughness and is this is likely due to insufficient distribution of the added carbon. Improved processing, to better distribute the added carbon and TiB$_2$, as well as better accounting of the native carbon and surface boron oxide content, will be the focus of future work.

ACKNOWLEDGMENTS
The authors would like to thank Mr. Matthew J. Motyka (DSI/ARL) and Mr. Herbert T. Miller (ORISE/ARL) for their technical support.

REFERENCES
1. J.C. LaSalvia, M.J. Normandia, H.T. Miller and D.E. MacKenzie, "Sphere Impact Induced Damage in Ceramics: I. Armor Grade SiC and TiB$_2$", *Cer. Eng. Sci. Proc.*, **26**(7), 171-79 (2005)
2. J.C. LaSalvia, M.J. Normandia, H.T. Miller and D.E. MacKenzie, "Sphere Impact Induced Damage in Ceramics: II. Armor-Grade B$_4$C and WC", *Cer. Eng. Sci. Proc.*, **26**(7), 180-88 (2005)
3. T.J. Moynihan, J.C. LaSalvia and M.S. Burkins, "Analysis of Shatter Gap Phenomenon in a Boron Carbide/Composite Laminate Armor System", In *20th Int. Symposium on Ballistics*, Vol. II., eds. J. Carleone and D. Orphal, 1096-1103 (2002)
4. J.C. LaSalvia, M.J. Normandia, D.E. MacKenzie and H.T. Miller, "Sphere Impact Induced Damage in Ceramics: III. Analysis", *Cer. Eng. Sci. Proc.*, **26**(7), 189-98 (2005)
5. M. Chen, J.W. McCauley and K.J. Hemker, "Shock-Induced Localized Amorphization in Boron Carbide", *Science*, **299**[5612], 1563-66, (2003)
6. J.C. LaSalvia, R.C. McCuiston, G. Fanchini, M. Chhowalla, H.T. Miller, D.E. MacKenzie and J.W. McCauley, "Shear Localization in a Sphere-Impacted Armor Grade Boron Carbide", To be published: *Proc. of the 23rd Int. Symposium on Ballistics*, Tarragona, Spain, (2007)
7. R. Telle and G. Petzow, "Strengthening and Toughening of Boride and Carbide Hard Material Composites", *Mat. Sci.Eng. A*, **105/106**, 97-104 (1995)
8. L.S. Sigl, "Microcrack toughening in Brittle Materials Containing Weak and Strong Interfaces", *Acta. Mater.*, **44**(9) 3599-609 (1996)
9. L.S. Sigl and H. Kleebe, "Microcracking in B$_4$C-TiB$_2$ Composites", *J. Am. Ceram. Soc.*, **78**(9), 2374-80 (1995)
10. K.A. Schwetz, L.S. Sigl, J. Greim, and H. Knoch, "Wear of Boron Carbide Ceramics by Abrasive Waterjets", *Wear*, **181-183**, 148-55 (1995)
11. V.V. Skorokhod, M.D. Vlajic and V.D. Kristic, "Mechanical Properties of Pressureless Sintered Boron Carbide Containing TiB$_2$ Phase", *J. Mat. Sci. Let.*, **15**, 1337-39 (1996)
12. L.S. Sigl, "Processing and Mechanical Properties of Boron Carbide Sintered with TiC", *J. Eur. Ceram. Soc.*, **18**, 1521-29 (1998)
13. V.V. Skorokhod and V.D. Kristic, "High strength-high toughness B$_4$C-TiB$_2$ composites", *J. Mat. Sci. Let.*, **19**, 237-39 (2000)
14. V.V. Shorokhod and V.D. Kristic, "Processing, Microstructure, and Mechanical Properties of B$_4$C-TiB$_2$ Particulate Sintered Composites I. Pressureless Sintering and Microstructure Evolution", *Powd. Met. And Met. Ceram.*, **39**(7-8), 414-23 (2000)
15. V.V. Shorokhod and V.D. Kristic, "Processing, Microstructure, and Mechanical Properties of B$_4$C-TiB$_2$ Particulate Sintered Composites II. Fracture and Mechanical Properties", *Powd. Met. And Met. Ceram.*, **39**(9-10), 504-13 (2000)
16. S. Yamada, K. Hirao, Y. Yamauchi and S. Kanzaki, "High Strength B$_4$C-TiB$_2$ Composites Fabricated by Reaction Hot-Pressing", *J. Eur. Ceram. Soc.*, **23**, 1123-30 (2003)
17. K.F. Cai, C.W. Nan, M. Schmuecker and E. Mueller, "Microstructure of hot-pressed B$_4$C-TiB$_2$ thermoelectric composites", *J. Alloys and Com.*, **350**, 313-18 (2003)
18. T.S. Srivatsan, G. Guruprasad, D. Black, R. Radhakrishnan and T.S. Sudarshan, "Influence of TiB$_2$ Content on Microstructure and Hardness of TiB$_2$-B$_4$C Composite", *Powd. Tech.*, **159**, 161-67 (2005)

19. T.S. Srivatsan, G. Guruprasad, D. Black, M. Petraroli, R. Radhakrishnan and T.S. Sudarshan, "Microstructural Development and Hardness of TiB_2-B_4C Composite Samples: Influence of Consolidation Temperature", *J. Alloys Comp.*, **413**, 63-72 (2006)

ELECTRO-CONDUCTIVE ZrO_2-NbC-TiN COMPOSITES USING NbC NANOPOWDER MADE BY CARBO-THERMAL REACTION

S. Salehi, J. Verhelst, O. Van der Biest, J. Vleugels
Department of Metallurgy and Materials Engineering, Katholieke Universiteit Leuven, Kasteelpark Arenberg 44, B-3001 Heverlee, Belgium

ABSTRACT

Niobium carbide (NbC) has a high melting point (3600°C), good electrical conductivity and high hardness. These properties encourage the investigation of zirconia-based niobium carbide composites aiming at achieving a composite with good fracture toughness, high hardness and at the same time a sufficient electrical conductivity to allow electro-discharge machining (EDM).
Since the commercially available NbC powders are relatively coarse (d_{50} = 1-3 µm), a NbC nanopowder was prepared by carbo-thermal reduction of Nb_2O_5. The synthesis time, temperature, and C/Nb_2O_5 ratio have been optimized to obtain a NbC powder with extremely fine grain size and minimum residual carbon and oxide impurities. Carbon black and Nb_2O_5 were mixed in different ratios and thermally treated at 1000-1450°C for 0.5-10 h. NbC powder with a d_{50} grain size of 185 nm was obtained from a powder mixture with a carbon/niobium oxide ratio of 1.88, thermally treated in vacuum (10^{-5} to 10^{-3} Pa) at 1350°C for 5 hours. ZrO_2-NbC-TiN (65:35:5 vol %) composites based on the synthesized NbC powder could be fully densified by means of hot pressing at 1450°C for 1 hour at 28 MPa, resulting in a composite with an excellent hardness of 16 GPa, a fracture toughness of 5 $MPa.m^{0.5}$ and an electrical resistivity of 0.8 mΩ.cm.

INTRODUCTION

Yttria-stabilized ZrO_2 (Y-TZP) has a high toughness and excellent strength but the hardness is modest and the material is electrically insulating. Transition metal carbides and nitrides on the other hand have high melting points and high chemical stability combined with a good electrical conductivity. Combining both phases is expected to result in ceramic composites with high hardness, toughness and enough electrical conductivity to be electro-erodable, avoiding expensive grinding operations for complex shaping of components and increasing the possibility of mass production and manufacturing cost reduction.
ZrO_2-based electro-conductive ceramic composites with TiN, WC, TiCN, or TiB_2 have been reported in literature [1-4]. At present, NbC is chosen as secondary phase to establish electrical conductivity since NbC has an excellent hardness of about 20 GPa and a low electrical resistivity of 35×10^{-3} mΩ.cm [5]. Commercially available NbC powders however are relatively coarse (d_{50} = 1-3 µm). Therefore, the possibility to synthesize NbC powder with minimum grain size and impurities by means of carburization is explored. Reported methods for NbC synthesis are high-temperature solid-state reactions of niobium metal or niobium oxides with carbon [6-7] or solid-gas reactions such as the reaction of $NbCl_5$ with hydrocarbons or the carburizing of oxide by methane-hydrogen gas mixtures [6]. In this work, carburization of Nb_2O_5 with carbon black was chosen because of its simplicity. Zirconia-based composites are processed from the optimum self-synthesized NbC powder and the mechanical properties are compared with those of commercial NbC powder based composites.

EXPERIMENTAL PROCEDURE

In the carburizing method, a mixture of Nb_2O_5 and carbon black is heated at elevated temperature in an inert atmosphere to form NbC according to equation (1) and (2) [8]. Nb_2O_5 (CBMM, Brazil) and carbon black (Grade 4, Degussa, Germany) were mixed with a C/Nb mol ratio of 3.50 and 3.76. An overview of all starting powders used is given in Table 1.

50 grams of fully formulated powder mixtures were mixed on a multidirectional mixer (type T2A, Basel, Switzerland) in ethanol in a polyethylene container of 250 ml for 48 h at 60 rpm. 250 grams WC-Co milling balls (diameter 3 mm, Ceratizit grade MG15) were added to the containers to break the agglomerates in the starting powders. The suspensions were dried in a rotating evaporator at 80°C. After additional drying for 3 h at 70°C, the powder mixture was sieved (315 mesh).

3 g of the sieved powder was placed in a SiC crucible (Haldenwanger, Germany) and covered with a SiC lid. The container was inserted in the sintering unit of a hot press (Model W100/150-2200-50 LAX, FCT Systeme, Rauenstein, Germany) in vacuum (~10^{-5} to 10^{-3} Pa) with a heating and cooling rate of 50 and 20°C/min respectively. Three parameters, i.e., the C/Nb mol ratio, temperature and dwell time were optimized.

The minimum test temperature was selected based on the Gibbs free energy of eq. (1) and (2), (eq. 3 and 4). Based on these calculations, NbC is expected to be formed at 850 °C (eq. 1) or 1030 °C (eq. 2) assuming a CO or CO_2 partial pressure of 10^{-4} Pa. Since the vacuum varies during the process and the kinetics of the reaction are not taken into account, a minimum temperature of 1150°C was used for experimentation.

$$1/2\ Nb_2O_5 + 7/2\ C \leftrightarrow NbC + 5/2\ CO \qquad (1)$$

$$1/2\ Nb_2O_5 + 9/4\ C \leftrightarrow NbC + 5/4\ CO_2 \qquad (2)$$

$$\Delta G_1 = 521.2 - 421.84 \times 10^{-3}\ T + 20.78 \times 10^{-3}\ T\ \times \ln(P_{CO})\ ^{[8]} \qquad (3)$$

$$\Delta G_2 = 313.0 - 208.10 \times 10^{-3}\ T + 10.39 \times 10^{-3}\ T\ \times \ln(P_{CO_2})\ ^{[8]} \qquad (4)$$

The ΔG's of reactions 1 and 2, calculated based on the recorded pressure and temperature profile, are presented in Fig. 1. Since the vacuum pump is continuously working during the process, it is assumed that the pressure inside the chamber above 600-700°C is only due to CO and CO_2 gases. In order to facilitate the calculations, the partial pressure of CO or CO_2 was considered to be half the recorded pressure. Although reactions 1 and 2 already start during the heating up at about 600°C and 860°C, a maximum reactivity was observed at 1350°C, as revealed by the significant vacuum pressure drop due to CO and CO_2 formation.

Table 1. Summary of the starting powders

	Producer	Grade / Impurities	Particle size
NbC	Treibacher (Austria)	0.25% O – 0.13% free C	1.18 µm (FSSS)
Nb_2O_5	CBMM* (Brazil)	1.5% mainly Ta and Ti	0.5-5 µm (d_{50} = 1.04 µm)
C	Degussa (Germany)	grade special 4	agglomerated nanopowder
ZrO_2	Tosoh (Japan)	TZ-0, TZ-3Y or TZ-8Y	37 nm
TiN	Kennametal (USA)	Jet-milled	1.03 µm
Al_2O_3	Baikowski (France)	Grade SM8	0.6 µm

*Companhia Brasileira de Metalurgia e Mineração

Electro-Conductive ZrO$_2$-NbC-TiN Composites Using NbC Nanopowder

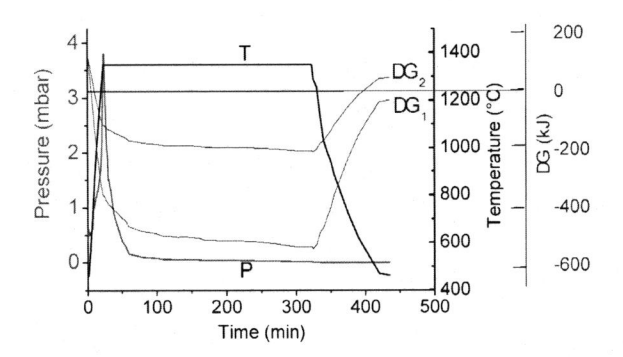

Fig. 1. Calculated ΔG$_1$ (eq. 1) and ΔG$_2$ (eq. 2) at the actual temperature and pressure during thermal treatment for 5 h at 1350°C.

The resulting NbC powder and composite microstructures were investigated by SEM (XL-30FEG, FEI, Eindhoven, The Netherlands) equipped with an X-ray energy dispersive analysis system (EDS, EDAX, Tilburg, The Netherlands) and X-ray diffraction (Seifert 3003 T/T, Ahrensburg, Germany) using Cu Kα radiation (40 kV, 40 mA). The average grain size of the powders was determined from SEM micrographs according to the linear intercept method [9], using Image-pro Plus software (Media Cybernetics, Silver Spring, Maryland). A minimum of 200 grains was counted for each powder grade.

The self-synthesized NbC powders were mixed with the proper amounts of ZrO$_2$ (grades TZ-0, TZ-3Y or TZ-8Y, Tosoh, Japan) and TiN (Jet milled, d: 1.03 μm Kennametal, USA,) to make discs with a height of 5 mm and a diameter of 30 mm. The overall Y$_2$O$_3$ stabilizer content of the powder mixtures was adjusted by mixing of the appropriate ZrO$_2$ starting powders. TiN was added to some of the composites to consume the excess carbon from the carburizing process by the formation of TiCN. Al$_2$O$_3$ (grade SM8, Baikowski, France) was added as a ZrO$_2$ grain growth inhibitor and sintering aid to all composites. The powders were mixed in the same way as the Nb$_2$O$_5$ and carbon black mixture. The powder was inserted into a graphite die coated with BN to separate the powder mixture from the graphite die/punch setup. The discs were hot pressed at 1450 or 1500°C for 1 hour under a load of 30 MPa using the same equipment as used for carburization, with a heating and cooling rate of 50 and 20°C/min respectively.

The Vickers hardness, HV$_{10}$, of the composites was measured (Model FV-700, Future-Tech Corp., Tokyo, Japan) with an indentation load of 10 kg. The indentation toughness, K$_{IC}$, was calculated from the length of the radial cracks originating in the corners of the Vickers indentations according to the formula proposed by Anstis et al. [10]. The density of the samples was measured in ethanol, according to the Archimedes method (BP210S balance, Sartorius AG, Germany). The reported hardness and toughness values are the mean and standard deviation obtained from five indentations. The electrical resistivity of the composites was measured on ground rectangular bars according to the 4-point contact method using a Resistomat (TYP 2302 Burster, Gernsbach, Germany). The bars were cut from the sintered discs by means of EDM and subsequently ground. The elastic modulus was measured using the resonance frequency method

[11], measured by the impulse excitation technique (Grindo-Sonic, J.W. Lemmens N.V., Leuven, Belgium).

RESULTS AND DISCUSSION
NbC powder synthesis

Nb_2O_5-carbon black starting powder mixtures with a C/Nb mol ratio of 3.00, 3.50 and 3.76 were heated to 1000-1450°C for a predefined time and the resulting powders were studied by SEM and XRD. Representative SEM micrographs of the carburized powders are shown in Fig. 2, whereas the corresponding XRD patterns are presented in Fig. 3. Although the conversion of the starting powder mixture into NbC is predicted to be thermodynamically favorable under the experimental conditions (see Fig. 1), niobium oxide and carbon can still be identified in the powder heated for 10 h at 1150°C, as shown in Fig. 2.b and Fig. 3. The conversion is much more pronounced after 30 min at 1350°C and almost complete after 5 h at 1350°C or 30 min at 1450°C. The NbC grain size after 30 min at 1450°C however is already substantially larger than after 5 h at 1350°C, as shown in Fig. 2.e and d. The NbC powder obtained from a Nb/C = 3.5 starting powder heated for 30 min at 1350°C, 5 h at 1350°C and 30 min at 1450°C had a d_{50} of 185, 225 and 339 µm and a d_{90} of 335, 490 and 1000 µm respectively. The synthesized NbC powders are substantially finer than the commercial powder, shown in Fig. 2.a.

The XRD patterns of the NbC powders obtained from starting powders with a different C/Nb ratio are compared with the commercial NbC powder in Fig. 3. The residual niobium oxide and carbon content of the carburized powders is low.

The NbC_x stoichiometry is reported to substantially influence its hardness. NbC shows a maximum hardness of ~25GPa at a $Nb_{0.8}C$ stoichiometry, whereas the hardness decreases to 21 GPa when modifying the stoichiometry to NbC [5]. For non-stoichiometric NbC_x, Brown [12] proved that a number of carbon vacancies in the NbC lattice introduce a corresponding change in the inter-planar distance, which is reflected in the lattice parameter, a. The x value of the NbC_x stoichiometry can be calculated from the lattice parameter, a, according to [12,13]:

$$a = 4.09847 + 0.71820\,x - 0.34570\,x^2 \qquad (5)$$

The lattice parameter was obtained according to the method suggested by Suryanarayana and Grant [14]. The stoichiometry of the NbC powder obtained from Nb_2O_3/C (carbon black) powder mixtures with a C/Nb mol ratio of 3.50 and 3.76 heated for 5h at 1350°C is $NbC_{0.70}$ and $NbC_{0.72}$ respectively. The composition of the commercial powder was measured to be $NbC_{0.96}$.

ZrO_2-NbC-TiN Composites

ZrO_2-NbC-TiN composites were hot pressed from two NbC powder grades (D and E), obtained from starting powders with a Nb/C mol ratio of 3.50 (D) and 3.76 (E) thermally treated for 5 h at 1350°C.

Representative SEM micrographs of the 3 mol% yttria stabilized ZrO_2-NbC-TiN (65:30:5) and the 4 mol% yttria stabilized ZrO_2-NbC-TiN (50:45:5) composites are shown in Fig. 4.a and b respectively. The presence of TiCN, the dark phase in Fig. 4.a as revealed by EDAX analysis, confirms the formation of TiCN by the reaction of TiN with the residual carbon. The white phase in Fig. 4.a was identified as WC, originating from the WC-Co milling medium. The gray matrix phase is yttria-stabilized ZrO_2 and the bright phase is NbC_x. Flake like residual carbon can be observed in Fig. 4.b, despite the addition of 5 vol% TiN.

XRD results, presented in Fig. 5, indicate that 5 vol% TiN is enough to consume the residual carbon in a composite with 30 vol % NbC. Moreover, 4.5 mol% yttria stabilizer was not enough to stabilize the t-ZrO$_2$ phase in the 50 vol% ZrO$_2$ in a composite without TiN addition, as revealed by the spontaneous degradation of the material. It seems that the presence of the residual carbon inhibits the stabilization of the tetragonal ZrO$_2$ phase. This observation is confirmed by the relatively high monoclinic ZrO$_2$ content in the samples without TiN addition, as shown in Fig. 5.

The mechanical properties of the different material grades are summarized in Table 2. For comparative reasons, a ZrO$_2$-NbC (60/40) disc was made using the commercial NbC powder (grade C). The hardness of the composites without TiN addition is very low because of the residual graphite and the micro cracks due to spontaneous transformation. The hardness of the composites made from self-synthesized NbC powder is higher than for the commercial powder based material. The reason is the smaller NbC grain size of the self-synthesized powder and the higher intrinsic phase hardness due to a higher hardness stoichiometry.

The relative density of the non-degrading composites was measured to be 99 %. The theoretical density was calculated based on a density of 6.05, 7.85, 5.43 and 2.90 g/cm^3 for ZrO$_2$, NbC, TiN and Al$_2$O$_3$ respectively.

Electro-Conductive ZrO$_2$-NbC-TiN Composites Using NbC Nanopowder

Fig. 2. SEM micrograph of commercial NbC powder (a), Nb$_2$O$_5$+C with a C/Nb = 3.5 heated for 10 h at 1150°C (b), 30 min at 1350°C (c,f), 5h at 1350°C (d) and 30 min at 1450°C (e).

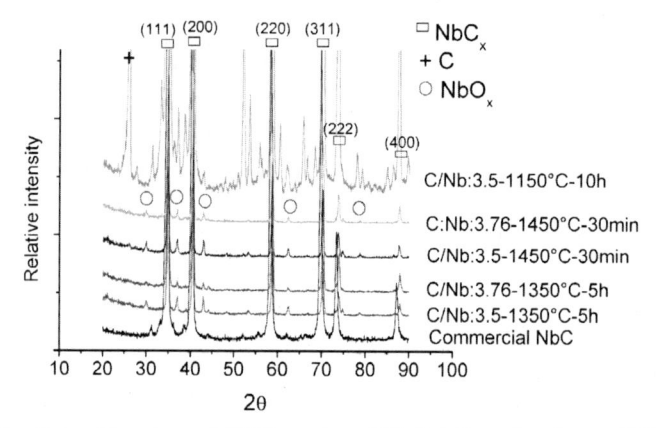

Fig. 3. XRD patterns of the commercial NbC powder and Nb$_2$O$_5$ + C powder mixtures (C/Nb = 3.50 or 3.76) heated for 10h at 1150°C, 5 h at 1350°C or 30 min at 1450°C.

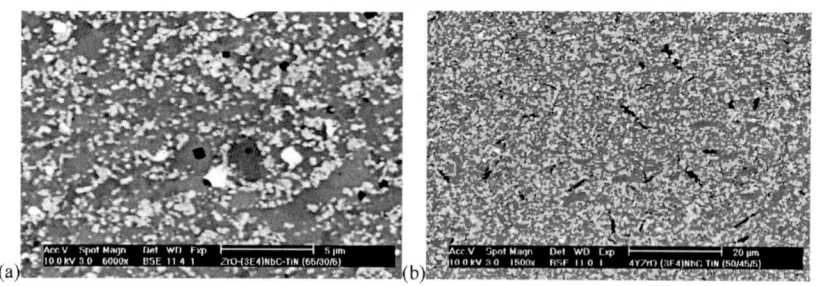

Fig. 4. Microstructure of the ZrO$_2$-NbC-TiN (65:30:5 vol%) (a) and (50:45:5 vol%) (b) composites. (ZrO$_2$ = gray matrix, TiCN = dark, NbC = bright, WC = white, alumina or carbon = black).

Table 2. Elastic modulus, hardness and indentation toughness of the investigated composites.

NbC* [vol %]	TiN [vol %]	Yttria content [mol %]	Density [%]	H_v [GPa]	K_{IC} [MPa.m$^{0.5}$]	E [GPa]
40 E	0	3.0	Cracked	8.78 ± 0.11	5.2 ± 0.3	260
30 E	5	3.0	99	15.82 ± 0.13	4.9 ± 0.2	257
45 E**	5	3.0	99	16.15 ± 0.08	2.9 ± 0.2	277
50 D	0	3.0	Cracked	10.89 ± 0.04	6.1 ± 0.1	273
45 D	5	4.0	99	15.99 ± 0.10	3.7 ± 0.3	277
50 E	0	4.5	Cracked	10.89 ± 0.05	5.9 ± 0.1	273
60 C	0	2.0	99	13.64 ± 0.07	5.4 ± 0.1	284

*E, D and C refer to the NbC grade used: E = starting C/Nb ratio of 3.5, D = starting C/Nb ratio of 3.76 and C is the commercial NbC powder
**Material grade sintered at 1450°C whereas all the others are sintered at 1500°C

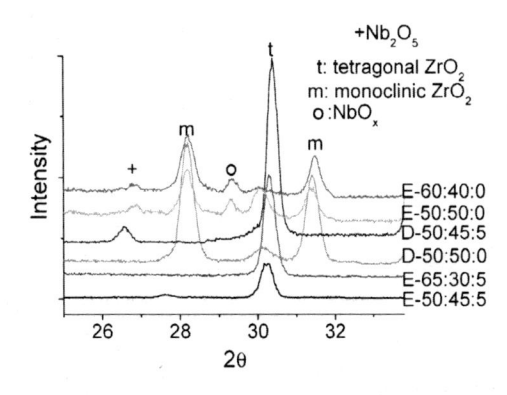

Fig. 5. XRD patterns of the SPS ZrO$_2$-NbC-TiN composites.

The resistivity of the 3 mol % yttria stabilized ZrO$_2$-NbC-TiN (65:30E:5) and (50:45E:05) was measured to be 0.81 and 2.49 mΩ.cm respectively, which is suitable for EDM machining [15,16]. This was confirmed by the fact that the blank specimens used for the E modulus determination were machined by means of wire-EDM (Robofil 240cc, Charmilles, Switzerland).

SUMMARY

The optimal carbothermal synthesis parameters to produce nanometer sized NbC powder from a mixture of Nb$_2$O$_5$ and carbon black in vacuum (~ 10^{-4} Pa) at 1350°C were found to be a reaction time of 5 hrs and a starting C/Nb ratio of 3.76. The resulting NbC powder has a d$_{50}$ grain size of 185 nm and a NbC$_{0.7}$ stoichiometry.

The addition of the carburized NbC powder to an Y-ZrO$_2$ matrix resulted in spontaneous transformation and composite degradation due to the presence of residual carbon. Adding 5 vol% TiN powder allowed to stabilize the Y-TZP phase in the composite, due to the reaction of TiN and carbon forming TiCN. The resulting Y-TZP ZrO$_2$-NbC-TiN composites combine an excellent Vickers hardness of 16 GPa, which is higher than for the commercial NbC powder based composite, with a moderate toughness of 3-5 MPa.m$^{1/2}$. All processed ZrO$_2$-NbC-TiN composites had a sufficiently high electrical conductivity allowing EDM machining.

ACKNOWLEDGEMENTS
This work was performed in the framework of the MONCERAT project supported by the Commission of the European Communities within the FP6 Framework under project No. STRP 505541-1.

REFERENCES
[1] S. Salehi, O. Van der Biest and J. Vleugels, "Electrically conductive ZrO$_2$-TiN composites", *J. Europ. Ceram. Soc.* 26, [15] (2006) 3173-3179.
[2] G. Anné, S. Put, K. Vanmeensel, D. Jiang, J. Vleugels, O. Van der Biest, "Hard, tough and strong ZrO$_2$-WC composites from nanosized powders", *J. Europ. Ceram. Soc.*, 25, [1] (2005) 55-63.
[3] J. Vleugels, O. Van Der Biest, "Development and characterization of Y$_2$O$_3$-stabilized ZrO$_2$ (Y-TZP) composites with TiB$_2$, TiN, TiC, and TiC$_{0.5}$N$_{0.5}$", *J. Amer. Ceram. Soc.*, 82 (10) (1999), 2717–2720.
[4] . B. Basu, J. Vleugels and O. Van Der Biest, "Processing and Mechanical properties of ZrO$_2$-TiB$_2$ composites", *J. Europ. Ceram. Soc.*, 25 (2005) 3629-3637.
[5] Hugh O.Pierson, Noyes Publications, "Handbook of Refractory carbides and nitrides"-ISBN 0-8155-1392-5, p. 67 & 70
[6] Liang Shi, Yunle Gu, Luyang Chen, Zeheng Yang, Jianhua Ma and Yitai Qian, "A co-reduction synthesis of superconducting NbC nanorods" *J. Phys. Condens. Matter*, **16** (2004) 8459-8464
[7] Shiro shimada, Tadashi Koyoma, Kohei Kodaira and Taru Mastushita, "Formation of NbC and TaC by solid-state reaction", *J. Mater. Sci.* 18 (1983) 1291-1296.
[8] Jian-Bao Li; Gui-Ying Xu; Sun, E.Y.; et al. "Synthesis and morphology of niobium monocarbide whiskers", *J. Amer. Ceram. Soc.* 81 [6] (1998), 1689-91.
[9] M.I. Mendelson, "Average grain size in polycrystalline ceramics", *J. Amer. Ceram. Soc.*, 52 (1969) 443-446.
[10] Anstis, G. R., Chantikul, P., Lawn, B. R. and Marshall, D. B., "A critical evaluation of indentation techniques for measuring fracture toughness", *J. Amer. Ceram. Soc.*, 64 (1998) 533-538.
[11] ASTM Standard E 1876-99, "Test Method for Dynamic Young's Modulus, Shear Modulus, and Poisson's ratio for Advanced Ceramics by Impulse Excitation of Vibration", ASTM Annual Book of Standards, Philadelphia, PA, 1994.
[12] D. Brown, "The chemistry of Niobium and Tantalum" pp. 553-622 in Comprehensive Inorganic chemistry, Vol. 3, Edited by J. C. Bailer Jr., H. J. Edeleus, R. Nyholm, and A. F. Trotman-Dickenson. Pergamon Press, London, U.K., 1975.
[13] Storms.E.K.1967 Refractory materials Vol 2,ed J.L.Margrave (new York: Academic) P 65.
[14] C. Suryanarayana and M.G. Norton, "X-ray Diffraction: A Practical Approach", Plenum Press, New York/London (1998) .P 157.
[15] König, W., Dauw, D.F., Levy, G., U. Panten, "EDM - future steps towards the machining of ceramics", *Annals of the CIRP*, 37 (1988) 623-631.
[16] R.F. Firestone, "Ceramic Applications in Manufacturing", SME, Michigan, 1988, 133.

HIGH TEMPERATURE STRENGTH RETENTION OF ALUMINUM BORON CARBIDE (AlBC) COMPOSITES

Aleksander J. Pyzik, Robert A. Newman, and Sharon Allen

Core R&D, New Products
The Dow Chemical Company
Midland, Michigan 48674

ABSTRACT

Aluminum Boron Carbide composites (AlBC) represent a family of light weight materials where multi-phase compositions are produced by reacting boron carbide with liquid aluminum. Even though these composites are characterized by very attractive room temperature properties, many applications that could benefit from light weight materials also require high temperature stability. This work has shown that AlBC composites exhibit high strength above the melting temperature of aluminum due to the strongly interconnected multiphase ceramic network that can be protected from further oxidation by formation of a dense alumina layer underneath the original material surface. This layer is a result of the combined oxidation behavior of boron carbide, binary and ternary reaction phases and aluminum metal. Once formed, it protects the AlBC composite material from further oxidation and also inhibits migration of aluminum to the surface. The strength of AlBC composite fabricated from 75 vol% boron carbide preforms that were infiltrated with 6061 Al alloy and heat-treated at 800°C in air showed significant improvement over the same starting material that was neither heat-treated nor oxidized. The room temperature strength of ~450 MPa was maintained until 900°C. Above this temperature, strength was reduced, but was still above 350 MPa at 1200°C. This is significantly higher than the strength of hot pressed boron carbide exposed to the same oxidation conditions and tested at comparable temperatures.

INTRODUCTION

Aluminum Boron Carbide composites (AlBC) represent a family of light weight materials where multi-phase compositions are produced by reacting boron carbide (B_4C) ceramics with liquid aluminum[1-3]. These composites are characterized by very attractive room temperature properties, such as high hardness (1400-1600 kg/mm^2), high Young's modulus (320-360 GPa), high strength (500 MPa) and excellent wear resistance[4,5]. However, many applications that could benefit from light weight materials also require high temperature stability. It is well known[6] that boron carbide itself oxidizes readily above 600°C forming volatile B_2O_3. This progressive reaction leads to significant changes in the composition and structure of the surface region and consequently has a negative effect on the properties and serviceability of boron carbide components at high temperature. However, it has also been reported that the addition of Al and Si dopants improve the oxidation resistance of boron carbide, enabling the material to better retain strength at higher temperatures[7]. AlBC contains both aluminum and small amounts of silicon. In contrast, however, to hot pressed boron carbide, AlBC composite is a complex, multiphase material which contains the reaction products of B_4C and Al: AlB_2, $AlB_{24}C_4$ and $Al_{3-4}BC$. These phases, plus residual B_4C and Al determine the rate of surface oxidation and subsequent high temperature properties. The change in the chemistry and microstructure, due to the oxidation, should have a direct effect on the surface region and as a result on the flexure strength of the material. Therefore, the strength measurement can be used as an indication of the magnitude of surface changes occurring. The purpose of this paper is two-fold; (1) to evaluate the high temperature strength of AlBC composites as a function of starting composition and processing conditions, and (2) to determine the oxidation mechanism in AlBC composite material.

EXPERIMENTAL PROCEDURES
The boron carbide powder ESK 1500s (bimodal) was used in this work. This powder contained 0.67 wt% oxygen, 0.04 wt% nitrogen and 21.27 wt% of carbon. The main impurities were 150 ppm of Si, 250 ppm of Fe and 75 ppm of Ti. The average particle size was 3.8 microns and the surface area was 6.84 m^2/g. In order to prepare boron carbide preforms, boron carbide powder was dispersed in distilled water. The suspensions were ultrasonically agitated and pH was adjusted by using NH$_4$OH to about 7. Slips were aged for 24 hours by rolling them in plastic containers, de-gassed and cast into 4x4 inch forms on a plaster of Paris mold. The densities of dried greenware were between 74 and 75 vol% of theoretical. The infiltration of boron carbide with 6061 aluminum metal was conducted under 100 millitorr of vacuum at 1160°C for 1 hour. After infiltration, dense AlBC coupons were machined into appropriate test specimens. These test samples were then heat-treated in air and argon under specific schedules described in later sections.

Crystalline phases were identified by collecting x-ray diffraction data with a Phillips diffractometer using CuKα radiation and a scan rate of 2 degrees per minute. The chemistry of all phases was confirmed by electron microprobe analysis of polished cross-sections using a CAMECA CAMEBAX electron probe. The accuracy in the determination of elemental composition was better than 3% of the amount present. Microstructures were characterized by using a Nikon Epiphot inverted optical microscope. . The flexure strength of the AlBC composites were measured in four point bending using the ASTM method C1161 for measuring flexure strength of advanced ceramic materials[8]. The specimen size was 3 x 4 x 45 mm. Flexure strength was determined from 8-10 measurements. The upper and lower span dimensions were 20 and 40 mm, respectively. All specimens were broken using a crosshead speed of 0.5 mm/min. Tests were conducted between room temperature and 1300°C under argon and air atmospheres. The heating rate was 5 degrees per minute. After reaching the desired temperature, a 30 minute equilibration period was instituted before application of load.

RESULTS AND DISCUSSION
Microstructure And Composition Of AlBC Composites
Aluminum Boron Carbide Composites are produced by reacting boron carbide with liquid aluminum to form various binary and ternary phases. The reaction temperature(s) and residence time(s) at these temperature(s) determine the quantity and chemistry of these phases. The reactions between B$_4$C and Al start below the melting temperature of aluminum; however, because the metal diffusion rate is very slow, the reaction time required to observe first phases (by x-ray diffraction) exceeds hundreds of hours. The kinetics of chemical reactions increases in the presence of the liquid metal phase. The low temperature phases that form above 660°C are Al$_{3-4}$BC and AlB$_2$. Both of these phases incorporate large amounts of metal, leading to rapid depletion of available Al. Around 970°C[9] the AlB$_2$ phase undergoes peritectic decomposition. In the absence of AlB$_2$, small amounts of Al$_3$B$_{48}$C$_2$ (β-AlB$_{12}$) and AlB$_{24}$C$_4$ (AlB$_{10}$) were also detected, but the overall quantity of reaction phases formed during post heat-treatment decreases. In fact, if dense AlBC samples are held for 10 hours at 700 °C versus 1000°C, the former heat

treatment produces an AlBC part with a significantly lower amount of unreacted free aluminum. Processing above 1000°C results in a moderate increase in the formation of the Al$_{3-4}$BC phase and a rapid increase in the formation of the AlB$_{24}$C$_4$ phase, especially in the temperature range of ~1150-1200°C. The AlB$_{24}$C$_4$ phase is deficient in Al and has higher hardness than either AlB$_2$ or Al$_{3-4}$BC. There are several other reaction phases possible in Al-B-C system, but under controlled heat-treatment conditions, these three are the most dominant.

The AlBC composite that was used for this oxidation study contained 74-75 vol% of starting boron carbide and about 18-20 vol% residual metal after infiltration was completed. The remaining 5-7% of the composite contained Al-B$_4$C reaction phases. The microstructure of this particular material had particles of boron carbide dispersed in a continuous matrix of aluminum alloy. Particles showed some necking, mostly due to a few percent of reaction phases forming between grains. After additional heat treatment the amount of ceramic phases was increased to 90-98% depending on the specific heat treatment conditions. An optical micrograph of typical AlBC bulk microstructure is shown in Figure 1.

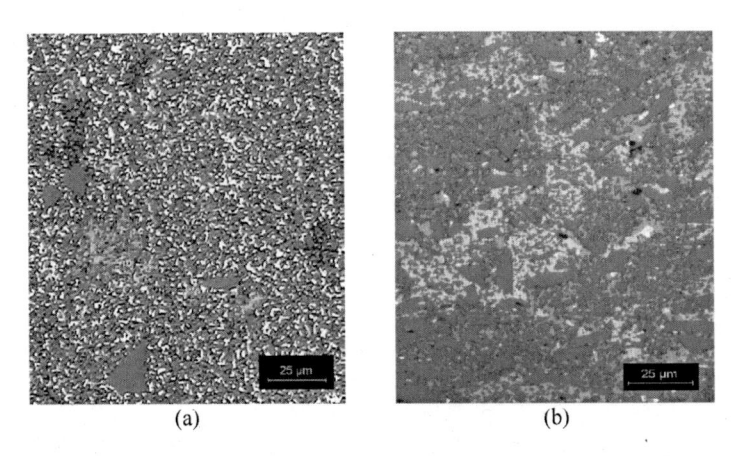

Figure 1. Microstructure of AlBC composite fabricated by infiltration 75 vol% boron carbide preform with 6061 Al alloy (a) directly after infiltration and (b) after additional heat-treatment at 800°C.

Strength of AlBC at Elevated Temperature
Strength was measured in argon and air from room temperature to 1200°C. The strength of AlBC tested in argon is shown in Figure 2 (as Set 1). It is degraded as a function of temperature from 550 MPa at 20°C to about 200 MPa at 1200°C. This is typical behavior that should be expected from a metal matrix composite with a continuous metal phase. Since testing is done in argon there is no damage coming from potential surface oxidation and grain boundary corrosion. The strength reduction is caused primarily by softening and melting of aluminum.

It is surprising however, that above the melting temperature of Al metal (660°C) the AlBC composite still maintained its mechanical integrity. We believe that this is due to the strength of the ceramic network itself. In fact, when boron carbide powder was mixed with a few percent of aluminum and the mixture was heated to form a fully reacted porous body (about 20% open porosity and no residual free metal) the high temperature strength of this material was very close to that shown for Set 1 in Figure 2. In a fully dense AlBC composite, the strength of the ceramic network can be further increased by producing additional binary and ternary B-C-Al phases during heat-treatment. For example, 100 hours of heat treatment of AlBC material at 800°C yields a reduction of free residual metal from 20% to only 5-8%. The effect of this heat-treatment on high temperature strength is presented in Figure 2, as Set 2.

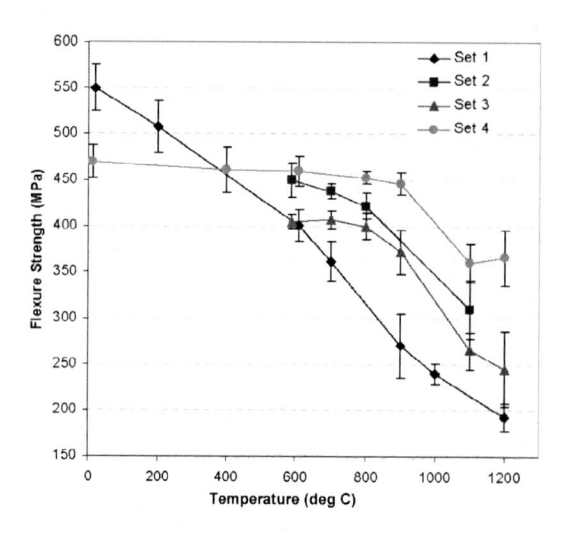

Figure 2. Strength of AlBC composites tested in argon (Set 1), tested in argon after additional heat treatment at 800°C for 100 hours (Set 2), tested in air (Set 3), and tested in air after additional heat-treatment (Set 4).

The low temperature strength of this ceramic-like material (Set 2) decreases, as shown in Figure 2 and reported elsewhere [9,10]. However, while the strength of the starting material (Set 1) continuously degrades vs. temperature, the strength of this 800°C heat-treated material (Set 2) is better retained above 400°C. The strength retention is good until 800°C. Above this temperature the strength curve of Set 2 changes its slope and declines to a value of ~320 MPa at 1100°C. Since all samples were tested under argon atmosphere, the potential effect of surface oxidation was eliminated and measured values correspond to the true strength of bulk AlBC material. The improved ability to maintain strength above the melting temperature of aluminum was mostly attributed to the stronger network of B-C-Al phases.

However, when the original AlBC material composition (represented by Set 1) was tested in air instead of argon, it showed a deviation from its original shape above 600°C. Above 800°C the strength values (Set 3) were closer to the heat-treated version (Set 2) than to starting AlBC composition (Set 1). Since residence time during testing at the given temperature was nearly identical for all samples, this difference cannot be due to changes in the bulk material composition. The observed difference is an indication of a positive contribution coming from the sample surface as a result of composite interaction with air. To verify this observation an additional set of AlBC test bars that was heat-treated at 800°C for 100 hours (identical to the Set 2 material) was also tested in air. The results are represented by Set 4 in Figure 2. Further improvement over Set 2 is evident. The composite strength remains virtually unchanged until 900°C and even at 1200°C the strength is still around 360 MPa (Fig. 2). Therefore, it became obvious that surface oxidation of AlBC composites containing 75 vol% of starting boron carbide contributed to the improvement of the overall material strength. This is in direct contrast to the oxidation of hot pressed pure boron carbide, where room temperature strength of ~375 MPa was reduced to 270 MPa when tested in air at 900°C.

It is important to keep in mind that strength values reported here for the oxidized bars represent the strength of specific AlBC test parts and not AlBC as a material. This is because strength was measured using parts that were first machined into 4-point bend bars, then exposed to air and tested. The same heat-treatment in air applied to the bulk material that was later machined into 4-point bend bars resulted in lower strength values, as presented in Figure 3. The Set 1A in this graph represents the strength of AlBC bars made from 75 vol% B_4C preforms that were infiltrated with aluminum and later oxidized at 800°C for 100 hours (the same as Set 4 in Figure 2). Set 2A shows the same material, however, that was first heat-treated at 800°C for 100 hours

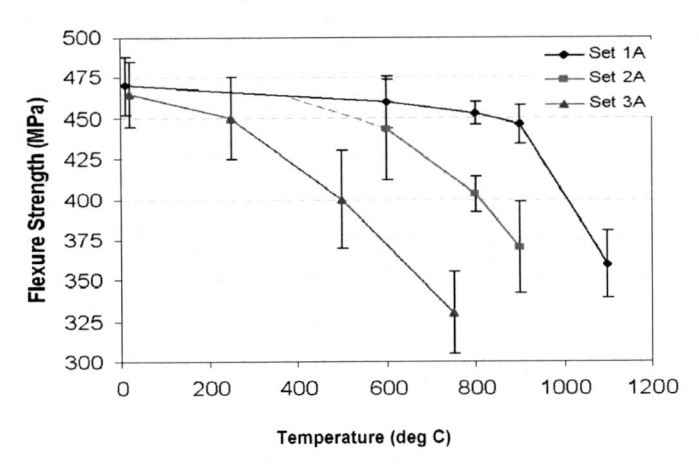

Figure 3. Strength of AlBC composites tested in air: Set 1A - 75 vol% B_4C preform Infiltrated and heat-treated at 800°C in air as 4-point bend bars; Set 2A - 75 vol% B_4C preform infiltrated and heat-treated at 800°C in air as bulk material that was then machined into bars; Set 3A - 62-63 vol% B_4C preform infiltrated and heat-treated at 800°C in air as bulk material that was later machined into bars.

as a single large part, then machined into test bars and tested. The initial portions of both curves overlap, but the advantage of long time surface oxidation becomes obvious at higher temperature.

Another factor that apparently contributes to high temperature strength is the initial content of boron carbide. The data for Set 3A (Figure 3) shows high temperature strength of AlBC that was prepared from 62-63 vol% B4C preforms, infiltrated with aluminum and reacted under conditions giving the final total amount of ceramic phases (and as a result total amount of unreacted aluminum) the same as in material represented by Set 2A of Figure 3. In both cases, the heat-treatment was finished before machining material into test bars. The difference is substantial; demonstrating that the amount of boron carbide in the starting preform is of primary importance and cannot be replaced by the formation of in-situ binary and ternary reaction phases. The positive effect of higher boron carbide content can be explained by the smaller B4C grain size versus the typical reaction phases, and by the fact that AlBC materials with high boron carbide content also tend to contain higher amounts of AlB24C4 reaction phase. This phase, as explained in the next section, can be important in slowing penetration of the oxidation layer.

Effect Of Long Term Oxidation On AlBC Strength And Surface Chemistry
In order to determine long-term oxidation effect on strength, four-point bend bars of AlBC composite (containing 75 vol% of starting B4C) were heated in air using the schedule outlined below. The first set (A) was exposed to 800°C for 50 hours followed up by 900°C for 200 hours. The second set (B) was heat-treated at 800°C for 50 hours and then at 1100°C for 50 hours. Finally, the third set (C) experienced the same conditions as set B plus an additional oxidation step at 1300°C for 5 hours. After heat-treatment was completed, all bars were broken at 900°C. The flexural strength of set A was 410 ± 25 MPa, set B was 320 ±18 MPa and set C was 277 ± 63 MPa. Even though AlBC test bar sets A and B were exposed to relatively high temperatures, they have higher strength value than unreacted original AlBC material tested at 900°C. In order to examine the reason for this behavior, all three types were examined microscopically. The bulk composition of the first two materials was similar; the main reaction phases were identified as AlB2, AlB24C4 and Al3-4BC. The bulk composition of set C had AlB24C4 and Al3-4BC but also about 5-8 % Al4C3 and only minor amounts of AlB2. All three materials revealed formation of an oxidation layer underneath the original sample surface. This layer varied in thickness (20-35 mm) and consisted of the two sub-layers, the inner and outer, as shown in Figure 4.

The inner layer(s) in all samples appeared to be dense and consisted primarily of alumina with small amounts of boron (6% in sample set A, 4% in sample set B and 1% in sample set C). The average thickness of this layer was about 2-3 microns (ranging from 1-8) in sample set A and 8 microns (5-25 microns) in sample set C exposed to 1300°C. The outer layer in AlBC composite subjected to oxidation at 900°C appeared to be quite porous and its composition was identified as primarily gamma alumina. The outer layer in the 1100°C heat-treated sample set B appears to be mostly dense, similar to its inner layer. This layer has about 2% of residual boron. Finally, the outer layer in sample set C that was exposed to heat-treatment at 1300°C consists of friable and porous alpha alumina as shown in Figure 5.
The inner layer(s) in all samples appeared to be dense and consisted primarily of alumina with small amounts of boron (6% in sample set A, 4% in sample set B and 1% in sample set C). The average thickness of this layer was about 2-3 microns (ranging from 1-8) in sample set A and 8 microns (5-25 microns) in sample set C exposed to 1300°C. The outer layer in AlBC composite subjected to oxidation at 900°C appeared to be quite porous and its composition was identified as primarily gamma alumina. The outer layer in the 1100°C heat-treated sample set B appears to be mostly dense, similar to its inner layer. This layer has about 2% of residual boron. Finally, the outer layer in sample set C that was exposed to heat-treatment at 1300°C consists of friable and porous alpha alumina as shown in Figure 5.

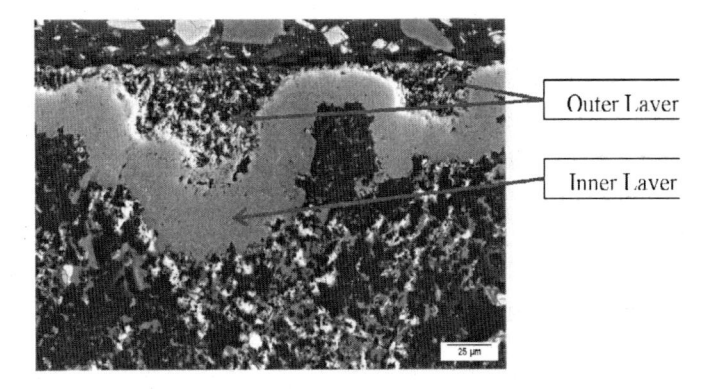

Figure 4. Oxide layer formed during oxidation of AlBC composite surface (red dashed line) consists of two sub-layers, the inner and outer, which vary in thickness, density and composition depending upon oxidation conditions. This particular cross-section was taken from material heat-treated at 900°C.

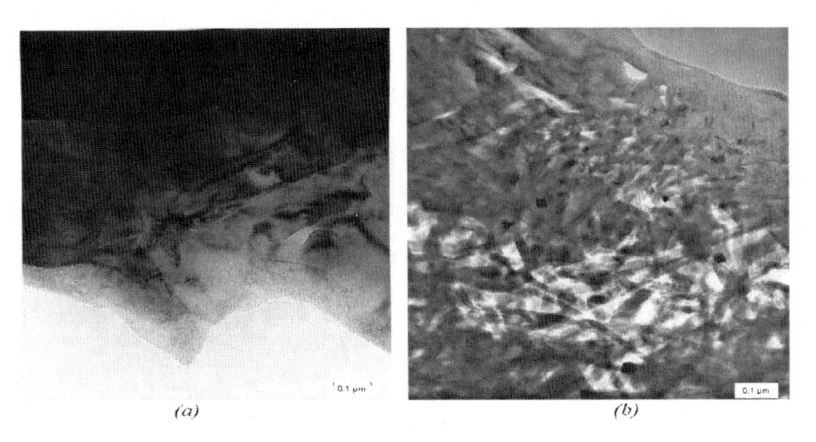

Figure 5. The structure of outer layers in AlBC samples oxidized at (a) 1100°C, and at (b) 1300°C (by Transmission Electron Microscopy).

The formation of the oxidation layer is very slow below the melting temperature of aluminum. However, above 660°C it grows rapidly to about 1-2 microns thickness at 700°C and to about 5-10 microns at 800°C. This thickness does not change even when exposed to 800°C for 500 hours. When AlBC is heated above 800°C, the oxidation layer grows thicker (~30 microns at 900°C /100 hours).

High Temperature Strength Retention of Aluminum Boron Carbide (AlBC) Composites

The uneven depth and waviness of the oxidation layer seemed to be dependent upon the local composition underneath the surface. The areas which had a majority of $AlB_{24}C_4$ (or $AlB_{24}C_4$ + B_4C) exhibited only very shallow oxide penetration, and formed an alumina layer that was fully dense (see Figure 6). This may be due to the fact that this ternary phase is typically surrounded by free aluminum. Free aluminum oxidizes rapidly during heating, forming a protective oxide layer which would prevent further penetration within these regions containing the $AlB_{24}C4$ phase. The other possibility is that $AlB_{24}C_4$ has better oxidation resistance than the other phases present, and this contributes to slower penetration. Deep penetration of the oxide layer was associated with regions rich in B_4C, AlB_2 and $Al_{3-4}BC$. The alumina layer in these regions consisted of two very distinctive inner and outer layers, as discussed previously. We hypothesize that in these regions the oxidation front moves until it encounters sufficient amounts of free aluminum to form a continuous alumina barrier coating. The observed difference in rate of barrier formation between different regions of AlBC composite can explain different strengths, as discussed in Figure 3. AlBC composites that are reacted to form multiple ceramic phases and then exposed to oxidation do not have a sufficient amount of free aluminum to form a dense and continuous protective barrier layer. The oxidation damage is deeper and the underneath surface

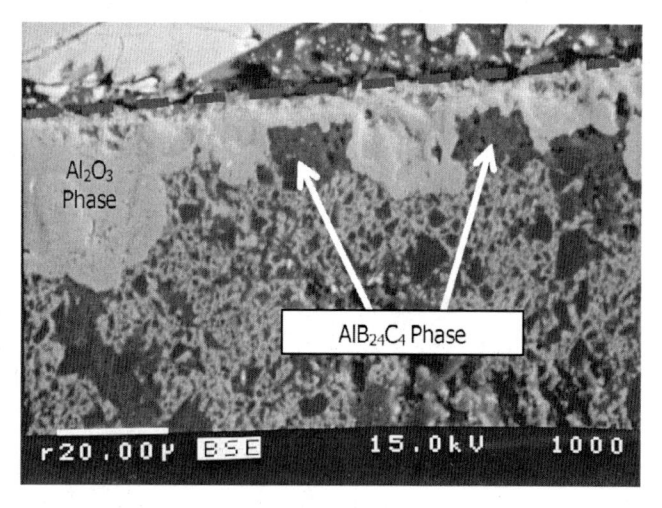

Figure 6. The uneven depth and waviness of the oxidation layer appears to depend on the local composition underneath the material surface (denoted by the red dashed line).
The areas which have majority of $AlB_{24}C_4$ phase exhibit shallow oxide penetration and the layer of alumina is fully dense. Deeper penetration of outer and inner alumina layers is observed in regions containing B_4C, $Al_{3-4}BC$, and AlB_2.

zone left behind the oxidation front is porous. As a result, the oxidation progresses as a function of time and the overall weight gain is higher than in samples containing more aluminum, as shown in Figure 7. However, it should be noted that the oxidation rate of both samples is still lower than pure boron carbide.

The protective oxidation treatment needs to be applied in the early stages of post-infiltration processing when more aluminum is still available. The formation of the protective oxidation layer was very slow below the melting temperature of aluminum. Above 660°C, however, the layer which initially consisted of B_2O_3 and Al_2O_3 grew rapidly to reach a thickness of 1-2 microns at 700°C and about 5-10 microns at 800°C after 100 hours of oxidation. When there is enough of free aluminum available, the coating is dense and the thickness remains almost constant, even after 500 hours exposure to air at 800°C. When as-infiltrated AlBC composite is oxidized at 900°C, the average thickness of the oxidation layer is ~30 microns after 100 hours.

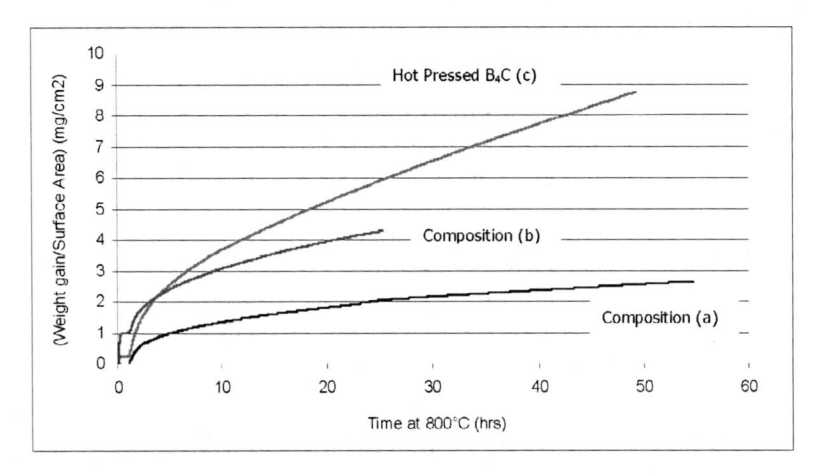

Figure 7. Weight gain at 800°C due to the oxidation of AlBC composites & hot pressed B_4C. (a) AlBC with 20-25 vol% of free aluminum; (b) AlBC composite with 5-10 vol% (c) of free aluminum; (c) weight gain due to oxidation of hot pressed B_4C.

CONCLUSIONS

Aluminum boron carbide composites exhibit high strength well above melting temperature of aluminum due to the strongly interconnected multiphase ceramic network that is protected from further oxidation by the formation of a dense alumina layer beneath the material surface. This barrier layer is a result of the combined oxidation behavior of boron carbide, binary and ternary aluminum-boron-carbon phases and aluminum metal. Once formed, it protects the AlBC composite from further oxidation and also inhibits the migration of aluminum to the surface. The protective alumina layer consists of two sub-layers, inner and outer. Their combined thickness depends on the oxidation conditions and may vary from a few microns at 700°C to 30 microns above 900°C. The continuity of this protective layer depends on the availability of free aluminum in the AlBC composite. The best results were obtained when heat-treatment to produce in situ aluminum-boron-carbon phases was combined with simultaneous surface oxidation and when the starting material composition was rich in aluminum. It appears that the most critical ceramic phases for the overall improvement of the oxidation resistance and high temperature strength are boron carbide and the ternary $AlB_{24}C_4$ phase.

The depth of the oxidation zone varies, depending on local material composition. Typically, the oxide layer is positioned deeper beneath the original surface when the local region consists of a mixture of B_4C, AlB_2 and $Al_{3-4}BC$ phases. In regions rich in $AlB_{24}C_4$ phase, the alumina layer is

situated just beneath the original part's surface. More work is needed, however, to determine the exact reasons for this behavior.
The strength of AlBC composite fabricated from 75 vol% dense boron carbide preform infiltrated with 6061 Al alloy and heat-treated at 800°C in air showed significant improvement over the same starting material that was neither heat-treated nor oxidized.

A strength of 450 MPa was extended up to 900°C, versus strength of ~370 MPa for AlBC composite that was not pre-oxidized, as shown in Figure 2. Also, a strength >350 MPa was measured at 1200°C, versus strength of <250 MPa for AlBC composite that was not pre-oxidized. This is higher than the strength of hot pressed boron carbide or any other aluminum-based MMC exposed to the same oxidation conditions and tested at comparable temperatures.

ACKNOWLEDGEMENTS
The authors would like to thank, Nick Shinkel, Amy Wetzel and Tom Collins for sample fabrication, characterization and mechanical testing; and John Blackson for analytical and microstructural characterization.

REFERENCES
[1] A.J Pyzik, I.A. Aksay, M. Sarikaya; Microdesigning of Ceramic-Metal Composites, in Ceramic Microstructures 86', ed. by J.A. Pask and A.G. Evans, Plenum Press, New York, Materials Science Research, v. 21, pp. 45-54, 1987.
[2] D.C. Halverson, A.J. Pyzik, I. A. Aksay, W.E. Snowden; Processing of Boron Carbide – Aluminum Composites, J. American Ceramic Society, 72, (5), 775, 1989.
[3] I. A. Aksay, D.M. Dabbs, J.T. Staley, M. Sarikaya; Bioinspired Processing of Ceramic Metal Composites, Third Euro-Ceramics, ed. P. Duran and J. F. Fernandez, vol. 1, pp.405-418, 1993.
[4] A. J. Pyzik, D. R. Beaman; Al-B-C Phase Development and Effects on Mechanical Properties of B_4C/Al Derived Composites, Journal American Ceramic Society, 78,(2), pp.305-12, 1995.
[5] A.J. Pyzik, R.A. Newman, A. Wetzel and E. Dubensky, Composition Control in Aluminum Boron Carbide Composites, Proceedings of 30th International Conference on Advanced Ceramics And Composites at Cocoa Beach, Fl, The American Ceramic Society, 2006.
[6] G.N. Makarenko; Borides of the IVb Group, in Boron and Refractory Borides, edited by V.I. Matkovich, Springer-Verlag Berlin Heidelberg New York, 1977, pp. 310-330.
[7] Yu. G. Gogotsi, V.P. Yaroshenko, F. Porz, Oxidation Resistance of Boron Carbide-Based Ceramics, Journal of Materials Science Letters, 11, 308-10, 1992.
[8] ASTM testing standard.
[9] J.C.V. Bouix, G. Gonzalez, C. Esnouf; Chemical Reactivity of Aluminum with Boron Carbide, Journal of Materials Science, 32, 4559-4573, 1997.

Environmental Effects

CORROSION RESISTANCE OF CERAMICS IN SULFURIC ACID ENVIRONMENTS AT HIGH TEMPERATURE

C.A. Lewinsohn, H. Anderson, and M. Wilson
Ceramatec Inc.
Salt Lake City, UT, 84119

A. Johnson
University of Nevada – Las Vegas
Las Vegas, NV

The Sulfur-Iodide (SI) process has been investigated extensively as an alternate process to generate hydrogen through the thermo-chemical decomposition of water.[1] The success of these high temperature processes is dependent on the corrosion properties for the materials of construction. Super-alloys are often considered because of their ability to be manufactured into heat exchangers and reactors by traditional fabrication methods. The creep and oxidation properties of these metals remain problematic due to these extreme temperatures (900°C) and corrosive environments. However, ceramic materials have been noted for their excellent corrosion resistance. In cooperation with the DOE and the University of Nevada, Las Vegas (UNLV), ceramic based micro-channel decomposer concepts are being developed and tested by Ceramatec, Inc.. In order to assess the viability of ceramic materials, extended high temperature exposure tests have been made to characterize the degradation of the mechanical strength and estimate the recession rates due to corrosion. The results of these corrosion studies will be presented with additional analysis including surface and depth profiling using high resolution electron microscopy.

INTRODUCTION

One potential problem with the utilization of the sulfur-iodine thermochemical cycle to produce hydrogen is that the final reaction in the cycle involves the decomposition of sulfuric acid at elevated temperatures.[2] This final step is a potential obstacle, because it creates an environment that varies significantly from most corrosive environments used to test the corrosion resistance of materials. Previous corrosion resistance studies on ceramics have been conducted in numerous environments including: combustion, gaseous N_2-H_2-CO, coal slag, air, dry and wet oxygen and even O_2-H_2O-CO_2 gaseous environments.[3,4,5,6] Despite the fact that several corrosion studies have been conducted, few studies involved exposure to environments of high temperature sulfuric acid decomposition. As a result, little is known about which materials could withstand such a harsh environment. This lack of knowledge regarding materials compatibility with decomposing sulfuric acid is an obstacle since this type of an environment must be endured and, more importantly, contained during a portion of the hydrogen production process.

In 1981, Irwin and Ammon of Westinghouse Electric Corporation conducted a materials screening program to identify potential structural materials for an acid vaporizer. Candidate materials, which included both metals and ceramics, were exposed to sulfuric acid, H_2SO_4, at temperatures ranging from 361°C to 452°C for 250-hour increments up to 1000 hours to assess corrosion compatibility with the process stream. Irwin and Ammon found that silicon and materials containing significant amounts of silicon, such as silicon carbide and silicon nitride, have the greatest resistance to attack by boiling sulfuric acid.[7] Fernanda Coen-Porisini of the

Commission of the European Communities JRC Ispra Establishment performed corrosion studies in sulfuric acid at 800°C in 1979.[8] These corrosion tests were mostly conducted on metal samples, but two of the tests also used a few ceramic samples. Both of these tests showed that alumina, mullite, and zirconia, which were the few ceramics tested, were unchanged or only displayed a deposited coating on the surface while all metals tested displayed considerable to severe corrosion.

Materials testing more relevant to the actual application of interest in this study, housing the thermochemical production of hydrogen, was conducted by T. N. Tiegs of Oak Ridge National Laboratory in 1981.[2] Tiegs identified a number of candidate materials from which he chose the leading materials for testing. The candidate materials were reported in decreasing order of preference as follows: SiC, MgO, MAS, Al_2O_3, Si_3N_4, sialon and BeO. The actual materials selected for testing, though, are SiC, Sialon, MgO, $ZrO_2(MgO)$ and $ZrO_2(Y_2O_3)$. These selected materials were then tested in an environment designed to simulate the decomposition of sulfuric acid. From the analysis of the test specimens following their exposure, Tiegs identified silicon carbide as the best material at 1000 and 1225°C in the simulated sulfuric acid decomposition environment. Tiegs recommended further testing for the SiC materials at conditions more representative of an actual sulfuric acid decomposition environment, that is, at temperatures of 800 to 900°C and pressures up to 3 MPa.

Ishiyama et al. report the percent weight change and corrosion rate of samples resulting from 100 hours of exposure to high-pressure boiling sulfuric acid. As seen from the results, SiC was the most corrosion resistant followed by Si-SiC and then by Si_3N_4. Also in Ishiyama's overall rating of the materials after 1000 hours of exposure the three above mentioned materials were listed as all being the least affected by the long exposure.

Based on the findings of these studies, silicon carbide and silicon nitride seem to be the most appropriate candidates for heat exchanger/decomposer materials and warrant further corrosion testing. This paper will describe the results from a series of corrosion tests to investigate the influence of the concentration of sulfuric acid and temperature on the corrosion of candidate ceramic materials for compact microchannel heat exchangers. Weight change, surface analysis, microscopy and mechanical testing were used to evaluate the effects of corrosion on the candidate materials. Alumina samples were included in the test matrix to verify the superior corrosion resistance of silicon-based materials and to compare their behavior with a material without silicon.

METHODS
1 – Experimental Setup

The corrosive environments for this exposure testing are selected to more closely mimic the decomposition environment of sulfuric acid and the expected conditions in the actual application of interest. The corrosion test setup consists of a long quartz tube partially housed inside a split tube furnace. The long quartz tube itself holds three large quartz cups and three small quartz cups as displayed in Figure 1. Starting at the top is a large quartz cup filled with quartz chips which acts as an evaporator and gas preheater. Below the evaporator cup sit the three small cups that hold the samples. Below the three sample cups are two large condenser cups, the top of which is filled with Zirconia media and the bottom with SiC media. The long quartz tube is capped on top by a solid Teflon manifold with a pliable Teflon gasket so that the sulfuric acid vapor and decomposition products stay trapped in the tube. This manifold is fitted with gas (air/oxygen) feed and a liquid (sulfuric acid) feed. In addition, the condensate is collected and disposed of in an appropriate waste barrel.

ASTM Standard C 1161-02C bend bars were scribed, weighed and randomly positioned within three small sample cups for sulfuric acid exposure. The simulated conditions were 60% H_2SO_4, 30% H_2O and 10% air at 900°C. These sample cups were then loaded into the quartz tube and the furnace was heated up to 900°C with flowing argon gas. Once at temperature, the simulated sulfuric acid environment was attained by switching over to air or oxygen from argon and by dripping in the acid solution. At predetermined intervals (100, 200, 500 and 1000 hours), samples were removed, weighed and fractured according to ASTM Standard C 1161-02C procedures. Table I describes the materials that were tested.

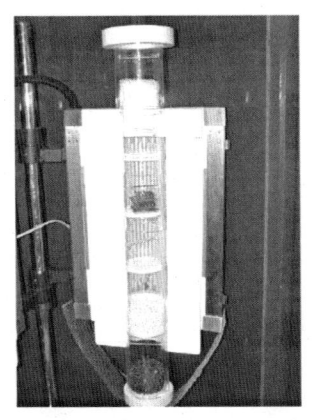

Figure 1. Sulfuric Acid Exposure Rig.

Table I

Material	Fabrication Process	Vendor	Identifier
Silicon Nitride	Hot Pressed	Ceradyne	SN-HP
Silicon Nitride	Gas Pressure Sintered	Ceradyne	SN-GP
Silicon Carbide	Pressureless Sintered	Morgan	SiC-PS
Silicon Carbide	Tape Laminated	Ceramatec	SiC-LS

RESULTS AND DISCUSSION

As seen in Figure 2, all of the materials experienced an increase in weight as a result of exposure to a sulfuric acid decomposing environment at 900 °C. The weight gain was so small that large standard deviations were often obtained with the average weight changes, and this reduces the confidence in determining the weight change between the different exposure times of 100 hours, 500 hours and 1000 hours. Spalling of corrosion products occurred whilst handling specimens of hot pressed silicon nitride after 1 000 h leading to an apparent low in weight.

As seen in Figure 3, there was no drastic drop in the average flexural strength of any of the materials with 1000 hours of exposure. The strengths of the silicon-based materials, on the other hand, were shown to have changed slightly over the 1000 hours of exposure as seen in Figure 3. Specifically, the flexural strength of SN-HP, SiC-LS and SiC-PS increased over the first 100 hours of exposure and then remained constant. The flexural strength of SN-GP showed a similar trend, but didn't increase until the 100 to 500 hours exposure time frame. This slight increase in strength of the silicon-based materials is believed to be a result of the blunting of surface defects caused by exposure to 900°C for expended time periods and the accumulation of silica on exposed surfaces.

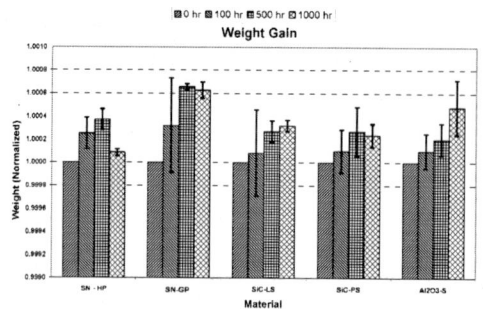

Figure 2. Weight Gains for 1000 hr Exposure Samples

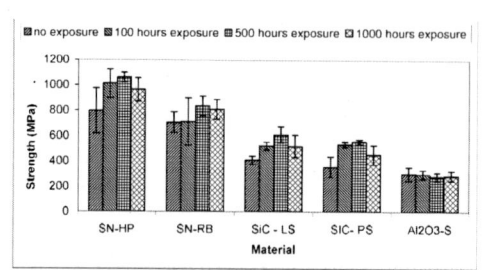

Figure 3. Strength Change for 1000 hr Exposure Samples

SEM images, shown in Figure 4, of unexposed and exposed samples after 1000 hours for each of the silicon-based materials, indicate the presence of a minor amount of reaction product. In an attempt to identify the reaction product, EDS and XPS analyses, were conducted on an unexposed bar and a bar after each of the different exposure times. XPS analysis confirmed that the reaction product was SiO_2 (Figure 5). Semi-quantitative EDS analysis, Figure 6, shows that the amount of oxygen on the exposed surfaces increases with exposure time. This results imply that the presence of sulfuric acid did not prevent the formation of silica on the surfaces of the silicon based materials that were tested and that this silica reaction product healed surface flaws leading to higher flexural strength values after corrosion testing. Future efforts will include performing thermodynamic calculations to verify whether the presence of sulfuric acid vapor changes the activity of the reactive species and to determine whether the microstructural observations agree with equilibrium phases or reactions limited by kinetics.

SUMMARY

In summary, silicon-based ceramics offer potential for application in sulfuric acid decomposition environments, such as those required in the thermochemical generation of hydrogen using the Sulfur-Iodide process. The effect of exposure to sulfuric acid vapour, in air, at high temperature on several silicon-based ceramics was investigated. These exposures caused silica films to form on the surfaces of the materials, due to oxidation, and subsequently an apparent increase in the flexural strength of the materials.

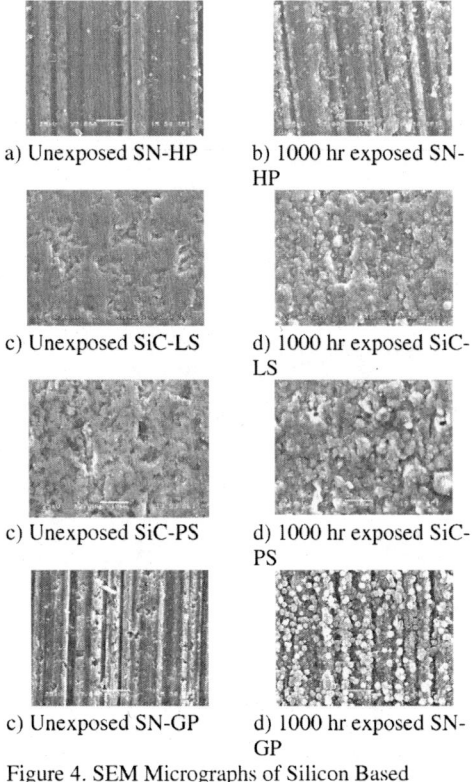

a) Unexposed SN-HP

b) 1000 hr exposed SN-HP

c) Unexposed SiC-LS

d) 1000 hr exposed SiC-LS

c) Unexposed SiC-PS

d) 1000 hr exposed SiC-PS

c) Unexposed SN-GP

d) 1000 hr exposed SN-GP

Figure 4. SEM Micrographs of Silicon Based Ceramics (2000x)

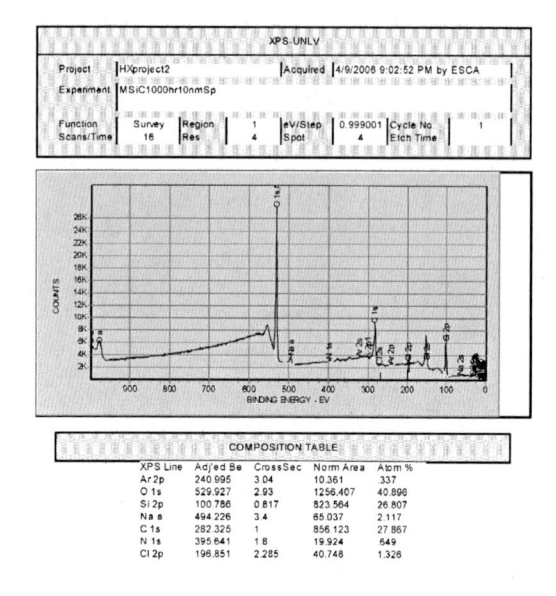

Figure 5. XPS analysis of SiC specimen after 1000 h of testing.

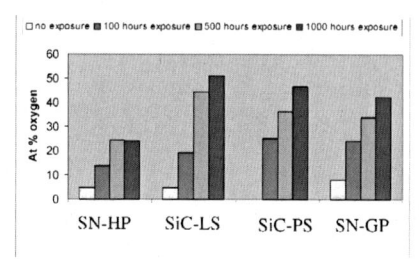

Figure 6. Variation in Presence of Oxygen on Exposed Surfaces of Silicon Based Materials

REFERENCES

[1] *Nuclear Hydrogen R&D Plan*. Department of Energy, Office of Nuclear Energy, Science and Technology. (2004, March). Retrieved January 17, 2006, from http://www.hydrogen.energy.gov/pdfs/nuclear_energy_h2_plan.pdf

[2] Tiegs, T. N. (1981, July). *Materials Testing for Solar Thermal Chemical Process Heat*. Metals and Ceramics Division, Oak Ridge National Laboratory. Oak Ridge, Tennessee. ONRL/TM-7833, 1-59.

[3] Jacobson, Nathan S. (1993). *Corrosion of Silicon-Based Ceramics in Combustion Environments.* J. Am. Ceram. Soc. 76 [1], 3-28.
[4] Opila, Elizabeth J., Smialek, James L., Robinson, Raymond C., Fox, Dennis S., Jacobson, Nathan S. (1999). *SiC Recession Caused by SiO_2 Scale Volatility under Combustion Conditions: II, Thermodynamics and Gaseous-Diffusion Model.* J. Am. Ceram. Soc. 82 [7], 1826-34.
[5] Vaughn, Wallace L. and Maahs, Howard G. (1990). Active-to-Passive Transition in the Oxidation of Silicon Carbide and Silicon Nirtide in Air. J. Am. Ceram. Soc. 73 [6], 1540-43.
[6] Jacobson, Nathan S., Opila, Elizabeth J. and Lee, Kang N. (2001). *Oxidation and corrosion of ceramics and ceramic matrix composites.* Current Opinion in Solid State and Materials Science 5, 301-309.
[7] Irwin, H.A., Ammon, R. L. *Status of Materials Evaluation for Sulfuric Acid Vaporization and Decomposition Applications.* Adv. Energy Syst. Div., Westinghouse Electric Corp., Pittsburg, PA, USA. Advances in Hydrogen Energy (1981), 2(Hydrogen Energy Prog., Vol. 4), 1977-99.
[8] Coen-Porisini, Fernanda. *Corrosion Tests on Possible Containment Materials for H2SO4 Decomposition.* Jt. Res. Counc., ERATOM, Ispra, Italy. Advances in Hydrogen Energy (1979), 1(Hydrogen Energy Syst., Vol. 4), 2091-112.

ANALYZING IRRADIATION-INDUCED CREEP OF SILICON CARBIDE

Yutai Katoh, Lance Snead, and Stas Golubov
Materials Science and Technology Division
Oak Ridge National Laboratory
1 Bethel Valley Road
Oak Ridge, TN 37831

ABSTRACT

Irradiation creep, which is among the major lifetime-limiting mechanisms for nuclear structural materials, is stress-driven anisotropic plastic deformation occurring in excess of thermal creep deformation in radiation environments. In this work, experimental irradiation creep data for beta-phase silicon carbide (SiC) irradiated at intermediate temperatures is analyzed using a kinetic model with an assumed linear-coupling of creep strain rate with the rate of self interstitial atom (SIA) absorption at SIA clusters. The model reasonably explains the experimentally observed time-dependent creep rate of ion-irradiated SiC and swelling evolution of ion- and neutron-irradiated SiC. Bend stress relaxation behavior during irradiation was then simulated using the developed model to examine the experimental data obtained by neutron irradiation experiments. Recommended directions of future experiment are provided to further verify and improve the models and assumptions in this work.

INTRODUCTION

Creep deformation and rupture are major lifetime-limiting mechanisms for high temperature materials for structural applications. In nuclear environments, irradiation creep is additive to thermally activated creep deformation [1]. Irradiation creep is anisotropic plastic deformation driven by external or internal stresses, and is induced or enhanced by irradiation in excess of thermal creep deformation. In many cases, irradiation creep is caused by the preferred nucleation of planar defect clusters and/or preferred absorption of supersaturated point defects at edge dislocations in favor of stress relaxation [2]. Because irradiation creep is typically not very sensitive to temperature, irradiation creep tends to dominate at relatively low temperatures. For structural ceramics and ceramic composites for nuclear applications, irradiation creep is potentially one of the most critical unresolved issues [3]. In this work, limited experimental irradiation creep data for beta-phase silicon carbide (SiC) is analyzed using a classical kinetic rate theory model.

MODEL DESCRIPTION

In mean-field rate theory of defect evolution in irradiated materials, time derivatives of the Frenkel defect concentrations are expressed by following equation.

$$\dot{C}_{i,v} = G_{i,v} - \mu_R D_i C_i C_v - D_{i,v} C_{i,v} \sum S^{i,v} \tag{1}$$

where G is the generation rate, μ_R is the recombination coefficient, D is the diffusivity, S is the sink strength of various extended defects, and subscripts i and v denote self interstitial atom (SIA) and vacancy, respectively. Recombination coefficient is given by $\mu_R = 4\pi r_R/\Omega$, where r_R is the recombination radius and Ω is the atomic volume. Equation (1) assumes $D_i \gg D_v$. In SiC, vacancies are practically immobile below ~800°C, as reported by various fundamental experiments [4, 5]. Recent *ab-initio* calculation studies report minimum migration energies of ~3.5 eV and ~2.4 eV for C and Si vacancies, supporting the extremely limited vacancy mobility at such temperatures [6]. For intermediate temperature, at which only SIAs are practically mobile, Eq. (1) can be rewritten as

$$\dot{C}_i = G - \mu_R D_i C_i C_v - D_i C_i \sum S_i \quad (2)$$

$$\dot{C}_v = G - \mu_R D_i C_i C_v \quad (3)$$

when $G = G_i = G_v$ is assumed.

Irradiation-produced defects observed in SiC by transmission electron microscopy after irradiation at ~300 - ~1000°C are small (typically a few nanometer-diameter) dislocation loops and even smaller unidentified defect clusters which likely are loops also. These clusters and/or loops are believed to be of interstitial-type, because of the lack of vacancy mobility. For simplicity, all the defect clusters are treated as extrinsic dislocation loops, and the loop number density is assumed to be the temperature-dependent constant. These are important assumptions because dislocation loops may not be the defects which are primarily responsible to the macroscopic radiation effects such as swelling and irradiation creep [7]. Now, because thermal dissociation of SIA clusters may be inhibited due to high SIA formation energy (>~6 eV [8]) in SiC, the time derivative of the average loop radius r_l is determined only by SIA influx and is given by:

$$\dot{r}_l = D_i C_i / b \quad (4)$$

where b is the magnitude of Burgers vector. Additionally, microstructural sinks other than SIA clusters are ignored, because typical mean inter-spacing of SIA clusters are <~10 nm, which is more than one order smaller than typical subgrain sizes in chemically vapor-deposited SiC. Thus, total sink strength for freely migrating SIAs is given as:

$$\sum S_i \approx S_l^i = 2\pi Z_l^i r_l N_l \quad (5)$$

where Z_l^i, taken as 1 in this work, and N_l are loop-interstitial bias factor and loop number density, respectively.

Mechanisms commonly used to explain irradiation creep phenomena of metallic materials are stress-induced preferred loop nucleation or alignment (SIPN), stress-induced preferred absorption of SIAs at dislocations (SIPA), and preferred absorption-enabled glide (PAG) [2, 9]. Both SIPN and SIPA require only SIA clusters and migrating SIAs to operate, hence are potential operating mechanisms in SiC at temperatures of current interest. Because there are no glissile dislocations observed in SiC irradiated at relatively low temperatures, PAG can not explain irradiation creep in SiC. Models of SIPN and SIPA both predict linear stress dependence of creep rates for metallic materials, which is in accordance with most experimental observations for metals [2].

In SiC irradiated at <~1000°C, configurations of SIA clusters have not been characterized sufficiently to discuss kinetic models of irradiation creep [10]. Moreover, because the growth rate of tiny SIA clusters is so slow (cluster sizes remain a few nanometers at 7.7 dpa and 800°C), it is very difficult to differentiate SIPN and SIPA even if either is the primary operating mechanism. However, for either mechanism, as the first order approximation, it will be reasonable to assume the creep rate proportional to the total SIA influx to SIA clusters, thus

$$\dot{\varepsilon}_{ic}/\sigma = k_{ic} D_i C_i S_I^i \tag{6}$$

assuming linear stress dependency is maintained, and k_{ic} is a coupling coefficient for creep rate and SIA absorption at SIA clusters.

Minimum migration energies for C and Si SIAs in cubic SiC were estimated to be ~0.5 and ~1.4 eV, respectively, by a recent *ab-initio* calculation [6]. Because of the expected high diffusivities (relative to evolutions of other types of defects), the SIA concentration can be treated as in pseudo-equilibrium with vacancies and other sinks. The C and Si Frenkel defects do not have to be differentiated with this pseudo-equilibrium treatment and vacancy immobility, when antisite defects are neglected. During most of the time of interest (>~10^{-3} dpa), $\mu_R C_v \gg S_I^i$ is true, then

$$C_i \approx G/\mu_R D_i C_v \tag{7}$$

$$\dot{C}_v \approx S_I^i G/\mu_R C_v \tag{8}$$

From Eqs. (4), (5), (7), and (8), the following relations are derived.

$$C_i \approx \frac{1}{D_i}\left(\frac{Gb}{9\pi N_I \mu_R t^2}\right)^{1/3} \tag{9}$$

$$C_v \approx \left(\frac{3Gt}{\mu_R}\right)^{2/3}\left(\frac{\pi N_I}{b}\right)^{1/3} \tag{10}$$

$$r_I \approx \left(\frac{3Gt}{\mu_R \pi N_I b^2}\right)^{1/3} \tag{11}$$

The approximated analytical result presented here is consistent with the general analytical solution for the kinetics of low-temperature irradiation processes in metals provided by Golubov

Analyzing Irradiation-Induced Creep of Silicon Carbide

[11]. With the analysis presented here, Eqs. (6), (9), and (11) indicate that the irradiation creep rate is proportional to the -1/3 power of time or dose. Additionally, if low temperature swelling of SiC is assumed to be accounted solely by vacancies, Eqs (10) and (11) imply that irradiation creep rate is in inverse proportion to the square root of swelling. This also suggests that, under a constant stress, a linear compliance that explicitly couples irradiation creep strain and swelling may exists.

RESULTS AND DISCUSSION

Computational analysis and comparison with experimental data

In Fig. 1, calculated evolution of SIA absorption rate at SIA clusters, $D_i C_i S_i^i$, is presented. Parameters used for this calculation are listed in Table. 1. The -1/3 power relationship is achieved at a dose of $\sim 10^{-3}$ dpa, approximately when a recombination-dominant condition ($\mu_R C_v \gg S_i^i$) is established, and continues until onset of the saturation behavior. A dose of $\sim 10^{-3}$ dpa is also typical before which SIA cluster nucleation through a homogeneous nucleation process nears completion. The calculation here incorporated cascade re-solution terms for SIA cluster growth and SIA production, which enables the saturation behavior of defect accumulation that is widely observed in experiments. The effect of cascade re-solution becomes apparent between 0.1 and 1 dpa in this calculation, as shown in Fig. 1. Equations (2)-(4) were slightly modified to take the effect of stoichiometric constraint for SIA clusters into consideration.

Table. 1. Parameters used in calculation.

Parameter (*Symbol*)	Value	Unit	Reference
Recombination radius (r_R)	7.5x10^{-10}	m	[12]
Diffusivity pre-exponential for SIA (D^0_i)	1.58x10^{-6}	m^2/s	[12]
SIA migration energy (E^m_i)	0.5	eV	[6]
SIA loop number density (N_l)	2.2x10^{25} x exp(-0.00404 *T*)	m^{-3}	[10]
Displacement rate (*G*)	1x10^{-6}, 1x10^{-3}, 3x10^{-6} (n, Si^{2+}, D$^+$)	dpa/s	-
Fraction of Si displacement to total displacement (f_{Si})	0.24, 0.32, 0.44 (n, Si^{2+}, D$^+$)	-	[13]

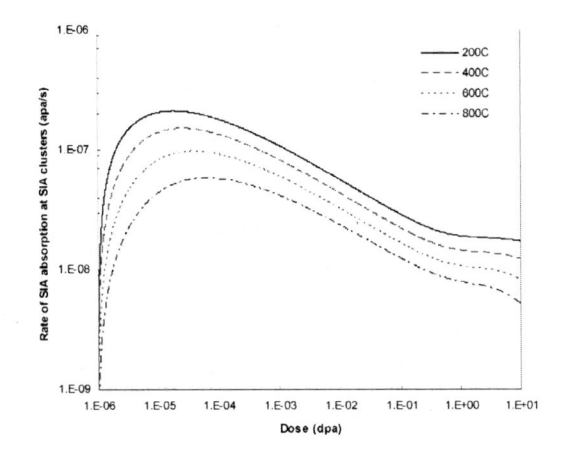

Fig.1. Calculated dose-dependent evolution of rate of SIA absorption at SIA clusters in beta-SiC at intermediate temperatures.

In Fig. 2, swelling in high-purity, chemically vapor-deposited SiC during Si ion irradiation [14] is plotted against dose. It is seen that the magnitude of swelling approximately follows the 2/3 power law in a broad temperature range below ~800°C until saturation occurs(<~0.3 dpa). In Fig. 3, dose dependence of swelling in the identical material during neutron irradiation in HFIR [15] is plotted. It is shown that the 2/3 power law also fits the neutron data reasonably at relatively low doses (<~0.1 dpa), but the calculation did not sufficiently reproduce the experimentally observed temperature dependence. These observations support adequacy of the present defect accumulation model to a limited extent.

In Fig. 4, calculated irradiation creep strains in SiC during 14 MeV deuteron irradiation are presented as a function of dose, along with experimental data replotted from the torsional irradiation creep data for SCS-6 CVD SiC fibers by Scholz [16]. The creep strain appears to be proportional to approximately 2/3 power of dose, which is in agreement with the analytical model prediction. In this calculation, the coupling coefficient of creep rate with SIA absorption rate at SIA clusters, k_{ic} in Eq. (6), was taken to be 3×10^{-3} MPa^{-1} to fit the experimental data, when diffusivity D_i is normalized to dpa. Fig. 4 also shows that the relative strains at different irradiation temperatures are in good agreement between experiment and calculation, which further support the model. The discrepancy between experiment and calculation at 800°C may be due to non-linear stress dependence or slightly anomalous creep behavior recorded only at this temperature in Scholz work.

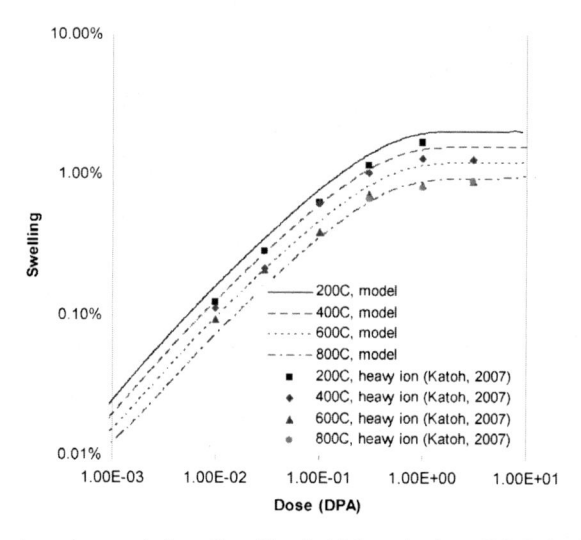

Fig. 2. Dose dependent evolution of swelling in high-purity beta-SiC during heavy ion irradiation: comparison of model and experiment. Experimental data by Katoh [14].

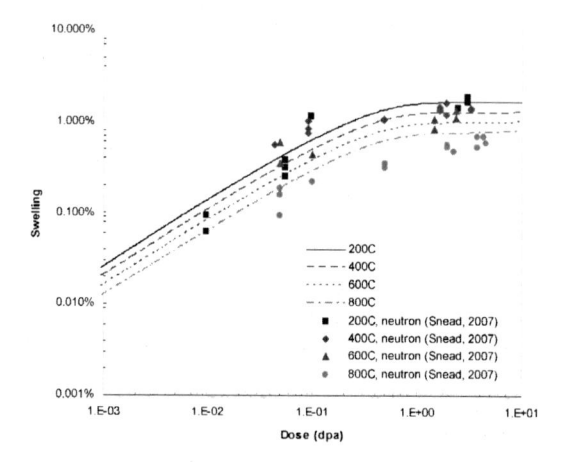

Fig. 3. Dose dependence of swelling in high-purity beta-SiC during neutron irradiation: comparison of model and experiment. Experimental data by Snead [15].

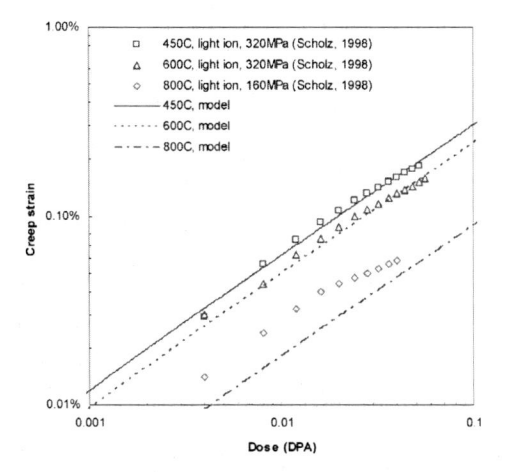

Fig. 4. Dose dependence of creep strain in beta-SiC during light ion irradiation: comparison of model and experiment. Experimental data by Scholz [16].

Analyzing neutron irradiation creep data

In a companion paper, bend stress relaxation (BSR) irradiation creep experiment was performed using thin strip specimens of high purity CVD SiC [17]. Detailed descriptions of experimental technique development and results are found elsewhere [18]. In Table 2, BSR irradiation creep data for CVD SiC are summarized. Additionally, result from an experiment by Price [19] was interpreted and included. These data are plotted in Fig. 5, in which calculated bend stress relaxation behavior of SiC based on the model described by Eqs. (2)-(6). The coupling coefficient k_{ic} was taken to be 3.5×10^{-4} MPa^{-1} to fit the experimental data for the intermediate temperatures (below 800°C). The fitted coupling coefficient was approximately one order smaller for the neutron-irradiated high purity CVD SiC than for the deuteron-irradiated SCS-6. This difference may be due to the effect of displacement cascade for neutron irradiation, difference in loading mode (flexural in the neutron irradiation vs. torsional shear in the deuteron irradiation), difference in absolute stress levels coupled with potentially non-linear stress dependence of creep rate, and/or difference in material quality. On the other hand, a constant irradiation creep compliance reasonably explain the experimental result at high temperatures (>1030°C). Obviously, more comprehensive sets of experimental data will be necessary to further examine adequacy of the models and various assumptions involved.

Table 2. Summary of neutron-irradiation creep data.

Material	T_{irr} °C	Fluence dpa[1]	Reactor	Initial / final bend stress[2], MPa	Creep strain ×10^{-4}	BSR ratio[3] m	Ref.
Rohm&Haas CVD SiC	640	3.7	HFIR[4]	87 / 36	1.12	0.42 (+0.03/-0.04)	[18]
	700	0.7		102 / 72	0.63	0.72 (+0.04/-0.06)	
	1030	0.7		86 / 61	0.52	0.73 (+0.04/-0.06)	
	1080	4.2		101 / 8	2.10	0.08 (+0.05/-0.06)	
Rohm&Haas CVD SiC	400	0.6	JMTR[5]	82 / 60	0.41	0.77 (+0.11/-0.19)	[18]
	600	0.2		81 / 57	0.49	0.73 (+0.10/-0.14)	
	600	0.6		81 / 46	0.75	0.58 (+0.10/-0.15)	
	750	0.6		80 / 55	0.53	0.71 (+0.11/-0.15)	
GA CVD SiC	780	7.7	ETR[6]	n/a[7]	n/a	0.36 (±0.03)[8]	[19]
	950	7.7				0.73 (+0.04/-0.06)[8]	
	1130	7.7				0.08 (+0.05/-0.08)[8]	

[1]Displacement per atom. [2]Maximum stress in a flexurally deformed sample. [3]Ratio of final-to-initial stresses. [4]High Flux Isotope Reactor, Oak Ridge National Laboratory. [5]Japan Materials Test Reactor, Japan Atomic Energy Agency. [6]Engineering Test Reactor, Oak Ridge National Labortoy. [7]Not available. [8]Calculated from linear-averaged creep compliance data.

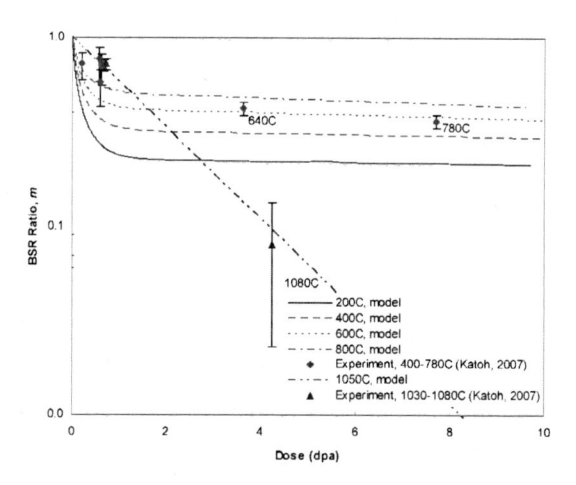

Fig. 5. Dose-dependent evolution of irradiation creep BSR ratio, m, of beta-SiC during neutron irrdiation: comparison of model and experiment. Experimental data by Katoh [18] and Price [19].

CONCLUSIONS

Simple models of irradiation-induced dimensional instability of silicon carbide at intermediate temperatures were developed based on assumptions of linear-coupling between creep rate and SIA absorption rate at SIA clusters, as well as between swelling and matrix vacancy concentration. The model predicted that both swelling rate and creep compliance are proportional to the -1/3 power of dose under constant temperature and dose rate, until saturation swelling takes place. These predictions are consistent with published experimental data for beta-phase CVD SiC on swelling under neutron and self ion irradiation and on creep under light ion irradiation. Moreover, BSR irradiation creep behavior was simulated using the developed model to confirm a reasonable agreement with the data obtained by neutron irradiation experiments.

To further examine adequacy of the models and assumptions, and to narrow the ranges of uncertain parameters, it is essential to add credible experimental data on both irradiation creep and swelling, particularly at different temperatures and dose levels within an intermediate temperature (~200 - ~800°C) / low (~1 dpa)-to-intermediate (~10 dpa) dose regime. It is also necessary that the stress dependence of irradiation creep rate is determined. Moreover, to understand the physical processes of irradiation creep in SiC, configurations of defect clusters and the influence of stress on microstructural development need to be further clarified. Result of this analysis will be utilized to design experimental matrices for future irradiation creep study on silicon carbide presently being planned in US and international fusion advanced materials programs.

ACKNOWLEDGEMENT
This research was sponsored by the Office of Fusion Energy Sciences, US Department of Energy under contract DE-AC05-00OR22725 with UT-Battelle, LLC.

REFERENCES
[1]   K. Ehrlich, J. Nucl. Mater. 100 (1981) 149-166.
[2]   J.R. Matthews and M.W. Finnis, J. Nucl. Mater. 159 (1988) 257-285.
[3]   Y. Katoh, L.L. Snead, C.H. Henager, A. Hasegawa, A. Kohyama, B. Riccardi and J.B.J. Hegeman, J. Nucl. Mater. (2007) doi:10.1016/j.jnucmat.2007.03.032.
[4]   H. Itoh, N. Hayakawa, I. Nashiyama and E. Sakuma, J. Appl. Phys. 66 (1989) 4529-4531.
[5]   A. Kawasuso, H. Itoh, N. Morishita, T. Ohshima, I. Nashiyama, S. Okada, H. Okumura and S. Yoshida, Applied Physics A 67 (1998) 209-212.
[6]   M. Bockstedte, A. Mattausch and O. Pankratov, Phys. Rev. B 68 (2003) 205201-1-17.
[7]   Y. Katoh, S. Kondo and L.L. Snead, J. Nucl. Mater. (submitted).
[8]   F. Gao, E.J. Bylaska, W.J. Weber and L.R. Corrales, Phys. Rev. B 64 (2001) 245208-1-7.
[9]   L.K. Mansur, Philosophical Magazine A 39 (1979) 497-506.
[10] Y. Katoh, N. Hashimoto, S. Kondo, L.L. Snead and A. Kohyama, J. Nucl. Mater. 351 (2006) 228-240.
[11] S.I. Golubov, Phys. Met. Metall. 60 (1985) 7-13.
[12] R.A. Johnson, J. Nucl. Mater. 75 (1978) 77-84.
[13] F. Gao and W.J. Weber, Phys. Rev. B 63 (2000) 054101-1-7.
[14] Y. Katoh, H. Kishimoto and A. Kohyama, J. Nucl. Mater. 307 (2002) 1221-1226.
[15] L.L. Snead, W. Cuddy, D. Peters, Y. Katoh, G.A. Newsome and A.M. Williams, J. Nucl. Mater. (submitted).
[16] R. Scholz, J. Nucl. Mater. 258-263 (1998) 1533-1539.
[17] Y. Katoh and L.L. Snead, Ceram. Eng. and Sci. Proc. 26 (2005) 265-272.
[18] Y. Katoh, L.L. Snead, T. Hinoki, S. Kondo and A. Kohyama, J. Nucl. Mater. (2007) doi:10.1016/j.jnucmat.2007.03.086.
[19] R.J. Price, Nucl. Tech. 35 (1977) 320-336.

PHYSICO-CHEMICAL REACTIVITY OF CERAMIC COMPOSITE MATERIALS AT HIGH TEMPERATURE: VAPORIZATION AND REACTIVITY WITH CARBON OF BOROSILICATE GLASS

Sebastien WERY and Francis TEYSSANDIER
PROMES CNRS UPR8521 – Tecnosud – Rambla de la thermodynamique – 66100 Perpignan, FRANCE

ABSTRACT
Both fundamental and experimental work has been carried out to improve our understanding of the physicochemical behavior of $B_2O_3{}_{(c)}$ as a function of several parameters: temperature (400 to 1400°C), total pressure (0.01 to 30 atm), water vapor content in the gaseous phase (0 to 28%) and microstructure of carbon substrate (vitreous and porous graphite).
Thermodynamic calculations were carried in two chemical systems, (B-H-O and Si-B-H-O) to determine *(i)* the parameters responsible for a high vaporization rate of $B_2O_3{}_{(c)}$ (especially the role of the water vapor), *(ii)* the major gaseous species formed at equilibrium and *(iii)* the influence of silica on $B_2O_3{}_{(c)}$ vaporization.
The interactions of $B_2O_3{}_{(l)}$ with carbon substrates according to experimental parameters (T, %H$_2$O, size of the drop and substrate roughness) were studied by the sessile drop method.

I. INTRODUCTION

Ceramic matrix composites (CMC), such as $C_{(f)}$ / $C_{(m)}$ and $SiC_{(f)}$ / $SiC_{(m)}$, are increasingly used in aeronautic and aerospace fields for the manufacturing of several parts such as nozzle, combustion chamber and exhaust cone. Thanks to the combination of good thermo-physical properties (low density, refractoriness, chemical inertia, etc.) and good mechanical resistance, these materials can replace superalloys in engines: they permit to increase their efficiency, reduce the size of the cooling systems and accordingly its weight and decrease NO_x emissions [1,2]. To resist to the severe operational conditions (high temperature, high pressure, high gas velocities and corrosive atmosphere), self healing matrix has been developed by including boron in the CMC structure. Thus, above 600°C, a vitreous phase ($B_2O_3{}_{(l)}$) protects the CMC against oxidizing agents by filling in the cracks formed during thermo-mechanical stress and so, ensure a higher lifetime to the component [3,4]. However, this oxide is very sensitive to water vapor, which is responsible for its volatilization [5,6,7,8].
In the present work, we focused both on the thermochemical behavior of $B_2O_3{}_{(c)}$ and B_2O_3-$SiO_2{}_{(c)}$ and on the wetting properties of $B_2O_3{}_{(l)}$ on carbon substrates using the sessile drop method. The purpose is to improve our understanding of the B_2O_3 reactivity in CMC.

II. FUNDAMENTAL AND EXPERIMENTAL PROCEDURES

II.1   Thermodynamic calculation

COACH and GEMINI2 softwares were used for the thermodynamic calculation and MODDE to display the response surface hence determined. Three hypotheses were assumed for the thermodynamic calculations: closed, isotherm and isobar system. The influence of the following parameters has been studied: *(i)* 400 < T(°C) < 1400; *(ii)* 0.01 < P(atm) < 30 atm; *(iii)* gaseous species: x. H$_2$O + (1-x). O$_2$.

II.2   Experimental procedure

The sessile drop method was undertaken in a hot wall reactor to follow the wetting behavior of $B_2O_3{}_{(l)}$ according to the experimental parameter. On an ideal surface, the wetting of a liquid on a solid is characterized by the contact angle θ defined by Young's equation (where σ_{lg}, σ_{sl} and σ_{sg} respectively correspond to the liquid-gas, solid-liquid and solid-gas surface energies) [9]:

$$\cos\theta = \frac{\sigma_{sg} \cdot \sigma_{sl}}{\sigma_{lg}} \quad \text{(Equation 1)}$$

Considering Young's equation, each solid-liquid-gas system corresponds to one contact angle. However, experiments show that the triple line can be stopped not only at the ideal contact angle, θ_Y, but also around it, with an advanced (θ_a) and/or a receding angle (θ_r), because of the defects neglected in the ideal case of Young: substrate roughness [10], chemical heterogeneity [11], etc.

The hot wall set up is composed of *(i)* a tubular kiln (maximum temperature 1100°C, heating ramp of 5°C/min), *(ii)* three mass flowmeters (Brooks) to control the atmosphere and *(iii)* a numerical camera to observe the drop profile. The temperature is regulated by a thermocouple K and the device is isolated from the ground vibrations by four spring stands.
The sample is located in the centre of the kiln and the drop is visualized through quartz portholes. Both the contact angle of the drop (θ ±2°) and the surface of the drop base (SDB, interface with substrate) are measured during experiments. B_2O_3 (s) powder (Sigma Aldrich 99,99%) was melted at the focus of a solar concentrator located at Odeillo (2kW) to obtain spherical drops. Vitreous carbon substrates (Carbon Lorraine V25) were mechanically polished with diamond paste down to 1μm and ultrasonically cleaned in ethanol. The roughness was measured by AFM (NT-MDT of SEMA). For the porous substrates (Carbone Lorraine 2123), the drop was deposited directly on the rough surface.
All the experiments were carried out with a total flow of 8l/h. The duration of each experiment (up to 30h for some samples) depends on the time necessary to reach equilibrium. The cooling down of the sample was realized by conduction-convection when the electrical alimentation of the furnace had been turned off. The morphological observations and chemical analyses were realized by optical microscopy, Scanning Electron Microscopy (SEM, Hitachi S4500) coupled with Energy Dispersive Spectroscopy (EDS, Noran-Vantage), Environmental Scanning Electron Microscopy (ESEM, Philips XL30) and Wavelength Dispersive Spectroscopy (WDS, Cameca SX100).

III. THERMODYNAMIC INVESTIGATION OF THE B-H-O SYSTEM

The purpose of this study is to determine the thermochemical behavior of B_2O_3 (c) according to operating conditions. We focused on: *(i)* the nature of the gaseous species formed according to the experimental parameters, *(ii)* the determination of the experimental conditions that favor the vaporization of B_2O_3 (c) and *(iii)* the thermochemical stability of B_2O_3 (c) considering all the species in the B-H-O system (gaseous and condensed).

III.1   Nature of the gaseous species

The nature and the amount of the gaseous species were determined by thermodynamic calculations starting from 1 mole of B_2O_3 (s) in a gaseous mixture composed of Ar, O_2 and H_2O (72/18/10) under atmospheric pressure. The variance of the system is equal to 3. Three intensive parameters were chosen: T, P and a ratio (H/(2O-3B)) that is representative of the gas phase composition and independent of the initial amount of B_2O_3 (s) introduced in the system.
The main gaseous $H_xB_yO_z$ (g) species at equilibrium according to the temperature range were:
T < 950°C: $H_3B_3O_6$ (g) and H_3BO_3 (g),

$950°C < T < 1200°C$: $H_3B_3O_{6(g)}$, $H_3BO_{3(g)}$ and $HBO_{2(g)}$,
$T > 1200°C$: $HBO_{2(g)}$.
Note: All the other boron bearing species (BO, BO_2, etc.) are present in negligible amount as compared to $H_xB_yO_{z(g)}$ ones.

III.2 Solubility of $B_2O_{3(s)}$ in the gas phase (the only condensed species considered is $B_2O_{3(s)}$)

Several calculations were performed with an initial amount of $B_2O_{3(s)}$ equal to 1 mole and an initial gas phase composed of $Ar + O_2 + H_2O$ $(0.72/x/(1-x))$ so that $x.O_2 + (1-x).H_2O = 0,28$. The ratio of $B_2O_{3(c)}$ vaporized is represented by an intensive parameter:

$$\%B_2O_{3(vaporized)} = \frac{(n_{B_2O_{3(s)}\ initial} - n_{B_2O_{3(s)}\ final})}{n_{gas}} \cdot 100 \quad (Equation\ 2)$$

Where $n_{B2O3\ (s)}$ (initial and final) respectively are the number of $B_2O_{3(s)}$ moles introduced in the system and at equilibrium, and n_{gas}, is the number of moles of gaseous phase at equilibrium.
Thanks to 3D graphic representations, the strong influence of the H_2O content on the vaporization of $B_2O_{3(c)}$ is displayed in figure 1: the higher the amount of water vapor, the higher the vaporization ratio of $B_2O_{3(c)}$ because of the formation of $H_xB_yO_{z(g)}$ species. The set of parameters corresponding to the highest $B_2O_{3(c)}$ vaporization ratio is: $P = 30$ atm, $T = 400°C$ and $\%H_2O = 28\%$

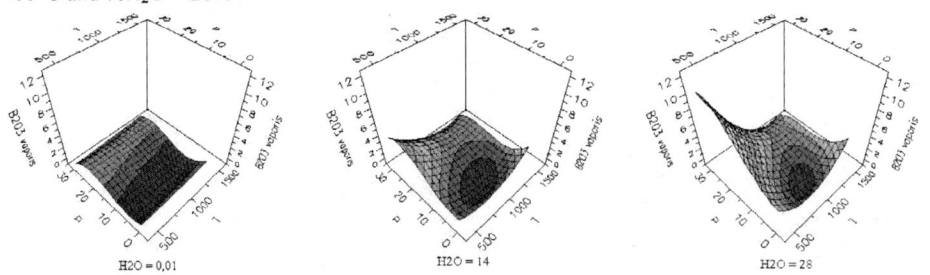

Figure 1: $B_2O_{3(c)}$ vaporization ratios according to Temperature, total Pressure and $\%H_2O$ in the gas phase.

III.3 Thermodynamic stability of $B_2O_{3(c)}$ (Both $B_2O_{3(c)}$ and $HBO_{2(s)}$ are taken into account in the condensed phase)

The $B_2O_{3(c)}$ vaporization ratios of are very different when the $HBO_{2(s)}$ species is introduced in the calculation. At low temperatures (400°C to 600°C), the $HBO_{2(s)}$ species "quench" the water molecules in the condensed phase, thus reducing their amount in the gas phase and the $B_2O_{3(c)}$ vaporization ratio. The maximum vaporization ratio is in this case 4% instead of 12.5% when $HBO_{2(s)}$ is not taken into account. Comparing figures 1 and 2, one can easily judge the influence of the $HBO_{2(s)}$ phase which significantly reduces the amount of vaporized $B_2O_{3(g)}$. The set of parameters corresponding to the highest $B_2O_{3(c)}$ vaporization ratio is now: $P = 0.1$ atm, $T = 1400°C$ and $\%H_2O = 28\%$. One can also notice the existence of a second maximum at 800°C / 30 atm that corresponds to a $B_2O_{3(s)}$ vaporization ratio of 2.75%. It is interesting to note that this second domain corresponds to the formation of $H_3B_3O_{6(g)}$ and $H_3BO_{3(g)}$ species, although the main one corresponds to $HBO_{2(g)}$. As a conclusion, when both $B_2O_{3(c)}$ and $HBO_{2(s)}$ condensed species are taken into account in the calculation, the water vapor still favors the vaporization of $B_2O_{3(c)}$, but its influence is limited in the temperature range 400 to 600°C because of $HBO_{2(s)}$ formation.

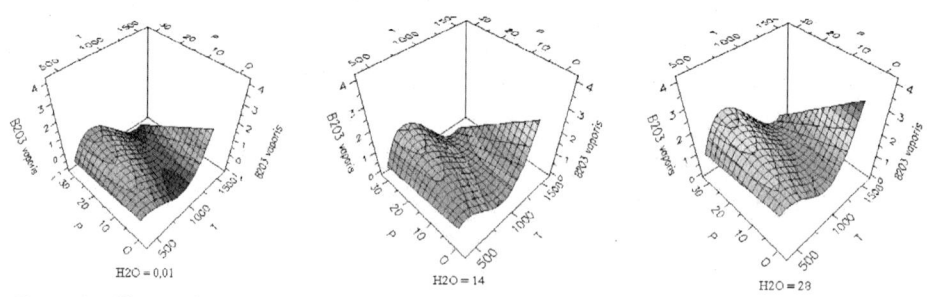

Figure 2 : Thermochemical stability of B_2O_3 (c) according to Temperature, total Pressure and %H_2O in the gas phase.

IV. THERMODYNAMIC INVESTIGATION ON THE B_2O_3-SiO_2 SYSTEM

B_2O_3 and SiO_2 mix in all proportions to form a glass. Using a Calphad thermodynamic approach, we modeled interactions between these two phases to determine the influence of silica on B_2O_3 (c) vaporization. Variation of the Gibbs free energy of formation of borosilicate glass is given in the literature as a function of composition[13]. However these data are referred to $B_{2/3}O$ and $Si_{1/2}O$ and are derived from lattice stabilities that are different from those of the SGTE bank used in our thermodynamic calculation. We accordingly modeled the Gibbs free energy of formation of borosilicate glass as a function of its composition using the Redlich-Kister [12] formalism. Gibbs free energy of mixing is given by:

$$\Delta G^{mix} = \Delta G^{xs} - T.\Delta S^{id} \quad \text{(Equation 3)}$$

Where $\Delta S^{id} = R.\sum_i (x_i.\ln x_i)$ is the ideal entropy of mixing and ΔG^{xs} the excess enthalpy of mixing. For a binary system A-B, the excess Gibbs free energy function is:

$$\Delta G^{xs} = x_A.x_B.f(x_A,x_B) \quad \text{(Equation 4)}$$

Where $f(x_A,x_B)$ is a Redlich-Kister polynomial equation.
To describe the temperature and composition dependence, we used the same polynomial equation, already used by Baret and al.[13]:

$$\Delta G^{xs} = [A + B.T + C.(1-2x)^2].x.(1-x) \quad \text{(Equation 5)}$$

A, B and C coefficients were calculated so that the polynomial equation reproduces the SiO_2-B_2O_3 phase diagram:
The B_2O_3-SiO_2 phases diagram thus established is in good agreement with the one experimentally determined by Rockett and al.[14] which is considered as the reference.

Two compositions (0.5 SiO_2 + 0.5 B_2O_3 and 0.9 SiO_2 + 0.1 B_2O_3) were chosen to look at the influence of the SiO_2-B_2O_3 interaction on the B_2O_3 (c) vaporization. Like in the previous calculation, the initial gaseous mixture is composed of Ar + O_2 + H_2O (0.72/x/(1-x)) so that x.O_2 + (1-x).H_2O = 0.28. Both B_2O_3 (c) and HBO_2 (s) are taken into account in the condensed phase.

Like B_2O_3, the B_2O_3-SiO_2 chemical system is very sensitive to the H_2O content: the higher the amount of water vapor pressure, the higher the vaporization of B_2O_3 (c). This trend is exemplified in figure 3. The sets of parameters corresponding to the highest B_2O_3 (c) vaporization ratio are the same as for the B-H-O system: P = 0.1 atm, T = 1400°C, %H_2O = 28%, and P = 30 atm, T = 800°C, %H_2O = 28% and correspond to the same preponderant gaseous species.

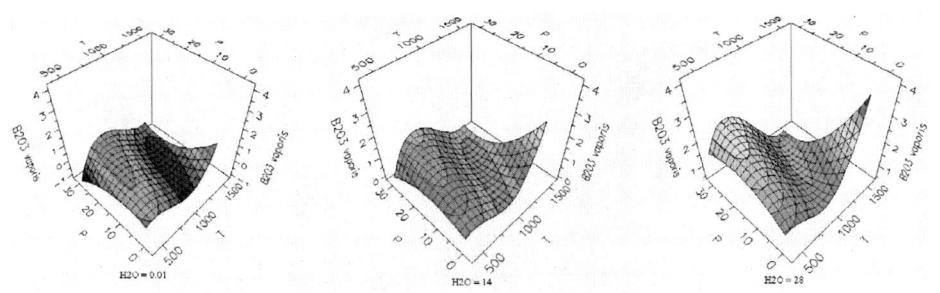

Figure 3 : Variation of the B_2O_3 (c) vaporization ratio as a function of Temperature, total Pressure and water vapor content

V. INTERACTION BETWEEN LIQUID B_2O_3 AND VITREOUS CARBON

V.1 Experiments in pure argon atmosphere

The influence of drop size, roughness and temperature has been studied under argon atmosphere. We first verified that two drops of different sizes (9.1 and 16 mg) gave both the same spreading kinetics and final contact angles, confirming that they were small enough to neglect gravity. Considering the influence of roughness, we studied at 600°C four samples of different roughness controlled by the grain size of the polishing paste (30, 14, 3 and 1µm). According to the Wenzel model, we observed that the contact angle diminishes when the roughness increases (in these experiments the contact angle ranges between 50° and 60°). All further vitreous carbon substrates were polished until 1µm.

V.1.1 Influence of temperature

The contact angle (θ) and the surface of the drop base (SDB), which is the interface between the drop and substrate, have been studied according to time and temperature. For all the curves related to the time dependence, time "0" corresponds to the beginning of the level temperature: the variations of θ and SDB taking place during the temperature ramp are not displayed. The variations of the contact angle represented in figure 4 are characteristic of a reactive wetting process.

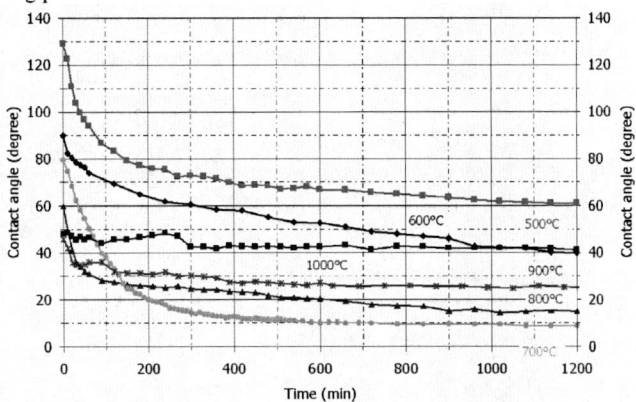

Figure 4: Influence of temperature on the spreading kinetics.

The reactivity between liquid B_2O_3 and vitreous carbon is very temperature dependant. This is revealed by the variation of the contact angle as a function of temperature presented in figure 5. The minimum that is observed at 700°C ($\theta = 9°$) is indicative of an enhanced reactivity between these two materials at this temperature.

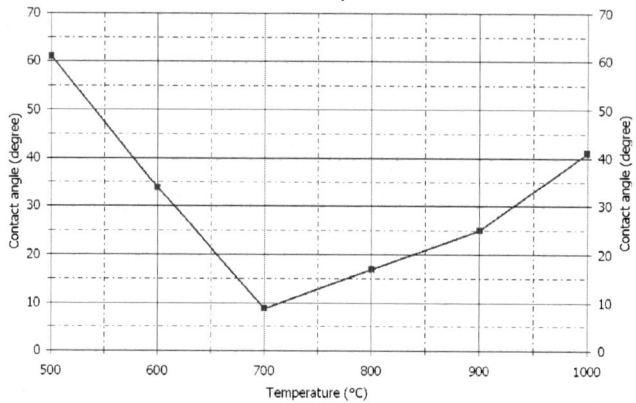

Figure 5: Variation of the contact angle with temperature.

V.1.2 Chemical and morphological analyses

The post mortem observation of the liquid-solid interface revealed a sharp interface between the solidified drop and the substrate and the presence of carbon in the drop. First composition analyses by EDS or WDS do not indicate that carbon detected in the solidified drop is includes in compound or solution $B_xC_yO_z$ phase. In contrast, these analyses suggest that small carbon grains coming from the substrate are "dissolved" into the drop. The C-B-O phase diagram is not well-known and further analyses are necessary to confirm this point. The different pictures furthermore reveal the presence of cracks at the triple point that probably result from thermal expansion mismatch.

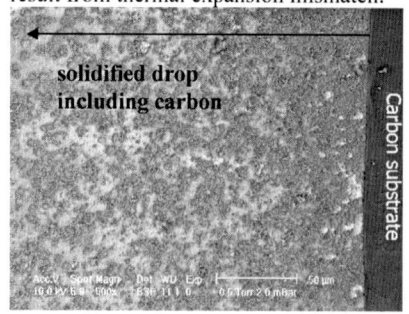

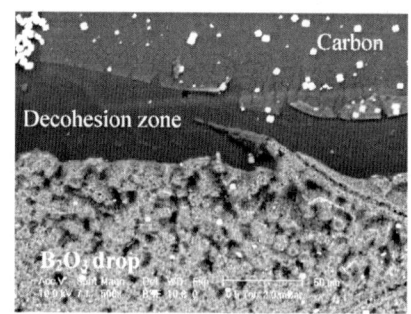

Figure 6: cross section observation of samples prepared at 600°C and 700°C.

V.2 Experiments under oxidizing atmosphere

We studied the variation of θ and SBD according to time and temperature. The experiments were carried out under $Ar + O_2$ and $Ar + H_2O$ atmospheres (respectively 72 and 28 Vol %). As expected, the gaseous corrosion kinetics of carbon under O_2 is very fast above 800°C and most of the carbon substrate has disappeared after 3h. In contrast to experiments carried out under argon atmosphere, the contact angle decreases when temperature increases (figure 7). Above 800°C, in the presence of H_2O, B_2O_3 is entirely vaporized after 200 min. Accordingly, the equilibrium value of the contact angle is difficult to determine. It is lower than 5° but it was not possible to determine if perfect wetting is reached. Above 800°C, in the presence of O_2 the same behavior is observed. However, the experimental conditions are worse due to corrosion of the carbon substrate.

The lack of wetting observed at 500°C under H_2O is due to the adsorption of water molecules on the reactive sites of carbon that prevent reactivity between the drop and carbon substrate.

As a general trend, O_2 is more efficient than H_2O to promote the reactive wetting of carbon by B_2O_3. It is interesting to note that the enhanced corrosion of the carbon surface by O_2 induces the formation of holes. These holes are responsible for imbibition of liquid from the drop when the drop spreads and the decrease of the surface of the drop base at T > 700°C (figure 8).

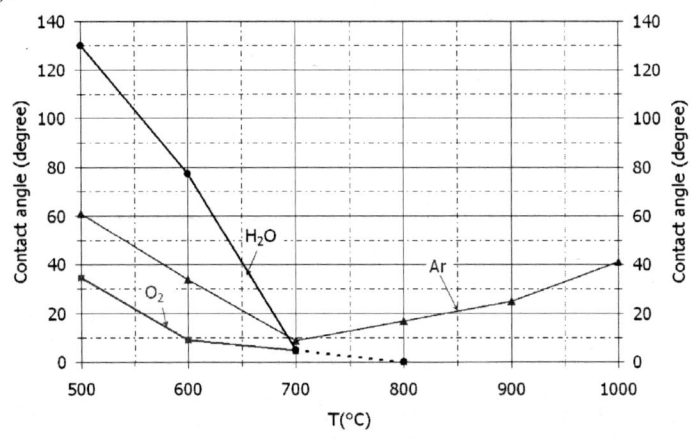

Figure 7: Variation of the contact angle according to atmosphere and temperature

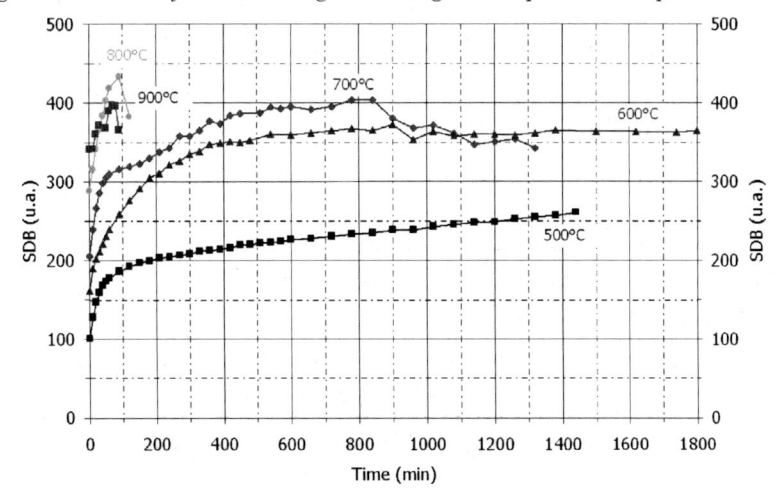

Figure 8: Variation of the drop spreading according to time and temperature.

VI. INTERACTION BETWEEN LIQUID B₂O₃ AND POROUS GRAPHITE

The wetting of a porous medium is usually divided into three steps[16]: *(i)* growth of the drop base until a maximum surface of contact is reached, *(ii)* steady state where the drop base remains unchanged and *(iii)* shrinkage of the drop base[16]. These variations result from two competing processes: the drop spreading over saturated porous layer and the imbibition of the liquid from the drop into the pores. The different steps of imbibition follow the law of Washburn and Middlemann[16]. The porous substrate (Carbone Lorraine 2123) contains 23% volume ratio of open pores. As determined by BET measurements, all the pores are smaller than 200μm and two maxima of the size distribution are observed at 5μm and 0.1μm.

The wetting of B_2O_3 (l) drops on porous graphite is similar to that observed on vitreous substrate. The same minimum contact angle is observed at 700°C (figure 8) revealing stronger chemical interactions at this temperature. The presence of pores nevertheless induces some differences in the spreading kinetics. Imbibition of the substrate by liquid B_2O_3 was checked by post mortem analysis of the substrate composition. These analyses reveal that imbibition takes place under oxidizing gas phase but not under pure argon. In the absence of oxidizing agent, the weak chemical interaction between carbon and liquid B_2O_3 are responsible for bad wetting conditions that do not allow filling the pores by capillary effects. In contrast, O_2 improves the wetting, thus favoring the penetration of pores by liquid B_2O_3.

The already described successive imbibition steps can be easily distinguished in figure 10. In the presence of H_2O and at 500°C, the same non-wetting phenomenon that was previously observed on vitreous substrate is observed. Above 600°C, both imbibition in the substrate and vaporization of the drop contribute to the decrease of the surface of the drop base, which corresponds to the third step of the imbibition process.

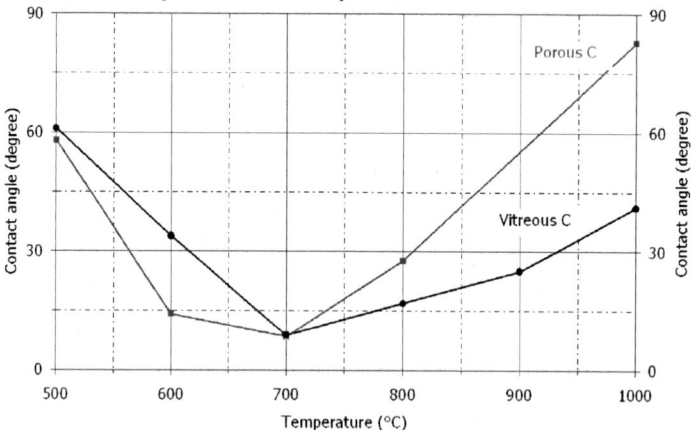

Figure 8 : Variation of the contact angle as a function of temperature.

As in the case of vitreous carbon, WDS and EDS analyses reveal the presence of carbon in the solidified liquid drop. The same hold true for these observations.

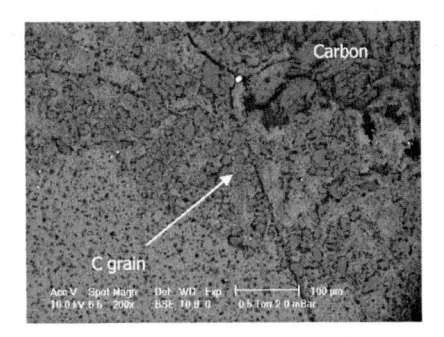

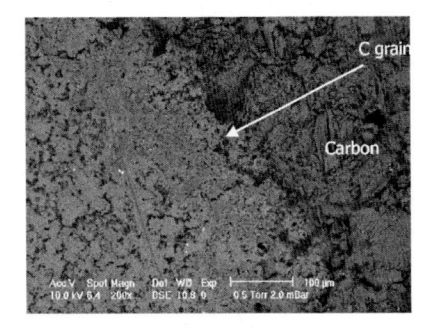

Figure 9 : Samples prepared at 800°C and 1000°C during 10h under pure argon.

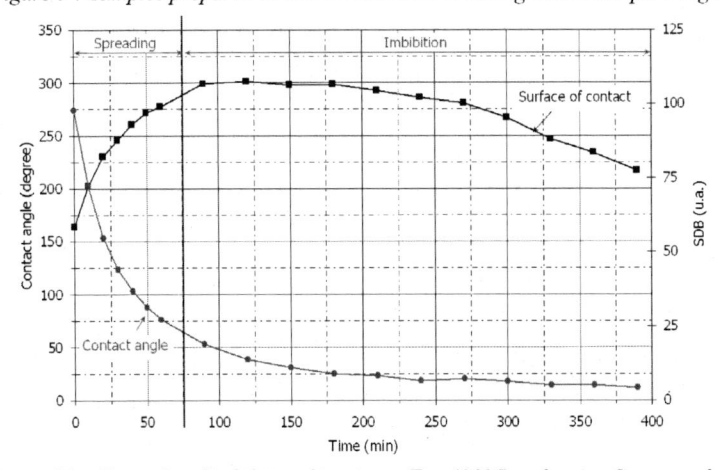

Figure 10 : Example of imbibition kinetics at T = 600°C under Ar+O₂ atmosphere.

CONCLUSION

The influence of temperature (400-1400°C), total pressure (0.01-30 atm) and water vapor content (0-28%) on the vaporization of B_2O_3 has been calculated under the thermodynamic equilibrium assumption in the B-H-O and B-Si-H-O systems. We determined the main gaseous $H_xB_yO_{z\,(g)}$ species formed at equilibrium as a function of the temperature range:
- for T < 950°C: $H_3B_3O_{6\,(g)}$ and $H_3BO_{3\,(g)}$,
- for 950°C < T < 1200°C: $H_3B_3O_{6\,(g)}$, $H_3BO_{3\,(g)}$ and $HBO_{2\,(g)}$,
- for T > 1200°C: $HBO_{2\,(g)}$.

We underlined the strong influence of water vapor pressure on the $B_2O_{3\,(c)}$ vaporization and determined the domains where this phenomenon preferentially takes place. We showed that the $HBO_{2\,(s)}$ species "quench" the water molecules in the condensed phase at low temperatures (400°C to 600°C), thus reducing the vaporization ratio of $B_2O_{3\,(c)}$ (12.5% to 4%). By modeling the SiO_2-B_2O_3 phase diagrams we concluded that the presence of SiO_2 does not significantly influence the $B_2O_{3\,(c)}$ vaporization.

We investigated the wetting behavior of B_2O_3 (l) on carbon substrates by the sessile drop technique.
We verified that the drop size has no influence on the final contact angle providing that capillary effect is predominant and that the variation of the contact angle follows Wenzel's model.
We have shown that under argon, the contact angle is minimal at 700°C, whatever the carbon substrate. On porous graphite substrate and in oxidizing atmosphere, both imbibition phenomenon of liquid B_2O_3 into the pores and its vaporization contribute to the decrease of the drop base surface at long interaction time, though under pure argon no imbibition takes place.
It is important to note that in the absence of water vapor the vaporization rate of liquid B_2O_3 remains low even at high temperature. The typical weight loss at the end of an experiment is smaller than 2%. This experimental observation is in accordance with our thermodynamic calculations.

REFERENCES

[1] R. NASLAIN, Design, preparation and properties of non-oxide CMC for application in engines and nuclear reactors: an overview, *Composites Science and Technology*, 2004, n°64, p. 155-170.

[2] S. SCHMIDT, S. BEYER, H. KNABE, H. IMMICH, R. MEISTRING and A. GESSLER, Advanced ceramic matrix composite materials for current and future propulsion technology application, *Acta Astronautica*, 2004, n°55, p. 409-420.

[3] F. LAMOUROUX, S. BERTRAND, R. PAILLER, R. NASLAIN and M. CATALDI, Oxidation-resistant carbon-fiber-reinforced ceramic-matrix composites, *Composites Science and Technology*, 1999, n°59, p. 1073-1085.

[4] R. NASLAIN, A. GUETTE, F. REBILLAT, R. PAILLER, F. LANGLAIS and X. BOURRAT, Boron-bearing species in ceramic matrix composites for long-term aerospace application, *Journal of Solid State Chemistry*, 2004, n°177, p. 449-456.

[5] W. CHUPKA, M. INGHRAM and R. PORTER, Mass Spectrometric Study of Gaseous Soecies in B-B2O3 system, *Journal of Physics and Chemistry*, 1956, vol. 25, p. 498-501.

[6] J.R SOULEN, P. STHAPITANONDA and J.L. MARGRAVE, Vaporization of Inorganic Substances: B_2O_3, TeO_2 and Mg_3N_2, *Journal of Physics and Chemistry*, 1955, vol. 59, p. 132-136.

[7] N.S. JACOBSON, S. FARMER, A. MOORE and S. SAYIR, High Temperature Oxidation of Boron Nitride: I, Monolithic Boron Nitride, *Journal of the American Ceramic Society*, 1999, vol. 2, n°82, p. 393-398.

[8] N.S. JACOBSON, G.N. MORSCHER, D.R. BRYANT and R.E. TESSLER, High-temperature Oxidation of Boron Nitride: II, Boron Nitride Layers in Composites, *Journal American Ceramic Society*, 1999, vol. 82, n°6, p. 1473-1482.

[9] T. YOUNG, Philos. Trans. R. Soc. (London), 1805, vol. 95, p. 65-87.

[10] R.N. WENZEL, Resistance of solid surfaces to wetting by water, *Industrial and Engineering Chemistry*, 1936, vol. 28, n°8, p. 988.

[11] A.B.D CASSIE and S. BAXTER, Wettability of Porous surfaces, *Trans. Faraday Soc.*, 1944, vol. 40, p. 546-551.

[12] O. REDLICH and A.T. KISTER, Algebraic representation of thermodynamic properties and the classification of the solutions, *Industrial and Engineering Chemistry*, 1948, vol. 40, n°2, p. 345-348.

[13] G. BARET, R. MADAR and C. BERNARD, Silica-based oxide systems, I. Experimental and Calculated Phase Equilibria in Silicon, Boron, Phosphorus, Germanium, and Arsenic Oxide Mixtures, *J. Electrochem. Soc.*, 1991, vol. 138, n°9, 2830-2835.

[14] T.J. ROCKETT and W.R. FOSTER, Phase relations in the system boron oxide-silica, 1965, *Journal of the American Ceramic Society*, vol. 48, n°2, p. 75-79.

[15] N. EUSTATHOPOULOS, M.G. NICHOLAS and B. DREVET, Wettability at high temperature, *Pergarmon Materials Series*, 2005.

[16] V.M. STAROV, S.A. ZHDANOV, S.R. KOSVINTSEV, V.D. SOBOLEV and M.G. VELARDE, Spreading of liquid drops over porous substrates, *Advances in Colloid and Interface Science*, 2003, vol. 104, p.123-158.

Acknowledgement
This work was funded by the joint research program "CPR DDV CMC" between SAFRAN-SNECMA PROPULSION SOLIDE, DGA and CNRS.

IRRADIATION EFFECTS ON THE MICROSTRUCTURE AND MECHANICAL PROPERTIES OF SILICON CARBIDE

MENARD Magalie[1]; LE FLEM Marion[1]; GELEBART Lionel[1]; MONNET Isabelle[2]; BASINI Virginie[3]; BOUSSUGE Michel[4]

1. CEA, Atomic Energy Commission, Nuclear Energy Division, Nuclear Material Department 91191, Gif Sur Yvette, France.
2. CIRIL (Ion-Laser Interdisciplinary Research Center)-GANIL, BP5133 14070, Caen, France.
3. CEA, Atomic Energy Commission, Nuclear Energy Division, Fuel Department, 13108 St-Paul-lez-Durance, France.
4. ENSMP/CDM, UMR CNRS 7633, BP 87, 91003 Evry, France.

ABSTRACT

Silicon Carbide (SiC) exhibits good thermomechanical resistance and high thermal conductivity at high temperature. As it is also compatible with fast neutron spectrum, SiC is then a candidate for structural applications in some nuclear reactors of the next generation, such as Gas Cooled Reactors. However, at high temperatures (T>1000°C), the simultaneous effect of stress and irradiation emphasizes the creep of SiC and induces changes in mechanical properties, depending on the temperature and the fluence. In the scope of a wide program aiming to select materials for these applications, ion irradiations have been performed to simulate neutron-induced damage in ceramics, and evaluate its consequences on microstructure and mechanical properties. The changes in microstructure and mechanical properties of commercial grade of α-SiC induced by surface irradiations with 95 MeV Xe ions at 400°C are investigated. Irradiations resulted in damages affecting a thickness of about 10 µm, for fluences ranging from $3.0.10^{14}$ to $3.6.10^{15}$ ions/cm². Raman spectroscopy analysis revealed that irradiations produced homonuclear Si–Si bonds and disordered phase of crystalline SiC between 3.10^{14} and $3.6\ 10^{15}$ ions/cm² fluence. These microstructural modifications contributed to a macroscopic swelling estimated by measuring the step height between the irradiated and virgin areas. Between 3.10^{14} and $1.2\ 10^{15}$ ions/cm² of fluence, the step height increases from 47 nm to 83 nm, then stabilizes with increasing fluence. Elastic modulus of α-SiC did not seem to be significantly affected by irradiation. Hardness of α-SiC exhibited an increase of 15% in the near-surface region of the samples (up to 6 µm depth at fluence of $3.6\ 10^{15}$ ions/cm²) then a decrease of 60% at the damage peak range that was attributed to a continuous buried disordered layer. Further mechanical tests such as creep and bending tests are planned on irradiated SiC, thanks to a dedicated device allowing characterisation of very thin beams.

INTRODUCTION

Silicon carbide (SiC) and more specifically SiC/SiC composites are selected for structural applications in next generation of nuclear reactors (Gaz Cooled Reactors), due to high-temperature thermomechanical properties and nuclear compatibility. In spite of significant results provided by Fusion or Generation IV programs, there is still a lack of data on the experimental feedback concerning the physical and mechanical properties of irradiated SiC. In this paper, heavy ion-beam irradiation is chosen to simulate neutron irradiation damages in polycrystalline hexagonal SiC (α-SiC, Xe, 95MeV). Because of the potentially high temperature applications of

SiC (400-1000°C in working conditions), irradiation was realized at 400°C. Five fluence levels were reached between 3.10^{14} and $3.6.10^{15}$ ions/cm^2). First, characterization of the microstructure performed by Raman spectroscopy is presented. Volume change, hardness and elastic modulus measured by profilometry, nanoindentation and acoustic microscopy are also reported. Results are discussed according to previous results and perspectives are presented in term of further mechanical characterisations.

EXPERIMENTAL

The investigated SiC grade deals with monolithic α-SiC Hexoloy SA, commercial product manufactured by Saint Gobain. The Hexoloy SA material was pressureless sintered from submicronic SiC powder with boron (0.7 wt%) and carbon (0.5 wt%) aids to increase its density. The material exhibits uniform micron-scale pores, and a bulk relative density of approximately 98% (3.21 g/cm^3). X-Ray Diffraction analysis confirmed the crystal structure is hexagonal, specifically the 6H polytype with traces of 4H polytype, commonly referred to as α-SiC. The microstructure of the material consists in equiaxial fine-grains of 5–8 μm in diameter. Before irradiation experiment, the surface of the samples is prepared by mechanical polishing up to ¼ μm.

Ten samples (24 × 4 × 2 mm^3) were irradiated at GANIL (National Large Accelerator of Heavy Ions) (Caen, France) [1], under high vacuum, with 95 MeV ^{129}Xe^{23+} ions. The irradiation was performed at 400°C thanks to a copper sample holder heated by resistivity. The temperature was monitored by a thermocouple placed between the sample and the metallic sheet which maintained the sample on the heating sample holder. In order to avoid beam heating, the ion flux was limited to 4.10^9 ion.cm^{-2}.s^{-1}. A metal plate protected the half of the surface samples during irradiation allowing measurements of macroscopic swelling by surface profilometry. Five fluence levels were obtained with an ion beam current close to 1 μA: 3.10^{14}, 6.10^{14}, $1.2.10^{15}$, $1.5.10^{15}$ and $3.6\ 10^{15}$.ions/cm^2. The damage profile as a function of depth from the surface has been characterized in terms of displacements per atom (dpa) for the different doses by TRIM code SRIM-2003 (Stopping and Range of Ions in Matter [2]) (Figure 1). Radiation damage calculations for SiC were made using threshold displacement energies of 35 eV for Si and 20 eV for C [3,4], leading to doses of 0.1, 0.2, 0.4, 0.5 and 1.2 dpa in the first 5μm of homogeneous damaged SiC.

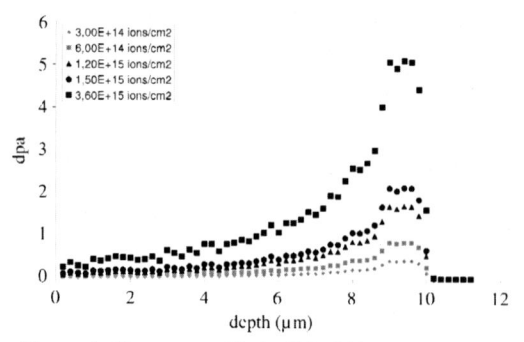

Figure 1 : Damage profile in SiC (TRIM simulations).

Raman scattering were recorded at room temperature in a backscattering geometry using a Jobin-Yvon T64000 spectrometer coupled with an Olympus microscope that contains an X–Y –Z stage. The microscope stage could be adjusted with an accuracy of 0.5 µm along the optical axis. The 514.5 nm line of an argon ion laser was focused on a 1 × 1 µm^2 spot and collected through a 100× objective with a numerical aperture value of 0.9. The laser intensity was kept around 30 mW to avoid any heating of the sample.

The modulus of elasticity E and hardness H were measured by Berkovitch nanoindentations. On the normal surface of the ion beam, they were determined for loadings ranging from 100mN to 700mN (i.e. penetration depth of 0.450µm-1.3µm) corresponding to the asymptote of the E=f(displacement) and H=f(displacement curves)). For characterisation on the cross section E and H were determined for low loadings between 3mN and 20mN (penetration depth of 0.07-0.2µm) allowing small print size and then good spatial precision along the damage profile. The change in E was also evaluated by acoustic microscopy at the frequency of 160 MHz allowing the analysis of the average properties of material in 10 µm depth. Nanoindentations were also conducted on lateral surfaces of highest irradiated sample to evaluate the change in hardness and elasticity modulus along the damage profile in depth.

The quantitative analysis of swelling was performed with a profilometer. Five scans of 2 mm in length were realized on each specimen over the borderline between an irradiated and a virgin area.

RESULTS
Structure of irradiated SiC :

Fig. 2 illustrates Raman spectra of an unirradiated specimen and the as-irradiated specimens. The unirradiated specimen spectrum shows four single strong peaks between 700 and 1000 cm^{-1}, which correspond to peaks from crystalline α-SiC 6H polytype [5-8].

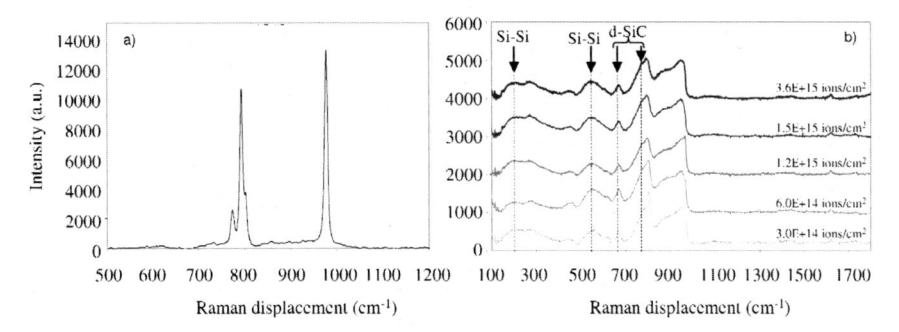

Figure 2 : Raman spectra of virgin 6H-SiC (a) and Xe-ion irradiated 6H-SiC at 400°C (b) indicating peaks of Si-Si homonuclear bonds (*Si-Si*) and disordered SiC (*d-SiC*).

After ion irradiation, α-SiC shows similar Raman spectra at the five fluence levels (Figure 2b). A decrease in intensity and broadening of the crystalline SiC peaks is observed after irradiation. Furthermore, additional peaks at 200 and 560 cm^{-1} and at 680 cm^{-1} occurs: they can be respectively attributed to the existence of amorphous Si (i.e. Si-Si homonuclear bonds) and SiC disordered network as already highlighted by [9-16]. The formation of Si-Si homonuclear bonds within the SiC network is observed from 3.10^{14} ions/cm^2, i.e. 0.1 dpa according to TRIM simulations (Figure 1). The peak around 800 cm^{-1} is attributed at highly disordered SiC. Anyway, the crystalline feature of α-SiC is still very present.

Swelling :
Irradiated SiC samples exhibit irradiation-induced volume modification characterized by the formation of a step border line between irradiated and virgin areas of the sample surface. Figure 3 shows the step height measured by a surface profilometer as a function of fluence of Xe. The step height grows with increasing fluence from 47 to 83 nm between 3.10^{14} and $1.2.10^{15}$ ions/cm^2 and stabilizes for higher fluence levels.

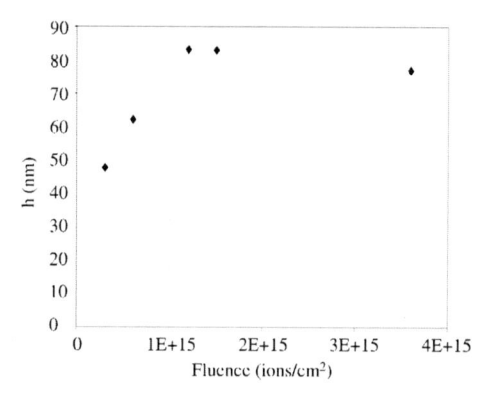

Figure 3 : Step height between irradiated and unirradiated areas of SiC as a function of fluence for Xe irradiations (95 MeV), at 400°C.

Elastic modulus and hardness:
Figure 4 presents the average values of elastic modulus E and hardness H with a 95% confidence level determined by nanoindentation and acoustic microscopy as a function of fluence. These two techniques are in good agreement and suggest that E is not affected by these irradiations (the small variation must not be significant according to experimental uncertainty). The average values of H show an increase of about 15% between 3.10^{14} and 6.10^{14} ions/cm^2 and saturation for the higher fluence levels.

Hardness and elastic modulus were also measured along the irradiation damage profile for irradiated SiC at the highest fluence (3.6 10^{16} ions/cm^2). Measurements were performed on the lateral surface from the free surface up to a distance close to that ion penetration depth (~ 10 µm). The results are presented on Figure 5 in parallel to estimated ion penetration depth. At the very surface of the sample, hardness is in agreement with the results shown on the Figure 4, i.e. it exhibits an increase of 15% with regard to the virgin sample. Moreover, hardness values show a plateau from surface to 5 µm in depth, suggesting homogeneous damages up to 5 µm in thickness. Beyond this, a significant decrease in hardness is observed: in the implanted ion area, the hardness is 65% lower than in the non irradiated sample. The elastic modulus does not seem to exhibit any significant evolution throughout the specimen. However, the scattering of the results is wider in the implanted ion zone and small load (3mN) leads to very low values of elastic modulus supporting a microstructural change in SiC.

Furthermore, micro-indentation showed that crack propagation processes are modified by irradiation. Irradiated specimens show a spalling of the surface which is more pronounced than for the non-irradiated specimens. As shown on Figure 6, irradiated SiC samples also exhibit shorter propagation cracks.

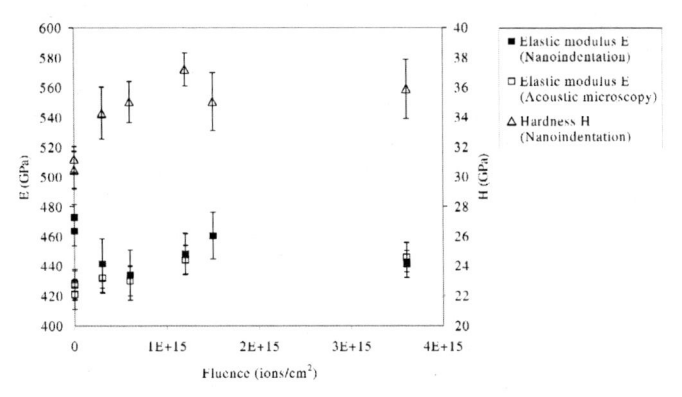

Figure 4 : Elastic modulus E and Hardness H of SiC as a function of fluence for Xe irradiations (95 MeV), at 400°C.

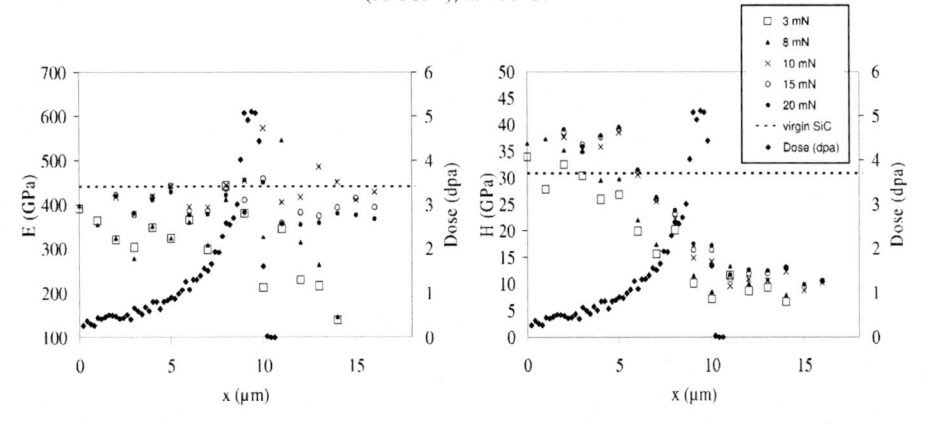

Figure 5 : Elastic modulus E and hardness H at various indentation loads as a function of depth and irradiation damages of SiC for fluence of 3.6 10^{16} ions/cm^2 (95 MeV Xe, 400°C). Dashed line corresponds to virgin-SiC E and H

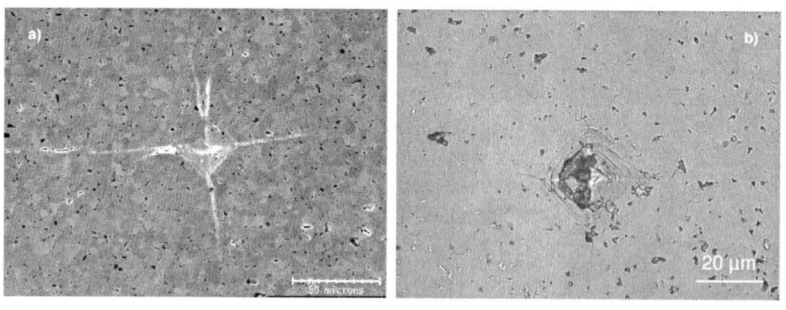

Figure 6 : Vickers micro-indentations (load of 2 kg) on virgin (a) and irradiated SiC to $3.6.10^{15}$ ions/cm^2 of fluence) (b).

DISCUSSION :

Effect of ion irradiation on microstructural damages in α-SiC examined by Raman spectroscopy is characterized by a decrease in the crystalline peak intensity and by the appearance of additional bands corresponding to the formation of homonuclear Si–Si bonds within the SiC network and to a disordered SiC structure. According to Bolse *et al.*, the disordered SiC structure is characterized by bond angle distortions [11] evidenced by the presence of bands around 500–600 cm^{-1}. This structure results in a short-range order and almost conserves the overall sp3 bond structure where each silicon atom is surrounded approximately by four C atoms and vice versa [17-19]. From previous studies, the disordered/distorted structure observed in this work can be defined as a transient state between the crystalline state and amorphous state [9,11]. At room temperature, this state vanishes with increasing fluence leading to total amorphous SiC. Sorieul *et al.* indicate that an irradiation at 400°C results in the stabilization of this transient state and a limitation of damage accumulation owing to the enhancement of the dynamic annealing. The increase in the irradiation temperature extends this stability of the disordered/distorted state up to irradiation dose of 26.4 dpa for α-SiC single crystals [9]. Furthermore, the formation of the band at 580 cm^{-1} associated with the disorder SiC vibration does not depend on the ion (Au or Xe) and consequently results from irradiation effects.

The observed increase in the hardness after Xe ion irradiations is in agreement with works by Parks et al. and Nogami *et al.* on β-SiC irradiated with Si ions and neutrons at temperatures above 400°C; besides, the authors observe a decrease in the elastic modulus after irradiations [20,21]. In the present work, the stability in elastic modulus may be due to the taking into account of greater volume of virgin material in determination of elastic modulus by indentation than that of hardness. A significant decrease in hardness with regard to virgin SiC is measured in the implanted zone area. This may be due to the presence of amorphous SiC softer than crystalline SiC according to previous results of irradiations performed at room temperatures [22-24].. In particular, McHargue *et al.* showed that the hardness of amorphous SiC was 40% of the virgin crystalline SiC hardness (Cr (260 KeV) ion irradiations) [22].This would be in agreement with the hardness values but further studies are necessary to confirm this hypothesis. In particular, since no amorphous SiC should appear after irradiation at 400°C (whatever the dose [24]), this decrease may be due to Xe implantation

The decrease in the crack length induced by indentation of SiC may corroborate the fractographic examinations performed by Katoh *et al.* on β-SiC after neutron irradiations [25]. An increase in the cleavage resistance was observed in large grains which could induce an improvement of the apparent toughness. Nevertheless, ion irradiations performed in the present work induce the formation of disordered subsurface phases and then, subsequent heterogeneous swelling: this may induce compressive stress state at the very surface of samples and could explain the significant spalling observed for irradiated samples.

To avoid such damage gradient throughout the specimen, highly energetic heavy-ion irradiations resulting in more homogeneous damage in larger depth are planned. These experiments should allow macroscopic mechanical characterizations, via creep bending tests on thin specimens.

CONCLUSION

Microstructure, swelling, elastic modulus and hardness of α-SiC was investigated after heavy ion irradiations (Xe, 95MeV) at 400°C. The results confirmed that at this temperature, SiC is not completely amorphous but also exhibits distorted SiC, this microstructure being stable beyond $3.0.10^{14}$ ions/cm^2 fluence level. The material seemed to be homogeneously damaged on 5µm in depth, allowing to highlight a 15% increase in hardness and an apparent increase in toughness, in agreement with previous work. Nevertheless, a probable heterogeneous swelling together to a thin damaged thickness suggested strong compressive stress at the surface of SiC. That is why more energetic heavy ions will be used to perform irradiation up to ~100µm in depth. These irradiations will also allow to perform creep bending tests on irradiated SiC, thanks to a dedicated device which requires thin samples (500µm in thickness). The correlation between surface and bulk characterisation will be done and mechanical behaviour laws of irradiated SiC will be precised (test of virgin SiC in progress).

ACKNOWLEDGEMENTS

Authors are very gratefull to Mr Lionel Gosmain who performed Raman spectroscopy analysis and Mr. Tabarant Michel and Mr. Pierre Forget for profilometry and nano-indentation measurements. They also want to thank Mr. Jean-Marc Costantini and Mr. Xavier Kerbiriou of helpful discussions on irradiation effects.

REFERENCES
[1] http://www.ganil.fr
[2] Ziegler F., Biersack J.P., Littmark U., The stopping and range of ions in solids, New York, 1985.
[3] Devanathan R., Weber W.J., Displacement energy surface in 3C and 6H SiC, Journal of Nuclear Materials, 278, Issue 2-3, 258-265, 2000
[4] Lucas G., Pizzagalli L., Comparison of threshold displacement energies in α-SiC determined by classical potentials and ab initio calculations, Nuclear Instruments and Methods in Physics Research Section B: Beam Interactions with Materials and Atoms, 229, Issues 3-4, 359-366, 2005
[5] Feldman D W, Parker J H Jr, Choyke W J and Patrick L, Phys. Rev., 170, 698–704, 1968

[6] Burton J.C, Sun L., Pophristic M., Lukacs S.J., Long F.H., Feng Z.C. and Ferguson I.T., J. Appl. Phys., 84, 6268–73, 1998
[7] Burton J.C, Sun L., Long F.H, Feng Z.C and Ferguson I.T, Phys. Rev. B, 59, 7282–4, 1999
[8] Nakashima S. and Harima H., Phys. Status Solidi a 162, 39–64, 1997
[9] Sorieul S., Costantini J.M., Gosmain L., Thomé L. et Grob J.J., Raman spectroscopy study of heavy-ion-irradiated alpha-SiC, J. Phys. Condens. Matter. 18, 5235-5251, 2006
[10] Zwick A, Carles R, Multiple-order Raman scattering in crystalline and amorphous silicon, Phys. Rev. B, 48, 6024-6032, 1993
[11] Bolse W., Conrad J., Rödle T. and Weber T., Surf. Coat. Technol., 74/75, 927–31, 1995
[12] Mélinon P., Blase X., Kéghélian P., Perez A., Ray C., Pellarin M., Broyer M., Si-C bonding in films prepared by heterofullerene deposition, Phys. Rev. B, 65, 125321, 2002
[13] Tuinstra F., Koening J. L., Raman Spectrum of Graphite, J. Chem. Phys., 53, Issue 3, 1126-1130, 1970
[14] Gilkes K. W., Sand H. S., Batchelder D. N., Robertson J., Milne W. I, Direct observation of sp^3 bonding in tetrahedral amorphous carbon using ultraviolet Raman spectroscopy, Appl. Phys. Lett., 70, Issue 15, 1980-1982, 1997
[15] [14] Merkulov V I, Lannin J S, Munro C H, Asher S A, Veerasamy V S and Milne W I 1997 Phys. Rev. Lett. UV Studies of Tetrahedral Bonding in Diamond like Amorphous Carbon 78, 4869-4872, 1997
[16] Ferrari A C and Robertson J 2000 Phys. Rev. B, Interpretation of Raman spectra of disordered and amorphous carbon, Phys. Rev., B 61,14095, 2000
[17] Finocchi F., Galli G., Parrinello M. and Bertoni C. M., Phys. Rev. Lett. 68, 3044–7, 1992
[18] Kelires M, Phys. Rev. B, 46, 10048–60, 1992
[19] Yuan X. and Hobbs L. W., Nucl. Instrum. Methods Phys. Res. B, 191, 74–82, 2002
[20] Park K.H., Kondo S., Katoh Y., Kohyama A., Mechanical Properties of β-SiC After Si- and Dual Si + He-Ion Irradiation at Various Temperatures, Fusions Science and Technology, 44, 455-459, 2003
[21] Nogami S., Hasegawa A., Snead L. L., Indentation fracture toughness of neutron irradiated silicon carbide, Journal of Nuclear Materials, Vol. 307-311, Issue 2, 1163-1167, 2002
[22] McHargue C. J., Williams J.M., Ion implantation effects in SiC, Nuclear Instruments and methods in Physics Research, B, 80-81, 889-894, 1993
[23] Ishihara M., Baba S., Takahashi T., Arai T., Hayashi K., Fundamental thermomechanical properties of SiC-based structural ceramics subjected to high energy particle irradiations, Fusion Engineering and Design, 51-52, 117-121, 2000
[24] Weber W. J., Wang L. M., Yu N., Hess N.J., Structure and properties of ion-beam-modified (6H) silicon carbide, Materials Science and Engineering A, 253, issue 1-2, 62-70, 1998
[25] Katoh Y., Snead L.L., Mechanical Properties of Cubic Silicon Carbide after Neutron Irradiation at Elevated Temperatures, Journal of ASTM International (JAI), Vol. 2, Issue 8, 13, 2005

OXIDATION OF ZrB₂-SiC: COMPARISON OF FURNACE HEATED COUPONS AND SELF-HEATED RIBBON SPECIMENS

S.N. Karlsdottir, J.W. Halloran
Materials Science and Engineering Department, University of Michigan
2300 Hayward St.Ann Arbor, Michigan, 48109-2136, USA

F. Monteverde, A. Bellosi
ISTEC, Institute of Science and Technology for Ceramics, National Research Council
Via Granarolo 64, Faenza, 48018, Italy

ABSTRACT

The oxidation of a ZrB₂-15vol%SiC composite was compared for specimens heated with a conventional furnace to a self-supportive self-heated ribbon specimen resistively heated in a novel table-top apparatus. Cyclic oxidation experiments at 1600°C in ambient air were performed on the ZrB₂-SiC samples with the two methods. A complex multi-layer oxide scale was observed for both cases. The structure of the oxide scale of the ZrB₂-SiC sample tested in the table-top apparatus was similar to the oxide scale formed by standard furnace oxidation testing. For both methods the oxide scale consisted of a SiO₂-rich surface layer and an inner ZrO₂ layer. Additionally, in the self-heated ribbon specimen a SiC-depleted region was observed below the ZrO₂ layer. The thicknesses of the oxide scales were estimated, 27 μm for the ribbon method and c. 30 μm for the furnace heated. The two different methods give comparable results for the ZrB₂-15vol%SiC composite. A conclusion can be drawn that the ribbon method is an alternate option to conventional furnace oxidation.

INTRODUCTION

Interest in Ultra-high temperature ceramics (UHTC) has increased significantly in recent years due the growing demand for materials for reusable thermal protection system and other components for future generation of hypersonic aerospace vehicles.[1] Newest designs of hypersonic vehicles require materials capable of operating in extreme reentering environment, such as oxidizing atmosphere at high temperatures (>1600°C) and corrosive gases at high velocities.[2] Among UHTC materials are the refractory borides; ZrB₂ and HfB₂, which have extremely high melting temperatures, greater than 3000°C, high thermal conductivity, high hardness and retained strength and chemical stability at elevated temperatures.[3] These properties also make them potential candidates for a variety of other high-temperature structural applications such as furnace elements, plasma arc electrodes, and molten metal crucibles.[4]

The ZrB₂- and HfB₂- based composites have been identified in the past by Hoffman[5-6] followed by groups led by Kaufman[7] to be the most promising materials for hypersonic applications. Tripp and Graham et al.[8,9] showed that the oxidation resistance of ZrB₂ can be greatly improved by the addition of SiC. The addition of SiC to ZrB₂ has also been reported to increase densification and oxidation resistance by Monteverde and Bellosi et al.,[10-12] Chamberlain and Fahrenholtz et al.,[13] and others.[14-16] ZrB₂- SiC composites form a complex oxide scale after oxidation at elevated temperatures. The oxide scale has been reported to be composed of a refractory oxide skeleton and amorphous glass components which both provide the oxidation resistance at high temperatures.[11-16] The oxide scale formation and oxidation behavior of the composite has though not fully been explained, partially due to the very high temperatures needed for testing UHTC. This is difficult and expensive with the high temperature facilities and techniques available today. The difficulty and high expense exist for acquiring temperatures above the normal operating range of conventional resistance heated furnaces capable of operating in air. Typically, the limit for continuous operation is 1700°C. Higher temperatures require alternate furnace designs, such as ZrO_2 resistance or testing in arc jet facilities which is very expensive and complicated. Thus characterization of the oxidation behavior and physical properties of UHTC at high temperatures is a challenging task.

Here we compare the oxidation behavior of a ZrB_2-15vol%SiC composite tested with a furnace anneal with a self-heated, self-supported specimen tested in a novel table-top apparatus. The furnace testing was conducted at the Institute of Science and Technology for Ceramics, National Research Council, (CNR-ISTEC), in Faenza, Italy. The table-top apparatus was designed and built at the University of Michigan, Ann Arbor, USA. It is capable of testing UHTCs materials inexpensively at 900-2000°C by resistively heating a ~ 500 μm thick ribbon sample. Morphology and composition of a multi-layer oxide scale of the ZrB_2-SiC composite are reported. The formation of the oxide scale tested by the two methods is also discussed and compared.

EXPERIMENTAL PROCEDURE

Material Fabrication

The material was fabricated at ISTEC-CNR. A ZrB_2-15vol%SiC mixture of commercial powders was ball-milled in absolute ethyl alcohol for 1 day, using zirconia milling media. After drying in a rotating evaporator, the powder mixture was sieved through a mesh screen with 250 μm openings. The powder mixture was then uniaxially hot-pressed in vacuum using an inductively heated graphite die. Peak temperatures/dwell times/applied pressures were 1820 °C / 15 min /30 MPa, about 20°C/min average heating rate. The temperature was measured by means of an optical pyrometer focused on the graphite die. In Table 1, some basic properties of the composite are shown.

Table 1 Density (d), Young modulus (E), Vicker's hardness (HV1.0), 4-pt flexure strength (σ) (tests carried out under air), coefficient of thermal expansion (CTE), and thermal conductivity (K_{TH})

d (g/cm^3)	E (GPa)	HV1.0 (GPa)	CTE (10^{-6}/K) (25-1300°C)	σ (MPa) 25°C	σ (MPa) 1500°C	K_{TH} (W/m K) 25°C	K_{TH} (W/m K) 1300°C
5.6	480	17.7±0.4	7.1	795±105	255±25	62.5	66.2

The fabrication of the self-supporting ribbon specimens was done in two steps. Firstly, the bulk material was cut with a wire-EDM machine into 2.5×2.0×25 mm bars. Secondly, about 100 μm of material was removed by diamond grinding to remove a heat-affected zone that can form during wire-EDM machining. The bars are then reduced in thickness in the center with a mechanical grinder (220 grit –diamond wheel) to make a ribbon, with a thickness of 450 µm, see Fig. 1. The specimens for oxidation in the conventional furnace were coupons cut from bulk material with a wire-EDM to the dimension of 2.5 x 2.0 x 25 mm.

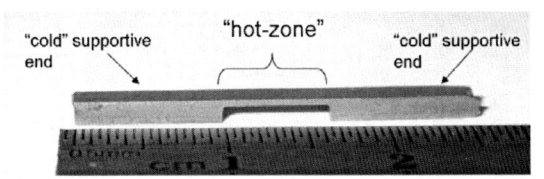

Fig. 1 Digital image of the self-supported, self-heated ribbon specimen.

Oxidation Testing

Cyclic oxidation of the ZrB_2-SiC specimens was performed at 1600°C in air using both methods.

Ribbon Specimen: The ribbon specimen is resistively heated by passing a modest current. The ribbon of the specimen (the hot-zone) can reach temperatures on the range of 900°C-2000°C, when a current is passed into the thicker ends of specimen. A table-top apparatus provides the current and controls the temperature of the specimen with a signal from a micro optical infrared pyrometer that is focused on the hot-zone of the specimen. Detailed descriptions of the design of the apparatus and its function are reported elsewhere.[17]

The specimen was tested in the apparatus for four cycles each for 5 min at 1600°C in stagnant ambient air with average heating rate of 690°C/min and a free cooling (705°C/sec). The fast cooling rate is due to the fast thermal response time of the specimen (7.97 ms at 1600°C), which is depends on the thickness of the specimen (450 µm) and its thermal diffusivity (1.41x10^{-5} m^2/s at 1600°C). The fast heating rate minimizes the oxidation effect on the specimen prior to the hold at 1600°C. The input power at peak temperature is about 90W. Figure 2 shows the temperature and the current profile of the specimen tested in the apparatus. With time the thickness of the oxide scale of the specimen increases and the un-oxidized interior becomes smaller, so less current is needed to maintain the temperature of the ribbon at 1600°C.

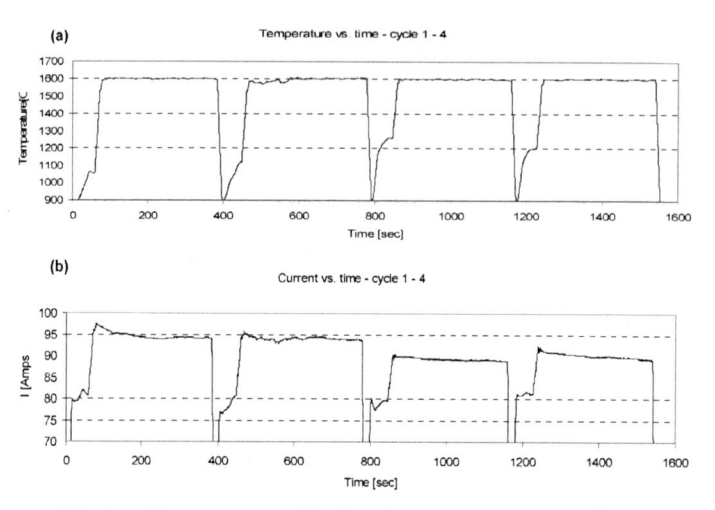

Fig. 2 (a) Temperature profile (b) and a current profile of the specimen tested in the apparatus at 1600°C for 4 cycles each for 5 min.

Furnace Heated Coupons: The composite was subjected to 4 cycled exposures at 1600 °C, 5 min each. The coupon with dimensions of 2.5 x 2.0 x 25 mm^3 was placed upon SiC supports, ensuring the minimal contact area between them. A bottom-loading furnace box, heated with MoSi$_2$ elements, and insulated with highly-porous Al$_2$O$_3$ fiber was used. Once slotted into the heated furnace chamber, a maximum time interval of 1 min was necessary to reach the set point of 1600°C.

For both the methods, prior to the testing the specimens were ultrasonically cleaned in acetone and dried. After oxidation, specimens were stored in moisture free desiccators to avoid any reaction of a B$_2$O$_3$ on the surface of the specimens. Cross-section of the oxidized specimen was prepared for microstructural analysis by non-aqueous polishing procedures down to 1μm finish. The composition and morphology of the multilayer oxide scale formed after the oxidation test was characterized by scanning electron microscopy (SEM), energy dispersive x-ray spectroscopy (XEDS), backscattering electron microscopy (BSE), and electron microprobe analyzer (EPMA). Analyses were done on the surface of the specimen and cross-section.

RESULTS AND DISCUSSION

A multi-layer oxide scale was formed on the ZrB$_2$-SiC samples during the cyclic oxidation with the two methods.

Ribbon Specimen: The surface of the ribbon specimen is covered with silica (SiO$_2$) rich glass. The glass contains small 2-5 micron dispersed zirconia (ZrO$_2$) particles and large 20-50 micron ZrO$_2$ regions. Figure 3 shows the surface view of the specimen and the corresponding XEDS analysis.

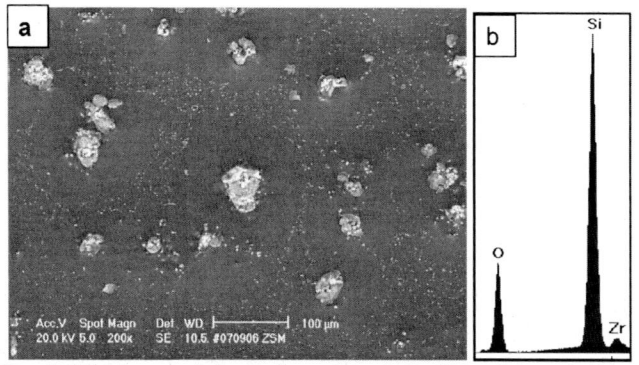

Fig. 3 SEM image of the surface of the ZrB$_2$-SiC self-heated ribbon tested 1600°C for 4 cycles (each 5 min), (a) Large ZrO$_2$ regions (white) in the SiO$_2$ glass matrix (dark grey), which has small ZrO$_2$ dispersed particles (white spots) (b) XEDS analysis of the surface.

The oxide scale formed on the ribbon specimen consists of three layers, which can be seen from the cross-section of the ribbon specimen shown in Fig. 4. The outermost layer is a silica (SiO$_2$) rich glass layer whereas the second layer of the oxide is a ZrO$_2$ layer. The thickness of the SiO$_2$ and ZrO$_2$ layers were measured to be on the average 8 μm and 9 μm. The third layer is a SiC-depleted region; c.a. 10 μm in thickness. The SiC-depleted layer is situated below the ZrO$_2$ layer and above the un-reacted bulk material (ZrB$_2$-SiC). Fig. 4 shows the three layers; SiO$_2$ (#1), ZrO$_2$ (#2) and SiC-depleted region (#3) and the un-reacted bulk material (ZrB$_2$-SiC).

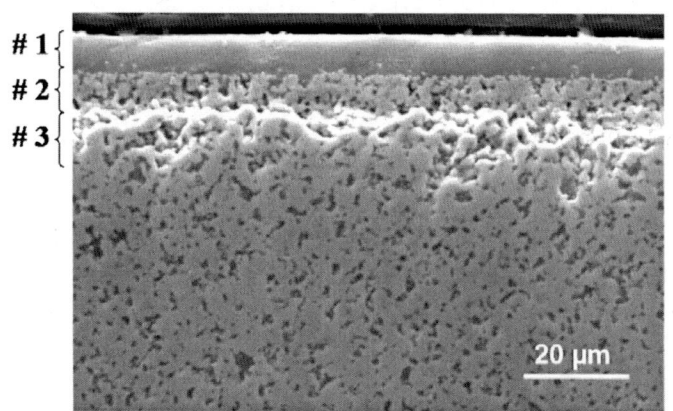

Fig. 4 SEM image of the cross-section view of the ribbon specimen showing the three layer structure of the oxide scale; SiO$_2$ (#1), ZrO$_2$ (#2) and SiC-depleted region (#3) and also the un-reacted bulk material (ZrB$_2$-SiC).

EPMA analyses were performed on the cross-section of the ribbon specimen shown in the BSE image in Fig. 5a. The elements: silicon (Si), oxygen (O), zirconium (Zr), and boron (B) were mapped with EPMA to obtain information about the distribution of elements in the oxide scale; the maps are shown in Fig. 5b-e. Boron (B) is not detectable with the SEM/XEDS used, thus the EPMA analysis were important in supplying information of any existed borosilicate glass or boron oxide (B_2O_3) in the oxide scale. B_2O_3 was not detected in the SiO_2 rich scale or in the ZrO_2 region with the EPMA, this is likely due to the small amount of the B_2O_3 existing in the oxide scale which could not be detected with the scanning speed (20ms) used for the EPMA analysis. The small amount of B_2O_3 is expected due to the high temperature (1600°C) that the specimen experienced. B_2O_3 is believed to start volatilizing extensively at temperatures above 1200°C for monolithic material.[11] Carbon (C) was not mapped due to the cross-section had to be coated with C for the EPMA analysis; other coatings are not suitable due to the light elements that were being analyzed.

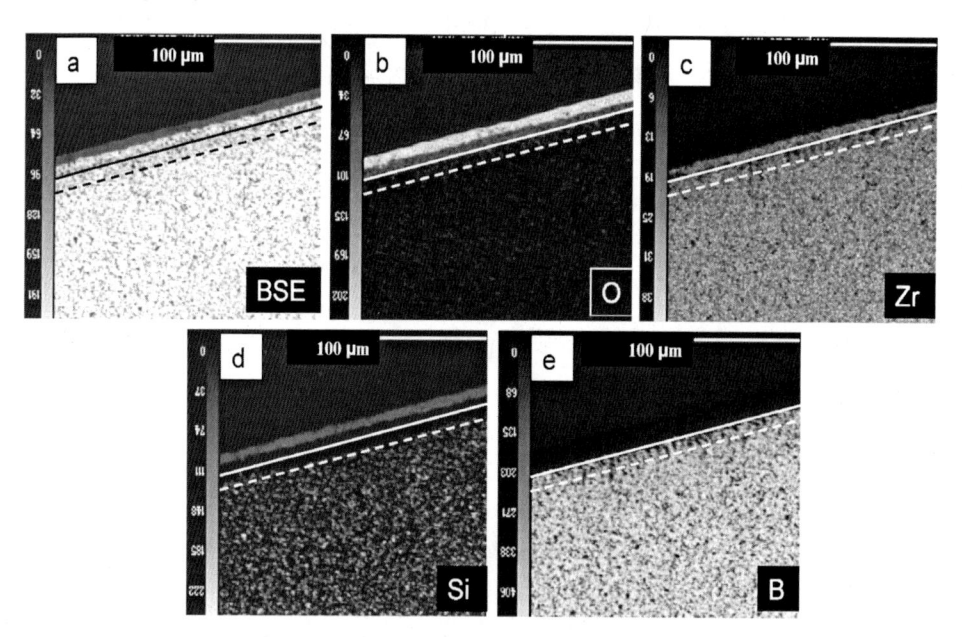

Fig. 5 (a) BSE image of the cross-section of the ZrB_2-SiC specimen tested in the apparatus b) O map (c) Zr map (d) Si map (e) B map. The solid lines outline the interface between the ZrO_2 layer and the SiC depleted region and the slashed lines outline the interface between the SiC depleted and the unreacted core.

The elemental maps (Fig. 5) show clearly that the first two layers of the oxide scale consist of ZrO_2 and SiO_2 formed during oxidation. For the third layer, the maps show evidence of Zr, B and possibly some O, but no Si. To gain more information about the distribution of the elements in the third layer, quantitative line analyses were done on the EPMA data. The line analysis is shown in Fig. 6 where the x-ray intensity ((cps) counts per second) vs. location in the oxide scale (μm) is shown for O, B, Zr and Si. The start of the line is labeled with A (i.e. 0 μm) and the end with B (i.e. 63 μm). From Fig. 6 we see that there indeed exists some O in the third layer along with Zr and B but there are no traces of Si. These results indicate that the third layer is a SiC depleted region. SiC- depleted intermediate layers have been previously observed on similar materials oxidized under different regimes [13, 18-20, 22], in some cases the oxidation of SiC leaving carbon [21]. The existence of oxygen in the third layer of the oxide scale suggests that the SiC-depleted layer consists of some ZrO_2 as well as ZrB_2 bulk material. Fahrenholtz et al.[22] reported that an interface separating a layer where ZrO_2 is dominant from a layer in which ZrB_2 is dominant may be located in SiC-depleted layers formed in oxidized ZrB_2-SiC composites.

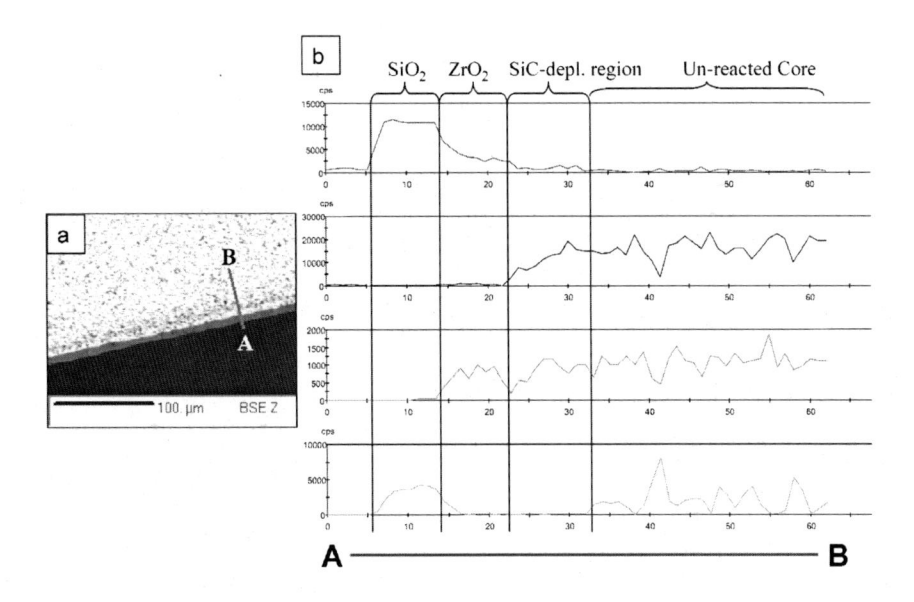

Fig. 6 Line analysis on EPMA maps (a) BSE image of the ribbon specimen showing the line where the intensity of the elements was recorded, A indicates the start of the line scan and B the end (b) graphs of the recorded intensity ((Cps) counts per second) vs. distance (μm) of the line scan shown in (a).

Oxidation of ZrB$_2$-SiC: Comparison of Furnace Heated Coupons and Self-Heated Specimens

Furnace Heated Coupon: The SEM micrograph in Fig. 7 illustrates changes in the virgin microstructure after the repeated exposures at 1600 °C. On top of the exposed faces, a glassy scale of silica has formed. The undulating trend of the thickness of such glassy product indicates that, owing to a diminished viscosity at the testing temperature, it may laterally flow out. Underneath, a partly porous scale which separates the outermost glassy scale from the as-sintered virgin bulk is basically constituted of zirconia grains embedded in glassy phase.

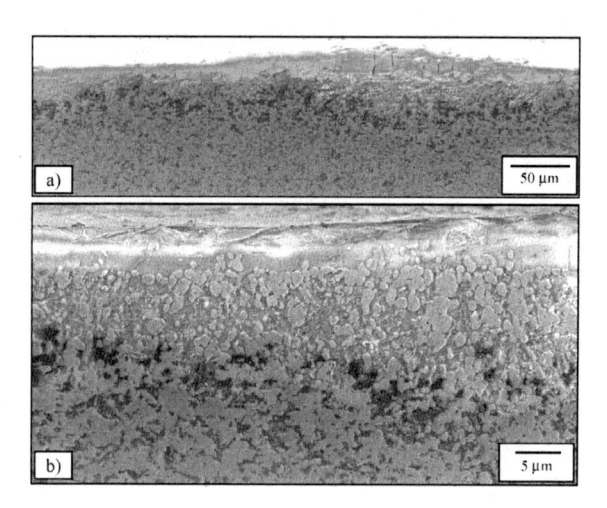

Fig. 7 SEM micrographs of the cross-section view of the furnace heated specimen showing the layered structure of the oxide scale.

COMPARISON OF FURNACE-HEATED AND SELF HEATED RIBBON SPECIMENS

The two testing methods are different. Firstly the size of the coupons (2.5 × 2 × 25 mm, V = 125.0 mm^3) tested in the conventional furnace are larger than the ribbon specimen (0.45×2.5×6.7 mm, V = 7.5 mm^3) tested in the apparatus, the volume of the furnace specimen being 16.7 times larger. Secondly, the furnace coupon is isothermally heated externally by convection in an enclosed environment, whereas the ribbon specimen is heated internally with a flowing current. The conditions are thus significantly different but the results from the microstructural and compositional analysis indicate that the oxidation behavior of the material tested at the two different conditions is rather similar. In fact, the comparison of SEM images in Figs. 4 and 7 verifies that the ribbon specimen (Fig. 4) and the furnace heated coupon (Fig. 7) both have a ZrO$_2$ layer underlying an external glassy layer with comparable thicknesses. However, no clear evidence of a third SiC-depleted layer was found in the furnace heated coupon, maybe consequently to the fast heating-up and cooling stages.

CONCLUSION

Cyclic oxidation experiments at 1600°C in ambient air were performed on ZrB_2-15vol%SiC specimens with conventional furnace and table-top apparatus. A complex multi-layer oxide scale was observed for both cases with a comparable thickness of the oxide scales. The oxide scale thickness of the ribbon specimen was measured to be on average 27 µm whereas for the furnace heated coupon ca 30 µm. The structure of the oxide scale formed during furnace oxidation has similarities to the oxide scale formed in the self-heated ribbon specimen. Both oxide scales have a ZrO_2 layer underlying an outermost SiO_2 surface layer. Additionally, in the self-heated ribbon specimen a SiC-depleted layer was observed below the ZrO_2 layer. The two different methods give comparable results for the ZrB_2-15vol%SiC composite. A conclusion can be drawn that the ribbon method is comparable to conventional furnace oxidation and could be used as an alternative option for high-temperature testing of UHTCs.

ACKNOWLEDGEMENT

The authors would like to thank the following for their contributions in support of this work: Carl Henderson for support in EPMA analysis, and Prof. Albert J. Shih and Jia Tao with EDM machining. We also would like to thank The Office of Naval Research for financial support under the grant N 0014-02-1-0034.

REFERENCES

[1] M. M. Opeka, I.G. Talmy, and J.A. Zaykoski, "Oxidation-Based Materials Selection for 2000°C + Hypersonic Aerosurface: Theoretical Considerations and Historical Experience," *J. of Mater. Sci.*, **39** [19] 5887-5904 (2004).

[2] P. Kolodziej, "Aerothermal Performance Constraints for Hypervelocity Small Radius Unswept Leading Edges and Nosetips," NASA Technical Memorandum 112204, July (1997).

[3] R. Reidel, Handbook of Ceramic Hard Materials, vol. 2, chapter: Transition Metal Boride Ceramics, pp. 881, Wiley-VCH, Weinheim, Germany (2000).

[4] C. Mroz. "Annual Mineral Review: Zirconium Diboride," *Am. Ceram. Soc. Bull.*, **74** [6] 165-166 (1995).

[5] L. Kaufman, "Boride Composites – A New Generation of Nose Cap and Leading Edge Materials for Reusable Lifting Reentry Systems," AIAA Advanced Space Transportation Meeting. *American Institute of Aeronautics and Astronautics*, AIAA 270-278 (1970).

[6] L. Kaufman and H. Nesor, "Stability Characterization of Refractory Materials under High Velocity Atmospheric Flight Conditions, Part I, Vol. I, Summary," Technical Report no. AMFL-TR-69-84, Air Force Materials Laboratory, Wright-Patterson Air Base, OH (1970).

[7] C. A. Hoffman, "Preliminary Investigation of Zirconium Boride Caramels for Gas-Turbine Blade Applicatons," NASA Technical Memorandum E52L15a, Lewis Flight Propulsion Laboratory, Cleveland, OH (1953).

[8] W.C. Tripp and H.C. Graham. "Thermogravimetric Study of Oxidation of ZrB_2 in Temperature Range of 800 Degrees to 1500 Degrees." *Journal of the Electrochemical Society*, **118** [7] 1195-1971 (1971).

[9] W.C. Tripp, H.H. Davis, and H.C. Graham, "Effect of SiC Addition on the Oxidation of ZrB_2," *Am. Ceram. Soc. Bull.*, **52** [8] 612-616 (1973).

[10] F. Monteverde, A. Bellosi and S. Guicciardi, "Processing and Properties of Zirconium Diboride-Based Composites," *J. Eur. Ceram. Soc.*, **22** 279-288 (2002).

[11] F. Monteverde and A. Bellosi. "Oxidation of ZrB_2-Based Ceramics in Dry Air," *Journal of the Electrochemical Society*, **150** [11] B-552-B559 (2003).

[12] F. Monteverde and A. Bellosi. "Development and Characterization of Metal-Diboride-Based Composites Toughened with Ultra-Fine SiC Particulates," *Solid State Sciences*, **7** [5] 622-630 (2005).

[13] A. Chamberlain, W. Fahrenholtz, G. Hilmas and D. Ellerby. "Oxidation of ZrB_2-SiC Ceramics under Atmospheric and Reentry Conditions," *Refractories Applications Transactions*, **1** [2] 1-8 (2005).

[14] M. M. Opeka, I.G. Talmy, E.J. Wuchina, J.A. Zaykoski and S.J. Causey. "Mechanical, Thermal, and Oxidation Properties of Refractory Hafnium and Zirconium Compounds," *J. Eur. Ceram. Soc.*, **19** 2405-2414 (1999).

[15] S. R. Levine, E.J. Opila, M. C. Halbig, J. D. Kiser, M. Singh and J. A. Salem. "Evaluation of Ultra-High Temperature Ceramics for Aeropropulsion Use," *J. Eur. Ceram. Soc.*, **22** 2757-2767 (2002).

[16] E.J. Opila, S.R. Levine and J. Lorincz. "Oxidation of ZrB_2-and HfB_2 based Ultra-High Temperature Ceramics: Effect of Ta Additions," *Journal of Materials Science*, **39** 5969-5977 (2004).

[17] S. N. Karlsdottir, and J.W. Halloran. (Accepted to the *J. Am. Ceram. Soc.*, April 2007).

[18] F. Monteverde, and A. Bellosi. "The Resistance to Oxidation of HfB_2-SiC Composite." *J. Eur. Ceram. Soc.* **25** 1025-1031 (2005).

[19] M. Gasch, D. Ellerby, E. Irby, S. Beckman, M. Gusman, & S. Johnson. "Processing, Properties and Arc Jet Oxidation of Hafnium Diboride/Silicon Carbide Ultra High Temperature Ceramics." *Journal of Materials Science* **39** 5925-5937 (2004).

[20] W. Fahrenholtz. "Thermodynamic Analysis of ZrB_2–SiC Oxidation: Formation of a SiC-Depleted Region." *J. Am. Ceram. Soc.*, **90** [1] 143-148 (2007).

[21] F. Monteverde "Beneficial effects of an ultra-fine SiC incorporation on the sinterability and mechanical properties of ZrB_2" *Appl. Phys. A* **82** 329-337 (2006)

[22] A. Rezaie, W.G. Fahrenholtz, and G.E. Hilmas, "Evolution of Structure during the Oxidation of Zirconium Diboride-Silicon Carbide in Air up to 1500C," *J. Eur. Ceram. Soc.* **27** [6] 2495-2501 (2007).

THE ROLE OF FLUORINE IN GLASS FORMATION IN THE Ca-Si-Al-O-N SYSTEM

Amir Reza Hanifi, Annaik Genson, Michael J. Pomeroy and Stuart Hampshire
Materials and Surface Science Institute
University of Limerick, Limerick, Ireland

ABSTRACT

Oxynitride glasses are found as grain boundary phases in silicon nitride ceramics. They are effectively alumino-silicate glasses in which nitrogen substitutes for oxygen in the glass network, and this causes increases in glass transition and softening temperatures, viscosities (by two to three orders of magnitude), elastic moduli and microhardness. Calcium silicate-based glasses containing fluorine are known to have useful characteristics as potential bioactive materials. Therefore, the combination of both nitrogen and fluorine additions to these glasses may give useful glasses with enhanced mechanical stability for use in biomedical applications.

This paper reports glass formation and evaluation of glass properties in the Ca-Si-Al-O-N-F system. Within the previously defined Ca-Si-Al-O-N glass forming region at 20 eq.% N, homogeneous, dense glasses are formed. However, addition of fluorine affects glass formation and reactivity of the glass melts and can lead to fluorine loss as SiF_4, and also nitrogen loss. As these gases evolve, bubbles are formed in the glass. The compositional limits for both dense and porous glass formation at 5 eq.% F have been mapped. At high fluorine contents under conditions when Ca-F bonding is favoured, CaF_2 crystals precipitate in the glass. The maximum fluorine content found in this system is 7 eq% at a composition $Ca_{28}Si_{51}Al_{21}O_{73}N_{20}F_7$. The role of the different cations in these oxyfluoro-nitride glasses is discussed.

INTRODUCTION

Oxynitride glasses form in silicate and alumino-silicate systems when a nitrogen containing compound such as Si_3N_4 dissolves to form a M-Si-O-N or M-Si-Al-O-N liquid at ~1600-1700°C and cools to form a glass (M is usually a di-valent [Mg, Ca] or tri-valent [Y, Ln] cation. Due to its ability to increase cross-link density, nitrogen raises the glass transition temperature, hardness, Young's modulus and the fracture toughness of these glasses and decreases the thermal expansion coefficient. Also nitrogen lowers the melting temperature and increases the viscosity of the melt which enhances glass forming ability and subsequent stability[1,2,3]. Al is known to expand the compositional range for glass formation in these systems, lower the melt temperature, increase nitrogen solubility and, in amounts of only a few atomic percent, suppress the phase separation observed in its absence[4,5,6].

Fluorine in silicate and alumino-silicate glasses acts as a powerful network disrupter which substitutes for bridging oxygen ions[7]. Fluorine reduces the glass transition temperature, viscosity and refractive index, aids crystallisation and increases the potential for phase separation[8]. The effect of fluoride addition on the structure of silicate or aluminosilicate glasses containing alkaline earth cations has been reported[7,9] and it has been shown that fluorine may bond to silicon as Si-F, to Al as Al-F, or to Ca as Ca-F. Fluorine loss (as SiF_4) occurs under conditions where formation of the Si-F bond is favoured. The bonding of fluorine to aluminum prevents fluorine loss from the glass melt and explains the reduction in the glass transition temperature[7]. For successful production of dense glasses in the Ca-Si-Al-O-N-F system, fluorine loss needs to be suppressed by means of careful design of chemical composition.

The aim of the current work was to explore a new generation of oxynitride glasses containing fluorine and to develop an initial understanding of the effects of composition on glass formation, structure and properties.

EXPERIMENTAL PROCEDURE

PREPARATION OF Ca-Si-Al-O-N-F GLASSES

The extent of the glass forming regions in various M-Si-Al-O-N systems (M = Ca, Mg, Y, etc.) has been studied previously[1-6,10] and represented using the Jänecke prism[1,2,10] with compositions expressed in equivalent percent (eq.%) of cations and anions[1,2,10] instead of atoms or gram-atoms. One equivalent of any element always reacts with one equivalent of any other element or species. For a system containing three types of cations, A, B and C with valencies of v_A, v_B, and v_C, respectively, then:

Equivalent concentration of A = (v_A [A])/(v_A [A] + v_B[B] + v_C[C]),

where [A], [B] and [C] are, respectively, the atomic concentrations of A, B and C, in this case, Si^{IV}, Al^{III} and Ca^{II}.

If the system also contains three types of anions, D, E and F with valencies v_D, v_E and v_F, respectively, then:

Equivalent concentration of D = (v_D [D])/(v_D [D] + v_E[E] + v_F[F]),

where [D], [E] and [F] are, respectively, the atomic concentrations of D, E and F, i.e. O^{II}, N^{III} and F^{I}.

To find the glass forming region in the Ca-Si-Al-O-N-F system, the previously reported glass forming region[2] of the Ca-Si-Al-O-N system at 20 eq.% nitrogen at 1700°C was used as the basis for further exploration with 5 eq.% F. Compositions within and outside this region were studied to evaluate the effects on glass formation at 1650°C of adding fluorine.

Changes in Ca:Al:Si cation ratios were made systematically in order to study the complete glass forming region at the fixed O:N:F ratio. Glasses were prepared from mixtures of silicon nitride powder (UBE) and CaF_2 (Aldrich) together with high purity (99.9%) SiO_2 (Fluka Chemika), Al_2O_3 (Sumitomo), CaO (Fisher) to give the required chemical composition in eq. % cations/anions.

Powder samples of total weight 10g were mixed using a magnetic stirrer in 50 ml isopropanol which was then evaporated using a hot plate. Compacts of 10 mm height and 20 mm diameter were pressed under 20 tonnes pressure. These were melted in a boron nitride lined graphite crucible under 0.1 MPa nitrogen at 1650°C for 1 h in a vertical Al_2O_3 tube furnace, after which the crucible was withdrawn rapidly from the hot zone to ambient temperature. The cooling rate was estimated to be ~100°C/min. The solid glass sample was removed and cleaned using an ultrasonic bath prior to further analysis.

CHARACTERIZATION OF GLASSES

X-ray diffraction analysis was used to determine if the samples were amorphous or crystalline and to identify the crystalline phases. Differential thermal analysis was carried out using a Stanton-Redcroft STA 1640 instrument. 50 mg of glass powder was heated at 10°C/min under 0.1 MPa nitrogen atmosphere. To find Tg_{onset} and Tg_{offset}, the first derivative was analysed and to find the Tg, the second derivative was analysed. Crystallisation and melting temperatures were obtained from the maxima in the peaks on the original DTA curve.

RESULTS AND DISCUSSION

GLASS FORMATION IN THE Ca-Si-Al-O-N-F SYSTEM

Fig. 1 shows the 75 compositions which were investigated to find the glass forming region at 20 eq% N and 5 eq% F. The cation compositions, which were selected from the amorphous and crystalline regions of the original Ca-Si-Al-O-N diagram, were changed systematically by 2-3 eq.% for each attempt. Fig. 2 shows the glass forming region found for the Ca-Si-Al-O-N-F system 20 eq.% N and 5 eq.% F at 1650°C based on the different compositions shown in Fig. 1. In the surrounding regions the different crystalline phases observed are also shown. All glasses were dense except for a region of Si-rich compositions where porous glasses were observed.

In the porous glass area of Fig. 2, there are bubbles on the surface in addition to the bubbles (pores) within the bulk of the glasses which is due to SiF_4 loss. Formation of SiF_4 is favored perhaps due to high Si:F ratios (>3) and low Al (6-15 eq.%) and Ca (13-25 eq.%) contents.

In some areas of this glass forming region inhomogeneous and phase separated glasses were found similar to other previously reported oxyfluoride and oxynitride glasses[7,11,12].

Different rules have been proposed to keep Al in 4-fold coordination in calcium alumino-silicate glasses, so combining with fluorine and preventing its loss[13,14]. When Al:Si ratio is <1, Al is in 4-fold coordination and Al may enter the glass network as $[AlO_4]^{5-}$ tetrahedra[13]. When Ca:Al ratio >1:2 and Al:Si ratio <1:1, Al is thought to be in 4-fold coordination as Ca ions can charge balance $[AlO_4]^{5-}$ tetrahedra[14]. When Al:Si ratio is >1 or Ca:Al is <1:2, then some of the Al ions

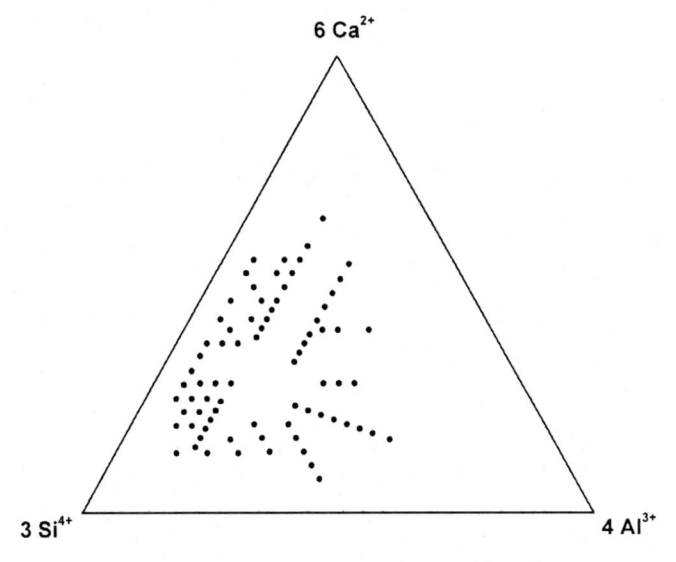

Fig. 1. Compositions studied (75) to investigate the glass forming region at 20 eq% Nitrogen and 5 eq% Fluorine in the Ca-Si-Al-O-N-F system.

The Role of Fluorine in Glass Formation in the Ca-Si-Al-O-N System

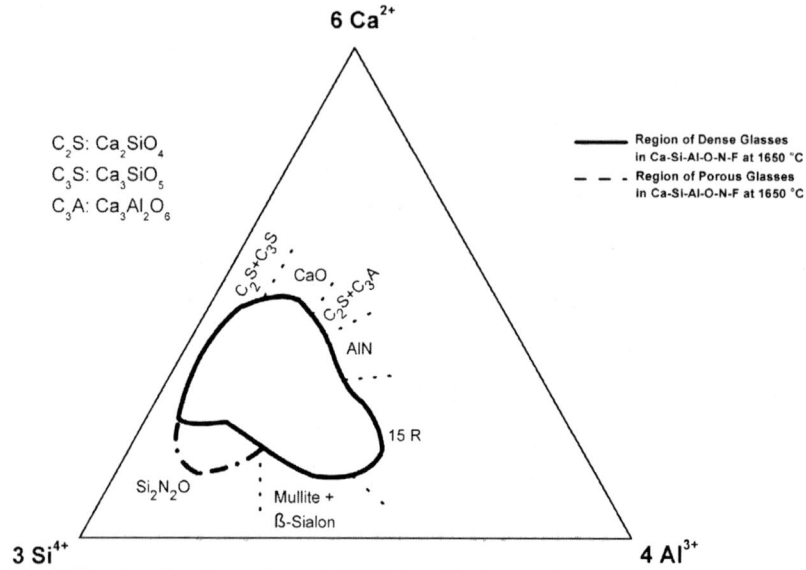

Fig. 2. The glass forming region at 1650°C found in the Ca-Si-Al-O-N-F system at 20 eq.% N and 5 eq.% F and the adjacent crystalline regions.

take up 6-fold coordination and so F tends to bond to Si which creates instability in the melt and loss of volatile SiF_4[14,15]. Fluorine loss can be suppressed by incorporation of sufficient basic network modifying oxide which in this system is CaO[7]. When there is sufficient Ca to form non-bridging oxygens, Si-F bond formation is prevented and, hence, no loss of SiF_4 can occur[16]. In addition, nitrogen increases the cross-linking in the network and, thus, can be bonded to either three Si atoms or only two[1] and still act as an effective bridging anion and therefore prevent formation of SiF_4.

In the case of Al, if Al:F = 1:1, then sufficient Al ions are present to satisfy the F ions present and form Al-F bonds. For 5 eq.% F, this requires 15 eq.% Al. Al-F bonds are favoured at Al > 15 eq.%, unless Al:Si >1 which leads to instability in the glass network and formation of Si-F bonds leading to possible SiF_4 bubble formation, resulting in porous samples. When Al:Si >>1, glass formation becomes difficult and calcium fluoride (CaF_2) is precipitated.

At < 15 eq.% Al (greater Ca content), a large area of dense glasses is observed which is evidence of the formation of Ca-F bonds since there is insufficient Al to satisfy all the F as Al-F bonds. While the samples are not porous, ~3% weight loss occurs in these glasses which means that N or F loss is still inevitable. At low Al contents (3-4%), samples were bloated and large weight losses observed. This is mainly due to the shift of the cation composition towards the $CaO-SiO_2$ binary which does not show a tendency for glass formation. Al-free Ca-Si oxynitride glasses are phase separated and addition of a few weight percentage of Al_2O_3 suppresses the phase separation in such systems[3].

The Role of Fluorine in Glass Formation in the Ca-Si-Al-O-N System

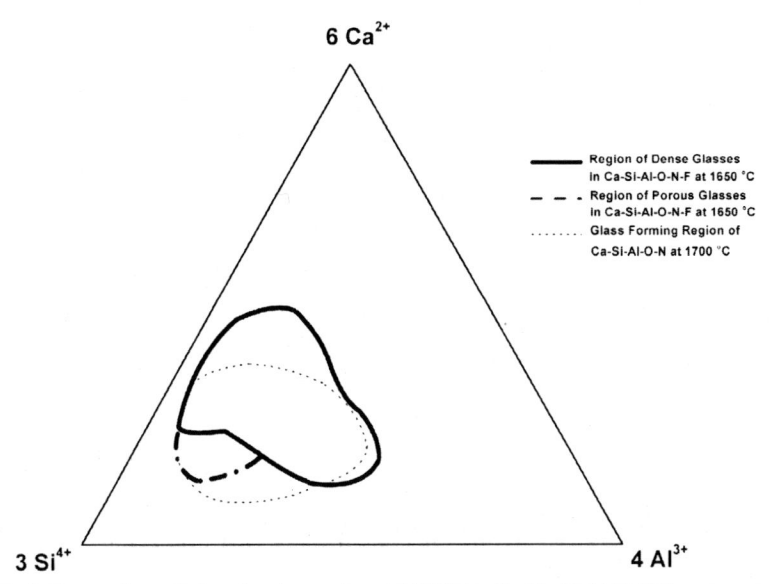

Fig. 3. Comparison of glass forming region at 1650°C in the Ca-Si-Al-O-N-F system at 20 eq.% N and 5 eq.% F and glass forming region at 1700°C in the Ca-Si-Al-O-N system at 20 eq.% N [2].

Fig. 3 compares the glass forming region at 1650°C in the Ca-Si-Al-O-N-F system at 20 eq.% N and 5 eq.% F (solid lines) with the glass forming region at 1700°C in the Ca-Si-Al-O-N system at 20 eq.% N (dotted line)[2]. The figure shows that, by addition of fluorine, the glass forming region has expanded towards more Ca-rich compositions. The glass forming region for the Ca-Sialon system at 1700°C is smaller than the area determined for Ca-Si-Al-O-N-F glasses at 1650°C and so it is clear that fluorine has a significant effect in expanding the glass forming region.

For a better understanding of the role of fluorine on thermal properties of glasses, some glasses with different cation ratios in the Ca-Si-Al-O-N-F system at 20 eq.% N and 5 eq.% F were selected. The melting points of these compositions were found from DTA analysis and compared with the liquidus temperatures of the same eq.% cation compositions found in the ternary of CaO-SiO_2-Al_2O_3 system[17]. Fig. 4 compares the liquidus temperatures for these oxide and oxyfluoronitride compositions. It is obvious that the addition of both nitrogen and fluorine reduces the liquidus temperatures of the high silica and alumina compositions by 100-250°C. At high SiO_2 and Al_2O_3 contents, there is not too much difference in the area of glass formation in both the Ca-Si-Al-O-N and Ca-Si-Al-O-N-F systems. However, there is a large expansion of glass formation in Ca-Si-Al-O-N-F system at 20 eq.% N and 5 eq.% F compared with the Ca-Si-Al-O-N system at 20 eq.% N, which is mainly due to the effect of nitrogen and fluorine together, as evidenced by the much lower liquidus temperatures of about 800°C at higher Ca contents. In the Ca-Si-Al-O-N system, high calcium compositions in this region are not fully melted and

The Role of Fluorine in Glass Formation in the Ca-Si-Al-O-N System

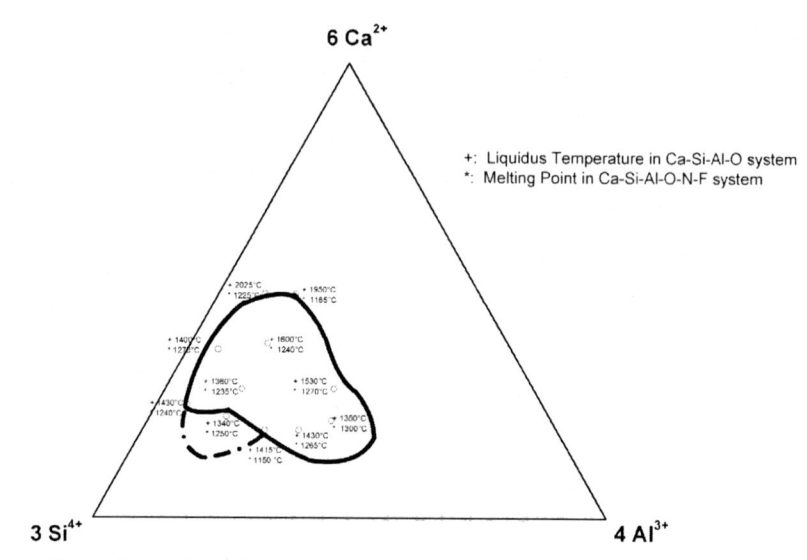

Fig. 4. Comparison of melting points of oxide and oxyfluoronitride compositions.

contain solid crystalline phases at 1700°C. Therefore the role of fluorine is to reduce the liquidus temperature and it also decreases glass transition temperature (see later) which may allow an extension of the melting range and so enhance glass formation.

In relation to melting temperatures, it was observed that by the addition of a defined amount of CaF_2 to Ca-Si-Al-O-N glasses and then remelting of the batch, a dense transparent Ca-Si-Al-O-N-F glass can be obtained at a temperature of 1500°C.

Fig. 5 shows regions which specify the homogeneity and the colour of the obtained glasses in this system. Within the area outlined in bold, dense homogeneous or inhomogeneous glasses or samples with few bubbles on the surface were obtained while in the area outlined by a dashed line, porous homogenous glasses were obtained. Most of the glasses in this system are gray or black in colour. The homogeneity of glasses is not only a function of composition and purity but also the melting conditions such as temperature or cooling rate; nitrogen pressure and pouring are important factors in this regard. Glasses in this system are mainly opaque but in some regions, thin slices of glass are transparent.

EFFECT OF FLUORINE AND NITROGEN ON GLASS STABILITY IN THE Ca-Si-Al-O-N-F SYSTEM

The effect of cation ratios and fluorine content on glass formation and crystallisation in the Ca-Si-Al-O-N-F system is shown in Table 1. It is clear that compositions which are amorphous at 5 eq.% F become crystalline when the amount of fluorine is increased. For example, for a cation composition of $Ca_{28}Si_{39}Al_{33}$, a dense glass is formed at 5 eq.% F while weight loss and precipitated CaF_2 occurs when 11 eq.% F is introduced.

The previous maximum N content glass in the Ca-Si-Al-O-N system was reported as 26 eq.% N, $Ca_{28}Si_{56}Al_{16}O_{74}N_{26}$[2]. After several attempts to find the maximum nitrogen content when

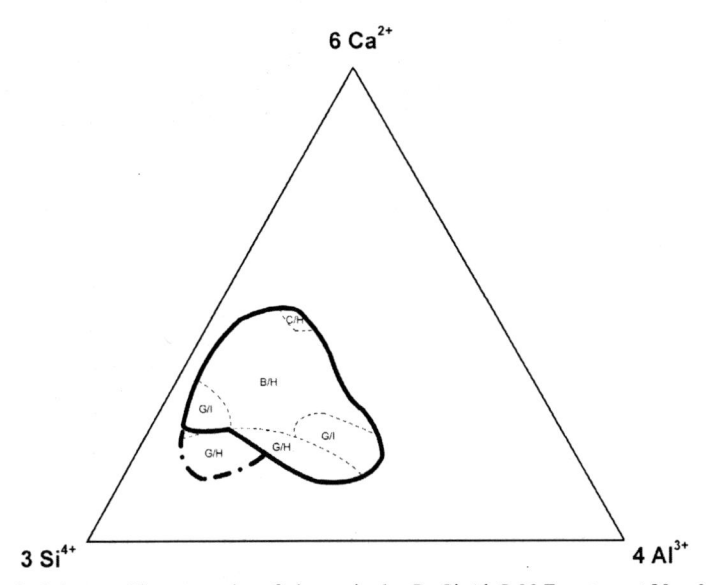

Fig. 5. Colour and homogeneity of glasses in the Ca-Si-Al-O-N-F system at 20 eq% N
and
5 eq% F. G = Gray, B = Black, C = Creamy, H = Homogeneous, I = Inhomogeneous

Table 1: XRD analysis of Ca-Si-Al-O-N-F compositions fired at 1650 °C showing amorphous or crystalline products obtained from different fluorine content samples.

Amorphous Composition	Crystalline Composition
$Ca_{28}Si_{39}Al_{33}O_{75}N_{20}F_5$	$Ca_{28}Si_{39}Al_{33}O_{69}N_{20}F_{11}$
$Ca_{23}Si_{47}Al_{30}O_{75}N_{20}F_5$	$Ca_{23}Si_{47}Al_{30}O_{70}N_{20}F_{10}$
$Ca_{28}Si_{45}Al_{27}O_{75}N_{20}F_5$	$Ca_{28}Si_{45}Al_{27}O_{71}N_{20}F_9$
$Ca_{19}Si_{57}Al_{24}O_{75}N_{20}F_5$	$Ca_{19}Si_{57}Al_{24}O_{72}N_{20}F_8$

5 eq.% fluorine is introduced into the system, it was found that not more than this level of nitrogen could be incorporated. At higher nitrogen contents, Si_3N_4 is precipitated which means that the network cannot dissolve more nitrogen. However, when a lower amount of fluorine (1 eq.% F) is added to the composition, nitrogen content was found to be ~40 eq.% N at the composition $Ca_{28}Si_{57}Al_{15}O_{59}N_{40}F_1$ without any crystallisation occuring.

Therefore it seems that fluorine affects the dissolution of nitrogen into the melt by lowering the liquidus temperature and increasing melting range and yet does not lead to crystallisation. Introduction of higher amounts of nitrogen into the glass should result in development of glasses with excellent mechanical properties and good chemical durability.

While oxide glasses can be formed containing up to 20 eq.% fluorine[7], it appears that the incorporation of a higher amount of fluorine is not possible in the presence of nitrogen. At 5 eq.% N, the maximum fluorine was found to be 10 eq.% F at $Ca_{28}Si_{57}Al_{15}O_{85}N_5F_{10}$, while at 10 eq.% N, the maximum was 8 eq.% F at $Ca_{28}Si_{48}Al_{24}O_{82}N_{10}F_8$ and at 20 eq.% nitrogen, the maximum

The Role of Fluorine in Glass Formation in the Ca-Si-Al-O-N System

fluorine was 7 eq.% F at $Ca_{22}Si_{57}Al_{21}O_{73}N_{20}F_7$ and $Ca_{28}Si_{51}Al_{21}O_{73}N_{20}F_7$. Therefore, it seems that by introducing more nitrogen, the amount of fluorine that can be incorporated reduces. Increasing the amount of fluorine results in CaF_2 precipitation and this can be explained by the fact that N occupies the corners of Si and Al tetrahedra and, as it does not tend to bond to the glass modifier, it forces F to attach to Ca. Increasing the amount of nitrogen results in this phenomenon occuring at lower amounts of fluorine.

Nitrogen as a cross linking agent, attaches to Si atoms and reduces the possibility of SiF_4 formation and, hence, stabilizes the glass composition. Fluorine has a significant effect on the solubility of nitrogen into the structure of Ca-Si-Al-O-N-F glasses and expands the glass forming region towards cation compositions that are more refractory with higher liquidus temperatures. Therefore, it can be concluded that the two anions have a complementary effect on glass formation in the Ca-Si-Al-O-N-F system.

GLASS TRANSITION AND CRYSTALLIZATION TEMPERATURES OF Ca-Si-Al-O-N-F GLASSES

Thermal properties of Ca-Si-Al-O-N-F glasses are summarized in Table 2. Addition of 5 eq.% fluorine to the oxide and oxynitride glasses leads to a sudden decrease in Tg by about 94°C. This is mainly due to the disruptive role of fluorine which substitutes for bridging oxygens and facilitates the motion of the structural units in the glass. The addition of 20 eq.% nitrogen increases the Tg of oxide and oxyfluoride glasses by about 38°C.

Both ions affect the motion of structural units in the glass. Cross-linked nitrogen keeps the units more rigidly fixed and, therefore, there is an increase in the thermal energy required for segmental mobility and thus an increase in Tg[18]. Fluorine has a more marked effect in terms of its disruptive role on thermal properties than the constructive role of nitrogen.

Both nitrogen and fluorine decrease the crystallisation temperature of oxide glass but fluorine is more effective in this regard. In oxide, oxynitride and oxyfluoride glasses, only one crystallisation peak is observed while in the oxyfluoronitride glasses, three peaks appear which are attributed to the multi-phase nature of the final crystalline glass-ceramics. The crystalline phases formed are: Gehelenite ($Ca_2Al_2SiO_7$), Calcium Fluoride (CaF_2), Cuspidine ($Ca_4Si_2O_7F_2$) and traces of Calcium Aluminum Oxide ($CaAl_4O_7$).

Table 2. Thermal properties of Ca-Si-Al-O-N-F glasses (varying anions).

Composition (eq%)	Tg$_{DTA}$ °C	Tc$_1$ °C	Tc$_2$ °C	Tc$_3$ °C	T$_m$ °C
$Ca_{28}Si_{57}Al_{15}O_{100}$	810	1090	-	-	1269
$Ca_{28}Si_{57}Al_{15}O_{95}F_5$	716	946	-	-	1206
$Ca_{28}Si_{57}Al_{15}O_{80}N_{20}$	848	1043	-	-	1261
$Ca_{28}Si_{57}Al_{15}O_{75}N_{20}F_5$	755	925	999	1059	1173

The first crystallisation temperature of this glass is lower than for both oxyfluoride and oxynitride glasses. This means that a combination of these two anions effectively reduces the crystallisation temperatures of calcium aluminosilicate glasses. Fluorine lowers the melting point of oxide and oxynitride glasses. Oxyfluoronitride glass has the lowest melting point in this series and this is mainly due to the presence of fluorine as a network disrupter.

CONCLUSIONS
1) A new generation of oxynitride glasses containing fluorine have been investigated.
2) The glass forming region in the Ca-Si-Al-O-N-F system at 20 eq.% N and 5 eq.% F is larger than the glass forming region at 20 eq.% N in the Ca-Si-Al-O-N system. Fluorine expands the range of glass formation in this oxynitride system.
3) Considerable reduction of liquidus temperatures by about 800°C at higher calcium contents shows that expansion of glass formation in the Ca-Si-Al-O-N-F system, compared with the Ca-Si-Al-O-N system, is mainly due to the effect of fluorine.
4) Fluorine facilitates the solution of much higher amounts of nitrogen into the melt than are possible in the Ca-Si-Al-O-N system.
5) Nitrogen and fluorine play a complementary role in glass formation in Ca-Si-Al-O-N-F system.

ACKNOWLEDGMENTS
The authors wish to acknowledge Science Foundation Ireland for financial support of this research and to thank colleagues in Materials and Surface Science Institute for their help and advice.

REFERENCES
[1] S. Hampshire, "Oxynitride Glasses, Their Properties and Crystallization- A Review," *J. Non-Cryst. Sol.*, **316**, 64-73 (2003).
[2] R. A. L. Drew, S. Hampshire and K. H. Jack, "Nitrogen Glasses," *Proc. Brit. Ceram. Soc.*, **31**, 119-132 (1981).
[3] R. E. Loehman, "Oxynitride Glasses," *J. Mater. Sci. Tech.*, **26**, 119-149 (1985).
[4] S. Hampshire, R. A. L. Drew and K. H. Jack, "Oxynitride Glasses," *Phys. Chem. Glass.*, **26** [5], 182-186 (1985).
[5] G. Leng-Ward and M. H. Lewis, "Oxynitride Glasses and Their Glass-Ceramic Derivatives," *Glasses and Glass-Ceramics*, ed. M. H. Lewis, Chapman and Hall, London, 106-155 (1990).
[6] R. E. Loehman, "Preparation and Properties of Oxynitride Glasses," *J. Non-Cryst. Sol.*, **56**, 123-134 (1983).
[7] R. Hill, D. Wood and M. Thomas, "Trimethylsilylation Analysis of the Silicate Structure of Fluoro-Alumino-Silicate Glasses and the Structure Role of Fluorine," *J. Mater. Sci.*, **34**, 1767-1774 (1999).
[8] A. Stamboulis, R. G. Hill and R. V. Law, "Characterization of the Structure of Calcium Alumino-Silicate and Calcium Fluoro-Alumino-Silcate Glasses By Magic Angle Spinning Nuclear Magnetic Resonance (MAS-NMR)," *J. Non-Cryst. Sol.*, **333**, 101-107 (2004).
[9] T. Maeda, S. Matsuya and M. Ohta, "Effects of CaF_2 Addition on the Structure of CaO-Al_2O_3-SiO_2 on Glasses," *J. Dent. Mater.*, **17** [2], 104-114 (1998).

[10]S. Hampshire, E. Nestor, R. Flynn, J. L. Besson, T. Rouxel, H. L. Lemrcier, P. Goursat, M. Sebai, D. P. Thompson and K. Liddell, "Yttrium Oxynitride Glasses: Properties and Potential for Crystallisation to Glass-Ceramics," *J. Euro. Ceram. Soc.*, **14**, 261-273 (1994).

[11]K. T. Stanton and R. G. Hill, "Crystallisation in Apatite-Mullite Glass–Ceramics As A Function of Fluorine Content," *J. Non-Cryst. Sol.*, **275**, 2061-2068 (2005).

[12]P. Jankowski and S. H. Risbud, "Formation and Characterization of Oxynitride Glasses in the Si-Ca-Al-O-N and Si-Ca-Al, B-O-N Systems," *J. Mater. Sci.*, **18**, 2087-2094 (1983).

[13]W. Loewenstein and M. Loewenstein, "The Distribution of Aluminum in the Tetrahedra of Silicate and Aluminates," *J. Amer. Mineral.*, **39**, 92-96 (1954).

[14]R. G. Hill, C. Goat and D. Wood, "Thermal Analysis of a SiO_2-Al_2O_3-CaO-CaF_2 Glass," *J. Amer. Ceram. Soc.*, **75** [4], 778-785 (1992).

[15]S. H. Risbud, R. J. Kirkpatrick, A. P. Taglialavore and B. Montez, "Sold-State NMR Evidence of 4-, 5-, and 6- Fold Aluminum Sites in Roller-Quenched SiO_2-Al_2O_3 Glasses," *J. Amer. Ceram. Soc.*, **70**, C10-C12 (1987).

[16]A. Rafferty, A. Clifford, R. Hill, D. Wood, B. Samuneva and M. Dimitrova-Lukacs, "Influence of Fluorine Content in Apatite Mullite Glass Ceramics," *J. Amer. Ceram. Soc.*, **83** [11], 2833-2828 (2000).

[17]E.M. Levin, C.R. Robbins, H. F. McMurdie, "Phase Diagram for Ceramists," *American Ceramic Society*, Columbus, OH, 220 (1964).

[18]M. J. Pomeroy, C. Mulcahy, S. Hampshire, "Independent Effects of Nitrogen Substitution for Oxygen and Yttrium Substitution for Magnesium on the Properties of Mg-Y-Si-Al-O-N Glasses," *J. Am. Ceram. Soc.*, **86** [3], 458-464 (2003).

WETTING AND REACTION CHARACTERISTICS OF AL_2O_3/SIC COMPOSITE
REFRACTORIES BY MOLTEN ALUMINUM AND ALUMINUM ALLOY

James G Hemrick
Oak Ridge National Laboratory
Oak Ridge, TN, USA

Jing Xu
West Virginia University
Morgantown, WV USA

Klaus-Markus Peters
Fireline TCON, Inc.
Youngstown, OH, USA

Xingbo Liu
West Virginia University
Morgantown, WV USA

Ever Barbero
West Virginia University
Morgantown, WV USA

ABSTRACT

The reactive wetting behavior in molten aluminum (Al) and Al alloy was investigated for alumina-silicon carbide composite refractory materials using an optimized sessile drop method at 900°C in a purified Ar-4% H_2 atmosphere. The time dependent behavior of the contact angle and drop geometry was monitored and the wetting kinetics were determined. The initial contact angle between the liquid Al/Al alloy and the refractory substrates was found to be greater than 90° and to gradually decrease with time. For two of the materials, it was found that the contact angles decreased to and angle less than 90° by the end of the two-hour test. For the third material, the contact angle was still greater than 90° at the conclusion of the two-hour test. The difference in wetting properties among the three types of refractories is attributed to their microstructural and compositional variations. The effect of magnesium in the molten Al alloy drops on the wetting kinetics and the reaction with the refractory substrates are also discussed. The results obtained provide important understanding of the wetting and corrosion mechanisms of alumina and silicon carbide materials in contact with molten aluminum.

INTRODUCTION

The work described in this paper was initiated to understand the wetting and reaction which occurs between alumina-silicon carbide composite refractories and molten aluminum or aluminum-magnesium alloy. This work is relevant to the aluminum melting and production industries, as refractory erosion and corrosion products can be sources of non-metallic inclusions and contamination of the aluminum, which leads to energy and production losses. Reactive metal (molten Al alloy) attack was explored, in which an Al drop wets, penetrates and reacts with a refractory substrate, degrading the refractory's properties and shortening its service life.

There is great interest in reducing the corrosion of refractory materials in the metallurgical processing industries as the refractory erosion and corrosion products can be sources of non-metallic inclusions and contamination of the aluminum, which leads to energy and production losses. One of the approaches taken to minimize the corrosion is to reduce the metal-ceramic contact surface by reducing the wettability between the ceramic and liquid metal. Many studies were performed to investigate the molten aluminum (Al) penetration in various ceramic materials. The majority of these studies addressed the penetration of SiO_2 substrates (silica glass or alumino-silicate ceramics) by molten Al[1-6]. Specific reports of exposure of SiC substrates to molten aluminum containing Mg were rarely found, but reference to the wetting of SiC by molten aluminum/silicon alloys[7-12] and copper alloy[13] were reported.

A linear dependence of the reaction layer thickness and composition with time was found for the interaction of SiO_2 substrates with molten Al.[1,5] Also, the wetting behavior was found to be dependent on the formation of a reaction zone by redox reactions and to consist of three regions with varying chemistries, dependent on the interdiffusion of Si^{2+}, Al^+, Al^{2+}, and Al^{3+}.[2] Further, during the reaction, it was found that Si is released into the liquid metal and diffuses toward the Al source[4], with his reaction occurring in up to five separate steps.[6]

Specific to refractory materials, the penetration of silica refractories by molten Al at 700-1000°C was studied by Brondyke using both traditional cup testing and immersion testing.[3] All tested commercial alumino-silicate refractories, used for aluminum melting applications, were found to be wetted and subsequently penetrated on exposure to molten Al. Results indicated that the problems associated with alumino-silicate refractories resulted from the penetration of molten Al, side-wall build-up of oxide, and formation of corundum and metallic silicon due to reaction of Al with the Si and Si-bearing constituents. Subsequently there was an increase in volume of penetrated product, which would lead to generation of tensile stress due to aluminum oxide build-up caused by oxidized aluminum and its alloy components around the metal line. The presence of tensile stresses would ultimately cause cracking in the refractories. Additionally, dissolution of Si occurred in the molten Al with the penetration rate controlled by diffusion of Al and Si through the aluminum oxide.

In this work the wettability and the corrosion of different compositions of SiC-Al_2O_3-Al-Si composite refractory materials by liquid Al/Al-Mg systems were investigated. Wetting was studied by the sessile drop method, while corrosion was studied through the characterization of the solid–liquid interfaces using spectroscopic methods. From results of this testing, the wetting and corrosion mechanisms are examined.

EXPERIMENTAL
Materials
    The composite refractory materials studied (supplied by Fireline TCON Inc) contained a continuous microscopic network of interpenetrating microscopic scale ceramic and metallic phases. The presence of metallic phases provide significant improvement in toughness and damage tolerance, while the ceramic phases lead to high hardness and improved performance at elevated temperatures. These materials contain various ratios of silicon carbide, alumina, aluminum, and silicon as listed in Table 1 and possess microstructures as shown in Figure 1.

Table 1. Refractory Substrate Compositions

Composition (wt. %)	MC	TC	TQ
Silicon Carbide	54	50	/
Aluminum Oxide	38	35	70
Aluminum/Silicon	8	15	30

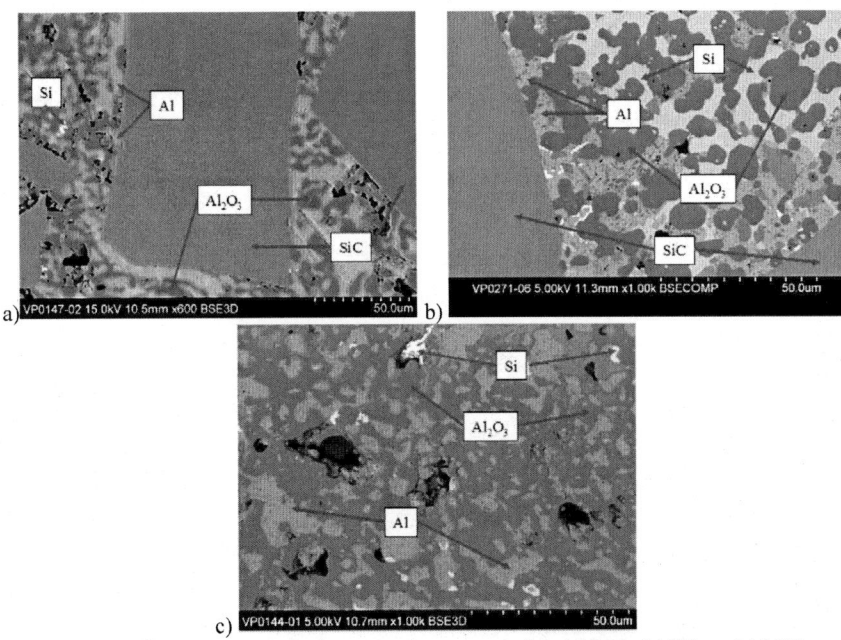

Figure 1. Microstructures of refractory substrate materials (a) MC, (b) TC, and (c) TQ.

The TCON materials evaluated in the present study are produced utilizing the displacement reactions shown in Equation 1.

$$3SiO_2 + 4Al \rightarrow 2Al_2O_3 + 3Si \quad \text{(base reaction)}$$
$$3SiO_2 + (phases) + 4Al \rightarrow 2Al_2O_3 + (phases) + 3Si^* \quad (1)$$
(*Si retained in final product is < 5%)

This transformation process is based on the direct oxidation of the metal (Al) by exposing the sacrificial silica preform to molten Al under carefully controlled conditions. Due to volumetric contractions caused by loss of silica and formation of alumina, void space is created as the reactions proceed (starting from the outside surfaces), and this space is completely filled with Al metal. The silicon by-product subsequently dissolves into molten Al with the resulting composite containing ≈70 wt. % (63 vol. %) Al_2O_3 and ≈30 wt. % (37 vol. %) Al-Si alloy. The process of

Wetting and Reaction Characteristics of Al$_2$O$_3$/SiC Composite Refractories

transforming a silica preform into a TCON component is efficient, in that there are minimal dimensional changes; the part's shape and size are essentially unchanged as the silica is converted to the alumina-aluminum alloy composite. Additionally, silicon carbide particles can be added to the composites to increase toughness, strength, wear resistance and thermal shock resistance.

Refractory plates (12 mm × 12 mm × 3 mm), which were cut from 25mm × 25 mm × 175 mm bars of originally supplied materials, were used as substrates in subsequent wetting studies. Aluminum pigs (99.99% pure) and aluminum alloy 5083, with composition given in Table 2, were machined down to cubes (5 mm × 5 mm × 3 mm) for melting during static testing or wire segments of 3 mm diameter which were melted to produce sessile drops during dynamic testing. Experiments were carried out in purified Ar - 4% H$_2$ (<1 ppm O$_2$) at 900°C and changes in contact angle between molten Al and refractory substrates were monitored with time using a high speed CCD camera.

Table 2. Chemical Composition of Aluminum Alloy 5083

Component	Al	Cr	Cu	Fe	Mg	Mn	Si	Ti	Zn
Wt. %	92.4 - 95.6	0.05 - 0.25	Max 0.1	Max 0.4	4 - 4.9	0.4 - 1	Max 0.4	Max 0.15	Max 0.25

Experimental Procedure

Static and dynamic sessile drop methods were used for the purpose of studying wettability. In the static (more conventional method), a solid cube of Al/Al alloy is placed on the refractory substrate and melted. A modified dynamic method was also employed in which molten Al/Al alloy was dropped on the refractory substrate by a heated delivery device. This modified method was developed to study the dynamic wetting behavior of the refractory/metal system as this approach has been suggested to be closer to the application conditions.[6] The apparatus used for both methods of sessile drop experiments is schematically illustrated in Figure 2.

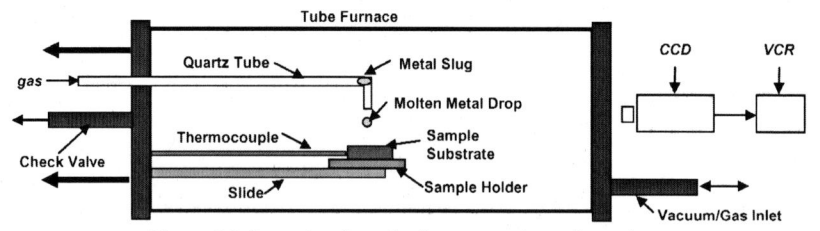

Figure 2. Schematic of sessile drop test unit configuration.

This apparatus consisted of a 33 kW horizontal circular infrared furnace, using an evacuating system with a rotary pump and refilling gas system supplying purified Ar-4% H$_2$. The quartz furnace chamber was enclosed on one end by a copper lid and slide device, which was used to move the experimental assembly inside the chamber. A small diameter quartz tube was also passed through the copper lid and extended to a location directly above the sample substrate where the tube was bent 90° and its diameter was reduced. This tube was used to contain the Al/Al alloy wire segment during heating and melting, which produced the molten metal drop for the dynamic test. Both sealed end caps of the furnace assembly contained quartz windows allowing a color CCD camera (640x480 resolution) to continuously monitor the experiments and record them. Three Type-S thermocouples (with ceramic sheath) were inserted into the horizontal quartz test chamber through the copper end plate for monitoring of the refractory substrate temperature (thermocouple in contact with substrate), molten metal drop temperature (thermocouple in contact with quartz tube), and the reaction temperature (thermocouple suspended in chamber near sample), respectively.

Before the experiment, the refractory substrate and the Al/Al alloy cube or wire segment were ultrasonically cleaned in acetone. The substrate was then carefully slid into the center of the horizontal chamber. For the static sessile drop method, a cube of Al/Al alloy was placed on top of the refractory substrate. The sealed chamber was evacuated to a vacuum of 1×10^{-6} Pa and then refilled with the purified Ar-4%H$_2$ gas. Following evacuation and refilling, the IR quartz chamber was heated to 900°C at a rate of 30°C per minute. The cube of metal was allowed to melt and the wetting behavior between the metal and refractory substrate was observed. The entire duration of the experiment was captured and recorded, from which the video still frames were extracted and analyzed to measure the contact angle changes with time.

For the dynamic sessile drop method, a wire segment of Al/Al alloy was placed into the quartz tube used for delivering molten metal to the substrate. This tube was inserted through the copper end plate into the IR chamber. The chamber was evacuated and refilled, in the same manner as for the static test. While the Al segment in the quartz tube was kept at the cold zone, the IR quartz chamber was heated up to 900°C at a rate of 30°C per minute. The furnace was allowed to stabilize for 20 minutes before the Al segment was slowly forced from the cold zone to the hot zone of the furnace where it was allowed to melt and pass through the vertical portion of the delivery tube as a molten drop onto the refractory test substrate. Upon delivery, the wetting behavior between the molten drop and the substrate were observed. Similar to the static test method, the entire duration of the experiment was captured and recorded, from which the video still frames were extracted and analyzed to determine the contact angle changes with the time.

At the end of each experiment, the substrate was removed from the furnace and prepared for examination by scanning electron microscope (SEM), energy dispersive spectroscopy (EDS), and electron microprobe. Further data analysis was carried out using axisymmetric-drop-shape-analysis (ADSA) software, by which the contact angle (CA), the drop base diameter (D), and drop height (H) were directly measured from the drop profiles.

RESULTS AND DISCUSSION

Sessile drop testing was carried out on three different composite refractory materials using two different molten metal drop materials. Contact angle and drop dimensional analysis for the TC composite refractory and molten Al system is discussed here as an example of the behavior exhibited by these materials. Figure 3 shows optical micrographs of a molten aluminum drop on the TC substrate during isothermal dwelling at 900°C, illustrating the changes in contact angle and drop geometry with time. The initial contact angle between the TC substrate and liquid Al was an obtuse angle of ≈125° ±1°. This value is similar to the value (120°C) reported by Z. Yu for an Al$_2$O$_3$ - Al system at 900°C[14]. The contact angle was found to gradually decrease to a value of 82° during a hold time of 65 minutes, indicative of the transition from non-wetting to wetting. The present experimental results are similar to the Al$_2$O$_3$ - Al system data reported previously[14,15], in which a steady value of 82° was observed after 40 minutes at 1000°C. Yet, in

Wetting and Reaction Characteristics of Al$_2$O$_3$/SiC Composite Refractories

the literature, the drop height continued to change with the dwell time until the end of the 120 minute experiment, when it reached an angle of 78°. This final state was considered to represent the onset of wetting and the build up of reaction equilibrium. Differences between the current experimental case and the literature cases are expected to be due to the differences in experimental temperature and substrate chemistries.

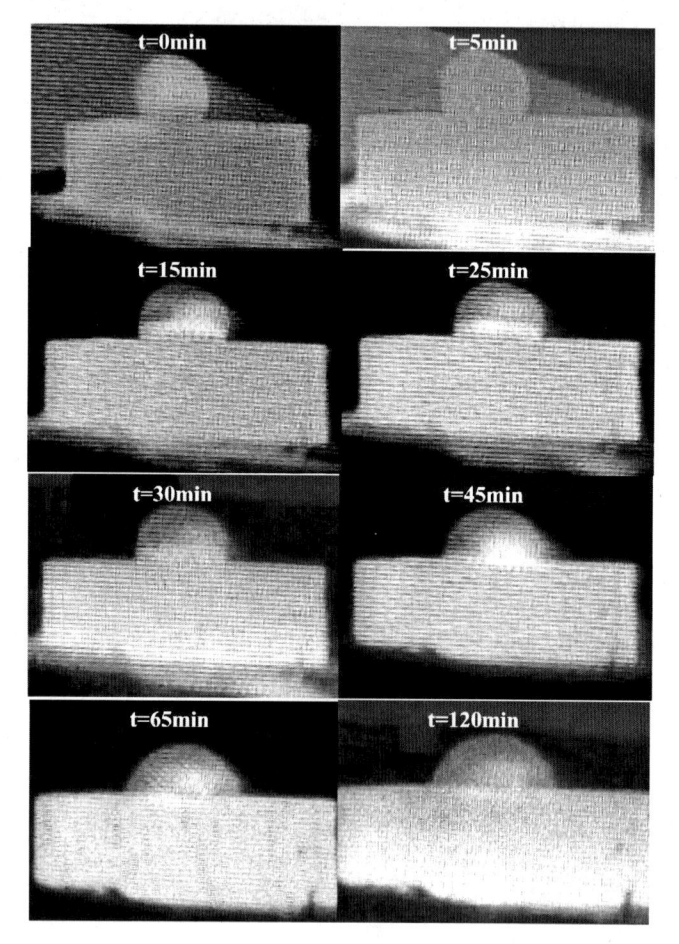

Figure 3. Wetting progress of molten Al droplet on TC substrate at 900°C.

Four parameters; namely contact angle (CA), drop volume (V), drop base diameter (D), and drop height (H), were plotted on a linear time scale revealing the wetting kinetics as shown in Figure 4. The change in contact angle characterized by the advance of the triple phase reaction (that where the solid substrate, liquid metal, and gaseous experimental environment are in contact) was due to the decrease in the drop height and/or the increase in the drop base diameter. The determinant factor was found to switch during the different sub-stages of the wetting process, leading to the maximum value of the drop volume being achieved after 10 minutes of contact. The wetting kinetics could be inferred by combination of the changes in contact angle with those in drop size[6,16]. After the initial rapid spreading stage (around 10 seconds) the interfacial front advanced quickly resulting in the steep slope found in the initial portion of the contact angle plot and lasting until the drop volume reached a maximum. After the initial stage, the decrease in drop height was dominant over the increase in drop base diameter, resulting in a decrease in overall drop volume. Finally, a steady state was attained at around 65 minutes, during which CA, V, D, and H remained unchanged.

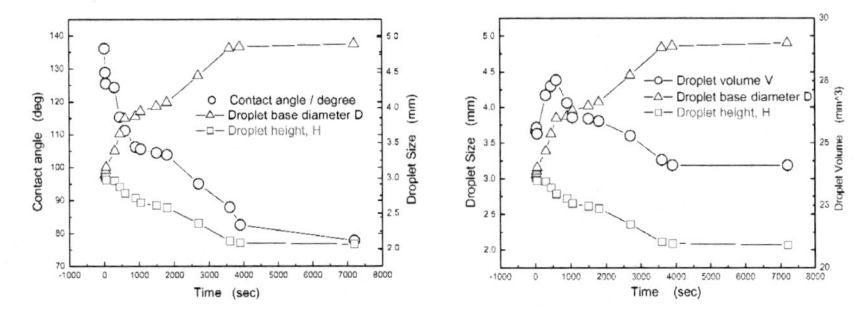

Figure 4. Variations in contact angle of TC substrate – liquid Al droplet size (base diameter and height) at 900°C.

Interfacial Morphology and Mechanism of Corrosion

Following completion of the experiments, polished cross section of samples were prepared using the standard metallographic procedures and the metal/refractory interface was examined by SEM/EDX as shown in Figure 5.

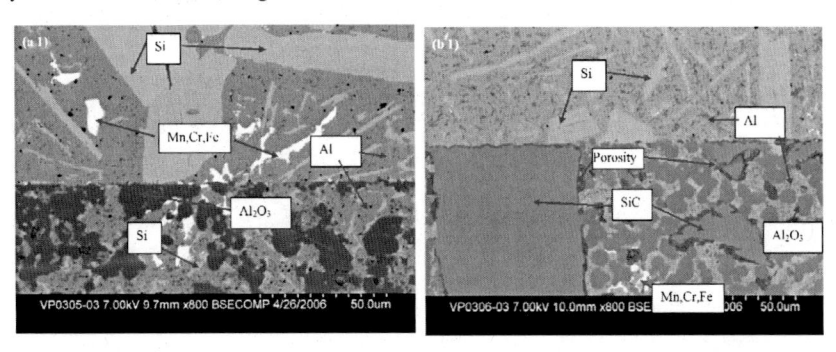

Wetting and Reaction Characteristics of Al₂O₃/SiC Composite Refractories

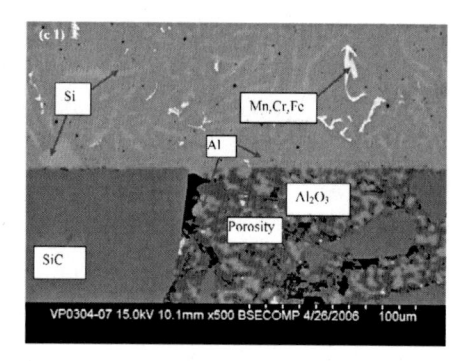

Figure 5. BSE photomicrographs and microprobe mapping of molten aluminum 5083 droplet cross section on (a) TQ, (b) TC, and (c) MC substrates at 900°C for 1 hour.

Interfacial regions for all three types of refractory showed a high concentration of silicon in the Al drop, indicating that the silicon diffused from the refractory substrate into the liquid Al drop during the experiment. The activation energy for such diffusion of silicon in molten aluminum has been estimated as 20 kJ/mol[5]. The silicon present has to be due to diffusion from the substrate, as the initial weight percent of silicon in the original alloy is less than 0.4%. This finding agrees with the observations that interfacial regions are the preferred position for the element transferring across the interface and into the other phase[17,18]. Low levels of Mn, Cr, and Fe were also detected at both sides of the interface.

During the wetting process, Al was found to diffuse into the pores of the refractory substrate. The diffused liquid Al joined with the existing residual Al metal present in the composite refractories, inhabiting both the pores as well as positions vacated by the diffusing silicon. The Al-Si-C phase diagram shows that thermodynamically, a three-phased monovariant equilibrium can be reached by reacting aluminum and SiC at temperatures above 612°C[19]. The equation for such a reaction is shown in Equation 2.

$$4Al_{(l)} + 4SiC_{(s)} \rightarrow Al_4SiC_{4(s)} + 3Si_{(s)} \qquad (2)$$

Another possible chemical reaction is the dissolution of SiC in the liquid metal as shown in Equation 3[20]:

$$SiC_{(s)} \rightarrow Si_{(M)} + C_{(s)} \qquad (3)$$

It was not possible to determine whether or not this reaction took place in our samples since both Si and C were introduced to the refractory surface during the specimen preparation process.

Effect of Molten Drop Material and Substrate Materials on Drop Morphology
Figure 6 shows back scattering SEM results of the TC substrate following wetting by pure molten Al and 5083 Al-Mg alloy drops after 2 hours at 900°C. Figure 7 shows back

scattering SEM micrographs of molten Al-Mg alloy drops on TQ, TC, and MC substrates after sessile drop experiments at 900°C for 1 hour.

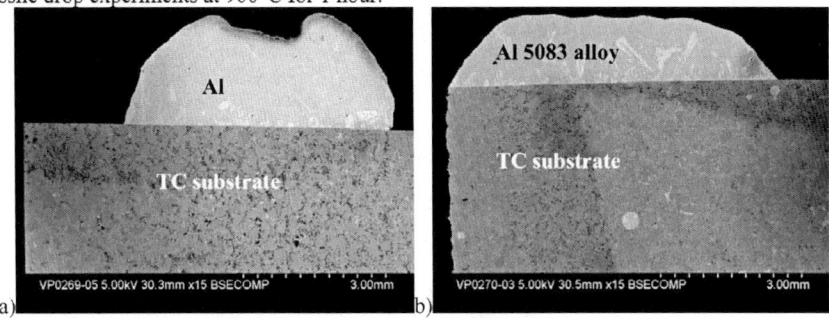

Figure 6. Back Scattering SEM photomicrographs of sections of TC substrate reacted with (a) molten Al droplet and (b) molten Al 5083 droplet after 2 hours at 900°C.

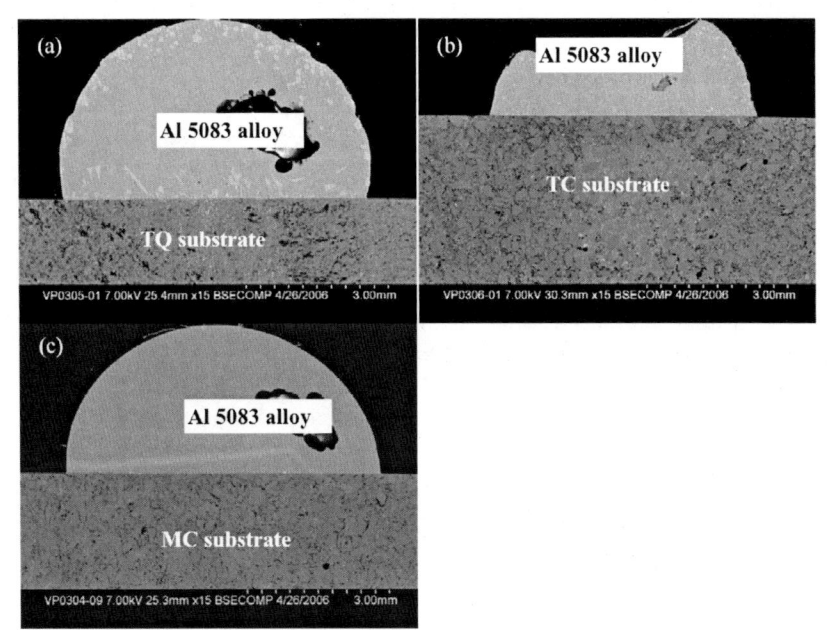

Figure 7. Back Scattering SEM photomicrographs of molten aluminum 5083 alloy droplets on (a) TQ, (b) TC, and (c) MC substrates after 1 hour at 900°C.

By comparing the wetting behavior of pure aluminum to that of the aluminum alloy, results showed that the Al alloy exhibited a larger drop base area and lower drop height, thus leading to a smaller contact angle than that obtained for pure Al. This contact angle difference, based on the drop dimensions after the wetting tests, suggests that the alloying magnesium promoted the wetting of the refractory with the molten metal drop. When heated to the testing temperature, reactive magnesium in the 5083 liquid alloy can readily diffuse across the interface into the substrate, reducing the magnesium content of the metallic drop.

Previous studies have shown that the addition of a reactive metal, i.e., Mg, to the liquid drop improves the wetting behavior and the interaction with the substrate.[21] The degree of wetting will depend on the chemical reactivity of Mg with the specific refractory substrate. In the case of TCON refractory materials, which contain SiC, Al_2O_3, Al and Si, the Mg can form silicides in the interface[22] or carbides with different stoichiometries. The microchemistry of these reaction products in the interface is important from the point of view of wettability. A pronounced $MgAl_2O_4$ phase was not found, which is consistent with previous results.[23]

Wetting Characteristics

The wetting mechanism is thought to occur as follows. In general, the wettability of a solid by a liquid is indicated by the contact angle. The contact angle, θ, between solid, liquid and gas/vapor is related by the Young–Dupre's equation[24] shown in Equation 4,

$$\cos\theta = \frac{\gamma_{sv} - \gamma_{sl}}{\gamma_{lv}} \qquad (4)$$

where γ_{lv} is the surface tension of the liquid metal, γ_{sv} is the surface energy of the solid, and γ_{sl} is the solid/ liquid interfacial energy. Based on the above equation, the contact angle, θ, can be decreased by increasing the surface energy of the solid, γ_{sv}, decreasing the solid/liquid interfacial energy, γ_{sl}, or by decreasing the surface tension of the liquid, γ_{lv}. It should be noted that as cos θ goes down, θ actually goes up. During heating, the silicon is believed to diffuse from the refractory substrate into the molten Al drop. The drop volume will therefore increase because of the silicon diffusion. Owing to the reactivity of the introduced silicon, the surface tension of the liquid aluminum, γ_{lv}, can be decreased by the adsorption of silicon on the interface and surface (see Figure 5) of the liquid. In the meantime, the solid/liquid interfacial energy, γ_{sl}, can also be decreased due to the enrichment of silicon on the interface. As a result, the contact angle, θ, will be decreased.

CONCLUSIONS

Reactive wetting tests were carried out in Ar-4%H_2 atmosphere using Al and Al-Mg alloys in contact with three types of alumina-SiC composite refractory materials. The interaction resulted in contact angle values which were initially obtuse, but changed to acute angles in most cases. These results were contrary to most commercially available refractory materials that are currently used in aluminum contact applications and that display acute contact angles for all test times. Some wetting and reaction mechanisms were discussed based on the identification of changes of contact angles and drops dimensions as well as the interfacial elemental mappings.

A strong diffusion of silicon from the substrate into the liquid metal drop was observed and silicon was noted to accumulate on the interface and surface area of the drop. Additionally, Al was found to move into the refractory.

The Mg found in the 5083 alloy apparently promoted the wettability of molten Al on refractory substrates as noted by the observations of changes in the wetting behavior.

ACKNOWLEDGEMENTS
The authors would like to acknowledge the contributions of Randy Parton, Randy Howell, and Donny McInturff of ORNL in support of the work described in this paper. The efforts of Edgar Lara-Curzio, HT Lin, and Jim Keiser in reviewing this document are also appreciated. This research was sponsored by the U.S. Department of Energy, Office of Energy Efficiency and Renewable Energy, Industrial Technologies Program, for the U.S. Department of Energy under contract DE-FC36-04GO14038.

REFERENCES
[1] A. Standage and M. Gani, *Journal of the American Ceramic Society*, **Vol. 50(2)**, 101-105 (1967).
[2] C. Marumo and Pask, *Journal of Materials Science*, **Vol. 12(2)**, 223-233 (1977).
[3] K. Brondyke, *Journal of the American Ceramic Society*, **Vol. 36(5)**, 171-174 (1953).
[4] E. Saiz and A. Tomsia, *Journal of the American Ceramic Society*, **Vol. 81(9)**, 2381-2393 (1998).
[5] W. Fahrenholtz, K. Ewsuk, R. Loehman, and P. Lu, *Journal of the American Ceramic Society*, **Vol. 81(10)**, 2533-2541 (1998).
[6] P. Shen, H. Fujii, T. Matasumoto, and K Nogi, *Metallurgical Materials Transactions A*, **Vol. 35A**, 583-588 (2004).
[7] S. Takahashi and O. Kuboi, *Journal of Materials Science*, **Vol. 21(7)**, 1797-1802 (1996).
[8] L. Mouradoff, P. Tristant, J. Desmaison, J. Labbe, M. Desmaison-Brut, and R. Rezakhanlou, *Key Engineering Materials*, **Vol. 113**, 177-186 (1996).
[9] A. Ferro and B. Derby, *Acta Metallurgica et Materialia*, **Vol. 43(8)**, 3061-3073 (1995).
[10] M. Pech-Canul, R. Katz, and M. Maklouf, *Metallurgical and Materials Transactions a: Physical Metallurgy and Materials Science*, **Vol. 31A(2)**, 565-573 (2000).
[11] V. Laurent, C. Rado, and N. Eustathopoulos, *Materials Science & Engineering A: Structural Materials: Properties, Microstructure and Processing*, **Vol. A205(1-2)**, 1-8 (1996).
[12] M. Shimbo, M. Naka, and I. Okamoto, *Journal of Materials Science Letters*, **Vol. 8(6)**, 663-666 (1989).
[13] J. Marin, L. Olivares, S. Ordonez, and V. Martinez, *Materials Science Forum*, **Vol. 416-418**, 481-486 (2003).
[14] Z. Yu, G. Wu, D. Sun, and L. Jiang, *Materials Letters*, **Vol.57**, 3111–3116 (2003).
[15] W. Jyng, H. Song, S. Park, D. Kim, *Metallurgical Materials Transactions B*, **Vol 27B**, 51-55 (1996).
[16] H. Fujii and H. Nakae, *Acta Metallugica et Materialia*, **Vol. 44**, 3567-3573 (1996).
[17] K. Rita, S. Veena, and S. Noel, *ISIJ International*, **Vol. 45**, 1261-1268 (2005).
[18] N.Sobczak, L. Stobierski, W. Radziwill, M. Ksiazek, and M. Warmuzek, *Surface and Interface Analysis*, **Vol. 36**, 1067-1070 (2004).
[19] J. C. Viala, P. Fortier, and J. Bouix, *Journal of Material Science*, **Vol. 25**, 1842-1850 (1990).
[20] G. Li, *Structure Ceramics Joining II*, 69 (1993).
[21] J. Marín, J. Lisboa, L. Olivares, P. Aguirre, R. Becerra, and G. Piderit, *Nucleotécnica*, **Vol. 15**, 19 (1995).
[22] Ternary Alloy Systems - Phase Diagrams, Crystallographic and Thermodynamic Data critically evaluated by MSIT, 139-145 (2004).
[23] P. Shen, H. Fujii, T. Matsumoto, and K. Nogi, *Acta Materialia*, **Vol 52**, 887-898 (2004).
[24] M Tagawa, *Journal of Japanese Society of Tribologists*, **Vol. 49(6)**, 499-500 (2004).

NDE and Test Methods

EVALUATION OF OXIDATION PROTECTION TESTING METHODS ON ULTRA-HIGH TEMPERATURE CERAMIC COATINGS FOR CARBON-CARBON OXIDATION RESISTANCE

Erica L. Corral, Alicia A. Ayala, and Ronald E. Loehman
Sandia National Laboratories
1001 University Blvd. SE, Suite 100
Albuquerque, NM, 87112

ABSTRACT

The development of carbon-carbon (C-C) composites for aerospace applications has prompted the need for ways to improve the poor oxidation resistance of these materials. In order to evaluate and test materials to be used as thermal protection system (TPS) material the need for readily available and reliable testing methods are critical to the success of materials development efforts. With the purpose to evaluate TPS materials, three testing methods were used to assess materials at high temperatures (> 2000 °C) and heat flux in excess of 200 Wcm^{-2}. The first two methods are located at the National Solar Thermal Test Facility (NSTTF) at Sandia National Laboratories, which are the Solar Furnace Facility and the Solar Tower Facility. The third method is an oxyacetylene torch set up according to ASTM E285-80 with oxidizing flame control and maximum achievable temperatures in excess of 2000 °C. In this study, liquid precursors to ultra high temperature ceramics (UHTCs) have been developed into multilayer coatings on C-C composites and evaluated using the oxidation testing methods. The tests will be discussed in detail and correlated with preliminary materials evaluation results with the aim of presenting an understanding of the testing environment on the materials evaluated for oxidation resistance.

INTRODUCTION

C-C composites are attractive candidates for next generation aerospace vehicles due to their high strength-to-weight ratio, resistance to extreme thermal shock, low coefficient of thermal expansion, and excellent strength retention and creep resistance over a wide temperature range. However, they oxidize in air at temperatures > 500 °C, thus rendering them useless for thermal protection in hypersonic atmospheric flights[1-3]. In order to advance the development of next generation TPS materials adequate and low cost testing and screening methods are necessary in order to rapidly evaluate and understand UHTC coating performance as thermal oxidation resistant materials on C-C composites. Currently, the most accepted test method for evaluating TPS materials oxidation performance is at an arc jet testing facility[4]. The availability, expensive and turn around time associated with a single arc jet test do not allow for rapid and iterative materials testing that are necessary to the development and understanding of TPS coating materials behavior. In this paper three testing methods will be discussed and evaluated for their potential use to evaluate the oxidation protection behavior of TPS materials. The first two testing methods are located at the National Solar Thermal Test Facility (NSTTF) at Sandia National Laboratories. They are the Solar Furnace Test Facility and the Solar Tower Testing Facility. They both use solar energy to control heat flux exposures onto the surface of UHTC coated C-C composite specimens. The third method was constructed in observation of ASTM E285-80, "Oxyacetylene Ablation Testing of Thermal Insulation Materials[5]." This test allows for laboratory scale testing that is reproducible, time efficient and serves to screen obviously poor

TPS materials. The three test methods are not capable of matching the high-altitude atmospheric flight conditions of an arc jet testing facility but do serve as valuable testing methods to evaluate C-C coatings at relatively low cost.

FLIGHT ENVIRONMENT and TPS MATERIALS

The flight environment of future re-entry vehicles of advanced hypersonic weapons will require materials that can withstand temperatures, depending on flight conditions, of up to 2700 °C[6]. Only very specialized materials can withstand this temperature, such as UHTCs[7, 8]. Examples of a couple UHTCs are boride materials like hafnium diboride (HfB_2) and zirconium diboride (ZrB_2), which have melting temperatures above 3027 °C. They also maintain tensile properties at high temperatures[9, 10] and provide oxidation resistance that allows them to be used in extreme environments like those associated with hypersonic flight, rocket re-entry and rocket propulsion[11]. In this paper multilayer UHTC coating on C-C discs will be discussed as they relate to their performance after solar furnace and solar tower testing.

In order to understand the flight environment during hypersonic flight conditions computation modeling has helped to gain an understanding of the variable heat flux controlled experiments that should be used while testing at the NSTTF. Using conventional aerothermodynamic techniques, at Sandia, an example of the cold wall heat flux conditions for three flight profiles are shown in Figure 1. The profiles represented are for boost glide hypersonic flight vehicles.

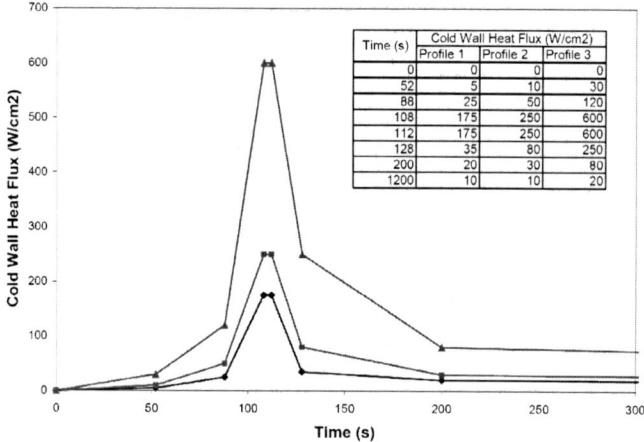

Figure 1. Aerothermal modeling indicates that peak heat flux temperatures used in hypersonic flights can reach up to 600 Wcm^{-2}.

The materials evaluated in this investigation are to serve as the outer layers of the thermal protection system. C-C composite discs and bulk graphite were tested to determine baseline material behavior for each testing method. Type CC139C carbon-carbon composite were supplied by HITCO Carbon Composites (Gardena, CA). Type CC139C is a 2D carbon-carbon composite comprising T300-3K fibers in an 8-harness satin weave, densified by chemical vapor infiltration. The density of CC139C is nominally 1.60 gcm^{-3} for the thickness used. Samples were cut at an angle to the plies to simulate a shingle layup. The TPS coatings tested were multilayer UHTC ceramic films of SiC, HfC, ZrB$_2$, or HfB$_2$. The coatings were applied by infiltration and dip coating onto the C-C composites in the form of slurries or polymeric liquid precursors.

OXIDATION TESTING FACILITY DESCRIPTIONS
Solar Furnace Facility

The Solar Furnace Facility uses a heliostat that tracks the sun to direct sunlight onto a mirrored parabolic dish. The focal point of the dish does not move therefore making it is simple to install experiments. The power level of the furnace is adjusted using an attenuator that works like a Venetian blind located between the heliostat and the dish. The small solar furnace test facility has a heliostat that is 95 m^2 and a dish that is 6.7056 m in diameter. Figure 2 shows a picture of the Solar Furnace Facility and where the parabolic mirror is in relationship to the position of the heliostat and attenuator. The furnace provides 16 kW total thermal power and time-dependent control of the thermal flux to a maximum of 800 Wcm^{-2}*. The flux is adjusted by a computer controlled attenuator that permits a time-varying flux profile. Samples measuring 1.59 cm in diameter and 0.64 cm thick were heated under a time varying profile to study the effect of the heating profile on the performance of the treatment. Specimen surface temperatures were monitored with pyrometers at 5.2 μm and 1.7 μm and the surface conditions observed via video and infrared cameras.

The advantage of testing at the Solar Furnace is the ability to easily control the variable heat flux profile while continuously monitoring the surface temperature of the specimen in real time. Currently, the test is open to air and the specimens are exposed to shear airflow, approximately 25 ms^{-1}. Each specimen is tested individually and the duration of each run is dependent upon cloud cover and can take up to 30 minutes. The maximum heat flux achieved per run is also dependent upon daily weather conditions. Figure 3 is an example of a Solar Furnace heat flux profile and the corresponding surface temperature measured on the specimen. The heating profile used was 0 to 50 Wcm^{-2} in 88 seconds, 50 to 175 Wcm^{-2} in 20 seconds, hold at 175 W cm^{-2} for four seconds, 175 to 80 W cm^{-2} in 16 seconds and finally from 80 to 30 Wcm^{-2} in 72 seconds. The heat flux profile shown in Figure 3 was designed based on the aerothermal modeling predictions provided by researchers at Sandia as seen previously in Figure 1.

Figure 2. The Solar Furnace Facility uses an attenuator located between the heliostat and the parabolic mirror dish to control the solar power level of the furnace.

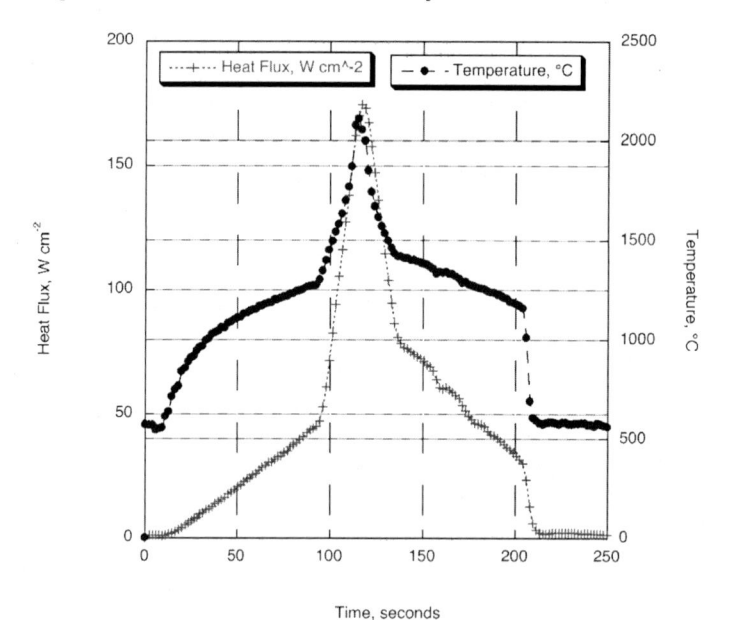

Figure 3. The Solar Furnace Facility can be used to design specific heat flux profiles and measure surface temperature of specimens in real time. The maximum heat flux reached during this test was approximately 175 Wcm^{-2} which corresponds to a maximum specimen surface temperature greater than 2000 °C.

Examples of preliminary test results for UHTC coated and un-coated C-C composites are shown in Table I. The percent mass loss was measured in order to evaluate the performance of the coating under variable heat flux conditions. These results suggest that the bare C-C discs that were filled with BC oxidize less than the un-coated C-C composites without filler. These tests were performed as validation runs for the testing facility and further experiments are in progress to further analyze the performance of the UHTC coated materials.

Table I. Percent mass loss measured after testing at the Solar Furnace Facility.

COATING/SUBSTRATE	% MASS LOSS
Un-coated /Bulk graphite	20.0
Un-coated/High Density C-C	23.0
Un-coated/High Density C-C BC Inhibited	16.0
SMP-10 SiC/Bulk Graphite	23.0
SiC-HfC-HfB$_2$/C-C	10.0
HfB$_2$-HfC-SiC/C-C	1.3
SiC-HfB$_2$/C-C	16.3
SiC-HfC/C-C	23.0
HfC-HfB$_2$/C-C	18.6

Solar Tower Testing Facility

The Central Receiver Test Facility, or Solar Tower Test Facility, can generate up to 5 MW of total thermal power and reach a maximum heat flux of 260 Wcm^{-2}. The illumination of the desired target area is up to about 2,800 m^2 with time-dependent control of the thermal flux. The Solar Tower operates using multiple heliostats to concentrate sunlight from a wide region. The advantage of the Solar Tower Facility is the peak heating region is large enough to heat many samples simultaneously or large structures. The Solar Tower Testing Facility and the array of heliostat are shown in Figures 4 and 5. The specimens are mounted onto a stainless steel holder, Figure 6, and then mounted on the tower as seen in Figure 7. The specimens tested measured 2.54 cm in diameter and 9.52 cm thick and were tested at fluxes of 150 Wcm^{-2} or 200 Wcm^{-2} for ten minutes. Figure 8 is an example of the heat flux profile used to test the UHTC coated C-C composites. Sample masses were recorded before and after testing and the back face specimen temperatures were recorded using thermocouples. The disadvantages of this test are the increased time it takes to perform each run, at twice the cost of the Solar Furnace, and the same weather related limitations to achieving maximum heat flux. The specimens shown in Figures 6 and 7 were used to determine testing conditions and validate the Solar Tower Facility as a suitable test method for TPS materials evaluation.

Evaluation of Oxidation Protection Testing Methods on Ultra-High Temperature Coatings

Figure 4. The Solar Tower Testing Facility is located at Sandia National Laboratories NSTTF.

Figure 5. The large array of heliostats focuses thermal energy from the sun onto a testing area at the top of the tower.

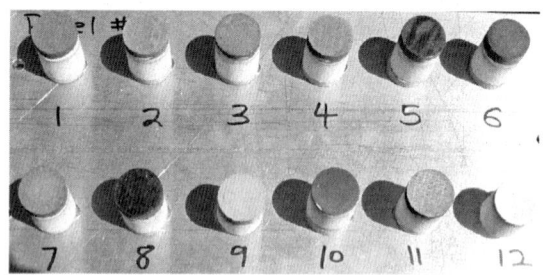

Figure 6. The UHTC coated C-C composites are mounted on a stainless steel holder before securing onto the tower assembly for testing. The specimens are approximately measuring 2.54 cm in diameter and 9.52 cm thick and are monitored for surface temperature only on the back face of each specimen.

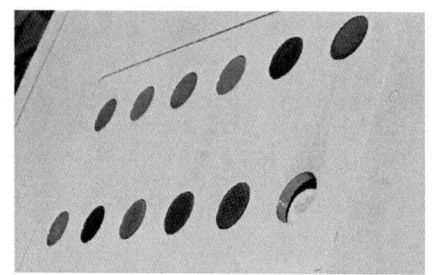

Evaluation of Oxidation Protection Testing Methods on Ultra-High Temperature Coatings

Figure 7. The UHTC coated C-C composite specimens are secured onto the tower assembly where all the specimens are exposed to the same heat flux at once. The peak heating area is approximately 2.800 m^2.

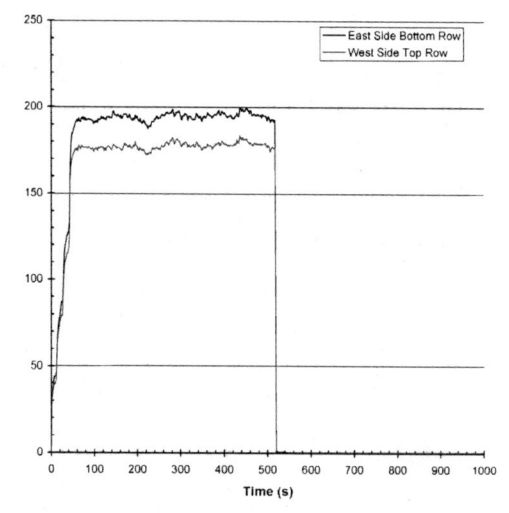

Figure 8. An example of a heat flux profile used to validate the testing Solar Tower Testing Facility for thermal oxidation testing of UHTC coated C-C composites. A heat flux of approximately 200 Wcm^2 was maintained for a duration of 400 seconds.

Oxyacetylene Ablation Testing Facility

The oxyacetylene ablation testing facility uses the combustion of oxygen and oxyacetylene gases to reach temperatures greater than 2000 °C at heating rates as high as 100 °C min^{-1}. The steady flow of hot gas provided by an oxyacetylene burner can be used to reliably heat under controlled laboratory conditions. The torch was set up in observation of ASTM E285-80, "The Standard Test Method for Oxyacetylene Ablation Testing of Thermal Insulation Materials.[5]" The use of this test method is to screen the most least oxidation resistant materials from further consideration. (Please note that since the combustion gases more closely resemble the environment generated in rocket motors, this method is more applicable to screening materials for nozzles than for aerodynamic heating.) The original intent of this test method is for screening poor materials from further consideration.

Figure 9 is a picture of the oxyacetylene ablation torch facility operating at approximately 1500 °C. The specimen shown in Figure 10 is that of an uncoated C-C composite filled with BC. The hot combustion gases are directed along the normal to the specimen until burn-though of the material is achieved. The ablation rate is determined by dividing the original thickness by the time to burn-through. The characteristics of the oxyacetylene heat source are neutral flame conditions and the velocity of reacted gases reach up to 25 ms^{-1}. A preliminary set of ablation rates were determined for un-coated C-C composites and are shown in Figure 11. The BC inhibited composites have a slightly lower ablation rate than compared to the conventional C-C composites without the filler.

Figure 9. The oxyacetylene ablation torch facility is shown operating at approximately 1500 °C. The surface temperature of the specimen is measured using a pyrometer and the back face specimen temperature is monitored by a high temperature material thermocouple.

Figure 10. A C-C disc filled with BC is shown during an oxyacetylene ablation test used to measure the un-coated material ablation rate at a given temperature.

Evaluation of Oxidation Protection Testing Methods on Ultra-High Temperature Coatings

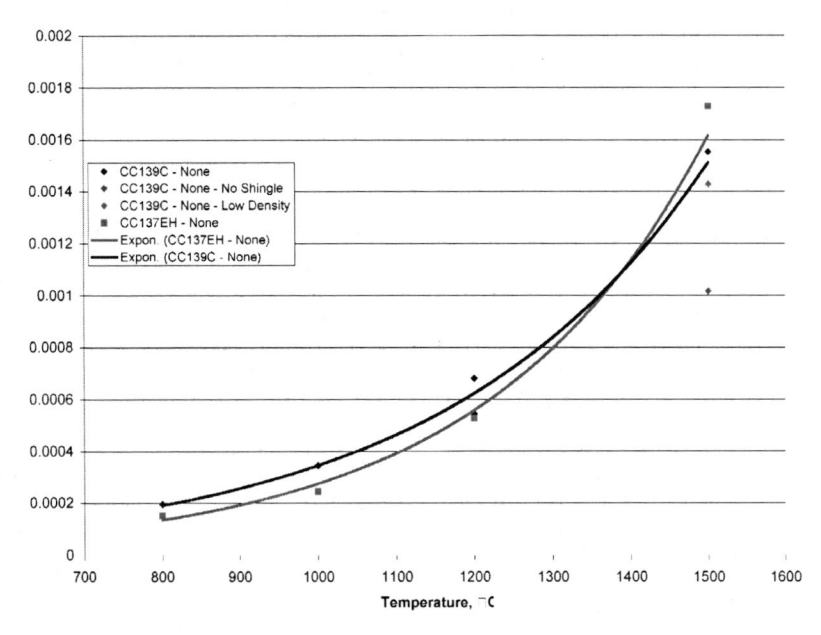

Figure 11. Ablation rates as a function of temperature were measured for un-coated C-C discs and un-coated C-C discs filled with BC. The filled composites have slightly lower ablation rates when tested at temperatures less than 1400 °C then tend to level of with the unfilled composites at temperatures greater than 1400 °C.

SUMMARY & CONCLUSION

In summary the use of low cost and reliable testing methods for the evaluation of TPS material performance is needed in order to advance the TPS materials development efforts and understanding of coating materials. The testing methods available at the NSTTF have shown to be useful tests for the evaluation of UHTC coating under controlled variable heat flux conditions. The advantages of testing at the Solar Furnace Facility are the accurate control of variable heat flux (> 400 Wcm^{-2}) and the ability to continuously monitor the specimen surface temperature (> 2000 °C). The advantages of testing at the Solar Tower Facility are the accurate control of variable heat flux profiles (> 200 Wcm^{-2}) and the large peak heating area. The advantage of having a screening test method such as, the oxyacetylene torch facility, allows for rapid evaluation of coating materials before testing at the NSTTF.

ACKNOWLEDGMENTS

The authors thank the Bryan D. Gauntt, Luke Boyer, Marlene Chavez, and Tom Westrich for their assistance in implementing the torch testing facility. Cheryl Ghanbari and Mike Edgar for testing at the NSTTF. Sandia is a multiprogram laboratory operated by Sandia Corporation, a Lockheed Martin Company, for the U.S. Dept. of Energy under Contract DE-AC04-94AL85000.

FOOTNOTES
*Please note that the maximum heat flux capability at the Solar Furnace Facility is expected to reach 800 Wcm^{-2} by early Spring 2007.

REFERENCES
1. McKee, D. W., Oxidation behavior and protection of carbon-carbon composites. *Carbon* **1987**, 25, (4), 551-557.
2. Ehrburger, P.; Lahaye, J., Characterization of carbon-carbon composites -I. *Carbon* **1981**, 19, 1-5.
3. Ehrburger, P.; Lahaye, J., Chracterization of carbon-carbon composites-II. *Carbon* **1981**, 19, 7-10.
4. Smith, J. A.; Kuntz, D. W.; Ayala, A.; Galloway, J. A.; Crane, N. B.; Emerson, J. A.; Loehman, R. E.; Daniel, S. D.; Zschiesche, D. *Arc jet results for coated and self coating TPS materials*; SAND2006-XXXX; Sandia National Laboratories: Albuquerque, 2006.
5. ASTME285-80, Standard test method for oxyacetylene ablation testing of thermal insulation materials. In *ASTM International*, 2002.
6. Savino, R.; Fumo, M. D. S.; Paterna, D.; Serpico, M., Aerothermodynamic study of UHTC-based thermal protection systems. *Aerospace Science and Technology* **2005**, 9 , 151-160.
7. Opeka, M. M.; Talmy, I. G.; Zaykoski, J. A., Oxidation-based materials selection for 2000°C hypersonic aerosurfaces: Theoretical considerations and historical experience. *Journal of Materials Science* **2004**, 39, 5887-5904.
8. Levine, S. R.; Opila, E. J.; Halbig, M. C.; Kiser, J. D.; Singh, M.; Salem, J. A., Evaluation of ultra-high temperature ceramics for aeropropulsion use. *Journal of the European Ceramic Society* **2002**, 22, 2757-2767.
9. Loehman, R.; Corral, E.; Dumm, H. P.; Kotula, P. *Ultra high temperature ceramics for hypersonic vehicle applications*; SAND 2006-2925; Sandia National Laboratories: June 2006, 2006; p 46.
10. Chamberlain, A. L.; Fahrenholtz, W. G.; Hilmas, G. E.; Ellerby, D. T., High-strength zirconium diboride -based ceramics. *Journal of the American Ceramics Society* **2004**, 87, (6), 1170-1172.
11. Chamberlain, A. L.; Fahrenholtz, W. G.; Hilmas, G. E.; Ellerby, D. T., Characterization of zirconium diboride ceramics for thermal protection systems. *Key Engineering Materials* **2004**, (264-268), 493-496.

NONDESTRUCTIVE EVALUATION OF SILICON-NITRIDE CERAMIC VALVES
FROM ENGINE DURATION TEST

J. G. Sun
Argonne National Laboratory
Argonne, IL 60439

J. S. Trethewey, N. S. L. Phillips, N. N. Vanderspiegel, J. A. Jensen
Caterpillar Inc.
Peoria, IL 61656

ABSTRACT
In this study, we investigated impact and wear damage in silicon-nitride ceramic valves that were subjected to an engine duration test in a natural-gas engine. A high-speed automated laser-scattering system was developed for the nondestructive evaluation (NDE) of 10 SN235P silicon-nitride valves. The NDE system scans the entire valve surface and generates a two-dimensional scattering image that is used to identify the location, size, and relative severity of subsurface damage in the valves. NDE imaging data were obtained at before and at 100 and 500 hours of the engine duration test. The NDE data were analyzed and compared with surface photomicrographs. Wear damage was found in the impact surface of all valves, especially for exhaust valves. However, the NDE examinations did not detect subsurface damage such as cracks or spalls in these engine-tested valves.

INTRODUCTION
Advanced ceramics are leading candidates for high-temperature engine applications that offer improved engine performance and reduced emissions. Among these ceramics, silicon nitrides (Si_3N_4) are being evaluated for valve train materials in automotive and diesel engines. Ceramic valves operate under high stress and temperature in a corrosive environment. Therefore, surface and subsurface damage induced by machining and engine operation testing must be characterized to ensure valve reliability.

Surface and subsurface defects and damages are known to significantly degrade fracture strength and fatigue resistance of Si_3N_4 ceramics (Ott, 1997). These defects/damages are normally in the form of microstructural discontinuities such as spalls, cracks, and voids and are typically within 200 μm under the surface. Because these ceramics are partially translucent to light, a laser-scattering method, based on the detection of cross-polarized optical scattering that originates mostly from the subsurface, can nondestructively and without contact measure variations in subsurface microstructure. By scanning the entire surface of a ceramic component and constructing a two-dimensional (2-D) scatter image, subsurface damage can be readily identified because they exhibit excessive scattering over the background (Sun et al., 1999).

In this study, a high-speed automated laser-scattering system was developed for nondestructive evaluation (NDE) of subsurface damage in Si_3N_4 ceramic valves (Sun et al., 2005, 2006). The NDE system scans the entire valve surface and generates a two-dimensional scattering image that is used to identify the location, size, and relative severity of subsurface damage. Laser-scattering NDE data were obtained for 10 SN235P Si_3N_4 valves that were subjected to a bench rig and an engine duration test. The data were analyzed and compared with surface photomicrographs.

AUTOMATED LASER-SCATTERING SYSTEM

An automated laser-scattering NDE system was developed to scan the entire surface of a valve. Because of the axi-symmetric surface profile, the scanning system requires a minimum of 4 degrees of freedom (i.e., 4 axes) to map the entire valve surface. Figure 1 is a schematic diagram of the experimental setup. A precision lathe unit is used to rotate and translate the valve, and a translation/rotation stage unit controls the position and orientation of the optical detection train so that the laser beam is always focused normally on the valve surface. One optical fiber is used to deliver the incident laser beam, another to bring the collected scattered light to a photomultiplier tube (PMT) detector. The data acquisition program controls the motion of all four stages to follow the profile of the valve surface, and synchronously collects laser-scattering data during a scan. The scanned data are plotted as a 2-D scatter image in which the scale at each pixel is proportional to the subsurface backscatter intensity recorded at the corresponding position of the valve surface (Sun et al., 1999, 2006). Because optical scattering is stronger at regions with subsurface defects and damage, such as cracks and pores, the 2-D scattering image can be used to identify the location, size, and relative severity of subsurface defects. The system can scan a complex valve head surface in one hour.

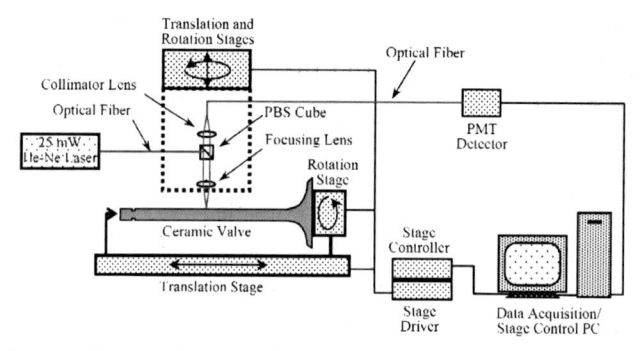

Fig. 1. Schematic diagram of automated laser-scattering system for valve scanning.

SILICON-NITRIDE CERAMIC VALVES

Ten SN235P Si_3N_4 ceramic valves, six intake (identified by I) and four exhaust (identified by E), have been run successfully for 500 hours in a Caterpillar natural-gas G3406 generator set at the National Transportation Research Center in Oak Ridge National Laboratory. Figure 2 is a photograph of these valves after 100-hr engine test. Each valve consists of a head section of varying radius and a stem section of constant radius. The end of the head section is a 4-mm-long conically shaped surface of constant slope (45°) to the axis. This surface contacts the matching surface of a metal valve-seat insert. Extending from the conical contact surface to the cylindrical stem surface is a curved smooth surface of varying radius (the fillet). During normal operation, the valve contact surface is in cyclic impact with a metal valve-seat insert. The impact stress propagates through the fillet into the stem, with the maximum principal tensile stress occurring in the region of the fillet. Therefore, material defects and processing (machining) damage in the entire valve surface and subsurface may affect valve reliability.

As seen in Fig. 2, all valves after the 100-hr engine test were covered by a coat of combustion byproduct. The coating is generally thicker on exhaust valves, especially for valves positioned at E3 and E7. However, the impact area of the valve seating surface is clean without any deposit. The impact area is not entirely within the conical contact surface; it lies around the edge between the contact and the fillet surface, as illustrated in Fig. 3. The coating affects the NDE result in two ways. First, because the scan starts from the top edge of the conical contact surface, the extra thickness of the coating makes it difficult for initial focusing. Second, the coating causes optical attenuation and scattering, so a thick coat will reduce the laser-scatter signal from the valve subsurface. Nevertheless, surface/subsurface damage is detectable under a thinner coating (e.g., at regions along the fillet radius towards the stem surface). Further, because the impact area was uncoated, the NDE data from this area are accurate and sensitive only to the surface/subsurface degradation. For the 500-hr engine-tested valves, the combustion coating was cleaned before the NDE examination. Therefore, the NDE scan results were not affected by this problem.

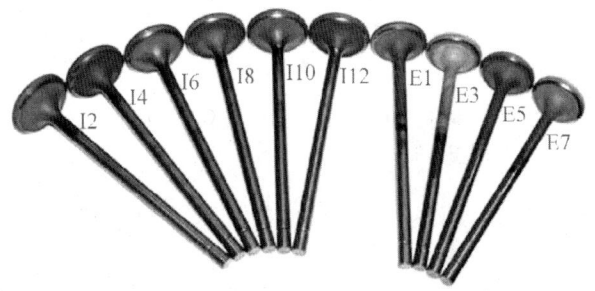

Fig. 2. Photograph of 10 SN235P Si_3N_4 valves after 100-hr engine test.

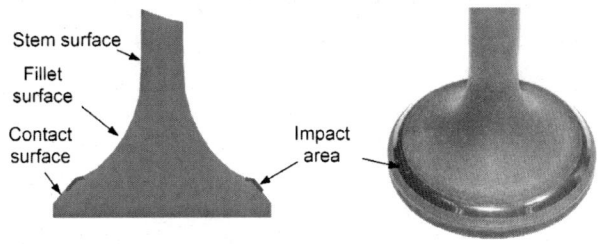

Fig. 3. Location of typical impact area on valve surface.

The laser-scattering NDE system (Fig. 1) was used to characterize surface and subsurface damage in the 10 SN235P valves shown in Fig. 2. These valves were machined by an initial coarse process and a finish machining process (Sun et al., 2006). Prior to the engine duration test, all valves were proof tested in a cyclic-impact bench rig for 20 hours. The laser-scatter NDE inspection did not find any damage in all valves after the rig tests. The engine test was planned for a total duration of 1000 hours. The valves were retrieved for nondestructive examinations at 100 and 500 hours of the engine operation. Typical NDE data for an intake and an exhaust valve at 100- and 500-hr engine operation are presented below.

EVALUATION OF SN235P VALVES AFTER 100-HR ENGINE TEST
A laser-scatter image of the valve-head surface of the SN235P intake valve I2 after the 100-hr engine test is shown in Fig. 4. Scan step size in both directions was ≈10 μm. The combustion coating induced higher scatter intensity in most of the valve-head surface, especially in the fillet surface. Within the impact area, shown in the enlarged NDE image, the scatter intensity in the conical contact surface is higher than that in the fillet surface (see edge location), which may indicate a difference of wear levels in the two surfaces. Nevertheless, no apparent damage was detected within the entire impact region.

Figure 5 shows the laser-scatter image of the SN235P exhaust valve E1 after the 100-hr engine test. The combustion coating on this valve is thin, so the valve surface underneath the coating is clearly imaged. The laser-scatter NDE data show generally higher scatter intensity in the impact area, indicating the presence of surface wear. From the enlarged NDE image, many individual spots of higher scatter intensity are detected; these spots are generally small (<50μm in size) and are the distributed porosities in the ceramic material (Andrews, 1999; Sun et al., 2005). The NDE data in the lower impact area (around the edge between contact and fillet surfaces) also exhibit several bands of lower scatter intensity, which could also be the result of nonuniform surface wear in that region. Despite these NDE indications of general surface wear from engine test, no subsurface damage such as cracks or spalls was detected for all exhaust valves. It should be noted that the NDE image data in Fig. 5 also display several gross features in the fillet and stem surfaces; these were not damage but due to surface coloration variation from nonuniform distribution of second phase material in the ceramic.

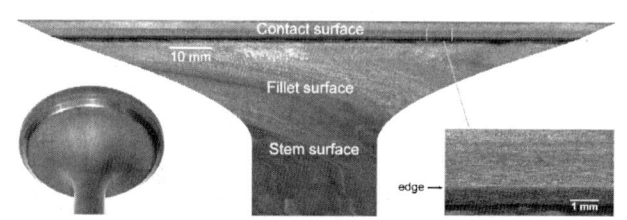

Fig. 4. Laser-scatter scan image of SN235P intake valve I2 after 100-hr engine test.

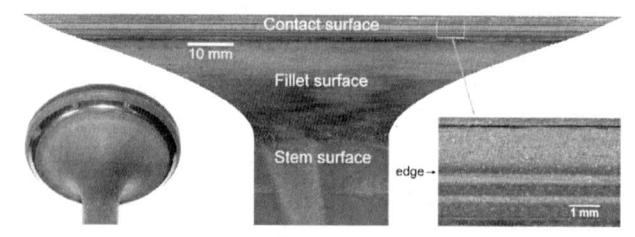

Fig. 5. Laser-scatter scan image of SN235P exhaust valve E1 after 100-hr engine test.

Figure 6 shows photomicrographs of typical impact surfaces in SN235P intake valve I2 and exhaust valve E1 after the 100-hr engine test. For intake valve I2, the original machining marks are still present, so the surface has little or no wear from the engine test. For exhaust valve E1, the machining marks have been worn off. The worn surface appears smooth (like polished surface), with distributed porosity spots and few subsurface damage marks (bright line) that are the remnants of the original machining damage. The difference in surface wear level between exhaust and intake valves is likely due to the temperature difference experienced by the valves in the engine operation. In general, the micrographic observations for these valves are consistent with those detected from the laser scatter NDE.

Intake valve I2   Exhaust valve E1

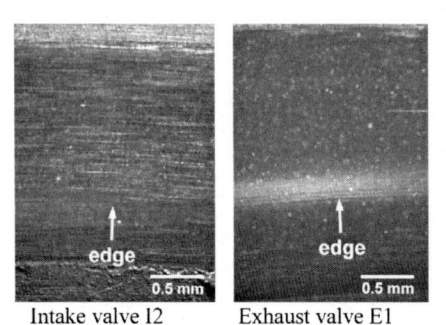

Fig. 6. Photomicrographs of impact surfaces of SN235P valves after 100-hr engine test.

EVALUATION OF SN235P VALVES AFTER 500-HR ENGINE TEST

Laser-scatter image of the SN235P intake valve I2 after the 500-hr engine test is shown in Fig. 7. The scatter intensity in the impact area is lower than that in the fillet and stem surface. Because damage is always shown with higher laser-scatter intensity, a lower scatter intensity in the impact surface is an indication of surface contamination. Within the impact area, shown in the enlarged NDE image in Fig. 7, no apparent subsurface damage was detected. The stripe pattern in the stem surface is due to the surface coloration variation.

Figure 8 shows the laser-scatter image of the SN235P exhaust valve E1 after the 500-hr engine test. Within the impact area, shown in the enlarged NDE image, many horizontal narrow bands of lower scatter intensity are detected. In addition, three wider bands of lower intensity that were present in the NDE data from the 100-hr engine test (see Fig. 5) are also observed. Because a lower laser-scatter intensity is not an indication of damage, the strong and complex NDE indications detected within the impact area should represent severe surface wear, not subsurface damage such as cracks or spalls.

Nondestructive Evaluation of Silicon-Nitride Ceramic Valves from Engine Duration Test

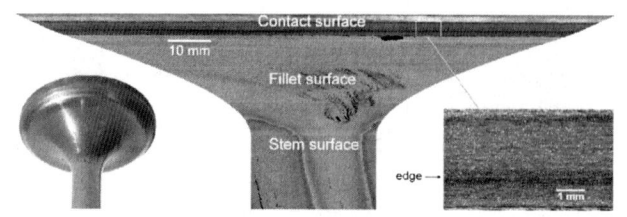

Fig. 7. Laser-scatter scan image of SN235P intake valve I2 after 500-hr engine test.

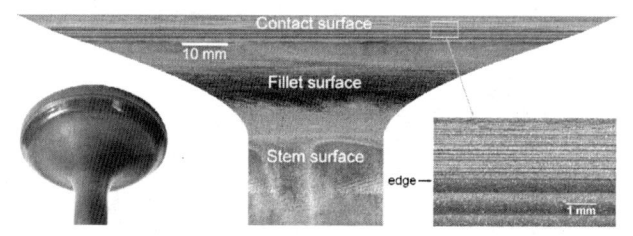

Fig. 8. Laser-scatter scan image of SN235P exhaust valve E1 after 500-hr engine test.

Figure 9 shows photomicrographs of impact surfaces of the intake valve I2 and the exhaust valve E1 after the 500-hr engine test. For intake valve I2, the machining surface mark has been worn off. The worn surface appears to be rough, making it susceptible for surface contamination as indicated from the NDE data (see Fig. 7). For exhaust valve E1, the impact surface displays considerable wear damage represented by narrow ridges and grooves that were detected in the NDE data (Fig. 8). A close examination of the metal valve-seat insert showed that the impacted insert surface was generally smooth, without the matching ridge/groove pattern. It is unclear how these ridges/grooves were generated. However, because the impact force was not normal to the valve surface, it is suspected that the impact shear stress may be responsible for the creation of these ridges/grooves.

Intake valve I2   Exhaust valve E1

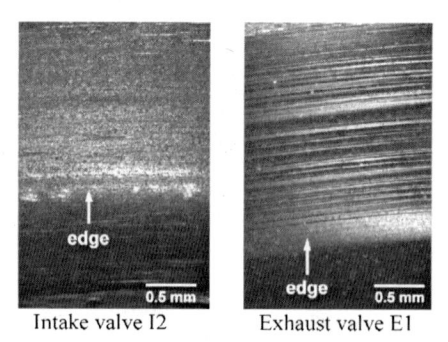

Fig. 9. Photomicrographs of impact surfaces of SN235P valves after 500-hr engine test.

CONCLUSION

A high-speed automated laser-scatter system was developed for nondestructive evaluation of 10 SN235P silicon-nitride ceramic valves that were subjected to an engine duration test. NDE imaging data of the valve-head surfaces for all valves were obtained at before and at 100 and 500 hours of the engine test. The NDE data were analyzed and compared with surface photomicrographs. In general, surface wear damage was found in the impact surface of all valves, especially for the exhaust valves. Considerable surface damage represented by narrow ridges/grooves was found within the impact surface of all exhaust valves after the 500-hr engine test. However, the NDE examinations did not detect subsurface damage such as cracks or spalls. These valves are currently being tested in the engine for another 500 hours. Results of this test will be presented in future publications.

ACKNOWLEDGMENT

This research was sponsored by the Heavy Vehicle Propulsion Materials Program, DOE Office of FreedomCAR and Vehicle Technology Program, under contract DE-AC05-00OR22725 with UT-Battelle, LLC.

REFERENCES

M. J. Andrews, 1999, "Life Prediction and Mechanical Reliability of NT551 Silicon Nitride," Ph.D. Thesis, New Mexico State University, Las Cruces, NM.

R. D. Ott, 1997, "Influence of Machining Parameters on the Subsurface Damage of High-Strength Silicon Nitride," Ph.D. Thesis, The University of Alabama at Birmingham, AL.

J. G. Sun, W. A. Ellingson, J. S. Steckenrider, and S. Ahuja, 1999, "Application of Optical Scattering Methods to Detect Damage in Ceramics," in *Machining of Ceramics and Composites*, Part IV, Chapter 19, Eds., S. Jahanmir, M. Ramulu, and P. Koshy, Marcel Dekker, New York, pp. 669-699.

J. G. Sun, J. M. Zhang, J. S. Tretheway, J. A. Grassi, M. P. Fletcher, and M. J. Andrews, 2005, "Nondestructive Evaluation of Machining and Bench-Test Damage in Silicon-Nitride Ceramic Valves," in *Ceramic Eng. Sci. Proc.*, eds. D. Zhu and W.M. Kriven, Vol.26, Issue 2, pp. 127-132.

J. G. Sun, J. M. Zhang, and J. S. Tretheway, 2006, "Nondestructive Evaluation of Machining and Bench-Test Damage in Silicon-Nitride Ceramic Valves," in *Ceramic Eng. Sci. Proc.*, eds. A. Wereszczack and E. Lara-Curzio, Vol.27, pp. 275-280.

MODEL OF CONSTRAINED SINTERING

Kais Hbaieb, Brian Cotterell
Institute of Materials Research and Engineering (IMRE)
3 Research link
Singapore, 117602

ABSTRACT

A simple model for constrained sintering of a film bonded to a thick substrate is developed. We assume that the strain rate in the film has four components: a free sintering strain rate, a thermal strain rate, an elastic strain rate and a viscous strain rate. The total strain rate in the film must follow the thermal strain rate in the substrate. Since the individual components of the strain rate can be large, we use logarithmic strain. A semi-empirical approach is adopted whereby the free sintering rate and uniaxial viscosity are expressed through mathematical functions. The expressions of these functions are determined by best fitting the functional curves to experimental data of unconstrained materials. The model developed provides a prediction of density profile and sintering stresses in constrained films.

INTRODUCTION

Constrained sintering is unavoidable in the processing of composite ceramic composites. Besides the intrinsic constrained sintering that takes place within the ceramic material due to the presence of agglomerates and inhomogeneities in green density, extrinsic constraint arises when different components undergo different sintering histories. In both cases, tensile stresses arise during sintering causing low final density and/or cracks.

Modelling constrained sintering can provide a helpful tool in predicting and characterizing sintering of complex green body. A simple linear viscous constitutive model is proposed to be sufficient to model the mechanical response of a sintered body[1]. Since the elastic strain is very small compared to densification strain at high temperature, many researchers neglect the elastic component of the strain[1-5]. In this paper we account for the elastic component of the strain as it is important at low temperature. Besides the elastic component, the total strain rate is composed of the free sintering strain rate, the viscous strain rate and the thermal strain rate. Although the model is simple, it requires determination of several thermo-mechanical properties, such as viscosity, viscous Poisson's ratio and free sintering strain rate. Among these three quantities, the free sintering strain rate is simplest to determine as it can be extracted from dilatometry shrinkage measurement. Various techniques are used to measure viscosity including sinter-forging[6], cyclic dilatometry[7], discontinuous sinter forging[8] and bending creep[9]. Among these various techniques cyclic dilatometry and discontinuous sinter forging are proved to be effective in minimizing anisotropy that would arise if a constant load is applied. In our work cyclic dilatometry is used.

The purpose of this work is to model sintering of a film deposited on a rigid pre-sintered thick substrate. The constraint applied by the substrate induces stresses in the film. We assume that both sintering strain rate and viscosity are functions of temperature and relative density of the sintered material. We develop a fully empirical relationship for both quantities. Using of these relationships for viscosity and free sintering strain rate in the governing equation of the constrained sintering model allows prediction of stresses in the film.

EXPERIMENTAL PROCEDURE

Model of Constrained Sintering

For both shrinkage and viscosity measurement yttria partially stabilized zirconia powder (3 mol. % YSZ) were pressed into cylindrical pellets of 8mm diameter. The thickness of the pellet varied from 5 to 6 mm. Both weight and dimensions of the pellet were measured before and after the experiment. Table I lists the weight, dimensions as well as the calculated green and final densities of the compact. The powder compact is heated at 2°C/min up to a temperature of 500°C, and thereafter at a heating rate of 3°C/min up to 1350°C. The temperature is held constant at 1350°C for one hour and subsequently reduced to room temperature at a rate of 3°C/min. For cyclic dilatometry a load of 150 g was uniaxially applied to the compact for 200s every 13 minutes. Both loading and unloading are instantaneous with temperature profile uninterrupted by the cyclic loading.

Table I. Calculated green and final density as well as dimensions of the samples before and after shrinkage for both shrinkage and cyclic loading tests

Sample	Before shrinkage			After shrinkage			Green Density [%]	Final Density [%]
	Weight	length	Diameter	Weight	Length	Diameter		
Shrinkage test	1.0095	5.36	8.06	0.9088	4.46	6.63	61.5	98.4
Cyclic dilatometry test	1.0066	5.4	8.05	0.9075	4.48	6.59	61.04	99

CONSTRAINED SINTERING MODEL

Model description
   A simple visco-elastic model is developed to model sintering of a film deposited on top of a pre-sintered rigid substrate. We assume that the strain rate in the film has four components: a free sintering strain rate, $\dot{\varepsilon}_f$, a thermal strain rate, $\dot{\varepsilon}_{fT}$, an elastic strain rate, $\dot{\varepsilon}_e$ and a viscous strain rate, $\dot{\varepsilon}_p$. When a thin film deposited on a pre-sintered thick substrate is sintered, the total strain rate in the film follows the thermal strain rate in the substrate, $\dot{\varepsilon}_{sT}$, so that

$$\dot{\varepsilon}_f + \dot{\varepsilon}_{fT} + \dot{\varepsilon}_e + \dot{\varepsilon}_p = \dot{\varepsilon}_{sT} \qquad (1)$$

Since the individual components of the strain can be large, we use logarithmic strain.

A fully empirical approach is taken and it is assumed that the actual functional form of the free sintering strain rate and viscosity is not important provided they can model the experimental data. In the following section we provide the procedure we follow to derive an empirical expression for free sintering rate and viscosity. The variation of the elastic modulus is only a weak function of the temperature and since the elastic strain rate is only important at low temperature we assume that the elastic modulus is only a function of the density and is given by the expression:

$$\frac{E-E_0}{E_d} = \frac{\rho-\rho_0}{1-\rho_0} \qquad (2)$$

where E_d is the elastic modulus of the dense film, E_0 is the green film average elastic modulus as measured by dilatometer at low temperature (below 1000C), ρ_0 is the green density and ρ is the relative density of the material. The Poisson's ratio is the elastic Poisson's ratio at initial density and 1/2 at final density. Since the range of the Poisson's ratio is small and its effect is not significant we assume that it varies linearly with density and thus it is given by the following expression:

$$\frac{v-v_e}{1/2-v_e} = \frac{\rho-\rho_0}{1-\rho_0} \qquad (3)$$

We do not differentiate between viscous and elastic Poisson's ratio. The rate of change of density is related to the volumetric strain rate, $\dot{\varepsilon}_v$, by

$$\dot{\rho} = -\rho\dot{\varepsilon}_v \qquad (4)$$

and the volumetric strain rate is given by

$$\dot{\varepsilon}_v = \frac{2\sigma(1-2v)}{E_p} + \frac{2\dot{\sigma}(1-2v)}{E} + 3\dot{\varepsilon}_f \qquad (5)$$

The constraint conditions causes formation of in-plane biaxial stresses in the film. Expressing the elastic and viscous strain components in terms of the biaxial in-plane stress, Equation 1 becomes

$$\frac{\sigma(1-v)}{E_p} + \frac{\dot{\sigma}(1-v)}{E} + \dot{\varepsilon}_f + \alpha_f\dot{T} = \alpha_s\dot{T} \qquad (6)$$

Where α_f, α_s and T are the thermal expansion coefficient of the film and substrate and the temperature, respectively,

Modeling free sintering rate

The shrinkage of ceramic powder compact (yttria partially stabilized zirconia, YSZ) was measured using a dilatometer and is plotted in Figure 1 versus temperature; the compressive strain is taken as positive. Engineering strain is calculated by dividing the displacement by the initial thickness of the specimen. Since the shrinkage can reach up to ~16%, the engineering

Model of Constrained Sintering

strain is converted to logarithmic strain. Strain rate is obtained from the experimental data by deriving strains relative to time. We assume that the strain rate is a function of temperature/time and density. Obviously in the fully dense state, when the relative density reaches 1, the free sintering rate must be zero. Thus, a suitable function to describe free sintering is

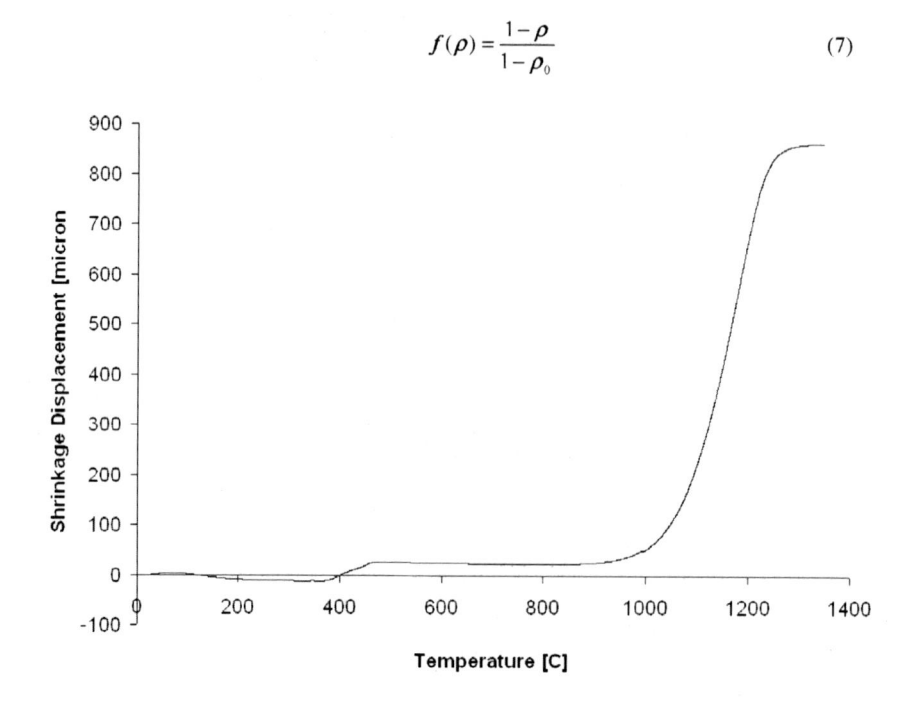

$$f(\rho) = \frac{1-\rho}{1-\rho_0} \qquad (7)$$

Figure 1. Typical shrinkage measurement for YSZ powder compact

At constant relative density, the sintering rate is well modeled by Arrhenius's equation. Hence the simplest empirical equation for free sintering rate is

$$\dot{\varepsilon}_f = Af(\rho)\exp\left(\frac{-B}{T}\right) \qquad (8)$$

where T is in Kelvin. However, the slightly more general equation

$$\dot{\varepsilon}_f = Af(\rho)[1 - Cf(\rho)]\exp\left(\frac{-B}{T}\right) \qquad (9)$$

can give a better fit to the experimental data.

Determination of the constants A, B and C from the shrinkage measurement data

In a free sintering experiment a compact of initial height, h_0, is heated steadily and the change in height, Δh, measured. The change in height is due to free sintering and thermal expansion. It is unlikely that the thermal expansion will be linear over the whole sintering temperature range, but for simplicity we assume it is linear. The thermal strain is small, but the free sintering strain is not. The free sintering strain rate is given by

$$\dot{\varepsilon}_f = \frac{1}{h}\frac{\Delta h}{\Delta t} - \alpha\frac{\Delta T}{\Delta t} \tag{10}$$

The free sintering strain can be found from the summation

$$\varepsilon_f = \log_e\left(\frac{h}{h_0}\right) - \alpha(T - T_0) \tag{11}$$

A good fit to the experimental data can be found if density is calculated from the final measured density. Thus the relative density is given by

$$\rho = \rho_f \frac{\exp(3[\varepsilon_f]_f)}{\exp(3\varepsilon_f)} \tag{12}$$

where ρ_f and $[\varepsilon_f]_f$ refer to the final density and free strain at final density, respectively.

A Fortran program is written to filter the experimental data to remove large fluctuations. This step is necessary especially where these data would cause the program to crash because the program attempts to take the log of a negative value. First the program calculates $\xi = \dfrac{\dot{\varepsilon}_f}{f(\rho)}$.
Over not too large a density range

$$\xi = A'\exp\left(\frac{-B'}{T}\right) \tag{13}$$

The temperature range 920-1050 ^{0}C has been found to be a suitable temperature range over which to determine the best values of A' and B'. The program then determine the constants C' and D' which give the best fit to the line

$$\frac{\xi}{A'\exp\left(\dfrac{-B'}{T}\right)} = C' + D' f(\rho) \tag{14}$$

It has been found that a range of f from 0.4 to 0.9 is a suitable range over which constants C' and D' can be determined. It is noted that C' is close to 1.0. Finally the two expressions are merged so that

$$A = A'C', \; B = B', \; C = \frac{D'}{C'} \tag{15}$$

Figure 2 shows a comparison between calculated strain rate and experimental data with the full line representing the modeling results.

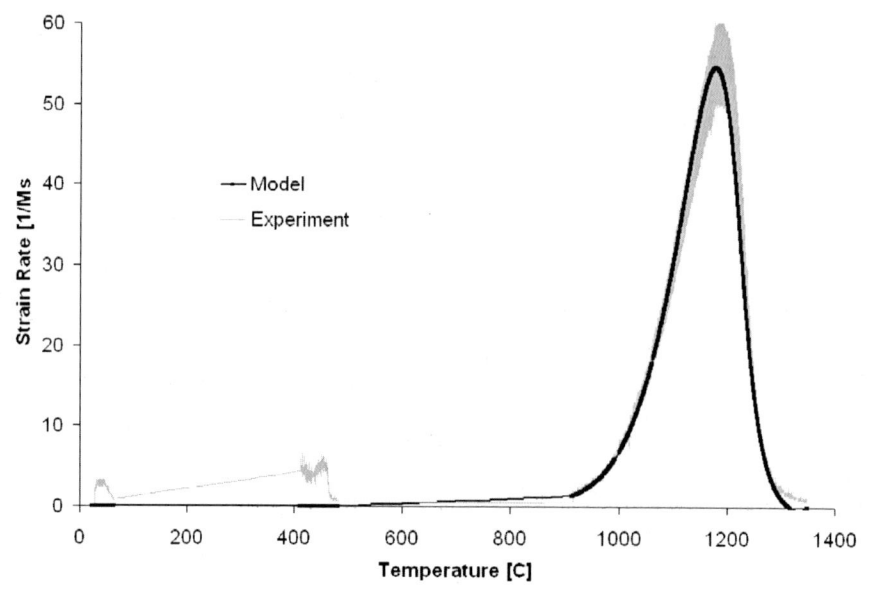

Figure 2. Measured and simulated free sintering strain rate versus temperature. Excellent fit is achieved by the proposed expression (Equation 9).

Modeling of viscosity

The viscosity is measured by applying cyclic loading on a powder compact for a finite time interval (200s in our case). The viscous strain rate is obtained by subtracting the free sintering strain rate from the total strain rate. The viscosity is determined by dividing the true stress applied on the sample by the viscous strain rate. One of the problems of interpretation of the data from the viscosity experiments is that part of the time the deformation is due to free sintering and part of the time due to free sintering plus viscous flow. Under viscous flow the diameter will get larger and under free sintering the diameter will get smaller. It is assumed that free sintering dominates so that if the axial strain is ε_z the diametrical strain is $\varepsilon_d = \varepsilon_z$. The total axial strain is given by

$$\varepsilon_z = \log_e\left(1 + \frac{\delta}{L_0}\right) \tag{16}$$

where δ is the axial displacement and L_0 the original length of the compact. The displacement at the end of the viscosity experiment was -910.28 μm. The original length of the compact was 5400 μm which gives a strain of -0.1846. The diameter measured after the experiment was 6.59 mm. However, this measurement was at room temperature. Assuming a coefficient of thermal expansion of 10.8×10^{-6} /deg C for YSZ powder, then the diameter at 1350°C would have been $6.59 \times (1+10.8 \times 10^{-6}(1350-23)) = 6.68$ mm. The original diameter of the compact was 8.05 mm, hence from the axial strain we can estimate the diameter at 1350°C to be
$D = D_0 \exp \varepsilon_d = 6.69$ mm, which is very close to the value obtained from the measurement of the diameter. Hence the assumption that $\varepsilon_d = \varepsilon_z$, i.e. isotropic sintering, seems reasonable.

The relative density should be calculated from the total strain less the thermal strain since thermal strain does not affect the porosity of the compact. The final total elongation less the thermal elongation is -985.53 μm. If the relative density is $\rho = 1$ at the end of the test, then the relative density of the original compact would be $\rho = \left[1 - \dfrac{985.35}{5400}\right]^3 = 0.5463$. The relative density of the compact obtained from measurement of its volume was .5503. Hence the assumption that $\rho = 1$ at the end of the test is reasonable. Since sintering is isotropic, we can use either diameter or sample thickness to measure thickness and thus also to determine the relative density of the compact.

Empirical expression for viscosity
The rate of viscous deformation depends on the reciprocal of the viscosity, E_p. Thus $1/E_p$ can be modelled with Arrhenius' equation

$$\frac{1}{E_p} = A' \exp\left(\frac{-B}{T}\right) f(\rho) \qquad (17)$$

where $f(\rho) = \dfrac{1-\rho}{1-\rho_0}$.

A more general empirical expression for viscosity is

$$E_p = A \exp\left(\frac{B}{T}\right)(1 + Cf(\rho))^{-1} \qquad (18)$$

Derivation of the constants A, B, and C
First we plot $\log_e(1/E_p)$ as a function of $1/T$ for temperatures greater than 800°C. Fitting a straight line to the results between $7.54 < 10,000/T < 8.86$ we obtain the relationship
$$\log_e(1/E_p) = 2.3664 - 14014/T \qquad (19)$$
In fitting a straight line one must make a judgement of the range used. Too large a range and the effect of the change in density will affect the results, too small a range and the inevitable scatter will affect the results.

The next step is to plot $(1/E_p)\exp(14014/T)$ as a function of $f(\rho)$ over the temperature range 800-1350°C and to find the line of best fit which in our case is

$$\frac{1}{E_p}\exp\left(\frac{14014}{T}\right) = 1.1628 + 10.923 f(\rho) \tag{20}$$

Thus the best fit to the expression of Eq. (18) is found to be

$$E_p = \frac{0.86\exp\left(\frac{14014}{T}\right)}{1+9.3937 f(\rho)} MPas \tag{21}$$

The viscosity at near full density is very sensitive to the density and it is almost impossible to model this high sensitivity. The viscosity versus temperature from the model and experimental results is plotted in Figure 3.

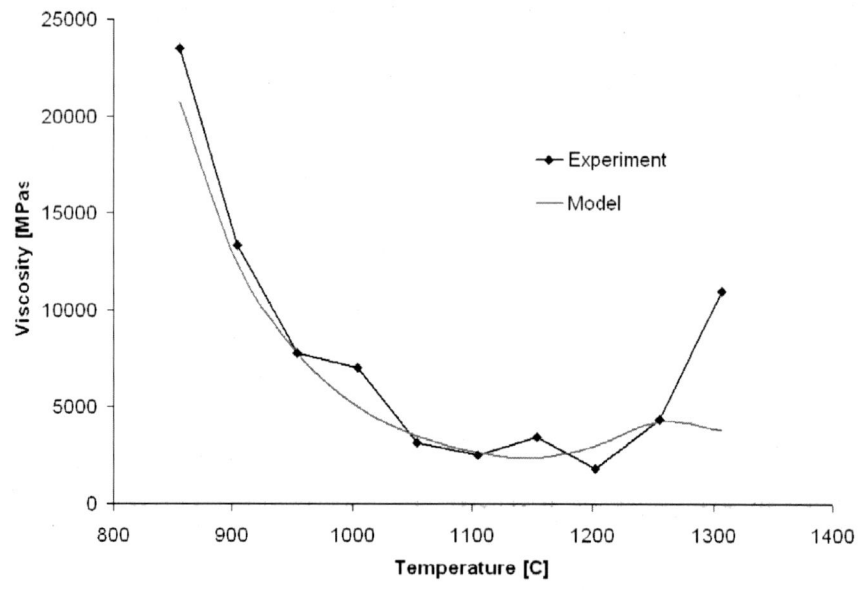

Figure 3. Experimental data for viscosity versus temperature as well as the simulated viscosity following equation 18.

Once the free sintering strain rate and the viscosity are determined, equation 6 can be used to predict the stresses in the constrained sintering film. A Fortran program is written to integrate the relevant equations. Figure 4 shows stress prediction for a fully constrained film under the same heating conditions as the pellets. The stresses predicted are very low and unlikely to cause cracks in the film, although damage initiation such as cavitational defects may be formed[7]. Previous works[7,10-11] have also reported low values for the constrained sintering stresses.

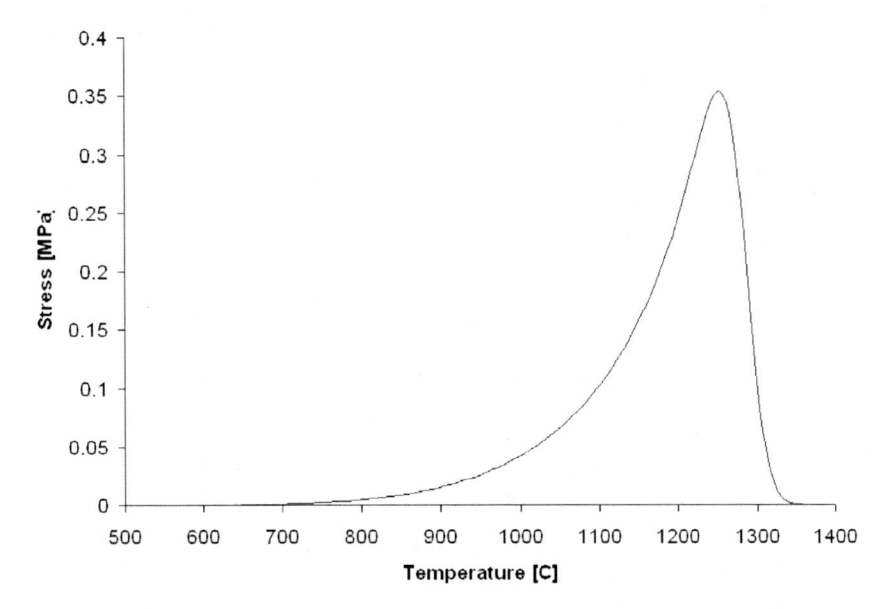

Figure 4. Stress prediction versus temperature in a fully constrained sintering film.

CONCLUSION

A simple visco-elastic model is formulated to predict stress in a fully constrained sintering film. Two experimental measurements are necessary for the model: free sintering rate and viscosity. These two quantities are well simulated by quasi-empirical equations. The predicted stresses are very low and unlikely to cause de-sintering cracks in the film, but they may induce cavitational damage.

REFERENCES

[1] R.K. Bordia and G. W. Scherer, On Constrained Sintering-I. Constitutive model for a sintering body; and II. Comparison of constitutive models, Acta Metall., 36, 2393-09 (1988).
[2] T.J. Garino, H.K. Bowen. Kinetics of Constrained Film Sintering, J. Am. Ceram. Soc., 73, 251-57 (1990).
[3] D. Green, P.Z. Cai and G.L. Messing. Residual Stresses in Alumina-Zirconia Laminates, J. Eur. Ceram. Soc., 19, 2511-17 (1999).
[4] A. Petersson and J. Agren. Constitutive behaviour of WC-Co materials with different grain size sintered under load, Acta Mater., 52, 1847-58 (2004).
[5] A. Mohanram, S.H. Lee, G.L. Messing and D.J. Green. A Novel Use of Constrained Sintering to determine the Viscous Poisson's Ratio of Densifying Materials, Acta Mater., 53, 2413-18 (2005).
[6] R.Z. Zuo, E. Aulbach and J. Roedel. Experimental determination of sintering stresses and sintering viscosities," Acta Mater., 51, 4563-4574 (2003).
[7] P.Z. Cai, D.J. Green and G.L. Messing. Constrained Densification of Alumina/Zirconia Hybrid Laminates .1. Experimental observations of processing defects, J. Am. Cer. Soc., 80, 1929-39 (1997).

[8]R.Z. Zuo, E. Aulbach and J. Rodel. Experimental determination of sintering stresses and sintering viscosities, *Acta Mater.*, **51**, 4563-74 (2003).

[9]S.H. Lee, G.L. Messing and D.J. Green. Bending creep test to measure the viscosity of porous materials during sintering, *J. Am. Ceram. Soc.*, **86**, 877-82 (2003).

[10]P.Z. Cai, D.J. Green and G.L. Messing. Constrained Densification of Alumina/Zirconia Hybrid Laminates .2. Viscoelastic Stress Computation, *J. Am. Ceram. Soc.,* **80**, 1940-1948 (1997).

[11]V.M. Sglavo, P.Z. Cai and D.J. Green. Damage in Al_2O_3 sintering compacts under very low tensile stress, *J. Mater. Sci. Lett.*, **18**, 895-900 (1999).

Fracture

STUDY OF FACTORS AFFECTING THE LENGTHS OF SURFACE CRACKS IN SILICON NITRIDE INTRODUCED BY VICKERS INDENTATION

Hiroyuki Miyazaki, Hideki Hyuga, Yu-ichi Yoshizawa, Kiyoshi Hirao and Tatsuki Ohji
National Institute of Advanced Industrial Science and Technology (AIST)
Anagahora 2266-98, Shimo-shidami, Moriyama-ku,
Nagoya 463-8560, Japan

ABSTRACT

Three possible factors affecting Vickers indentation results were assessed using silicon nitrides in order to clarify the resulting large scatter in calculated indentation fracture resistance, K_R, reported in the VAMAS round robin tests. The degree of misreading both crack and diagonal sizes depended on the operators, which brought about the variation of K_R from 4.9 to 5.3 MPa·m$^{1/2}$. Measurements with relatively new and used indenters gave the same crack lengths, indicating that indentation fracture resistance was insensitive to slight damage at the corner of the indenter. Optical microscopic observations of crack tips induced by the indentation confirmed little extension of cracks after the unloading of an indenter, which was reflected on the negligible difference in fracture resistance attained at 1 and 30 min after the indentation. It was suggested that the inconsistent outcomes of K_R from various laboratories that participated in the round robin tests were not originated from the latter two factors but mainly from the error in measuring the crack length.

INTRODUCTION

Because of the ease of applicability to small samples of material and simple testing procedure, the indentation fracture (IF) method has been widely used for evaluation of fracture resistance[‡], K_R, of ceramics since it was proposed by Lawn and his co-workers.[1] Determining the fracture resistance of small ceramic parts such as bearing balls is impossible without this method since alternative techniques have not been presented. However, IF method has not been without detractors. One of the most critical and practical issues is that in between-laboratory consistency is poor,[2,3] which was revealed by round-robin tests conducted in order to standardize the indentation fracture-resistance test for ceramics (e.g. VAMAS#3[4] and #9[5,6], etc.[7]). This issue is a major reason relegating the IF method as a subsidiary technique, thus limiting its usage to special purposes. Although a plausible explanation for the scatter among several laboratories was subjectivity of the operators in reading the crack length,[2,4-7] there have been few systematic studies about experimental factors affecting the resistance. Because of the industrial need for fracture resistance assessment of small ceramic parts such as silicon nitride bearing balls, it is important to eliminate the source of scatter by clarifying the influence of measuring conditions on crack length. In this paper, three conditions were studied using commercial silicon nitrides for ball bearings: the effects of the time after indentation, the condition of the corner of an indenter on the crack length, and the human factor.

[‡] Note the fracture resistance, K_R, as used here is not to be equal to fracture toughness K_{Ic}. K_R is an estimate of a material's resistance to cracking as introduced by an indenter. K_{Ic} is an intrinsic property of a material and is independent of test method.

EXPERIMENTAL PROCEDURE

Materials

Two commercial Si_3N_4 ceramics for ball bearings, TSN-03[¶] and EC-141[*], were used in this study. Some characteristics of these materials are summarized in Table 1. The bulk density was obtained by the Archimedes method as used in JIS R 1634.[8] The Young's modulus was obtained by the ultrasonic pulse echo method as used in JIS R 1602.[9] The Vickers hardness was calculated from the size of impressions with force of 196 N as used in JIS R 1607.[10] The fracture toughness was determined by the SEPB method according to JIS R 1607.[10]

Test Procedure

In order to check the difference in reading the crack length and diagonal size due to the operators, eight Vickers indentations were made on the polished surface of A samples by a single operator with the dwell time of 15 s. It was assumed that the weight of error in reading the crack length should become relatively larger as the size of crack was decreased. Thus, a relatively low indentation force of 98 N was selected to find the subtle difference due to the operators. The same impressions were measured successively by four persons with and without the experience of this test. The same optical microscope equipped with a hardness tester [§] was used by all operators. The magnification of both the eyepiece and the objective lens was 10 X. Bright field illumination was employed. All the above measurements were conducted in accordance with JIS R 1607 except the low level of force.[10] In the case of the study on the effect of the condition of edge of an indenter, a relatively new indenter and a used indenter with a flaw on the edge were employed. In order to detect the slight difference in both the surface crack length and the diagonal size due to the indenters, both TSN-03 and EC-141 samples were indented at relatively high force of 196 N with the two indenters. The dwell time was also 15 s. The surface crack and diagonal sizes were measured immediately after the unloading by the most experienced operator. Bright field images were observed with a 10 X eyepiece and a 50 X objective using a traveling microscope[¥]. The presence of the subcritical crack growth was examined by monitoring the crack tips introduced on the two samples using 50 X objective and bright field illumination. The indentation force and the contact time were 196 N and 15 s, respectively, and a relatively new indenter was used. The lengths of the indentation diagonals and surface cracks were measured with the measuring microscope 1 and 30 min after the unloading by the most experienced operator.

The effect of variation in the measured crack length on the fracture resistance was also estimated. In this study, the ratios of crack length to diagonal size were larger than 2.5 for both samples, indicating that the crack systems were median/radial ones.[11] Then, the equation for the median crack were used. The fracture resistance, K_R, was determined from the as-indented crack lengths as follows:

$$K_R = \xi(E/H)^n P c^{-3/2} \qquad (1)$$

[¶]Toshiba Materials Co., LTD., Yokohama, Japan. The year of manufacture was 2006.
[*]NGK Spark Plug Co., LTD., Nagoya, Japan. The year of manufacture was 2006.
[§]Model AVK-C2, Akashi Corp., Yokohama, Japan
[¥] Model MM-40, Nikon Corp., Tokyo, Japan

where ξ and n are material-independent dimensionless constants for Vickers-produced radial cracks, E and H are Young's modulus and the hardness, P is the indentation force, and c is the half-length of as-indented surface crack length. In this study, Miyoshi's equation was applied and the values of the constants were $\xi = 0.018$ and $n = 1/2$.[12] K_R was calculated for each indentation using the hardness value obtained for each impression.[10]

RESULTS AND DISCUSSION
Variation in the Reading the Crack and Diagonal Sizes Due to the Operators

Figure 1 shows an example of the indentation observed optically. Nearly all (98.8%) of the indentations in this study were symmetric. Table 2 shows the variation in the results of Vickers indentation test obtained by different operators. The differences in the crack lengths were ~ 10 μm, which was nearly equal to the standard deviation. When compared to the size of crack size ~ 270 μm, the percent of the variation was less than 4 %. It can be deduced that the error in reading the crack length by the inexperienced operator was not excessive. The scatter in the diagonal size due to the observer was ~ 4 μm and was twice as large as the standard deviations. There might be some misreading in diagonal size obtained by inexperienced operators since standard deviations of their data were larger than that of experienced person. Due to these differences in observed crack and diagonal sizes, the calculated fracture resistance ranged between 4.89 ± 0.26 and 5.30 ± 0.28 MPa·m$^{1/2}$, which showed that the human factor alone can cause some degree of the scatter.

In the case of the round robin tests, each participant used different measuring instruments with various modes of viewing. The length measurements were also likely to be affected by both equipments and observation mode. Thus it is rational to suppose that the error due to the human factor can be exacerbated by machine factors, resulting in significant disagreement in calculated K_R between participants as reported in the VAMAS report.[4-6]

Effect of the Condition of an Indenter Edge

Figure 2 shows SEM micrographs of corners of the relatively new and used indenters. The relatively new indenter had the sharp and straightforward edge, whereas a flaw was on the edge of the used indenter. Indentation results obtained with a used indenter are compared with that for relatively new indenter in Table 3. The sizes of crack and diagonal varied little between the two measurements using the relatively new indenter and used one, leading to the slight difference in the obtained fracture resistance. It appears that the fracture resistance measurements by IF technique was not so sensitive to the flaw on the edge of the used indenter. It is well known that the driving force for crack extension from indenters is the constraints of the indentation plastic zone.[1] In the model of Lawn et. al, the strain in the plastic zone is determined by the volume of the impression, but has nothing to do with the conditions of the corner of the indenter.[1] Therefore it is reasonable to assume that the small damage on the edge has slight influence on the crack length.

Effect of Time after Indenter Unloading on the Crack Length

Figure 3 shows the enlarged crack tips of TSN-03 sample 1, 10 and 30 min after indentation. The tip location could be assessed by using nearby tiny microstructural features as convenient location references. By comparing the tip locations, it can be seen that crack growth larger than several micrometers did not take place in the sample. The extension of crack tips 10 and 30 minutes after the indentation was hardly detected for the EC-141 as well. Then the slow crack growth, even if it occurred slightly, was negligible as compared with the crack length of c =~ 220 μm.

Indentation fracture resistance calculated by the data of 1 and 30 min after the indentation are presented in Figure 4. Since slow crack growth was not active in the samples, the K_R was hardly affected by the elapsed time.

It has been reported that the crack length for silicon nitride did not increase but decreased slightly (2.7%) 17 days after indentation.[2] Other reports in the literature indicated that silicon nitrides were not susceptible to environmentally-assisted slow crack-growth at room temperature.[13,14] Therefore, it appears that the time after completion of contact is not a controlling factor for the scatter of fracture resistance of Si_3N_4 in the round robin tests such as VAMAS #9.[4-7]

CONCLUSION

In order to clarify the source of errors in the measurements of indentation fracture resistance by the IF method, the dependence of measured surface crack length in Si_3N_4 on the observer were studied, as well as the effects of the time after indentation and the state of the corner of the edge. It was found that the observed sizes of crack and diagonal were dependent on the operator's experience, giving rise to the variation of K_R between 4.89 ± 0.26 and 5.30 ± 0.28 MPa·m$^{1/2}$. By contrast, K_R calculated from results using a used indenter was almost the same as that calculated from results obtained using a relatively new indenter, indicating that the slight deterioration in the corner of the indenter hardly affected the indentation results. Observation of the crack tips with an optical microscope revealed that the Si_3N_4 was insensitive to the environmentally-assisted subcritical crack growth. Hence the time after indenter unloading had little effect on K_R. From these results, it is assumed that the difficulty in finding the correct position of crack tips and impression corner is most likely to be major cause of errors in the fracture resistance evaluation by the IF technique.

ACKNOWLEDGMENT

This work has been supported by METI, Japan, as part of the international standardization project of test methods for rolling contact fatigue and fracture resistance of ceramics for ball bearings.

REFERENCES

[1] B. R. Lawn, A. G. Evans and D. B. Marshall, Elastic/Plastic Indentation Damage in Ceramics: The Median/Radial Crack system, *J. Am. Ceram. Soc.*, **63**, 574-81 (1980).
[2] G. D. Quinn, Fracture Toughness of Ceramics by the Vickers Indentation Crack Length Method: A Critical Review, *Ceram. Eng. Sci. Proc.* **27** [3] (2006).
[3] G. D. Quinn and R. C. Bradt, On the Vickers Indentation Fracture Toughness Test, *J. Am. Ceram. Soc.*, **90**, 673-680 (2007).
[4] D. M. Butterfield, D. J. Clinton and R. Morell, The VAMAS Hardness Round-Robin on Ceramic Materials, VAMAS report#3, National Physical laboratory, Teddington, Middlesex, United Kingdom, 1989.
[5] H. Awaji, T. Yamada and H. Okuda, Result of the Fracture Toughness Test Round Robin on Ceramics – VAMAS Project-, *J. Ceram. Soc. Jpn.*, **99**, 417-22 (1991).
[6] H. Awaji, J. Kon and H. Okuda, The VAMAS Fracture Toughness test Round-Robin on Ceramics, VAMAS report#9, Japan fine ceramic center, Nagoya, 1990.

[7] Report of preliminary investigation for standardization of fine ceramics, Japanese Fine Ceramics Association, 1998.
[8] JIS R 1634, Test Methods for Density and Apparent Porosity of Fine Ceramics," Japanese Industrial Standard, (1998).
[9] JIS R 1602, Testing Methods for Elastic Modulus of Fine Ceramics," Japanese Industrial Standard, (1995).
[10] JIS R 1607, Testing Methods for Fracture Toughness of Fine Ceramics," Japanese Industrial Standard, (1995).
[11] K. Niihara, R. Morena and D. P. H. Hasselman, Evaluation of K_{Ic} of Brittle Solids by the Indentation Method with Low Crack-to-Indent Ratios, *J. Mater. Sci. Lett.*, **1**, 13-16 (1982).
[12] T. Miyoshi, N. Sagawa and T. Sasa, Study of Evaluation for Fracture Toughness of Structural Ceramics, *J. Jpn. Soc. Mech. Eng.*, **A, 51**, 2489-97 (1985).
[13] K. D. McHenry, T. Yonushonis and R. E. Tressler, Low-Temperature Subcritical Crack Growth in SiC and Si_3N_4, *J. Am. Ceram. Soc.*, **59**, 262-63 (1976).
[14] A. Bhatnagar, M. J. Hoffman and R. H. Dauskardt, Fracture and Subcritical Crack-Growth Behavior of Y-Si-Al-O-N Glasses and Si_3N_4 Ceramics, *J. Am. Ceram. Soc.*, **83**, 585-96 (2000).

Table 1 Mechanical properties of the Si₃N₄ ceramics for ball bearings used in this study.

Sample	Bulk Density † (g·cm⁻³)	Young's modulus ‡ (GPa)	Vickers hardness ¶ (GPa)	Fracture toughness* (MPa·m^(1/2))
A	3.231 ± 0.004	304.6 ± 0.4	14.9 ± 0.2	5.55 ± 0.12
B	3.203 ± 0.001	309.6 ± 0.4	14.4 ± 0.2	5.52 ± 0.16

† The bulk density was obtained by the Archimedes method as used in JIS R 1634 with the sample size, $N = 3$.[8]   ‡ The ultrasonic pulse echo method as used in JIS R 1602 ($N = 3$).[9]   ¶ Hardness was obtained from the indentation with the force of 196 N according to JIS R 1607 ($N = 7 - 8$).[10]   * The SEPB method as used in JIS R 1607 ($N = 6$ for A sample and 10 for B sample).[10]   Uncertainties are one standard deviation.

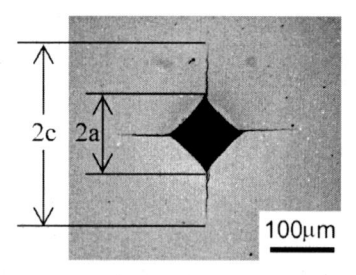

Figure 1. An example of Vickers indentation in Si₃N₄ (sample: A) produced with the indentation force of 98 N and dwell time of 15 s.

Table 2 The variation in the Vickers indentation outcomes due to the operator

Operator	Experience of measuring	Average diagonal size, 2a (μm)	Average crack size, 2c (μm)	Vickers hardness (GPa)	Fracture resistance (MPa·m^(1/2))
1	Enough	110.2 ± 1.2	277.3 ± 10.9	15.0 ± 0.2	4.89 ± 0.26
2	Plenty	110.5 ± 1.4	269.1 ± 6.4	14.9 ± 0.3	5.12 ± 0.18
3	None	114.0 ± 2.0	268.6 ± 9.0	14.0 ± 0.3	5.30 ± 0.28
4	None	113.3 ± 1.8	276.1 ± 11.3	14.2 ± 0.3	5.06 ± 0.24

Eight impressions were made by a single operator and were measured successively by four operators. The indentation force was 98 N with the dwell time of 15 s. Uncertainties are one standard deviation.

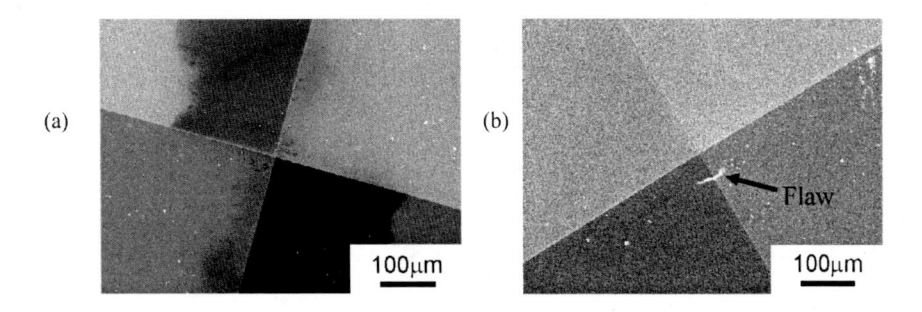

Figure 2. The enlarged corner of the indenter observed with SEM (a): relatively new indenter and (b): used one.

Table 3 Effect of the condition of indenter on the results of Vickers indentation test

Sample	Type of indenter	Average diagonal size, 2a (μm)	Average crack size, 2c (μm)	Vickers hardness (GPa)	Fracture resistance (MPa·m$^{1/2}$)
A	Relatively new	156 ± 2	431 ± 13	14.9 ± 0.2	5.05 ± 0.11
	Used	157 ± 1	429 ± 10	14.8 ± 0.2	5.10 ± 0.07
B	Relatively new	159 ± 2	448 ± 10	14.4 ± 0.2	4.89 ± 0.07
	Used	160 ± 2	452 ± 12	14.3 ± 0.1	4.85 ± 0.05

Eight impressions were made by using each indenter for each sample with the indentation force of 196 N and the dwell time of 15 s. Crack and diagonal sizes were measured by the most experienced operator. Uncertainties are one standard deviation.

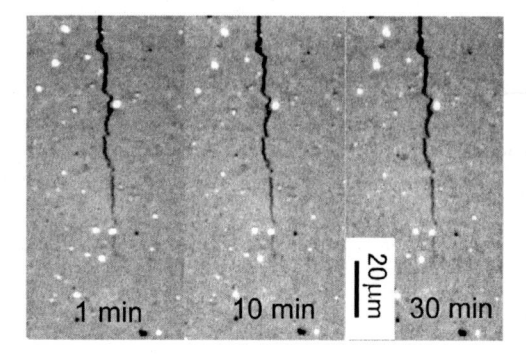

Figure 3. Observations of the crack tip 1, 10 and 30 min after the unloading with an optical microscope (Sample: TSN-03). A bright field illumination and 50 X objective were used.

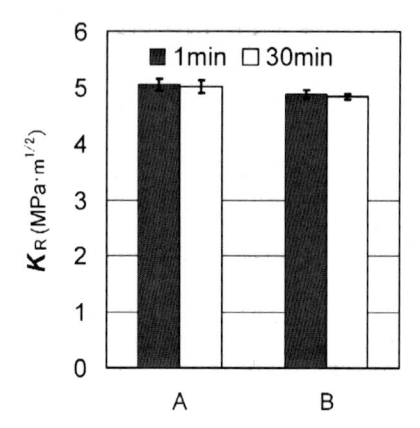

Figure 4. Effect of time after indentation on the fracture resistance for the two Si_3N_4 samples. Eight impressions were made for both samples with the indentation force of 196 N and the contact time of 15 s. Crack and diagonal sizes were measured by the most experienced operator. Error bars are ± 1 standard deviation.

STRENGTH RECOVERY BEHAVIOR OF MACHINED ALUMINA BY CRACK HEALING

Kotoji Ando, Wataru Nakao, Koji Takahashi
Department of Energy and Safety Engineering
79-5 Tokiwadai, Hodogaya-ku, Yokohama, 240-8501, Japan

Toshio Osada
Post-graduate student, Department of Energy and Safety Engineering
79-5 Tokiwadai, Hodogaya-ku, Yokohama, 240-8501, Japan

ABSTRACT
Machined alumina containing 20 vol.% SiC whiskers has been subjected to various heat-treatments. The effect on bending strength has been investigated as a function of crack-healing temperature and time. By heating at 1673 K for 10 h prior to testing, specimens fractured from sites other than those associated with the final machining process. Thus, it was concluded that the machining cracks were completely healed by this heat treatment. Moreover, the strength at elevated temperature of the healed samples exhibited similar properties to healed polished samples. It is proposed that the crack-healing treatment is a viable economic technique to be applied to structural ceramics.

INTRODUCTION

Ceramics have been used in various fields as machined components. In particular, alumina is expected to be employed in structural components operating at high temperatures in various atmospheres, because of its excellent mechanical properties and oxidation resistance. However, alumina has low fracture toughness, resulting in numerous random cracks due to machining. Final machining operations, such as polishing and lapping, together with nondestructive inspection, are generally performed to remove the non-acceptable flaws. Although these techniques are very costly, they cannot eliminate minute flaws. Thus, the reliability of the ceramic components cannot be secured [1]. If such cracks can be healed perfectly, the ceramics components can offer improved reliability without costly final machining.
The present authors have sintered alumina based materials reinforced SiC particles or SiC whiskers which have previously demonstrated an excellent crack-healing ability. Crack-healing is a function in which materials detect surface cracks by themselves, and then heal them all [2-12]. For example, a semi-elliptical surface crack introduced by indentation can be completely healed by high temperature heat-treatment in air. Depending on scale, cracks of this nature can reduce the bending strength of the materials by as much as 85 %. However, the crack-healed materials have demonstrated similar levels of strength and fatigue limit as the base material up to approximately 1573 K [9-12], since most crack-healed specimens failed from a base material.
Whether this function can also be useful for cracks caused by machining is a very interesting research subject. If the crack-healing is available to heal the cracks, improving the reliability and machining efficiency, and then the crack-healing allow the machining costs to save. On repair for cracks caused by machining, Matsuo et al. [13] reported that the machined monolithic alumina recovered its strength considerably by heat treatment. However, the material's strength recovery was not complete. Moreover, monolithic alumina dose not have the crack-healing ability; [8] thus, the phenomena of strength recovery due to heat treatment is not related to the crack-healing. Recently, Ando et al. [14] reported that heavily machined mullite/ SiC

composites could recover its strength considerably through the crack-healing process. However, it has not yet been clear whether the crack-healing is a useful technique to recover the strength of machined alumina.

In this study, alumina/ SiC whiskers composite was sintered and the crack-healing behavior for the cracks caused by machining was investigated. The composite has fracture toughness of 5.7 MPa·m$^{1/2}$, which was evaluated by SEPB method. This value was larger then that of monolithic alumina (K_{IC} = 3 - 4 MPa·m$^{1/2}$) [10]. Small bending specimens (22 × 4 × 3 mm) were made. These specimens were machined into various configurations so that they have various stress concentration factors to replicate those of components. The machined specimens were crack-healed under various conditions. The fracture stress of these specimens after crack-healing were evaluated systematically. From the obtained results, the crack-healing effect on the cracks was discussed.

EXPERIMENTAL
Material and crack-healing condition

The alumina powder (AKP-20, Sumitomo Chemical Co. Ltd., Tokyo, Japan) used has an average particle size of 0.5 μm. The SiC whisker (SCW, NO.1-0.8, Tateho Chemical Industries, Ako, Japan) used has length of 30 - 100 μm and diameter of 0.8 – 1.0 μm. For mixing the raw powder, the mixture of alumina powders and 20 vol.% SiC whiskers were blended well in Isopropyl alcohol for 12 h using alumina balls and an alumina mill pot. Rectangular plates (90 × 90 × 6 mm) were hot pressed in Ar at 2123 K under 40 MPa for 1 h. The sintered plates were cut into 3 × 4 × 22 mm rectangular bar specimens as shown in Figure 1. The specimens were polished to a mirror finish according to the JIS standards [16] on a tensile surface. The edges of the specimens were beveled 45° to prevent fractures due to edge-cracks. In this paper these specimens were called "smooth specimens". As shown in Figure 1, Semicircular groove, Thorough-hole and square groove were made at the center of the smooth specimens by using diamond-coated ball-drill. The other grinding conditions are listed in Table 1. For case of semicircular groove, maximum depth of cut by one pass (d) was 15 μm, because diamond grains suffered from damage significantly, resulting in decreased grinding efficiency.

The machined specimens were subjected to crack-healing treatment by heat-treating in air at 1373 - 1673 K for 1 or 10 h. The specimens are here called as machined specimen healed. Also the smooth specimens were crack-healed at 1573 K for 1 h in air. These specimens are defined as healed smooth specimen. Even the smooth specimen involved minute flaws such as surface-crack and surface-pore. The flaws can be completely healed by heat-treatment [10-12]. Thus, all the fractures of the healed smooth specimens initiate from largest embedded flaw such as pore. In this paper, therefore, the healed smooth specimens were treated as specimens without surface flaws.

Test method

All fracture tests were performed on a three-point bending system with a span of 16 mm as shown in Figure 1. The crosshead speed was 0.5 mm/min. The tests took place at room temperature and at elevated temperature. Nominal fracture stress (σ_{NF}) of machined specimens was evaluated by

$$\sigma_{NF} = \frac{M_F}{Z} \qquad (1)$$

where Z is the section modulus of the center of the machined specimens, and M_F is a bending moment as specimens fractured.

Bending strength, σ_B, of the semicircular grooved specimen is defined as the stress at the center of the semicircular groove evaluated from the load at failure, P and given by

$$\sigma_B = \alpha \sigma_{NF} \qquad (2)$$

where α is a stress concentrate factor. The value of Z and α were obtained from data book [17], and listed in Table 2. The value of α of square groove could not be obtained.

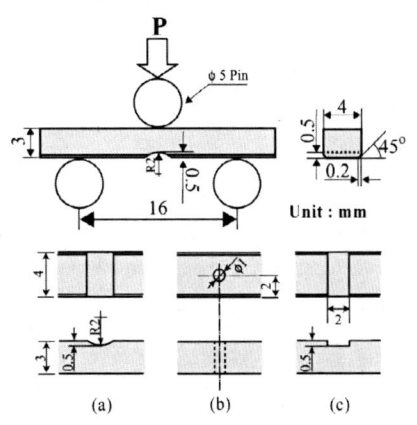

Figure 1 Schematic drawing of three-point loading system, test specimen size and machining figurations: (a) type A: semicircular groove, (b) type B: through-hole, (c) type C: square groove.

Table1 Grinding conditions

Specimen preparation	Semicircular groove	Through -hole	Square groove
Grindstone	$\phi 4$ #140 diamond electrocoated	$\phi 1$ #200 diamond electrocoated	$\phi 2$ #140 diamond electrocoated
Rotational speed (rpm)	5000	5000	5000
Table feed speed (mm/min)	100	2	140
Depth of cut by One pass (μm)	5,10 and 15	10	10

Table 2 Stress concentration factor and section modulus.

Shape of specimen	Semicircular groove	Through-hole	Square groove
Section modulus, Z [mm³]	4.2	4.5	4.2
Stress concentration factor, α	1.4	2.2	-

RESULTS AND DISCUSSION
Crack-healing behavior

Figure 2 shows the effect of the crack-healing conditions on the strength recovery behavior of the machined specimen. The shape of specimens was semicircular groove, and d = 10 μm. The symbols ◇ in left column and ● in right column show the bending strength (σ_B) of as-machined specimen and healed smooth specimen, respectively. The average σ_B of these specimens were found to be 514 MPa and 1015 MPa, respectively, indicative of an 100% strength recovery upon crack-healing.

The as-machined specimens always fractured from surface cracks caused by machining as shown in Figure 3 (a) and (b). The cracks introduced by machining are formed by coalescence of similar semi-elliptical cracks as shown in Figure 3. Moreover, the critical crack size differed extensively for each specimen, and therefore, σ_B also showed large scatter from 382 MPa to 609 MPa.

The symbols ◆ and □ in Figure 2 show the σ_B of machined specimen healed at elevated temperatures for 1 h and 10 h, respectively. The σ_B of healed specimens increases gradually up

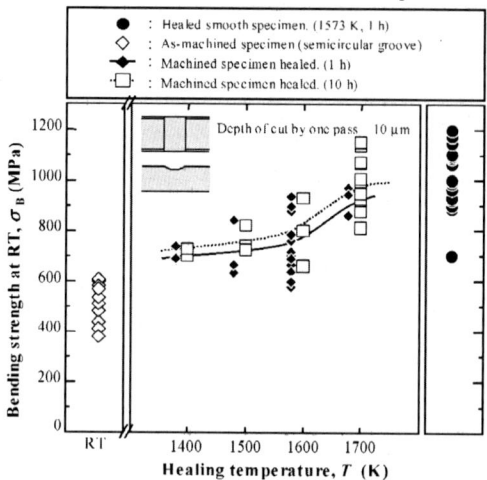

Figure 2 the effect of crack-healing conditions on the strength recovery behavior of machined specimen.

to 1573 K, above which it increases abruptly. The average σ_B has a maximum at healing temperature of 1673 K for 10 h. Moreover, the σ_B was almost equal to the fracture stress of healed smooth specimens. Figures 4 (a) and (b) show SEM images of surface of the machined specimens healed at 1573 K for 1 h and at 1673 K for 10 h, respectively. In this figure, the dark and the bright regions implied the oxidation products and alumina, respectively. Oxidation products were found to extend as the healing temperature and time increases. For the case of 1673 K - 10 h, a large amount of the products originating from SiC whiskers were confirmed to cover the whole cracks caused by machining surface, and thus the complete strength recovery appeared. From these results, the optimal crack-healing conditions for machining cracks was defined as heating at 1673 K for 10 h. However, it was found that these conditions are quite different from the optimal crack-healing conditions for indentation crack, which was reported previously as 1573 K for 1 h [10-12].

(a) (b)
Figure 3 SEM images of fracture surface of as-machined specimen:
(a) σ_B=609 MPa and (b) σ_B=382 MPa.

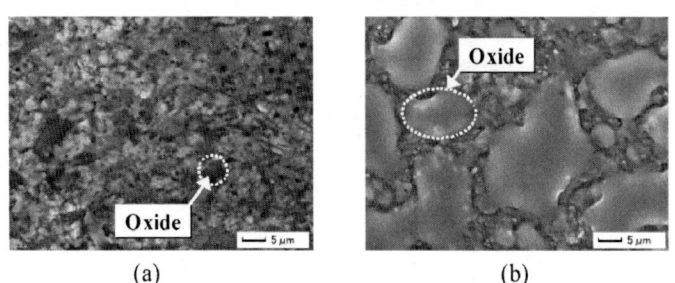

(a) (b)
Figure 4 SEM images of surface of machined specimen healed
(a) at 1573 K for 1 h and (b) at 1673 K for 10 h.

Statistical analysis of bending strength

Figure 5 shows the Weibull plots of the bending strengths of the as-machined specimen ($\Diamond$), the machined specimen healed at 1573 K for 1 h ($\blacklozenge$), the machined specimen healed at 1673 K for 10 h ($\Box$). Where d = 10 µm. Also the Weibull plot of σ_B of the healed smooth specimen ($\bullet$) was shown in this figure. A two parameter Weibull function is given by

$$F(\sigma_B) = 1 - \exp\left\{-\left(\frac{\sigma_B}{b}\right)^c\right\} \quad (3)$$

where b is a scale parameter, c is a shape parameter. Table 3 shows the scale parameter, b and the shape parameter, c of the four kinds of samples shown in Figure 5.

The values of b and c of the machined specimen healed at 1673 K for 10 h were 1026 MPa and 11.5, respectively. The values were almost equal to those of smooth specimen healed. Furthermore, it was found that the fracture initiation site of the machined specimen healed was the embedded flaw from SEM observations. From the result, it is, therefore, found that healing the cracks introduced by machining completely, the machined specimen healed at 1673 K for 10 h indicates the same fracture mechanism as the smooth specimen healed. Alternatively the machined specimen healed at 1573 K for 1 h had lower values of b and c than the smooth specimen healed, as this crack-healing condition was inadequate.

Another important aspect involved in this analysis is that the shape and maximum size of the machining cracks have large scatter. In the earlier studies [15], it was also found that the value of c of a specimen that had an indented crack was large (c =13.8), because these specimens always fracture from the same sized crack introduced by design. However, the values of b and c of the as-machined specimen were 549 MPa and 6.69, respectively. Thus, it was confirmed that crack-healing is useful for the cracks having a large scatter of its size and shape, which introduced by hard machining.

From these results, it can be concluded that the strength of machined specimens was completely recovered by crack-healing at 1673 K for 10 h. Therefore, the crack-healing treatment is an effective technique for increasing the reliability of machined alumina ceramics.

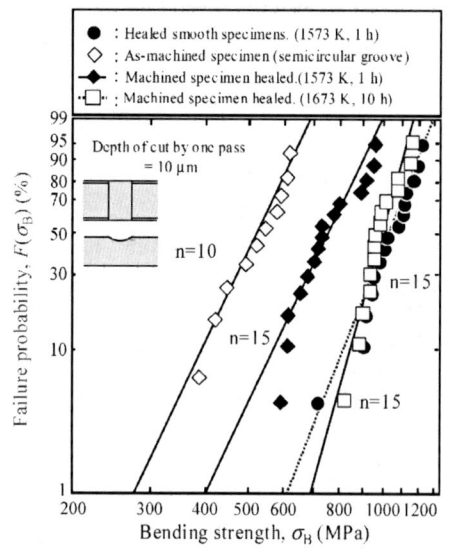

Figure 5 Weibull plot of bending strength.

Strength Recovery Behavior of Machined Alumina by Crack Healing

Table 3 Scale parameter and shape parameter.

	Scale parameter [MPa]	Shape parameter
Healed smooth specimen	1075	8.15
As-machined specimen	549	6.69
Machined specimen healed (1573 K, 1 h)	796	6.70
Machined specimen healed (1673 K, 10 h)	1026	11.5

Working limit of machining with crack-healing treatment.

Figure 6 shows the effect of depth of cut by one pass (d) on the σ_B of healed machined specimens containing a semi-circular groove. Also the data on as-machined specimens were included in figure 6. The symbols ◆ and □ show the σ_B of machined specimen healed at 1573 K for 1 h and at 1673 K for 10 h, respectively. Also the symbol ◇ shows the σ_B of as-machined specimen. The symbols ● and ○ in the left column show the σ_B of the healed smooth and

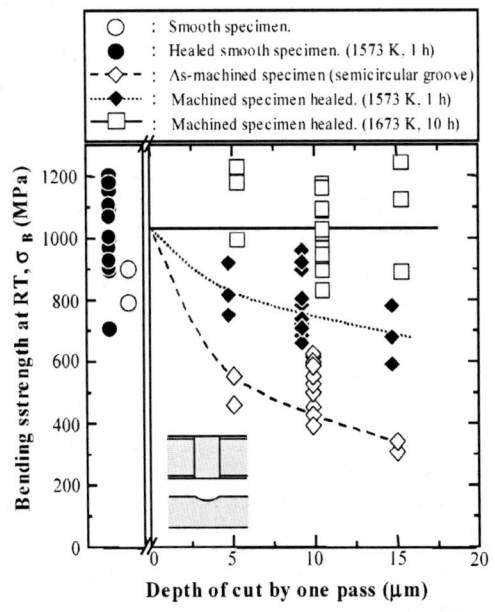

Figure 6 The effect of depth of cut by one pass on the local fracture stress of the machined specimens.

smooth specimen, respectively.

Strength Recovery Behavior of Machined Alumina by Crack Healing

Throughout the whole range of the d, a complete strength recovery was attained by crack-healing treatment at 1673 K for 10 h, because these average strengths were almost equal to that of healed smooth specimen. Thus, crack-healing is possible for relatively large cracks (The crack depth was approximately 60 μm) as shown in the figure 7, which was introduced through hard machining (for depth of cut up to 15 μm). Alternatively by heat-treatment at 1573 K for 1 h, only a limited strength recovery could be attained, even for the depth of cut by one pass as low as d = 5 μm.

Thus, it was concluded that the working limit of machining with crack-healing treatment was d = 15 μm.

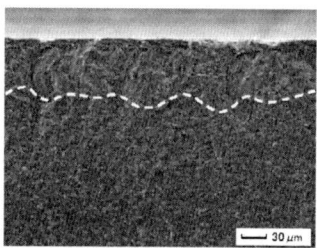

Figure 7 SEM images of fracture surface of as-machined specimen.
: d = 15 μm, σ_B = 297 MPa.

Effect of test temperature on the bending strength of machined specimens healed

Figure 8 shows the temperature dependence of the bending strength of the healed machined specimens (□). For comparison, the temperature dependences of the fracture stresses of the indented-crack-healed specimen (▲) and the healed smooth specimen (●) are also shown in this figure. The symbols with asterisks (*) indicate the crack-healed specimens fractured outside the healed pre-crack. These specimens indicate that the pre-crack was completely healed and the healed pre-crack exhibited the same level strength to the base material even up to 1573 K.[10-12]

The bending strengths (σ_B) of the healed machined specimens were almost constant in the temperature range from room temperature to 1373 K. The constant value of σ_B was found to be about 1000 MPa, and was a little higher than that of the crack-healed specimens and the healed smooth specimen at 873 K ~ 1373 K. This could be caused by the local plasticity at high temperature. At high temperature, alumina is able to exhibit limited plasticity. In this study, the fracture stress was obtained from the elastic solution. Thus, the obtained fracture stress was over estimated. Above 1373 K, the σ_B decreases markedly with the increase of the test temperature. However, it exhibited the same σ_B of the healed smooth specimen at 1573 K. Moreover, SEM observations confirmed that all of the machined specimens did not fracture from examples of the healed machining-cracks. Thus, it could be concluded that healed machining-cracks exhibited the same levels strength to the base material (Al_2O_3/SiC composite) up to 1573 K.

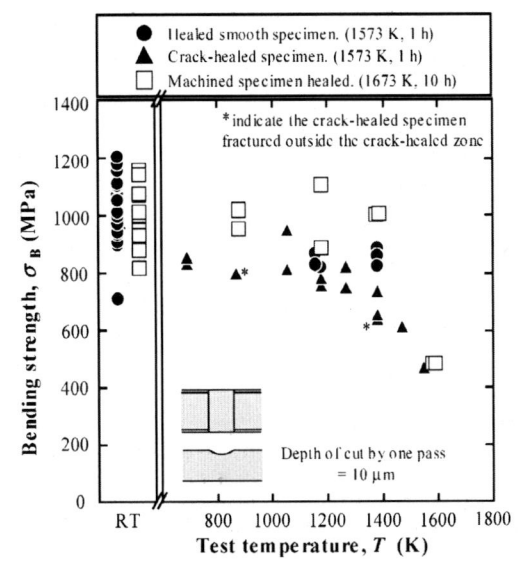

Figure 8 Temperature dependence of the local fracture stress

Strength recovery of various machined specimens by crack-healing.

Figure 9 shows crack-healing behavior of various machined specimens. The machined figurations were thorough-hole (type B) and square groove (Type C). The symbols ▼ and ▽ show the σ_{NF} of machined specimens healed at 1673 K for 1 h and 10 h, respectively. The symbols △ show the σ_{NF} of as-machined specimens. Also the effect on σ_{NF} of semicircular groove (type A), machined with the $d = 10$ µm, show in this Figure.

The average σ_{NF} of as-machined specimens type A, B and C were found to be 367 MPa, 458 MPa and 250 MPa, respectively. From comparison of σ_{NF} between type A and C, the σ_{NF} of the as-machined specimen (type C) was lower than that of semicircular groove. This results from the differences in the crack depth and stress concentration factor. On the other hand, the σ_{NF} of as-machined specimen (type B) was found to be the largest of all samples. The direction of the main cracks in the type B specimens were not perpendicular to the tensile stress applied by bending due to the relative orientation of the path of the diamond grains movement during machining of the through-hole.

The σ_{NF} of the healed machined specimens increases significantly, as healing-time increases. The σ_{NF} of the machined specimens healed at 1673 K for 10 h also show a maximum. The average σ_{NF} of type B and type C increase by 59 % and 173 % compared to those of as-machined specimens, respectively. From these results, it was found that the various cracks initiated by machining into various machined figurations, could be healed by heat-treatment at 1673 K for 10 h.

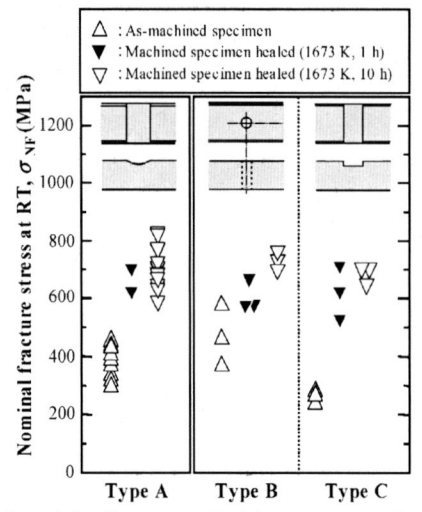

Figure 9 Effect of crack-healing on nominal fracture stress of various machined

CONCLUSION

In this study, alumina/SiC whisker composites were sintered and small bending specimens were made. These specimens were machined into various geometries to provide different stress concentration features to replicate those found in engineering components. The machined specimens were crack-healed under various conditions. The fracture stress of these specimens after crack-healing was evaluated. From the results obtained, the crack-healing effect was discussed. The main conclusions were as follows:

(1) The optimized crack-healing condition for the machining cracks was defined as 1673 K for 10 h. However, this condition is clearly different from that identified previously for indentation cracks.
(2) For depth of cut by one pass up to 15 μm, complete strength recovery was attained by crack-healing at 1673 K for 10 h.
(3) Healed machining-cracks exhibited the same strength as the base material up to 1573 K.
(4) All forms of crack, introduced by various machining operations, could be healed by exposure at 1673 K for 10 h.
(5) These results have demonstrated that crack-healing is a highly effective technique which may potentially reduce machining costs and improve the structural integrity of machined alumina.

REFERENCES

[1]W. Kanematsu, Y. Yamauchi, T. Ohji, S. Ito and K. Kubo, "Formulation for the effect of surface grinding on strength degradation of ceramics", *J. Ceram. Soc. Jpn.*, **100**, 775-779 (1992), (in Japanese).

[2]F. F. Lange and K. C. Radford, "Healing of surface cracks in polycrystalline Al_2O_3", *J. Am. Ceram. Soc.*, **53**, 420-1 (1970).

[3] C. F. Yen and R. L. Coble, "Spheroidization of tubular voids in Al_2O_3 crystals at high temperatures", *J. Am. Ceram. Soc.*, **55**, 507-509 (1972).

[4] J. Zhao, L. C. Stearns, M. P. harmer, H. M. Chan, G. A. Miller and R. F Cook, "Mechanical behavior of Al_2O_3-SiC "nanocomposite"", *J. Am. Ceram. Soc.*, **76**, 503-510 (1993).

[5] I. A. Chu, H. M. Chan and M. P. Harmer, "Machining-introduced surface residual stress behavior in Al_2O_3-SiC nanocomposite", *J. Am. Ceram. Soc.*, **79**, 2403-2409 (1996).

[6] A. M. Thompson, H. M. Chan and M. P. Harmer, "Crack healing and surface relaxation in Al_2O_3-SiC "Nanocomposite"", *J. Am. Ceram. Soc.*, **78**, 567-571 (1995).

[7] H. Z. WU, S. G. Roberts and B. Derby, "The strength of Al_2O_3/SiC nanocomposite after grinding and annealing", *Acta Mater.*, **46**, 3839-48 (1998).

[8]B. S. Kim, K. Ando, M. C. Chu and S. Saito, "Crack-healing behavior of monolithic alumina and strength of crack-healed member", *J. Soc. Mater. Sci. Jpn.*, **52**, 667-673 (2002), (in Japanese).

[9]K. Ando, B.S. Kim, M.C. Chu, S. Saito and K. Takahashi, "Crack-healing and mechanical behavior of Al_2O_3/SiC composites at elevated temperature", *Fatigu. Fract. Engin. Mater. Struct.*, **27**, 533-541 (2004).

[10]K. Takahashi, M. Yokouchi, S.K. Lee, K. Ando, "Crack-healing behavior of Al_2O_3 toughened by SiC whiskers", *J. Am. Ceram. Soc.*, **86**, 2143-2147 (2003).

[11]S.K. Lee, K. Takahashi, M. Yokouchi, H. Suenaga and K. Ando, "High temperature fatigue strength of crack-healed Al_2O_3 toughened by SiC whiskers," *J. Am. Ceram. Soc.*, **87**, 1259-1264 (2004).

[12]K. Ando, M. Yokouchi, S. K. Lee, K. Takahashi, W. Nakao and H. Suenaga, "Crack-healing behavior, high temperature strength and fracture toughness of alumina reinforced by SiC whiskers", *J. Soc. Mat. Sic., Jpn.*, **53**, 599-606 (2004), (in Japanese).

[13]Y. Matsuo, T. Ogasawara, S. Kimura, S. Sato, and E. Yasuda, "The effect of annealing on surface machining damage of alumina ceramics", *J. Ceram. Soc. Jpn.*, **99**, 384-389 (1991), (in Japanese).

[14]S.K. Lee, M. Ono, W. Nakao, K. Takahashi, and K. Ando, "Crack-healing behavior of mullite/SiC/Y_2O_3 composites and its application to the structural integrity of machined components", *J. Eur. Ceram. Soc.*, **25**, 3495-3502 (2005)

[15]M. Ono, W. Nakao, K. Takahashi, K. Ando, M. Nakatani. "A methodology to increase a strength and guarantee a reliability of a Al_2O_3/SiC composite ceramics component by crack-healing and proof testing" *J. high pressure inst. Jpn.*, **44**, 80-90, (2006), (in Japanese)

[16]Japan Industrial Standard R1601, Testing method for flexural srength of high performance ceramics, Japan Standard Association, Tokyo (1993).

[17]M.Nishida, Stress concentration. Morikita Publishing Co., Ltd., 572-574 (1967). (in Japanese).

MODELING CRACK BIFURCATION IN LAMINAR CERAMICS

K. HBAIEB[1], R.M. MCMEEKING* AND F.F. LANGE
Materials Department and *Department of Mechanical Engineering
University of California, Santa Barbara, California 93106

ABSTRACT
Crack bifurcation is observed in laminar ceramics. In such composites, alternating material layers have tensile and compressive residual stress, due to thermal expansion mismatch. The compressive stress ensures that crack growth leading to failure in the laminar system is mediated by threshold strength, but, in some cases, it also leads to bifurcation of the propagating flaw. The phenomenon of bifurcation takes place when the crack tip is propagating in the compressive layer, and occurs typically at a distance equal to a few laminate thicknesses below the free surface and beyond. The observation of this phenomenon is usually associated with the presence of edge cracking in the compressive layers of the laminar ceramic. The energy release rates for the straight and bifurcated cracks are calculated from the results of finite element computations and compared. When edge cracking is ignored, the crack is simulated as a through-thickness crack in an infinite body, and the energy release rate is used to predict crack deviation and bifurcation. Based on this, the finite element model successfully predicts bifurcation in only one material combination that was investigated in experiments. However, the experimental bifurcation takes place in two additional material combinations. When the effect of edge cracking is incorporated into the finite element simulations, the energy release rate calculations successfully predict the phenomenon of bifurcation in three material combinations, as observed in the experiments. Since no edge cracks are present in the fourth material combination tested experimentally, its lack of bifurcations is automatically predicted by the model.

INTRODUCTION
It has been demonstrated that laminar ceramics having residual stresses that are compressive and tensile in alternating layers exhibit threshold strength when subjected to bending[1]. The stresses in the compressive layers inhibit the otherwise unstable growth of cracks until the applied load exceeds a threshold. As a consequence, these systems possess the desirable characteristic of reliability under load (*i.e.* the absence of failure at low applied load) that is absent from ordinary ceramics. In some laminates[2-6] the cracks that cause failure propagate straight and the measured threshold strength is predicted well by theoretical and computational models for such cracks[1,2]. In other laminates, containing thick compressive layers and/or having high stress magnitudes in the compressive layers, crack bifurcation within these layers takes place[2-6] and the measured threshold strength lies above the level predicted for cracks that grow straight[1,2].

In this context, Rao *et al.*[5,6] investigated four composites. Each composite consisted of 5 alumina layers of thickness 550 μm having tensile residual stress, with 4 thin compressive layers in between them of thickness 55 μm and composed of mixtures of mullite and alumina. The residual stresses are caused by thermal expansion differences between the alumina and the alumina/mullite mixture. These composites are designated MXX according to the amount of mullite present where XX is the weight percentage of it in the compressive layer. In the experiments, pre-cracks approximately 300 μm in length were created in the middle of the central tensile layer by a Vickers indenter with a load of 5 kg. Specimens were then subjected to

Modeling Crack Bifurcation in Laminar Ceramics

4-point flexural bending, causing the cracks to propagate unstably and without bifurcation through the entire central tensile layer, and through about 10% of the two neighboring compressive layers, where the tips arrested. The load was then increased gradually until propagation of the cracks re-commenced, and the flaws then grew stably through the compressive layers until the tip (or tips) reached the next tensile layer or near to it. At that stage, unstable propagation of the cracks set in, and the specimens failed.

Three of the composites, M40, M55 and M70, *i.e.* those with higher residual stress magnitudes, exhibit crack bifurcation during their growth under load, as shown in Fig. 1. The fourth, M25, is absent such crack bifurcation. In those laminates exhibiting crack bifurcation during loading, there are also shallow edge cracks at the surfaces of the compressive layers that are produced spontaneously upon cooling after processing and firing of the composite. During cooling, the spontaneous edge cracks propagate down the mid-planes of the compressive layers to arrest in the interior, as depicted schematically in Fig. 2. In these specimens, crack bifurcation during loading does not take place at the free surface of the component, but instead occurs some distance below, close to the depth corresponding to the tip of the spontaneous edge crack. As can be seen in Fig. 1, the angle that each branch makes with the prior crack path is approximately 60° (actually 57° to 67°). The experiments have also confirmed that the propagation of the crack growing under load is stable and not dynamic, so that a monotonically increasing load is required to force the crack progressively through the compressive layer. In the composite designated M25, neither bifurcation of cracks growing under load nor spontaneous edge cracking takes place, and the crack growing under load is also stable. Thus experimental observation in the laminates composed of the materials designated M25, M40, M55 & M70 suggests that crack bifurcation under load occurs in a stable manner under the same conditions that favor the spontaneous creation of edge cracks during cooling, whereas the conditions that do not lead to spontaneous edge cracking during cooling also do not favor crack branching under load.

A purpose of the current paper is to investigate the reason why the cracks growing under load in some of the specimens tested by Rao *et al.*[5,6] (*i.e.* M40, M55 & M70) bifurcated, and why others (*i.e.* M25) did not. For this purpose, we carry out finite element simulations of cracks in laminar ceramics to understand their behavior, and to see if the maximal energy release rate hypothesis for crack deviation and bifurcation is consistent with the observed behavior. We present results that compare the energy release rates for bifurcated and straight cracks in an attempt to establish the conditions that cause crack branching. We first consider a crack propagating in a plane strain laminar ceramic strip subjected to the stress that arises at the tensile surface of a 4-pt flexural specimen. In the absence of a satisfactory explanation of crack bifurcation and branching based on the energy release rate for cracks in plane strain laminar strips, we then consider the effect of stresses arising in a finite thickness laminar ceramic body due to the presence of specimen free surfaces, and the spontaneous formation of edge cracks in the compressive layers. We find that when these stresses are taken into account in an approximate, 2-dimensional model, the energy release rate at the tip of the bifurcating crack growing under load can be correlated in a suggestive manner with the tendency for such crack branching.

Modeling Crack Bifurcation in Laminar Ceramics

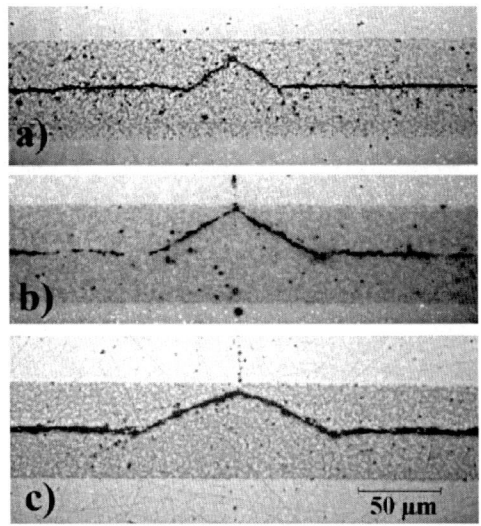

Fig. 1: Optical micrographs of crack branching for the materials: a) M40, b) M55 and c) M70. The crack propagating under load grows vertically downward and is arrested just inside the compressive layer, which has a slightly darker shade than the tensile layers. Under rising load, the crack then bifurcates and the branches grow stably as the load is increased. The branching does not occur at the free surface but some distance below it and has been revealed by removal of material. The horizontal dark line is the spontaneous edge crack, which grows down into the material as the surface is removed by grinding, causing some of the branched crack to open up so that it becomes visible in the micrograph.

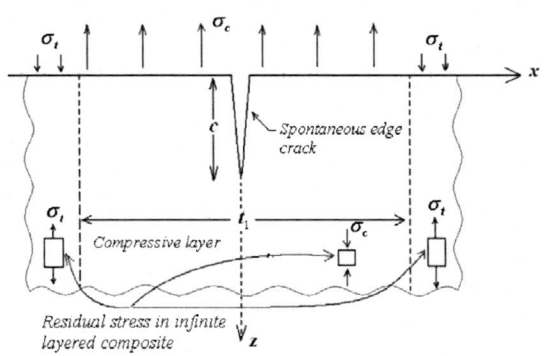

Fig. 2: A spontaneous edge crack along the mid-plane of a compressive layer of a laminar ceramic extending from the surface to the interior of the layer.

MODEL DESCRIPTION

Figure 3 depicts a geometry used in the analysis that contains a straight crack without branching or bifurcation. Due to symmetry, only one quarter of the whole specimen is modeled, specifically the segment illustrated in Fig. 3. The model is used to simulate the experiments of Rao et al.[5-6] carried out for the alumina-alumina/mullite composites, and, therefore, the layer dimensions and properties are the same as in the experiment. As in the experiment, the quadrant analyzed contains two compressive layers with thickness $t_1 = 55$ µm and two and a half tensile layers of thickness $t_2 = 550$ µm. The height of the model, h, is 5000 µm, and is therefore many times greater than the thickness of the tensile layers. The crack has its tip in the first compressive layer adjacent to the central tensile one (*i.e.* the leftmost layer in Fig. 3). The bending of the specimen causes tensile loading parallel to the vertical direction in Fig. 3, and the problem is analyzed as if it were in plane strain with tension parallel to the layers. The Young's modulus, E, and the coefficient of thermal expansion (CTE), α, of the tensile alumina and compressive alumina/mullite layer materials are presented in Table I, with the different values given for the compressive layers depending on their mullite content. In the simulations, we ignore any difference or variation in Poisson's ratio, ν, and set its value equal to 0.24 for the entire model throughout all analyses. Displacement boundary constraints are applied to the segment analyzed and shown in Fig. 3 so that, other than on the crack surface, no displacement occurs normal to the left and bottom edges. All edges of the model are free of shear traction and the right hand edge is free of normal traction as well. In addition, the crack surfaces are free of all traction.

Table I: Material properties of alumina and alumina/mullite layers.

	Young's modulus, E (GPa)	Coefficient of Thermal Expansion, α (10^{-6} C^{-1})
Alumina	401[13]	8.3[14]
Mullite/Alumina (M25)	347	7.75
Mullite/Alumina (M40)	319	7.37
Mullite/Alumina (M55)	293	6.94
Mullite/Alumina (M70)	268	6.46
Mullite	220[15]	5.3[16]

Thermal residual stresses are induced in the model by simulation of the cooling of the specimen to 1200°C below the stress-free temperature. The uniform in-plane stress near the crack plane in each layer that arises in the absence of the crack is given in Table II. After the thermal residual stresses have been induced in the model, a uniform traction is applied along the top edge to simulate the applied tension.

Table II: Residual stresses in alumina and alumina/mullite layers

	Alumina σ_r^2 (MPa)	Alumina/mullite σ_r^1 (MPa)
Mullite/Alumina (M25)	22.5	-282
Mullite/Alumina (M40)	35.2	-440.4
Mullite/Alumina (M55)	47.5	-594.4
Mullite/Alumina (M70)	59	-739

The models analyzed contain either a straight crack as depicted in Fig. 3 or a branched crack. Both geometries are depicted in detail of the crack tip region shown in Fig. 4. The crack is considered to have first arrested with its tip in the compressive layer so that its half length is a, slightly greater than the half width of the central tensile layer. Thereafter, under stable growth, the crack either extends straight so that its half length extends to $a+b$, or it bifurcates symmetrically in such a way that two branches of length b and included angle 2θ extend from the straight crack, as shown in Fig. 4.

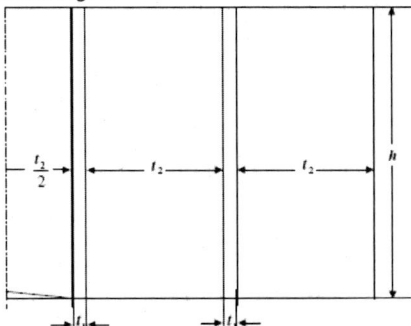

Fig. 3: Layered ceramic body analyzed in the plane strain calculations (not drawn to scale). Due to symmetry, only a quarter of a specimen is modeled. The crack is located along the bottom edge from the bottom left corner and its tip is in the compressive layer.

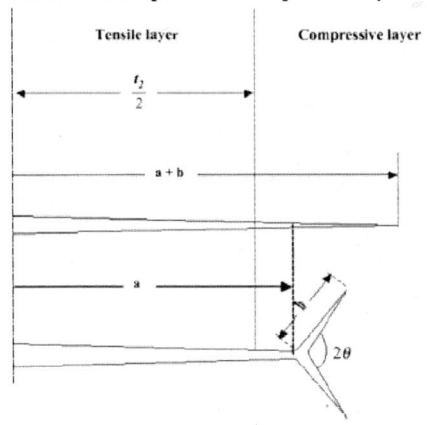

Figure 4: Detail of the model geometry at the crack tip.

Computations are carried out using the finite element code ABAQUS. Various checks were carried out to ensure accuracy of the results, both in terms of the fineness of the finite element model and the reliability of the results for the crack tip energy release rate, computed by the domain integral method in ABAQUS. The fineness of the finite elements is realized by making the finite elements very small. This exercise included comparison of the numerical results with those for a branched crack in an infinite body of homogenous material calculated by Vitek[8] using a dislocation model, and a check against results for a kinked crack provided by

Cotterell and Rice[9]. It was concluded from these studies that models and methods of sufficient accuracy are being employed for the computations described in this paper.

RESULTS FOR CRACKS SUBJECT TO TENSION AND THERMAL STRESS

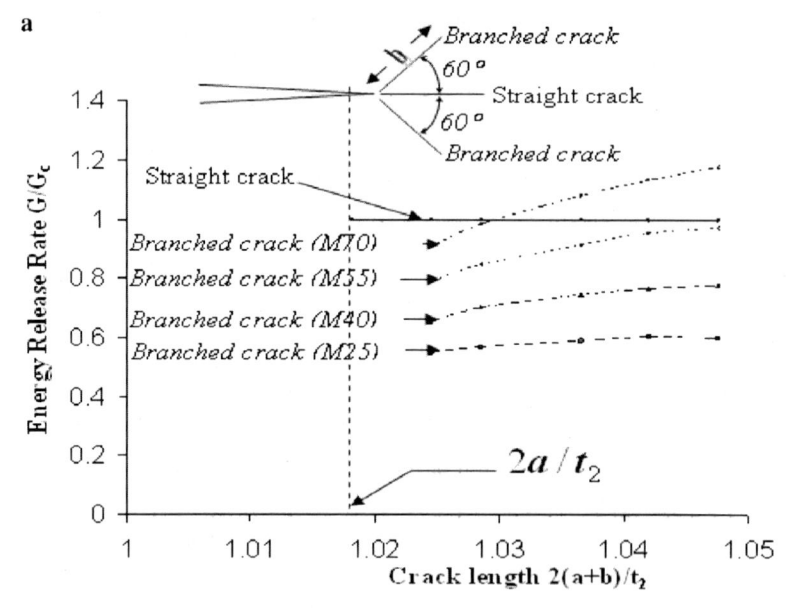

Figure 5a shows results for the energy release rate for straight and bifurcated cracks where $\theta = 60°$ for the crack branches. These results are obtained with a thermal stress present along with an applied tensile stress representing the effect of bending. No other external loads are present. The applied stress for the straight crack is chosen for each crack half length $a+b$ and for each compressive layer (M25, M40, M55 or M70) so that the energy release rate, G, at the crack tip is exactly equal to the toughness, G_c. This is achieved by adjusting the applied stress in each calculation for the straight crack until the Mode I crack tip stress intensity factor due to the combination of applied load and thermal stress is exactly equal to $2\,MPa\sqrt{m}$, which is assumed to be the plane strain fracture toughness $K_c = \sqrt{G_c E'}$ for the compressive layer no matter its composition. Note that $E' = E/(1-\nu^2)$. The same stress as used for a straight crack of half length $a+b$ to cause the crack tip energy release rate to equal the toughness is also applied to the bifurcated crack having the same value of $a+b$ (see Fig. 4). All branched cracks are assumed to emanate from a straight crack having the same half-length a.

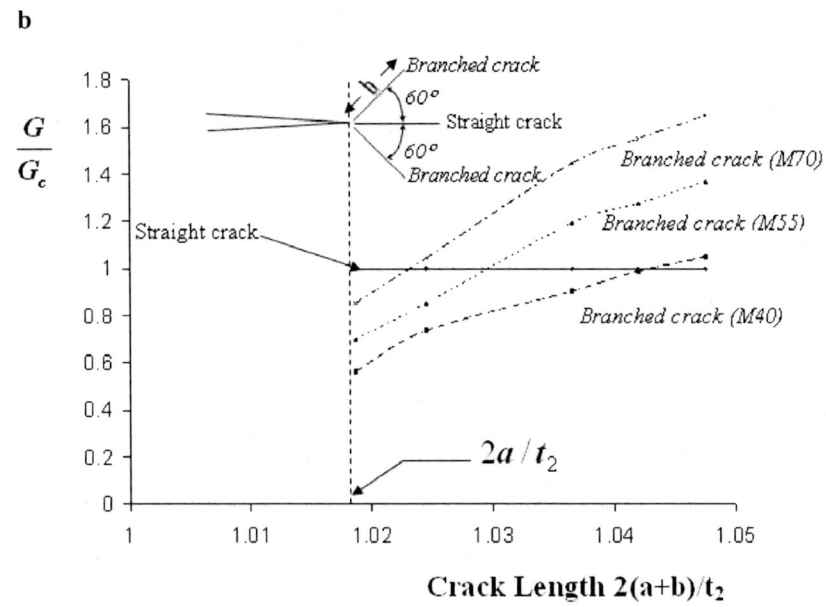

Modeling Crack Bifurcation in Laminar Ceramics

b

Figure 5: (a) The energy release rate versus crack length for straight and bifurcated cracks in composites M25, M40, M55 & M70 subject to thermal stress and applied load simulating the effect of bending. The branched crack having branches of length b extending from a crack of half length a is subjected to the same applied stress as a straight crack of half length $a+b$. The common value of a for the branched cracks is shown by the dashed vertical line. (b) Energy release rate results for straight and bifurcated cracks in the composite systems under consideration, M40, M55 and M70, where the stresses imposed on the cracks are due to applied loads, thermal stress and the effect of a free surface and spontaneous edge cracks.

Figure 5a shows the energy release rate results for different values of $a+b$. The results for all straight cracks are represented by the single straight line at $G/G_c=1$. As expected, the energy release rate at the tips of bifurcated cracks is smaller then that for a straight crack in all cases when the length of the branch, b, is very small. This feature is consistent with the results for branched cracks in homogeneous materials free of thermal stress, where the energy release rate for the branched cracks is always smaller than that for the straight crack[8]. However, it can be seen in Fig. 5a that the energy release rate at the tips of bifurcated cracks in heterogeneous composites, subject to applied loading and thermal stress, increases with the length of the branch relative to the value of the energy release rate for the straight crack. Indeed, for branches such that b/t_2 is greater than about 0.01, the energy release rate at the tips of the branched cracks in the M70 composite is larger than the energy release rate for a straight crack with the same value of $a+b$ and the same value of applied stress. It can be deduced from Fig. 5a that bifurcated cracks in a heterogeneous composite having thermal stress require a larger applied stress to cause the flaw to propagate compared to the stress required to propagate a straight crack in almost all cases studied. A lower stress for propagation of the bifurcated crack only occurs for long branches, as can be seen in Fig. 5a for the M70 composite.

However, the preceding paragraph appeals to a rather simplistic set of models regarding the deviation of cracks propagating in a brittle manner. Instead, we can consider a richer possibility discussed by Cotterell and Rice[9] in the context of homogeneous materials, and also used by He and Hutchinson[10] to explain and model the deviation and non-deviation of brittle cracks approaching an interface between dissimilar materials. In this model, there are a large number of microflaws in the brittle material, and they have a variety of sizes and orientations, or, as the crack grows, a number of branches are thrown out by the crack tip in random orientations and, within limits, to random extents. In either case, we can consider a straight crack with a number of incipient branches along its front with a variety of orientations and lengths. Given the results in Fig. 5a, the longer branches will have a larger energy release rate than segments of the crack that are still straight. Therefore, such long branches will extend rapidly, and propagation of the straight sectors will die out, with the branches running down the length of the crack front so that the whole flaw deviates and, due to symmetry, bifurcates. As noted above, a version of this model was presented by Cotterell and Rice[9], who pointed out that when the T-stress is positive for a straight Mode I crack, the lengthening of a deviation will give rise to an increasing energy release rate, thereby causing its unstable growth at the expense of its straight course. In this view, once a deviation becomes established and if it is long enough, the positive T-stress for the original straight crack keeps the branch growing, and in symmetric loadings such as here, bifurcation would take place. A model accounting for above in combination with simulation of the effect of the edge crack will be presented shortly.

EFFECT OF THE SPONTANEOUS EDGE CRACK

As mentioned in the introduction, in composites M40, M55 & M70, the growing crack propagates towards an edge crack in the compressive layer that, driven by the thermal residual stresses, has grown in spontaneously from the free surface. The situation is depicted in Fig. 6, which is a sketch of the free surface on the tensile side of the bending specimen showing both the growing crack and the spontaneous edge crack. The spontaneous edge crack appears first in specimens M40, M55 & M70 upon cooling after processing. After it has spontaneously appeared and grown due to the build up of residual stress during cooling, the edge flaw stops propagating with its crack tip energy release rate exactly equal to the toughness, G_c, so that there are high stresses remaining around its tip. The 3-dimensional state of stress around the tip will affect the growing crack in two different ways. The tensile stress parallel to the y axis (see Fig. 6), present due to plane strain constraint near the tip of the spontaneous edge crack, will assist the propagation of the growing crack in a direct way. However, of more importance to us, the high tensile stress parallel to the x axis associated with the Mode I stress field around the tip of the spontaneous edge crack will augment the T-stress experienced by a growing crack propagating straight. Thus, by the model of Cotterell and Rice[9], the presence of the spontaneous edge crack can destabilize straight propagation of the growing crack, leading it to deviate and, by symmetry, to bifurcate. It is notable that this effect would be strongest a distance c below the free surface, where c is the length of the spontaneous edge crack. This seems to be in accord with the observation of Rao et al.[5-6] that the growing crack does not deviate or bifurcate at the free surface, but instead does so some distance below the free surface, arguably adjacent to the tip of the spontaneous edge crack.

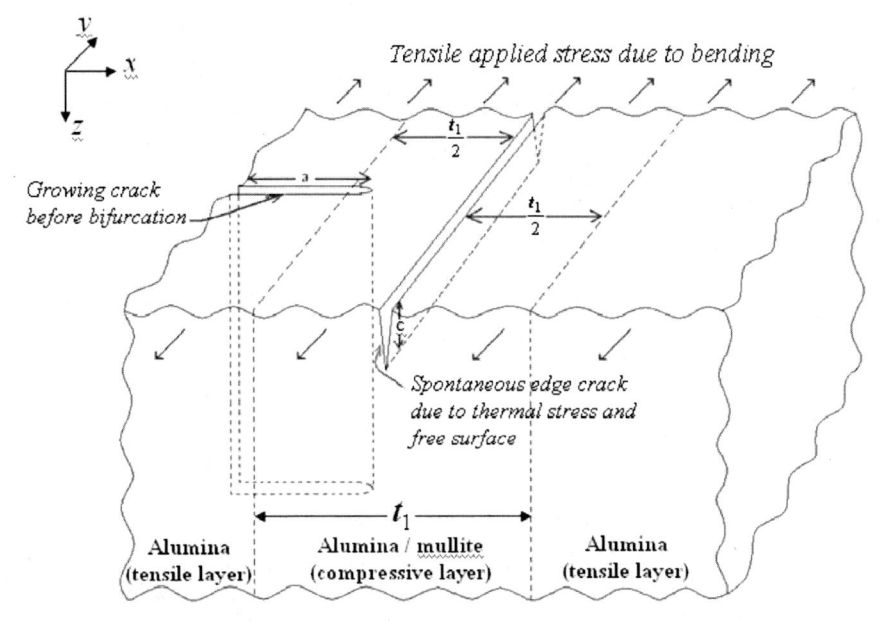

Figure 6: A three dimensional sketch showing both the growing crack and the spontaneous edge crack. Spontaneous edge cracks appear first when specimens M40, M55 & M70 are cooled after processing. They occur where the compressive layers have a free surface. The growing crack propagates from left to right until it just penetrates into the compressive layer.

SIMULATION OF THE SPONTANEOUS EDGE CRACK

The energy release rate G per unit crack area for a very shallow surface flaw extending by propagation along the center line of the compressive layer at the free surface is given by Ho et al.[10] and Lange et al.[11]

$$G = \frac{0.34 \sigma_c^2 t_1 (1-\nu^2)}{E_1} \qquad (1)$$

where E_1 is Young's modulus for the material in the compressive layer. Equation (1) shows that G depends not only on the magnitude of the compressive residual stress and the elastic properties of the compressive layer, but also on the layer thickness. To estimate the length of the spontaneous edge crack, and the stresses adjacent to it, a separate finite element calculation is carried out. The calculations of the edge crack length, for which the crack tip energy release rate equals the material toughness, yield the following values: 17μm for M40, 29μm for M55 and 46μm for the M70 material, all relative to a compressive layer thickness of 55 μm. Note that edge crack growth in the M25 composite is not possible because the critical condition for crack propagation is not met. That is, the value of the energy release rate given by Eq. (1) falls below the toughness of the material, and thus an edge crack cannot nucleate. Once the correct crack length is established in each composite, the stresses adjacent to the edge crack are extracted from the model. The stresses needed are the tensile components parallel to the x axis, i.e. σ_{xx}, and parallel to the y axis in Fig. 6, i.e. σ_{yy}, along a horizontal line passing just below the crack tip. Consequently, in the computations to be described below, the growing crack will be subjected to 3 stress fields, such that the effects of both the free surface of the specimen and the presence of the edge crack, as well as the applied load and the thermal stress are taken care of.

DESCRIPTION OF THE MODEL FOR SIMULATING THE EFFECT OF THE SPONTANEOUS EDGE CRACK

The computations described in 'MODEL DESCRIPTION' for which results have been given in Section 3 are now repeated but without remote applied load or thermal stress. Instead tractions are applied to the crack surface representing the effect of the applied loads, the thermal stress, the free surface and the spontaneous edge crack; the details of the calculations are given elsewhere[12]. The model depicted in Fig. 3 is used once more with elastic properties of the layers as given in Table I, and symmetry boundary conditions are applied to the leftmost and lower edges (except for the crack surface), as in the calculations described in 'MODEL DESCRIPTION'.

RESULTS OF SIMULATIONS FOR CRACKS SUBJECT TO APPLIED LOAD, THERMAL STRESS AND THE STRESS DUE TO THE PRESENCE OF A FREE SURFACE AND A SPONTANEOUS EDGE CRACK

Figure 5b shows the results for the crack tip energy release rate for the composite systems M40, M55 and M70 for both straight cracks and those that have bifurcated at an angle of $\theta = 60°$. It has been numbered Fig. 5b to expedite comparison with the equivalent results in Fig. 5a for the crack in the absence of edge flaws. Similar to the case of the plane strain crack subject only to applied load and thermal stress, the uniform applied stress is determined to be that required so that the crack tip energy release rate G for the straight crack of half length $a+b$ is equal to the compressive layer material toughness, G_c, and then the same uniform applied load is used for the branched crack having the same value for its effective length $a+b$. The results for the bifurcated crack indicate that at the same applied stress, small branches have a lower crack tip energy release rate than the straight crack with the same value of $a+b$, whereas longer branches have a larger crack tip energy release rate. In addition, the crack tip energy release rate for the bifurcated crack increases rapidly with the length of the branches, and exceeds the energy release rate for the straight crack when the branches are still relatively short. Fig. 5b also shows that in the case of M70, quite short branches are all that are needed to cause the energy release rate for the bifurcated case to exceed that of the straight flaw, whereas progressively longer branches are needed in the case of M55 and M40 respectively.

In summary, the results for the M40, M55 & M70 composites that were obtained with crack surface tractions designed to represent, in an approximate way, the total sum of the effects of the applied load, the thermal stress, the nearby free surface and the spontaneous edge crack do offer a rationale for crack deviation and bifurcation in these materials due to the presence of the spontaneous edge crack and its absence in the M25 composite. In the cases of the M40, M55 & M70 composites, when there is a free surface and a spontaneous edge crack present, the crack tip energy release rates for bifurcated cracks with relatively short branches exceed that for a crack that has continued to grow straight. Figure 5b indicates that this situation arises when the branch length b exceeds 13%, 4% and 2% of the compressive layer thickness respectively in the M40, M55 and M70 composite. Since the compressive layer thickness is 55 μm, this means that if b exceeds 7.4 μm, 2.1 μm and 1.3 μm respectively for the M40, M55 & M70 composite in the presence of a free surface and a spontaneous edge crack, the crack tip energy release rate for the bifurcated cracks will exceed that for the crack that has grown straight. Therefore, if we postulate that the compressive layers contain microflaws of random orientation and size up to a length of 7.5 µm, then by the model described here, we would conclude that the growing cracks in the M40, M55 & M70 composite would deviate and, due to symmetry, bifurcate when there is a free surface and a spontaneous edge crack present. Since the M25 composite does not in practice have spontaneous edge cracks, this material will not have deviation or bifurcation of the growing crack.

CONCLUSIONS

A phenomenon of crack bifurcation has been observed to take place in some laminar ceramics with alternating material layers having tensile and compressive residual stress. The phenomenon of bifurcation is experienced by cracks that propagate in specimens subjected to bending, and takes place when the crack tip is propagating in the compressive layer. Finite element calculations are carried out to explore the cause of these bifurcations. The calculations suggest that the growing crack only initiates a deviation of its path, and a bifurcation, under the influence of a spontaneous edge crack present in the compressive layers of the composite. These spontaneous edge cracks develop because of thermal stress and exist only in the M40, M55 and M70 composites, but are absent in the M25 material. Consequently, path deviation and bifurcation of the growing crack are absent in the M25 composite.

REFERENCES
[1] M.P. Rao, A.J. Sanchez-Herencia, G.E. Beltz, R.M. McMeeking and F.F. Lange, Laminar Ceramics that Exhibit a Threshold Strength, Science 286, 102-105 (1999).
[2] Oechsner, M., Hillman, C., Lange, F.F., Crack Bifurcation in Laminar Ceramic Composites. J. Am. Ceram. Soc. 79, 1834-1838 (1996).
[3] A.J. Sanchez-Herencia, C. Pascual, J. He and F.F. Lange, ZrO_2/ZrO_2 Layered Composites for Crack Bifurcation. J. Am. Ceram. Soc. 82, 1512-1518 (1999).
[4] A.J. Sanchez-Herencia, L. James, F.F. Lange, Bifurcation in Alumina Plates Produced by a Phase Transformation in Central, Alumina/Zirconia Thin Layers. J. Euro. Ceram. Soc. 20, 1297-1300 (2000).
[5] M.P. Rao. Laminar Ceramics that Exhibit a Threshold Strength. Ph.D. Dissertation, University of California, Santa Barbara (2001).
[6] M.P. Rao, F.F. Lange, Factors Affecting Threshold Strength in Laminar Ceramics Containing Thin Compressive Layers. J. Am. Ceram. Soc. 85, 1222-1228 (2002).
[7] K. Hbaieb, R.M. McMeeking. Threshold Strength Predictions for Laminar Ceramics with Cracks that Grow Straight. Mech. Mater. 34, 755-772 (2002).
[8] V. Vitek, Plane Strain Stress Intensity Factor for Branched Cracks. Int. J. Frac. 13, 481-501 (1977).
[9] B. Cotterell, J. R. Rice, Slightly Curved or Kinked Cracks. Int. J. Frac. 16, 155-169 (1980).
M. He, J.W. Hutchinson, Crack Deflection at the Interface between Dissimilar Elastic Materials. International J. Sol. Struc. 25, 1053–1067 (1989).
[10] S. Ho, C. Hillman, F.F. Lange, Z. Suo, Surface Cracking in Layers Under Biaxial, Residual Compressive Stress. J. Am. Ceram. Soc. 78, 2353-2359 (1995).
[11] F. F. Lange, M.P. Rao, K. Hbaieb, R.M. McMeeking. Ceramics that Exhibit a Threshold Strength. In: Proceeding of the PAC Rim IV Ceramic Armor Materials by Design Symposium, Maui, Hawaii, Ceramic Transactions of the American Ceramic Society, Columbus, OH (2001).

[12]K. Hbaieb, R.M. Meeking and F. F. Lange, Crack Bifurcation in Laminar Ceramic Having Large Compressive Stress. Int. J. Sol. Struc. (2007).
[13]G.A. Tracy and G.D. Quinn, Fracture toughness by the surface crack in flexure (SCF) method. Ceram. Eng. and Sci. Proceedings 15, 837-845 (1994).
[14]P. Aldebert and J.P. Traverse, α-Al_2O_3: A high temperature thermal expansion standard. High Temperature-High Pressure 16, 127-135 (1984).
[15]J.F. Bartolome and M. Diaz, J. Requena, J.S. Moya and A.P. Tomsia. Mullite/Molybdenum ceramic-metal composites. Acta Mater. 47, 3891-3899 (1999).
[15]K.N. Lee and R.A. Miller and N.S. Jacabson. New generation of plasma-sprayed mullite coatings on silicon carbide. J. Am. Ceram. Soc. 78, 705-710 (1995).

DELAYED FAILURE OF SILICON CARBIDE FIBRES IN STATIC FATIGUE AT INTERMEDIATE TEMPERATURES (500-800°C) IN AIR

W. Gauthier and J. Lamon
University of Bordeaux1
Laboratoire des Composites Thermostructuraux
3 Allée de la Boétie
33600 Pessac France

ABSTRACT

It has been established by a number of investigators that slow crack growth in ceramic materials leads to time dependence of strength. In the present paper, the lifetime of SiC Hi-Nicalon multifilament tows and single filaments is investigated in static fatigue at 500°C and 800°C in air. Experimental data show that lifetime of tows obeys the classical power law $t\sigma^n=A$ whereas that of single filaments depends on filament strength rank. A slow crack growth based model is proposed for lifetime predictions for single SiC filaments. The model involves the Paris law, statistical distributions of filaments strengths (Weibull model) and lifetimes at a given applied stress. It allowed SPT (Strength-Probability-Time) diagrams to be established for single filaments.

INTRODUCTION

Ceramic matrix composites (CMCs) are very attractive for high-temperature structural applications. Nowadays, CMCs are essentially used in space and defense applications. The control and prediction of CMC components lifetime is a crucial issue for future applications in civil aircrafts engines.
Unexpected failures of SiC/SiC composites and SiC Nicalon multifilament tows have been observed under low stresses at intermediate temperatures (500-800°C) [1,2]. Failure of woven SiC/SiC composites is controlled by the tows [3]. Recent studies have shown the sensitivity of SiC Nicalon tows to subcritical crack growth [2]. This paper focuses on the lifetime of Hi-Nicalon SiC fibres in static fatigue, at intermediate temperatures in air. Lifetime of multifilament tows and single filaments is analysed, first using an empirical model. Then, a subcritical crack growth model is proposed for single filaments.

Much research work has demonstrated the sensitivity of refractory materials to slow (subcritical) crack growth [4-5]. The word 'subcritical' means that the stress intensity factor at crack tip, K_I, is lower than the critical value, K_{IC}, which characterizes the effective fracture toughness of the material. As a result of slow crack growth, material strength is time dependent. Crack length, a, is related to stress intensity factor, K_I, by the following equation:

$$K_I = \sigma Y \sqrt{a} \qquad (1)$$

where σ is the remote applied stress and $Y = 2/\pi^{1/2}$ for a penny-shaped crack.
More generally in ceramics, crack velocity, v, is related to the stress intensity factor, K_I, by the empirical Paris law:

$$v = \frac{da}{dt} = A_1 K_I^n \qquad (2)$$

where t is time. Constants A_1 and n depend on material and environment.

EXPERIMENTAL PROCEDURE

Silicon carbide Hi-Nicalon fibres (Nippon Carbon, Tokyo, Japan) were tested. As-received multifilament tows containing 500 filaments, and single filaments of 15 µm average diameter were subjected to static fatigue at high temperature. Two specific testing devices were used for the tows (Fig.1-a) and the single filaments (Fig.1-b). The gauge length (25 mm) was located in the furnace hot-zone at a uniform temperature. Silica tubes were used to protect samples against possible pollution from furnace elements. They allowed environment control through a constant gas flow (N_2 / O_2). The test specimens were heated up to the test temperature before loading (heating rate ~ 20°C/min). Then, a dead-weight-load was hung slowly using the lifting system (this operation took < 10 s). Lifetime was captured automatically in the computer when specimen failed.

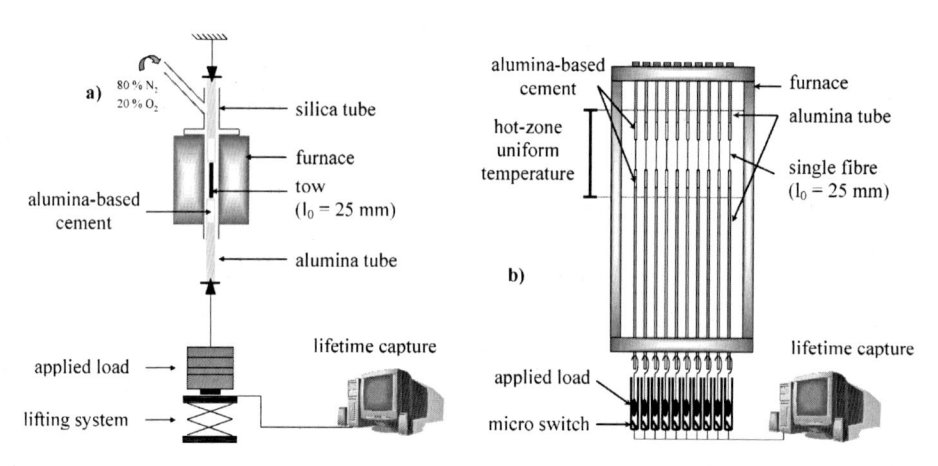

Figure 1. Static fatigue testing systems at high temperature for multifilament tows (a) and single fibre (b).

Sample ends were affixed within alumina tubes using an alumina-based cement. Special care was taken during test specimen preparation. A specific device for specimen preparation was used to ensure good alignment of tows and of filaments within the tows. For this purpose, a low load was applied on tows.

The applied stress was derived from the load, using the following equation which takes into account the individual fibre failures resulting from application of the load:

$$\sigma = \frac{c g \rho l N_0}{m [N_0 - N(c)]} \quad (3)$$

where c is the applied load, g is gravity constant, ρ is tow density, l is tow length and m is tow mass. N_0 is the initial number of unbroken fibres and $N(c)$ the number of broken fibres under the applied load c.

For single filament stress determination, each filament diameter was measured by laser diffraction. The Fraunhofer model was used to derive diameter from the diffrction spectrum:

$$d = \frac{2\lambda D}{i} \qquad (4)$$

where λ is the laser wave length, D is the distance between the fibre and the screen and i the distance between the two first minima of the diffraction spectrum.

RESULTS AND DISCUSSION

Static fatigue lifetime of Hi-Nicalon tows and single fibres
The static fatigue behavior of tows and single filaments at 500°C et 800°C is illustrated by the stress-lifetime data plotted on figure 2. Each data point corresponds to one test. For the tows, at least three tests were performed for a given applied stress. The amount of tests was much larger for the single filaments (>30 for each set of specimens).

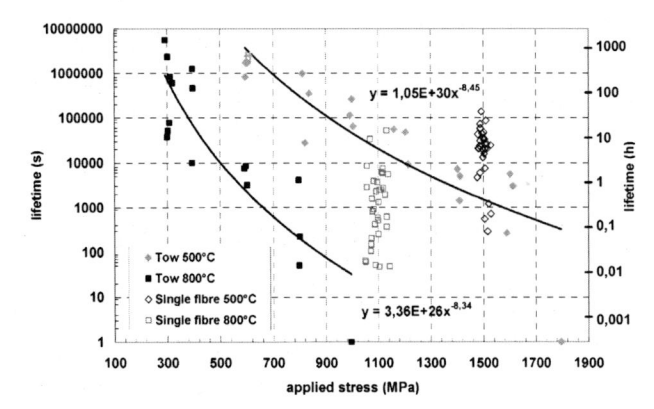

Figure 2. Lifetime of Hi-Nicalon tows and single filaments, in static fatigue at 500°C and 800°C.

Lifetime of tows decreases when temperature and applied stress increase. The lifetime-stress dependence is described by the following power-type law:

$$t\,\sigma^n = A \qquad (5)$$

where A and n are constants depending on material and environment.
This power-type law is often applied to nonlinear time-dependent phenomena such as slow crack growth induced by environment in ceramics. The following A and n values were estimated by regression from figure 2 data:

- 500°C: $n = 8,45$ and $A = 1,05.10^{30}$

- 800°C: $n = 8,34$ and $A = 3,36.10^{26}$

The stress exponent n seems to be temperature independent. It classically depends on material only. Constant A depends on environment (temperature). The stress exponent n was estimated at 8,4 ± 0,1 for the Hi-Nicalon tows. The fatigue behaviour of Hi-Nicalon tows is similar to that obtained previously on Nicalon ones [2] but the current n and A values are greater than those estimated for Nicalon tows ($n \approx 2,6$ and $A \approx 1.10^{12}$) [2].

A large amount of static fatigue tests were carried out on single filaments for each applied stress. Lifetimes exhibit an significant scatter (Fig. 2). Lifetime data were treated using the Weibull model.

Lifetime data obtained for a given applied stress were ordered from smallest to largest and probabilities $P_{(i)}$ were then assigned using the following estimator:

$$P_{(i)} = (i-0,5)/N_t \tag{6}$$

where i is the rank and N_t the total number of data.

Failure probability in static fatigue is given by the following equation:

$$P_\sigma(t) = 1 - \exp\left[-\frac{V}{V_0}\left(\frac{t}{t_0}\right)^{m_t}\right] \tag{7}$$

where V_0 is the reference volume ($V_0 = 1$ mm^3) and m_t and t_0 are the statistical parameters.

These parameters were estimated using linear regression:
- 500°C : $m_t = 0,64$ et $t_0 = 3,78.10^5$ s
- 800°C : $m_t = 0,98$ et $t_0 = 8,94.10^5$ s

$P_\sigma(t)$ represents the rank of the considered fibre in the lifetime distribution. The rank of a single fibre in the statistical distribution of strength data under conditions of catastrophic failure (i.e. in the absence of delayed failure) is referred to as α [6]. Assuming that these ranks are identical, it comes from equation (7):

$$P_\sigma(t) = 1 - \exp\left[-\frac{V}{V_0}\left(\frac{t}{t_0}\right)^{m_t}\right] = \alpha \tag{8}$$

The coincidence of both ranks implies that crack velocity is the same in all the fibres, since the critical flaw is the same under conditions of catastrophic failure and delayed failure. The above hypothesis is acceptable since all the filaments were subjected to identical test conditions (temperature, pressure, environment and applied stress). Assuming that the fatigue mechanism observed on tows is relevant to single filaments, equations (5) and (8) allow the determination of constant A (denoted $A(\alpha)$), relative to the considered single filament:

$$A(\alpha) = \sigma^n\, t_0 \left[\frac{V_0}{V}\ln\left(\frac{1}{1-\alpha}\right)\right]^{1/m_t} \tag{9}$$

Using relation (8), it is now possible to plot the lifetime behaviour of a single filament with respect to its rank (or failure probability) in the statistical distribution of strengths. Results are shown on figures 3 and 4 for Hi-Nicalon single filaments at 500°C and 800°C respectively.

Delayed Failure of Silicon Carbide Fibres in Static Fatigue at Intermediate Temperatures

These results are in good agreement with, first, the fatigue behaviour of tows, and second, previous fatigue results obtained on Nicalon tows [2]. Indeed, tows fail after a critical number of filaments have broken individually, and it appears that lifetime of tows corresponds to particular values of α = 5 % at 500°C and α = 4 % at 800°C. Then, the analysis of acoustic emission signals has shown that this value of α was about 1% to 8% at 600°C for Nicalon tows [2].

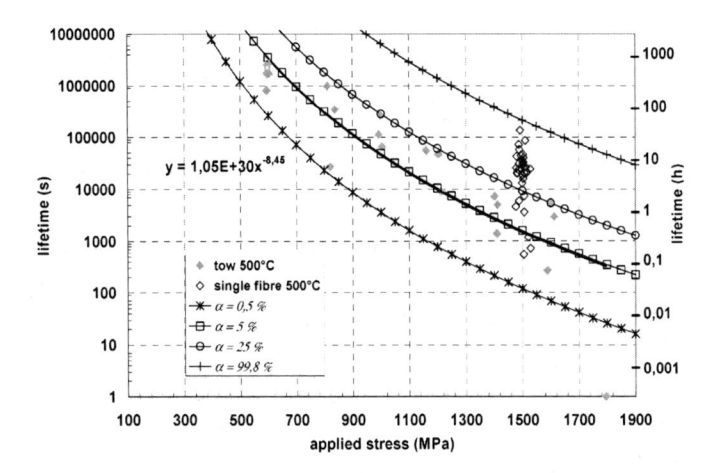

Figure 3. Lifetime of Hi-Nicalon single filaments at 500°C with respect to filament rank α, determined using the empirical model

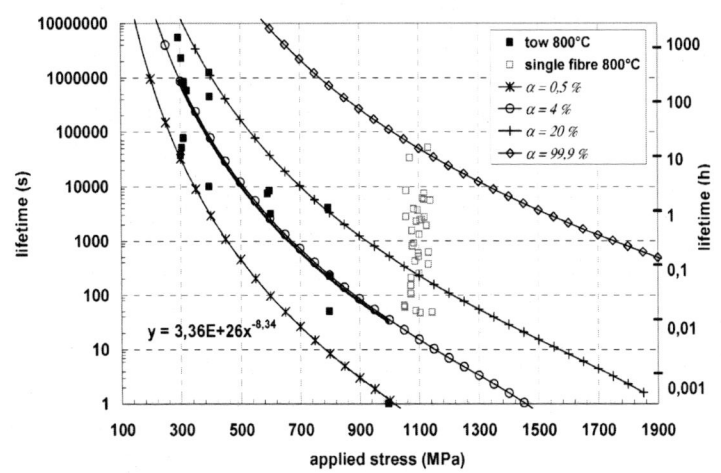

Figure 4. Lifetime of Hi-Nicalon single filaments at 800°C with respect to filament rank α, determined using the empirical model

Finally, lifetimes of tows display a reduced scatter. This scatter can be related to that which is usually observed on tows strength data, as discussed in a previous paper [6].

Subcritical crack growth model for single fibres

A subcritical crack growth model for lifetime prediction was proposed earlier for tows only [2]. In the present paper, the relevance of the model to single filaments and multifilament tows is discussed. The relation between single filaments and tow lifetimes is investigated for a better understanding of experimental results.

The subcritical crack growth model is based on the Paris law (Eq.2), which is usually employed to describe the slow propagation of a crack in ceramics or metals.

Under a constant stress, the lifetime of a single filament is the time for the critical flaw to grow from initial size C_j to critical length a_c:

$$t = \int_{C_j}^{a_c} \frac{da}{V} = \frac{2}{\sigma^n A_1 (n-2)} \left[\frac{C_j^{2-n/2}}{Y^n} - \frac{K_{IC}^{2-n} \sigma^{n-2}}{Y^2} \right] \tag{10}$$

where K_{IC} is the critical value of K_I (fracture toughness) and $Y = 2/\pi^{1/2}$ (Eq.1).
The initial flaw size C_j can be characterized by the strength of the fibre, σ_f, in absence of environmental effects (fast fracture conditions):

$$C_j = \frac{K_{IC}^2}{\sigma_f^2 Y^2} \tag{11}$$

The statistical distribution of fibre strengths σ_f is described by the Weibull equation:

$$P(\sigma) = \alpha = 1 - \exp\left[-\frac{V}{V_0}\left(\frac{\sigma_f}{\sigma_{0f}}\right)^{m_f}\right] \tag{12}$$

where V_0 is the reference volume ($V_0 = 1$ mm^3) and m_f and σ_{0f} the statistical parameters. Rearranging equations (11) and (12) into (10) gives the following lifetime-stress relationship:

$$t = \frac{2 K_{IC}^{2-n}}{Y^2 A_1 (n-2) \sigma^n} \left[\frac{\sigma_{0f}^{n-2}}{V_{m_f}^{n-2}} \left[\ln\left(\frac{1}{1-\alpha}\right)\right]^{\frac{n-2}{m_f}} - \sigma^{n-2} \right] \tag{13}$$

Experimental static fatigue results show that $t\,\sigma^n = A$, where A is a constant. As a consequence, the constant A_1 that appears in the Paris law can be derived from (13) for $\sigma \to 0$:

$$A_1 = \frac{2 K_{IC}^{2-n}}{Y^2 A(\alpha)(n-2)} \left[\frac{\sigma_{0f}^{n-2}}{V_{m_f}^{n-2}} \left[\ln\left(\frac{1}{1-\alpha}\right)\right]^{\frac{n-2}{m_f}} \right] \tag{14}$$

Substitution of equation (14) into (13) gives the following lifetime-stress relationship:

$$t = \frac{A(\alpha)}{\sigma^n} \left[1 - \left(\frac{\sigma}{\sigma_{0f}}\right)^{n-2} \left(\frac{V}{\ln\left(\frac{1}{1-\alpha}\right)}\right)^{\frac{n-2}{m_f}} \right] \qquad (15)$$

And finally, equation (13) gives the theoretical statistical lifetime distribution expressed in terms of filament rupture rank α under a constant stress σ:

$$\alpha = 1 - \exp\left[-\frac{V}{\sigma_{0f}^{m_f}} \left(\frac{t\sigma^n Y^2 A_1 (n-2)}{2 K_{IC}^{2-n}} + \sigma^{n-2} \right)^{\frac{m_f}{n-2}} \right] \qquad (16)$$

Discussion

The following filament characteristics are required for prediction of lifetime of a single filament and a tow using the proposed model:
  (i) statistical parameters m_f and σ_{0f};
  (ii) stress exponent n derived from the experimental law $t\sigma^n = A$ and
  (iii) constant A_1 and toughness K_{IC} for use of Eq. 13, or the empirical $A(\alpha)$ values (Eq. 9) for the use of Eq. 15.
Constant A_1 depends on $A(\alpha)$ (Eq.14). It is estimated from the experimental value of A determined on tows. The corresponding α values are estimated using Eq. 9.
It is now possible to compare the experimental lifetime distribution (Eq. 7) with the theoretical one (Eq. 16). The good agreement shown by the curves plotted on figure 5 supports the pertinence of the analysis. It confirms the correspondence between the lifetime and the strength rank $P_\sigma(t) = \alpha$, which reflects the dominant contribution of pre-existing flaws in delayed failure.

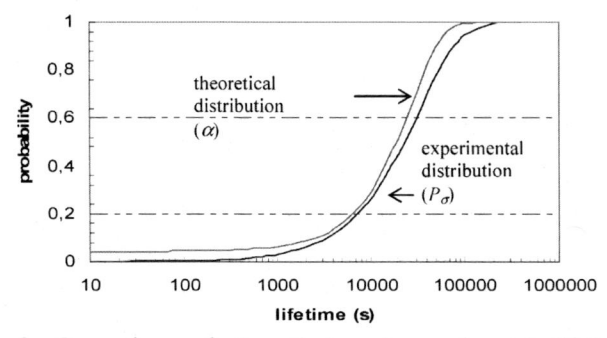

Figure 5. Comparison of theoretical and experimental lifetime distributions (temperature = 500°C).

Experimental results have shown that lifetimes of tows coincide with single filaments lifetimes for a particular value α_c of α (Fig. 3-4), which corresponds to the failure probability of the critical single filament (that one which causes failure of the tow).
At last, the multifilament tow average lifetime behaviour can be evaluated using equation (13) provided the tensile failure probability of the tow α_c is known. The

proposed model finally allows SPT (Strength - Probability - Time) diagrams to be constructed. These diagrams proposed by Davidge [7], are representative of ceramic materials because they relate lifetime, the applied stress and failure probability. These diagrams were established for Hi-Nicalon single filaments at 500°C and 800°C in figure 6.

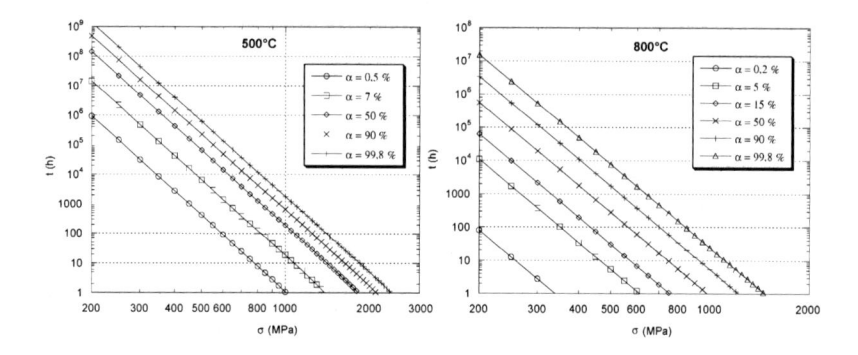

Figure 6. SPT diagrams (Strength - Probability - Time) for Hi-Nicalon single filaments at 500°C and 800°C

CONCLUSIONS

The delayed failure of SiC Hi-Nicalon single filaments and multifilament tows was observed in static fatigue at 500°C and 800°C. Single filaments and tows lifetimes obey the power law $t\sigma^n = A$, with $n = 8,4 \pm 0,1$. The constant A depends on temperature and on the failure probability α of the considered single filament. The lifetime of the tow corresponds to the critical single fibre lifetime within the tow, which is given by a particular value α_c of α. This critical value of α seems to depend on temperature. It may also be affected by fibres interactions leading to a multifilament tow weakening. The static fatigue behaviour of tows and single filaments was described using an empirical model. Then, a model based on subcritical crack growth was proposed. It allowed relevant lifetime predictions for single filaments and multifilament tows. Finally, SPT (Strength - Probability - Time) diagrams were established for Hi-Nicalon single filaments at 500°C and 800°C.

REFERENCES

[1] S. Bertrand, R. Pailler and J. Lamon, "Influence of strong fibre/coating interfaces on the mechanical behaviour and lifetime of Hi-Nicalon/(PyC/SiC)$_n$/SiC minicomposites", *J. Am. Ceram. Soc.*, **84**, 787-794 (2001).

[2] P. Forio, F. Lavaire and J. Lamon, "Delayed failure at intermediate temperatures (600°C-700°C) in air in silicon carbide multifilament tows", *J. Am. Ceram. Soc.*, **87**, 888-893 (2004).

[3] J. Lamon, "A micromechanics-based approach to the mechanical behaviour of brittle-matrix composites", *Composites Sciences and Technology*, 61, 2259-2272 (2001).

[4]R. J. Charles and W. S. Hillig, Symposium on Mechanical Strength of Glass and Ways of Improving It, Florence, Italy, 1962. Union Scientifique Continentale du Verre, Charleroi, Belgium, 1962.

[5]S. M. Wiederhorn, "Subcritical Crack Growth in Ceramics", *Fracture Mechanics of Ceramics*, Vol 2, 613-646, Edited by R. C. Bradt, D. P. Hasselman and F. F. Lange, Plenum Press, New York (1974).

[6]V. Calard and J. Lamon, "Failure of fibers bundles", *Composites Sciences and Technology*, **64**, 701-710 (2004).

[7]R. W. Davidge, J. R. McLaren and G. Tappin, "Strength-probability-time (SPT) relationships in ceramics", *Journal of Materials Science*, **8**, 1699-1705 (1973).

FRACTURE-TOUGHNESS TEST OF SILICON NITRIDES WITH DIFFERENT MICROSTRUCTURES USING VICKERS INDENTATION

Hiroyuki Miyazaki, Hideki Hyuga, Yu-ichi Yoshizawa, Kiyoshi Hirao and Tatsuki Ohji
National Institute of Advanced Industrial Science and Technology (AIST)
Anagahora 2266-98, Shimo-shidami, Moriyama-ku,
Nagoya 463-8560, Japan

ABSTRACT
In order to select a preferable equation for the indentation fracture (IF) technique which is necessary for practical evaluation of fracture resistance, K_R, of small and tough silicon nitride ceramics, two types of Si_3N_4 samples were prepared, one with almost flat R-curve behavior and the other with rising R-curve behavior, and their K_R were calculated using the IF method with various indentation loads ranging from 49 N to 490 N. By comparing the fracture toughness, K_{Ic}, estimated from the single-edge precrack beam (SEPB) and surface crack in flexure (SCF) methods with K_R from IF method using 4 different equations in the same region of crack depth, it was demonstrated that the Miyoshi's equation gave the closest value to K_{Ic} for Si_3N_4 ceramics with flat R-curve behavior. In contrast, K_{Ic} from SCF located middle of the data points of Niihara (higher side) and Miyoshi's equations (lower side) in the case of sample with rising R-curve behavior. These results clarified that there was no single IF formula which could produce a match with K_{Ic} regardless of the difference in microstructures of Si_3N_4 ceramics.

INTRODUCTION
Silicon nitrides are now widely used into tribological applications such as bearings for hard disks and mechanical seals in corrosive environments as the results of much investigation to improve their wear resistance through microstructural control. For such applications, it is very important to know the fracture toughness, K_{Ic}, of real parts themselves. However, tribological parts such as bearings generally have limited sizes, which makes it hard to assess the fracture toughness by the conventional standard testing methods such as single-edge precrack beam (SEPB), surface crack in flexure (SCF) and chevron-notched beam methods, etc. In that case, the use of the indentation fracture (IF) method is inevitable for evaluation of fracture resistance since no other methods applicable to miniature samples has been proposed so far. The IF method has been widely studied for determining fracture resistance of ceramics since it was proposed by Lawn and his co-workers.[1] The advantage of this method is applicability to the tiny specimens and the simplicity of the procedure. However, it possesses many weak points to be solved[2] and then IF method might be considered as a subside technique for fracture resistance evaluation.

One of the major confusions about this method is that many formulae have been proposed to figure out the fracture resistance from the surface crack length,[3-6] each leading to different values.[7-10] Consequently, the international market of ceramic bearing balls demands to select a IF equation suitable for Si_3N_4 ceramics in order to guarantee the fare trading. Although there are a few reports which tried to select the best equation for Si_3N_4 ceramics, the samples were limited to the one with the low K_{Ic} and flat R-curve behavior.[2,7,8] In the case of Si_3N_4 samples with R-curve behavior and/or higher K_{Ic}, it was reported that K_{Ic} from the SEPB or SCF methods was inconsistent with K_R from IF technique by some researchers.[2,11,12] Most of them attributed

the discrepancy to the difference in the crack length measured for both techniques.[11,12] Each researcher used a different IF equation and the verification of the equation has not been made.
Our previous study tried to choose the most proper equation for Si_3N_4 ceramics with and without R-curve behavior.[13] It was revealed that the Miyoshi's equation gave the closest value to K_{Ic} from SEPB test in the case of sample with flat R-curve behavior. For the Si_3N_4 with the rising R-curve behavior, however, we failed to select the best IF equation because of the lack of the information about the crack bridging length in the SEPB specimens. In this study, in order to overcome the disadvantage of SEPB technique, the surface crack in flexure (SCF) method as well as SEPB technique, were employed to give the reference value, since the crack length can be defined clearly in the SCF test. R-curve behavior of the two different Si_3N_4 ceramics were characterized by the IF method with various indentation loads. The abilities of four representative IF equations[3-6] to yield the same K_{Ic} as SCF method was judged in the same crack depth region.

EXPERIMENTAL PROCEDURE
Materials
Two types of Si_3N_4 ceramics were used in this study. One was prepared by us using quite small amounts of sintering additions and the other was Si_3N_4 for ball bearings commercially purchased from Toshiba company (TSN-03). For the former ceramic, α-Si_3N_4 powder was mixed with 1 wt% Al_2O_3 and 1 wt% Y_2O_3 in ethanol by using nylon-coated iron balls and a nylon pot for 24 h. The slurry was dried, and then passed through 125 mesh. The powders were hot-pressed at 1950°C for 2 h with an applied pressure of 40 MPa in a 0.9 MPa N_2 atmosphere. Hereafter, we designate the sample as 1A1Y. The machined samples were polished and plasma etched in CF_4 gas before microstructural observation by scanning electron microscopy (SEM). Figure 1 shows microstructures of the samples. It can be found that grain sizes of both samples were almost similar, whereas the grain boundary phase of TSN-03 was more abundant than that of 1A1Y.

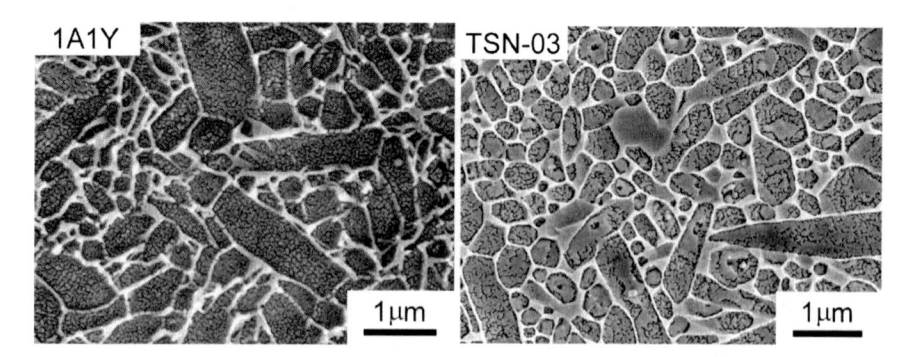

Figure 1. Microstructures of the Si_3N_4 sintered with 1 wt% Al_2O_3 and 1 wt% Y_2O_3 at 1950°C for 2 h (1A1Y) and Si_3N_4 for ball bearings from Toshiba company (TSN-03).

Test Procedure

Vickers indentations were made on the polished surface perpendicular to the hot-pressing axis. The indentation loads of 49, 98, 196, 294 and 490 N were chosen to vary the crack size over a broad range. The lengths of the impression diagonals, 2a, and sizes of surface cracks, 2c, were measured with a traveling microscope immediately after the unloading. Only indentations whose 4 primary cracks emanated straight forward from each corner were accepted. Indentations whose horizontal crack length differed by more than ~ 10% from the vertical one were rejected as well as those with badly split cracks or with gross chipping. Figure 2 shows examples of Vickers impressions, which were observed optically on TSN-03 samples with loads of 98 and 490 N, respectively. Nearly all the indentations were symmetrical and had well-developed shapes. Eight impressions were made at each load, and the total number of rejected one was only three for the TSN-03 samples. In the case of 1A1Y samples indented at loads below 490N, only one impression was unaccepted. When the load was increased up to 490 N, however, half of the indentations on the 1A1Y sample experienced serious chipping. The dark contrast around the plastic impression of 490 N in figure 2 was attributed to the significant swelling around the indentation.

In the case of median/radial crack system, the fracture resistance, K_R, was determined from the as-indented crack lengths as follows:

$$K_R = \xi (E/H)^n P c^{-3/2} \quad (1)$$

where ξ and n are material-independent, dimensionless constants for Vickers-produced radial cracks, E is Young's modulus and H is the Vickers hardness for the Miyoshi,[4] Niihara[3] and Ramachandran[6] equations, whereas H is the mean contact pressure (load over projected area) in the Anstis formula.[5] In this study, H was calculated for each indentation. P is the indentation load and c is the half-length of as-indented surface crack length. In the Niihara's equation,[3] $n = 2/5$, whereas $n = 1/2$ in those equations presented by Miyoshi[4], Anstis[5] and Ramachandran.[6] The values for ξ in the equations of Niihara, Miyoshi, Anstis and Ramachandran are 0.0309, 0.018, 0.016 and 0.023, respectively.

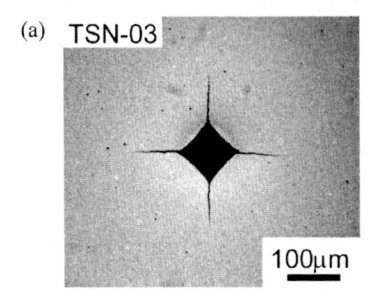

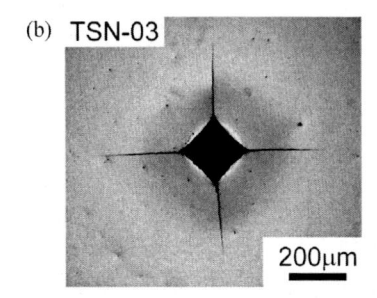

Figure 2. Optical image of the (a) 98 N indentation and (b) 490 N indentation on the TSN-03 samples.

For fracture toughness measurement, rectangular specimens (4 mm in width x 3mm in breadth x ~40 mm in length) were machined from each sintered sample. The SEPB test was performed according to JIS R 1607 with a pop-in crack depth of about 2 mm.[14] For the SCF measurements, the samples were precracked in accordance with the SCF method as described in ISO 18756.[15] For 1A1Y sample, Knoop indentations with the load of 98 and 490 N were conducted on the surface normal to the hot-pressing direction. In the case of TSN-03, the indentation loads were 196, 294 and 490 N. An amount of material removed by polishing was 5~6 times the depth of the impression for the lower indentation loads, whereas that for loads of 294 and 490 N was 7 ~ 8 times the depth of the impression since the lateral cracks became deeper for the indentations with the higher loads. Four-point bending strength was measured with an inner and outer span of 10 mm and 30 mm, respectively, and a crosshead speed of 0.5 mm/min. Precrack sizes were attained on the SEM micrographs of the fracture surfaces after the test. K_{Ic} was estimated from the crack size, the specimen dimensions, the fracture stress and the stress intensity factor, in accordance with ISO 18756. The average of fracture toughness for each load was calculated from 4 ~ 6 measurements which had acceptable crack morphology.

RESULTS

Table 1 summarizes mechanical property data measured by the present authors for the two Si_3N_4. The Young's modulus and hardness of 1A1Y sample were higher than those of TSN-03 sample, which was attributable to the small volume fraction of intergranular phase that is softer than pure Si_3N_4. By contrast, the fracture toughness for the 1A1Y sample was lower than that for TSN-03 sample. This phenomenon can be also explained by the difference in the volume fraction of sintering additives as follows. The poor intergranular phase in the 1A1Y sample is expected to result in a rigid bonding between the grains which suppress the crack bridging, whereas the bonding between grains in TSN-03 should be less tight than that for the 1A1Y due to the enough intergranular phase, which allow the crack bridging to occur more frequently, leading to the higher fracture toughness.

Figure 3 shows a typical profile of Vickers crack observed with an optical microscope on bend-specimen fracture surfaces of 1A1Y sample. The dark-gray half-annular contrast in figure 3 demonstrates that the crack was the median/radial type. A core zone with the shape of a spherical segment immediately below the indentation appears light in the center of the crack. The crack patterns produced at other loads exhibited median/radial system as well.
Median/radial cracks were also generated in TSN-03 samples. The ratio of as- indented crack length to the characteristic dimensions of the "plastic" impression, c/a, was larger than 2 in the range of the indentation load investigated for both samples, confirming that the cracks in this study were median-radial cracks.[10,16,17] Thus the equation (1) can be used for assessing the fracture resistance of the two Si_3N_4.

Table 1 Mechanical properties of the 1A1Y and TSN-03 Si_3N_4 samples

Property	1A1Y	TSN-03
Fracture toughness (MPa·m$^{1/2}$)	4.55±0.09	5.55±0.12
Young's modulus (GPa)	315.2±1.3	304.6±0.4
Hardness (GPa)*	16.3±0.2	14.8±0.2

Hardness was attained from the indentation with the load of 196 N.
Uncertainties are one standard deviation.

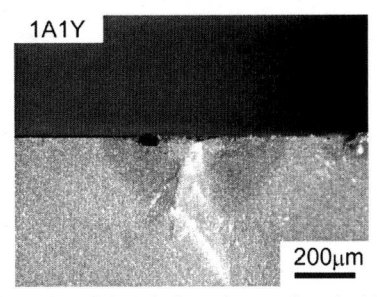

Figure 3. Crack profile under Vickers indentation produced with the load of 294 N in the 1A1Y sample.

The fracture resistances determined from the as-indented crack lengths, for the 1A1Y and TSN-03, are shown in figure 4, tables 2 and 3, as a function of the indentation load respectively. The fracture resistance of 1A1Y sample calculated by the Niihara's equation and that by the Ramachandran's equation was almost the same. All four equations showed little dependence on the indentation load. In the case of TSN-03, K_R from Niihara's and Ramachandran's equations gave the same value again. The increase in K_R with the load was more evident than that for 1A1Y sample, indicating the presence of rising R-curve behavior for this material.

It is well known that the increase in fracture resistance with crack extension (rising R-curve behavior) in Si_3N_4 is caused by the increment in shielding force which arises from the bridging of elongated grains in the sample.[18,19] Then, a parameter which represents the bridging length should be used as the X axis of a R-curve plot when K_R from IF method is compared with K_{Ic} from SCF. As shown in figure 3, the damage zones are located at the center of the cracks in the case of IF test, whereas the damage zones were removed in the SCF specimens (figure 5). It is reasonable to suppose that the crack depth in the SCF test specimen can be employed as the parameter of the size of crack bridging. However, it is inadequate to use the surface crack length c, as a parameter of the crack-bridging length since the bridging should only occur in the crack region excluding the core zone. In this study, crack depth is adopted as the X axis of the R-curve plots. The crack depth for the Vickers impression was defined as half of crack length minus half of diagonal size of plastic impression, $c-a$, because both of crack and core zone had half-penny shapes.

The K_R from IF technique as a function of crack depth are presented in figure 6, as well as K_{Ic} from both SCF and SEPB tests. For both sample, K_{Ic} from SCF agreed with the value from SEPB. In the case of 1A1Y sample, almost flat R-curve behavior was confirmed by the plots of K_R from the four equations, which is attributable to the poor crack bridging as discussed above. The Miyoshi's equation gave the closest outcome to K_{Ic} data from SCF and SEPB. By contrast, the data points of K_R for TSN-03 showed the rising R-curve behavior. K_R from the Miyoshi's equation resided lower side of the K_{Ic} and that from Niihara located upper side. Both of them kept almost the same distance from the K_{Ic} data from SCF. It is evident from the results that none of the four popular equations could produce the agreement with the K_{Ic} constantly.

Table 2. Vickers indentation fracture resistance results for the 1A1Y sample

Indentation Load (N)	Average Diagonal half length a (μm)	Average Crack half length c (μm)	K_{Ic} from SCF test (MPa√m)	Miyoshi K_R (MPa√m)	Anstis K_R (MPa√m)	Ramachandran K_R (MPa√m)	Niihara K_R (MPa√m)
49	36.6±0.3	90.6±3.2	-	4.39±0.08	3.76±0.07	5.61±0.10	5.62±0.10
98	51.9±0.4	140.3±5.2	4.60±0.07	4.59±0.16	3.93±0.14	5.87±0.20	5.88±0.19
196	74.7±0.7	221.0±6.4	-	4.72±0.10	4.04±0.09	6.03±0.13	6.03±0.13
294	90.6±0.8	292.9±10.5	-	4.60±0.10	3.94±0.09	5.88±0.13	5.88±0.13
490	120.3±1.9	420.2±18.0	4.72±0.18	4.59±0.10	3.93±0.09	5.87±0.13	5.83±0.12

Uncertainties are one standard deviation.

Table 3. Vickers indentation fracture resistance results for the TSN-03 sample

Indentation Load (N)	Average Diagonal half length a (μm)	Average Crack half length c (μm)	K_{Ic} from SCF test (MPa√m)	Miyoshi K_R (MPa√m)	Anstis K_R (MPa√m)	Ramachandran K_R (MPa√m)	Niihara K_R (MPa√m)
49	38.9±0.4	92.8±5.5	-	4.50±0.14	3.85±0.12	5.75±0.18	5.72±0.17
98	54.8±0.6	138.3±3.8	-	4.87±0.05	4.17±0.04	6.22±0.06	6.19±0.06
196	78.3±0.7	215.8±5.8	5.63±0.10	5.05±0.06	4.32±0.05	6.45±0.08	6.41±0.07
294	95.4±0.5	278.8±6.9	5.84±0.05	5.11±0.04	4.37±0.03	6.53±0.05	6.49±0.05
490	124.2±0.6	393.6±16.0	5.82±0.13	5.14±0.09	4.40±0.08	6.57±0.12	6.52±0.12

Uncertainties are one standard deviation.

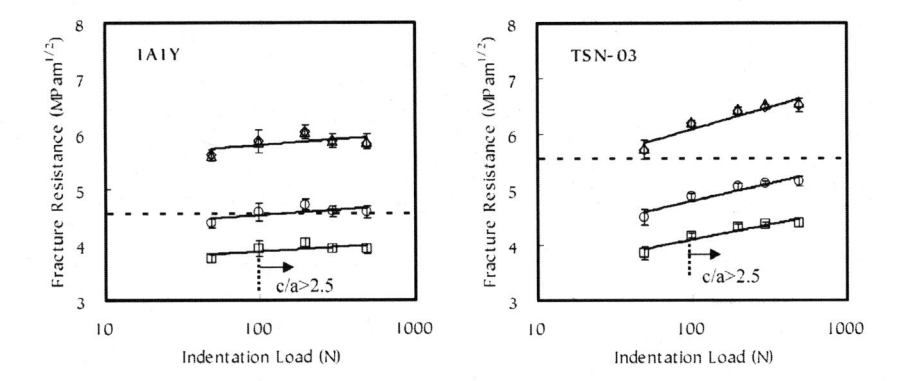

Fracture Toughness Test of Silicon Nitrides with Different Microstructures

Figure 4. Fracture resistance determined by IF method at various indentation loads for the 1A1Y and TSN-03 Si_3N_4 samples by using the equations of Miyoshi (○), Anstis (□), Ramachandran (⊔) and Niihara (◊). Dashed line indicates SEPB's value. Error bars are ± 1 standard deviation.

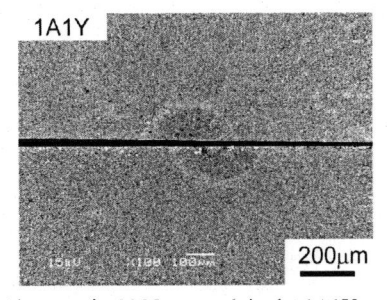

Figure 5. SEM image of a 98 N precrack in the 1A1Y sample.

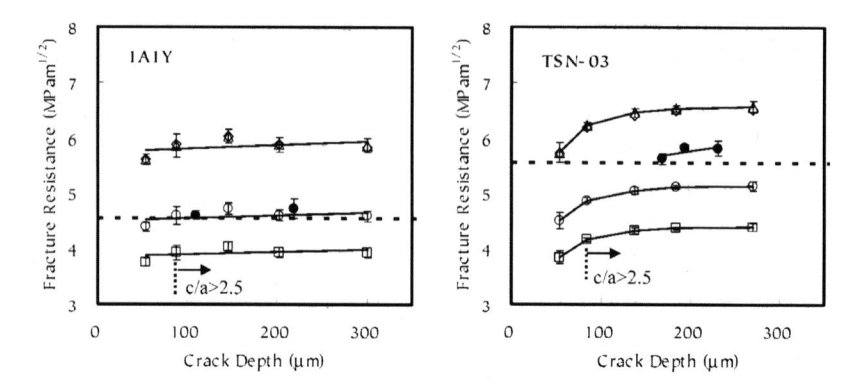

Figure 6. Fracture resistance as a function of crack depth determined by IF (open symbols) and SCF (closed symbols) techniques for 1A1Y and TSN-03 samples. ○: Miyoshi, □:Anstis, ⌷: Ramachandran, ◊:Niihara. Dashed line indicates SEPB's value. Error bars are ± 1 standard deviation.

DISCUSSION
Comparisons of the fracture toughness obtained by the SEPB and fracture resistance from IF method have been conducted by several researchers. In the case of the silicon nitride with the flat R-curve behavior, Awaji et al.[7] and Pezzotti et al.[20] determined K_{Ic} by various techniques using Si_3N_4 ceramics with relatively low fracture toughness, 2 ~ 6 $MPa·m^{1/2}$. They pointed out that K_R estimated by the Niihara's equation was higher than K_{Ic} by the SEPB's technique. Awaji also reported that K_R calculated with Miyoshi's equation was relatively near the value from the SEPB method, which is consistent with our results. Also, the roughly estimated K_R value of Miyoshi's equation from the data of Pezzotti (by multiplying the Niihara's value with an appropriate constant), exhibited a coincident with the value obtained from the other techniques. The consistency between K_R from the Miyoshi's equation and K_{Ic} from SEPB was also verified for Al_2O_3, SiC and Si_3N_4 by the round-robin test conducted to standardize the toughness testing method,[21] which allowed the adoption of IF method as the standard test technique in the JIS R 1607.[14] Quinn also reported recently that the closest outcomes were from the Miyoshi's equation using a standard reference material SRM 2100 prepared by NIST.[2] Thus, in the case of the Si_3N_4 with the almost flat R-curve behavior, K_R from Miyoshi's equation was the closest data among the four equations.

The best outcomes from the Miyoshi's equation seems to originate from the fact that the constant ξ was estimated from the quasi-theoretical analysis using FEM method with the measured values of both crack length and diagonal size of Si_3N_4 indented at 196 N.[4] By contrast, the value of ξ for the Ramachandran's equation came from the approximation using the simplified model.[6] It is reasonable to expect that the accuracy of the estimation from such approximation should be inferior to that of Miyoshi's estimation. In the case of the Anstis's and Niihara's equations, the values of ξ were the average using a host of miscellaneous materials such as glasses, Al_2O_3, B_4C and Si_3N_4, etc.[3,5] The difficulty in detecting the crack tips and the amount of post-indentation slow crack growth differ among these materials,[7,8] which would result in the inadequate values of ξ for Si_3N_4 ceramics. Accordingly, in the case of the Si_3N_4, it seems natural that the Miyoshi's equation could produce the nearest values to K_{Ic}.

From the theoretical point of view, only one IF formula should be applicable regardless of the difference in R-curve behavior since these formulae were derived from the similar models and share the form but only differ in some adjustable constants. Then, K_R from the Miyoshi's equation should also agree with K_{Ic} in the case of Si_3N_4 with the rising R-curve behavior. A possible explanation for the deviation of K_R(Miyoshi) from K_{Ic} can be presented as follows. The distortion near the boundary of damage zone may prevent the crack bridging. Then, the length of crack bridging for IF test may be smaller than c-a. In this case, the data points of K_R from the surface crack length are expected to shift to lower side in X axis. Whatever the reasons are, it seems difficult for the single IF equation to provide the same K_{Ic} as SCF technique for the samples with different R-curve behavior.

CONCLUSION

Because of the necessity for practical evaluation of fracture resistance, K_R, of small and tough Si_3N_4 samples such as ball bearings, K_R was evaluated by the IF method using the representative formulae for the two Si_3N_4 ceramics with different R-curve behavior. The K_R obtained from the four equations was compared to the fracture toughness, K_{Ic}, from the SEPB and SCF methods. K_R from the IF method for the Si_3N_4 with almost flat R-curve behavior was nearby to K_{Ic} from the SEPB and SCF tests when the Miyoshi's equation was used to the data for indentations whose c/a were larger than 2.5. This result was consistent with the previous report, which confirmed that the Miyoshi's equation was preferable for Si_3N_4 with the flat R-curve behavior. By contrast, in the case of rising R-curve material, K_{Ic} from SCF laid halfway between the data points of Miyoshi's and Niihara's equations. These results indicated that there was no single, microstructure-independent way of correlating the IF method with the SCF method. In conclusion, the K_R of different type of Si_3N_4 ceramics measured by the IF test were not adequate to predict the K_{Ic} which was obtained reliably by the standardized fracture toughness test such as SCF and SEPB.

ACKNOWLEDGMENT

This work has been supported by METI, Japan, as part of the international standardization project of test methods for rolling contact fatigue and fracture resistance of ceramics for ball bearings.

REFERENCES

[1] B. R. Lawn, A. G. Evans and D. B. Marshall, Elastic/Plastic Indentation Damage in Ceramics: The Median/Radial Crack system, *J. Am. Ceram. Soc.*, **63**, 574-81 (1980).
[2] G. D. Quinn, Fracture Toughness of Ceramics by the Vickers Indentation Crack Length Method: A Critical Review, *Ceram. Eng. Sci. Proc.* **27** [3] (2006).
[3] K. Niihara, R. Morena and D. P. H. Hasselman, Evaluation of K_{Ic} of Brittle Solids by the Indentation Method with Low Crack-to-Indent Ratios, *J. Mater. Sci. Lett.*, **1**, 13-16 (1982).
[4] T. Miyoshi, N. Sagawa and T. Sasa, Study of Evaluation for Fracture Toughness of Structural Ceramics, *J. Jpn. Soc. Mech. Eng.*, **A, 51**, 2489-97 (1985).

[5]G. R. Anstis, P. Chantikul, B. R. Lawn and D. B. Marshall, A Critical Evaluation of Indentation Techniques for Measuring Fracture Toughness: I, Direct Crack Measurements, *J. Am. Ceram. Soc.*, **64**, 533-38 (1981).

[6]N. Ramachandran and D. K. Shetty, Rising Crack-Growth-Resistance (*R*-Curve) Behavior of Toughened Alumina and Silicon Nitraide, *J. Am. Ceram. Soc.*, **74**, 2634-41 (1991).

[7]H. Awaji, T. Yamada and H. Okuda, Result of the Fracture Toughness Test Round Robin on Ceramics – VAMAS Project-, *J. Ceram. Soc. Jpn.*, **99**, 417-22 (1991).

[8]H. Awaji, J. Kon and H. Okuda, The VAMAS Fracture Toughness Test Round-Robin on Ceramics, VAMAS report#9, Japan fine ceramic center, Nagoya, 1990.

[9]E. Rudnayová, J. Dusza and M. Kupková, Comparison of Fracture Toughness Measuring Methods Applied on Silicon Nitride Ceramics, *J. DePhysiqueIV*, **3**, 1273-76 (1993).

[10]C. B. Ponton and R. D. Rawlings, Vickers Indentation Fracture Toughness Test Part 2, Application and Critical Evaluation of Standardized Indentation Toughness Equations, *Mater. Sci. Tech.*, **5**, 961-976 (1989).

[11]R. Choi and J. A. Salem, Crack-Growth Resistance of in Situ-Toughened Silicon Nitride, *J. Am. Ceram. Soc.*, **77**, 1042-46 (1994).

[12]J. Yang, T. Sekino and K. Niihara, Effect of Grain Growth and Measurement on Fracture Toughness of Silicon Nitride Ceramics, *J. Mater. Sci.*, **34**, 5543-48 (1999).

[13]H. Miyazaki, H. Hyuga, K. Hirao and T. Ohji, Comparison of Fracture Resistance Measured by IF Method and Fracture Toughness Determined by SEPB Technique Using Silicon Nitrides with Different Microstructures, *J. Eur. Soc. Ceram.*, **27**, 2347-54 (2007).

[14]JIS R 1607, Testing Methods for Fracture Toughness of Fine Ceramics," Japanese Industrial Standard, (1995).

[15]ISO 18756, Fine Ceramics (Advanced Ceramics, Advanced technical Ceramics) – Determination of Fracture Toughness of Monolithic Ceramics at Room Temperature by the Surface Crack in Flexure (SCF) Method, International Organization for Standards, Geneva, 2003.

[16]T. Lube, Indentation Crack Profiles in Silicon Nitride, *J. Eur. Ceram. Soc.*, **21**, 211-18 (2001).

[17]D. B. Marshall, Controlled Flaws in Ceramics: A Comparison of Knoop and Vickers Indentation, *J. Am. Ceram. Soc.*, **66**, 127-31 (1983).

[18]P. F. Becher, Microstructural Design of Toughened Ceramics, *J. Am. Ceram. Soc.*, **74**, 255-69 (1991).

[19] R. W. Steinbrech, Toughening Mechanisms for Ceramic Materials, *J. Eur. Ceram. Soc.*, **10**, 131-42 (1992).

[20]G. Pezzotti, K. Niihara and T. Nishida, Some Microstructural Conditions for Evaluating Fracture Toughness and R-Curve Behavior in Platelet-Reinforced Composites, *J. Testing and Evaluation*, **21**, 358-65 (1993).

[21]Report of preliminary investigation for standardization of fine ceramics, Japanese Fine Ceramics Association, 1998.

SELF-CRACK-HEALING ABILITY OF ALUMINA/ SIC NANOCOMPOSITE FABRICATED BY SELF-PROPAGATING HIGH-TEMPERATURE SYNTHESIS

Wataru Nakao, Yasuyuki Tsutagawa, Koji Takahashi, Kotoji Ando
Yokohama National University
79-5, Tokiwadai, Hodogaya-ku,
Yokohama, 240-8501, Japan

ABSTRACT

The improvement in the self-crack-healing rate at the lower temperature is tried by minifying SiC particles to nanometer-size. Alumina containing manometer-sized SiC particles composite was prepared by the following reaction,

$3(3Al_2O_3 2SiO_2) + 8Al + 6C = 13Al_2O_3 + 6SiC$.

The formed nanometer-sized SiC particles were mainly entrapped inside alumina grains. The grains size of the intra SiC particles ranged from 10 nm to 30 nm, corresponding to one tenth that of commercial SiC particles. The formed nanometer-sized SiC caused 250 K decrease to the temperature at which the complete strength recovery can be attained by self-crack-healing for 10h. Furthermore, the limit temperature for the bending strength of the alumina/ nanometer-sized SiC particles composite was found to be equal to that of the alumina containing commercial SiC particles composite. It is summarized that new alumina/ nanometer-sized SiC particles composite can be applied to structural components with self-crack-healing from 1223 K to 1573 K.

INTRODUCTION

To ensure the structural integrity of ceramic components, it is important to manage surface cracks. During service, structural ceramics have several cracking scenarios, e.g., crash, fatigue, thermal shock and corrosion. Low fracture toughness exhibited in most ceramics leads to high sensitivity to flaws. Therefore, these scenarios cause a large strength decrease. If self-crack-healing occurs as soon as cracks are introduced and gives cracked sample complete strength recovery, the structural integrity of ceramic component can be ensured over the whole lifetime.

Self-crack-healing must, of course, occur under service conditions to satisfy the above requirements. The present authors succeeded that alumina[1,2] and mullite[3-5] which are expected to apply at high temperatures in air have been endowed with self-crack-healing ability. The self-crack-healing ability is generated by only incorporating SiC. The phenomenon is driven by the oxidation of SiC and can occur at high temperature in air as soon as cracks are introduced. The reaction makes the space between the crack walls completely filled with the formed oxide, because the reaction includes 80.1 % volume increase in the condense phases. Furthermore, the oxidation includes a large exothermic heat. The reaction heat makes the formed oxide and the base material to react and melt once, resulting in the formation of the strong bonding between the reaction products and crack walls. However, self-crack-healing rate has still been insufficient. Actually in the case alumina containing 15 vol% commercial SiC particles[1], if the crack healing temperature is more than 1473 K, the indented crack having depth of 0.05 mm can be completely healed within 10 h, but the time above 100 h is required to do at 1273 K. The long crack-healing time is inadequate to ensure the structural integrity, because the crack introduced by first damage is possible not to be healed completely until the second damage.

In the present study, the improvement in the self-crack-healing rate at the lower temperature is tried by minifying SiC particles to nanometer-size. Nanometer-sized SiC particles have high specific surface. Thereby these would increase the crack healing rate. Furthermore, large number of SiC particles would cause large resistance to the glide deformations of alumina grain and alumina grain boundary. Alumina containing manometer-sized SiC particles composite was actually prepared by the following reaction,

$3(3Al_2O_3 2SiO_2) + 8Al + 6C = 13Al_2O_3 + 6SiC$ (1).

Zhang et al.[6] succeeded fabricating alumina containing ~50 nm SiC nanometer-sized particles composite by this method.

EXPERIMENTAL
Sample Preparation

In the present study, the reaction synthesis expressed as Reaction (1) was adapted to produce alumina/ nanometer-sized SiC particles composite. The synthesized composite has the SiC contents of 18.4 vol.%. Thus, the composite is abbreviated as AS18NP in this paper.

The used mullite (KM101, Kyoritsu Materials Ltd, Japan), aluminum (600F, Minaruko Ltd., Japan) and carbon powders (#4000B, Mitsubishi Chemical Ltd., Japan) have means of the particles size of 780 nm, 5400 nm and 20 nm, respectively.

Carbon powder was dispersed well in ethanol using ultra sonic vibration, prior to the mixing. Mullite and aluminum powders were added to carbon dispersed ethanol and the powders were mixed well via a Teflon pot and alumina balls for 24 h. To finish reaction (1), the dried mixed powder was heated at 1673 K for 5 h in a carbon crucible and Ar atmosphere. After that, the synthesized AS18NP powder was sintered via hot-press at 2073 K for 1 h in Ar. The sintered plate was cut into rectangular test specimens (Width = 4 mm, Length = 22 mm, Height = 3 mm).

Experimental Procedure

A semi-elliptical surface crack was made at center of the specimen surface by Vickers indenter. The introduced crack has surface length of 0.1 mm and aspect ratio of 0.9. The cracked specimens were crack-healed at temperatures from 1073 K to 1573 K for 1 h - 100 h. These specimens were called as "crack-healed specimen" in this paper.

All fracture tests were conducted on a three-point bending. The span was adapted to be 16 mm in order that the fracture is as possible as to occur from the crack-healed pre-crack. The test temperatures were room temperature and high temperatures from 1073 K to 1673 K. Microstructure and fracture initiation were investigated using a scanning electron microscopy (SEM).

Same experiments were conducted on the alumina containing 15 vol% commercial SiC particles, which has means particles size of 270 nm, composite. The composite is abbreviated as AS15P.

RESULTS AND DISCUSSIONS
Microstructure

Figure 1 shows the microstructure of the AS18NP. The grains size of the alumina matrix ranges from 1 μm to 10 μm. The formed nanometer-sized SiC particles were mainly entrapped inside alumina grains, as pointed by black arrow in Fig. 1. Alternatively, the submicron-sized SiC particles are located at the grain boundaries of alumina grains, as pointed by white arrow in Fig. 1. The grains size of the intra SiC particles ranges from 10 nm to 30 nm, corresponding to one tenth that of commercial SiC particles. The nanometer-sized SiC particles are beneficial for improving crack-healing ability.

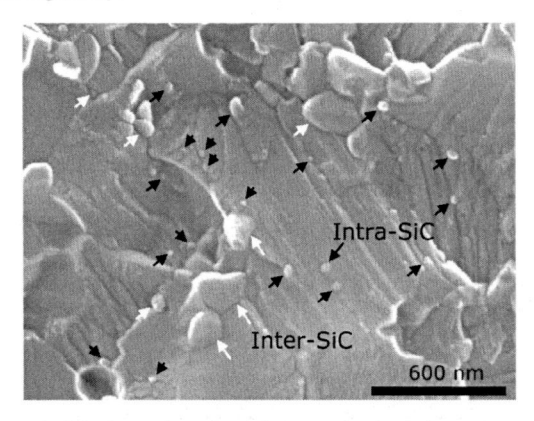

Figure 1 SEM image of the fracture surface of the synthesized alumina containing nanometer-sized SiC particles composite (AS18NP)

Crack-Healing Behavior

Figure 2 shows the bending strength of the crack-healed AS18NP as a function of crack-healing temperature. The crack-healing time is 10 h. Moreover, the left side column shows the bending strength of as-cracked specimens and the center-lined symbols indicate the specimens fractured from the pre-crack crack-healed. As crack-healing temperature increases, the bending strength increases. When the crack-healing temperature is more than 1223 K, the strength recovery is found to be saturated. Moreover, half of all specimens exhibiting the saturated strength recovery fractured from somewhere the healed pre-crack else, such as embedded flaw. This implies that crack-healing makes the pre-crack to erase completely. Therefore, the complete strength recovery is attained by heat-treatment at 1223 K for 10 h. In similarly, the composite was found to be able to heal the pre-crack completely by heat-treatment at 1373 K for 1 h.

For comparison, the data on AS15P is also shown as the open triangle in Fig. 2. The cracked strength and crack-healed strength of AS15P were found to be approximately equal to those of AS18NP. However, the complete strength recovery in AS15P was attained by crack-healing at the temperature of more than 1473 K for 10h. It was, therefore, confirmed that nanometer-sized SiC causes 250 K decrease to the temperature that self-crack-healing become active.

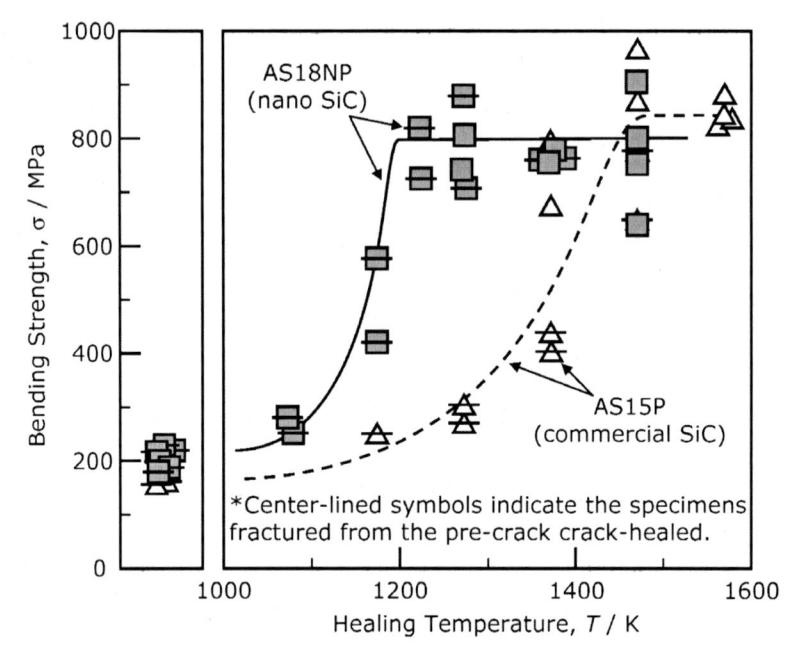

Figure 2 Bending strength of the AS18NP crack-healed for 10 h at several temperatures, with that of the crack-healed AS15P, in which the center-lined symbols indicate the specimens fractured from the pre-crack crack-healed.

High Temperature Strength

Figure 3 shows the temperature dependence of the bending strength of AS18NP crack-healed at 1473 K for 1 h in air. The strength decreases gradually as temperature increases. The means of the bending strengths are 850 MPa at RT, 650 MPa at 1273 K, 600 MPa at 1473 K, 500 MPa at 1573 K and 300 MPa at 1673 K. Furthermore, there are no outstanding plastic deformation behaviors from the all stress-strain curve. Therefore, the limit temperature for the bending strength of AS18NP would be at the least 1573 K. The similar behavior of high temperature was occurred in the crack-healed AS15P, which is shown as the open triangle in Fig. 3. Therefore, it was confirmed that the nanometer-sized SiC cannot gives further improvement in the limit temperature for the bending strength.

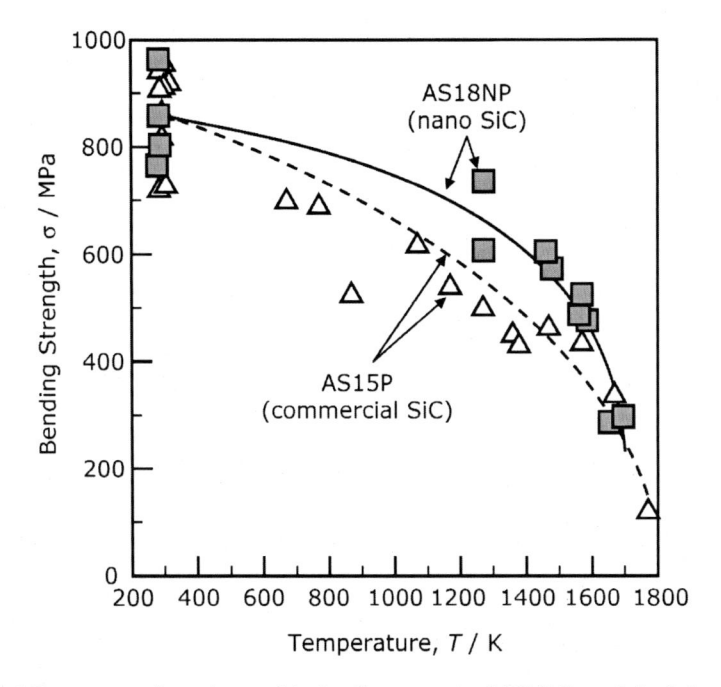

Figure 3 Temperature dependence of the bending strength of AS18NP crack-healed at 1473 K for 1 h in air, with that of AS15P

CONCLUSIONS

Using the following reaction synthesis,
$3(3Al_2O_3 2SiO_2) + 8Al + 6C = 13Al_2O_3 + 6SiC$,
the present authors prepared alumina composite containing nanometer-sized SiC particles whose grains size corresponds to one tenth that of commercial SiC particles. The formed nanometer-sized SiC caused 250 K decrease to the temperature at which the complete strength recovery can be attained by self-crack-healing for 10h. Furthermore, the limit temperature for the bending strength of the alumina/ nanometer-sized SiC particles composite was found to be equal to that of the alumina containing commercial SiC particles composite. It is summarized that new alumina/ nanometer-sized SiC particles composite can be applied to structural components with self-crack-healing from 1223 K to 1573 K.

REFERENCES

[1]K. Ando, B.S. Kim, M.C. Chu, S. Saito and K. Takahashi, Crack-healing and Mechanical Behaviour of Al_2O_3/SiC Composites at Elevated Temperature. Fatigue and fracture of Engineering Materials and structures. 2004, 27, 533-541

[2]K. Takahashi, M. Yokouchi, S.K. Lee, K. Ando. Crack-Healing Behavior of Al_2O_3 Toughened by SiC Whiskers. Journal of the American Ceramic Society. 2003, 86(12), 2143-2147

[3]M. C. Chu, S. Sato, Y. Kobayashi and K. Ando. Damage Healing and Strengthealing Behavior in Intelligent Mullite/SiC Ceramics. Fatigue and fracture of Engineering Materials and structures. 1995, 18(9), 1019-1029

[4]M. Ono, W. Ishida, W. Nakao, K. Ando, S. Mori and M. Yokouchi. Crack-Healing Behavior, High Temperature Strength and Fracture Toughness of Mullite/SiC Whisker Composite Ceramic. Journal of the Society of the Materials Science Japan. 2004, 54(2), 207-214

[5]W. Nakao, S. Mori, J. Nakamura, M. Yokouchi, K. Takahashi, and K. Ando. Self-Crack-Healing Behavior of Mullite/SiC Particle/SiC Whisker Multi-Composites And Potential Use for Ceramic Springs. Journal of the American Ceramic Society. 2006, 89(4), 1352-1357

[6]G. J. Zhang, J. F. Yang, M. Ando and T. Ohji. Reactive Hot Pressing of Alumina-Silicon Carbide Nanocomposite. Journal of the American Ceramics Society. 2004, 87(2), 299-301

THROUGH-LIFE RELIABILITY MANAGEMENT OF STRUCTURAL CERAMIC COMPONENTS USING CRACK-HEALING AND PROOF TEST

Kotoji Ando[1], Masato Ono[2], Wataru Nakao[1] and Koji Takahashi[1]
1) Department of Energy and Safety Engineering, Yokohama National University
2) Post-graduate student, Yokohama National University
79-5 Tokiwadai, Hodogaya-ku, Yokohama, 240-8501, Japan
andokoto@ynu.ac.jp (Kotoji Ando)

ABSTRACT

To overcome decrease in component's mechanical properties and reliability, crack-healing could be a very useful technology. In the present study, firstly, strength recovery of Al_2O_3/SiC specimen with a semicircular groove by crack-healing was investigated for improving the structural integrity of machined alumina and reducing machining costs. Secondly, using the crack-healing, a new methodology to guarantee the reliability of ceramic components before service [crack-healing + proof test] was proposed. Thirdly, if a crack initiated during service, reliability would be severely impaired. Therefore, if a material can heal a crack during service, and if the healed zone has enough strength at the temperature of healing, it would be very desirable for structural integrity.

From the above points of view, a new methodology to guarantee the structural integrity of ceramic components using in situ crack-healing was proposed and the usefulness is discussed using the test results in terms of crack-healing behavior and proof test theory proposed by authors.

INTRODUCTION

Structural ceramics are brittle and sensitive to flaws. As the results, the structural integrity of a ceramic component may be seriously affected. The followings can be the excellent methodology to overcome these problems [1]; (a) toughen the ceramic by fiber reinforcement, etc., (b) activate the crack-healing ability and heal a crack after machining. If a crack-healing ability [2-4] was used on structural components for engineering use, considerable advantages can be anticipated. With this motivation, the authors developed Si_3N_4 [5], mullite [6], alumina [7], and SiC [8] with very strong crack-healing abilities. To use these materials with a high degree of efficiency, the following topics should be studied systematically; (a) the effect of the healing condition on the strength of the crack-healed zone [5,6,8], (b) the maximum crack size which can be healed completely [9], (c) the high temperature strength of the crack-healed member [1,7-9], (d) the cyclic and static fatigue strength of the crack-healed member at elevated temperature [9,10], (e) a methodology to guarantee structural integrity of the ceramic component using the crack-healing ability [11]. Using these technologies, the reliability of ceramic components can be well guaranteed before service.

However, if a crack initiates during service, the reliability of ceramic components will decrease considerably depending on the crack size. These are two ways of overcoming this problem; (a) a periodic proof test to remove the components with non-acceptable flaws; (b) activating the in situ crack-healing ability and heal the crack which initiated during service.

Flow chart of a new methodology to guarantee the structural integrity of a ceramic component is shown in Fig. 1. The new concept consisted of the following three stages; (a) crack-healing under optimized conditions, (b) proof testing, and (c) in-situ (in-service) crack-

Through-Life Reliability Management of Structural Ceramic Components

healing [12]. By machining, many surface cracks will be induced and reliability will decrease considerably. Crack-healing under optimized condition causes complete healing of surface cracks and increment in reliability. Proof testing is required to ensure the fracture associated with the embedded flaws, because the embedded flaw cannot be healed. From Fig. 2 in which the effect of environment on the crack-healing behavior of and Al_2O_3/ 15 vol. % SiC particles [13] is shown, one can understand that the crack-healing cannot occur without oxygen. The ceramic components used at high temperature, the proof testing must be conducted at the operating temperature, but it is no reality. Recently, a new theory to explain the temperature dependence of minimum fracture stress guaranteed based on non-linear fracture mechanics was proposed [11] by Ando et al. as following equation,

$$\sigma_G = \frac{2\sigma_0^T}{\pi} \arccos\left\{ \left(\frac{K_{IC}^T}{K_{IC}^R}\right)^2 \left(\frac{\sigma_0^R}{\sigma_0^T}\right)^2 \left\{\sec\left(\frac{\pi\sigma_p^R}{2\sigma_0^R}\right) - 1\right\} + 1\right\}^{-1} \quad (1),$$

where the superscript T and R indicates the value at elevated temperature T and room temperature, respectively. By obtaining the temperature dependence of K_{IC} and σ_0, one can estimate σ_G. Thus, before service, the structural integrity of ceramic components can be confidently guaranteed using the concept; crack-healing + proof test. After service, if a crack initiated, structural integrity will decrease considerably depending on the crack size. However, if a material can heal a crack in service (that is to say, if a material has in situ crack-healing ability), it would be very desirable for structural integrity. Thus, for the whole lifetime, a new concept which may be called [crack-healing + proof test + in situ crack-healing] will be very desirable.

In this paper, in situ crack-healing behavior of alumina/SiC composite has been made systematically, and a new concept; [crack-healing + proof test + in situ crack-healing] is proposed. The usefulness of the new concept is discussed using crack-healing behavior of machined alumna specimen and previous test results in terms of crack-healing and proof test theory by authors.

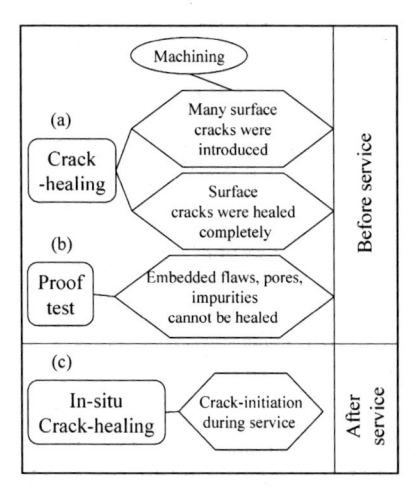

Figure 1 Flow chart of a new methodology to guarantee the structural integrity of ceramic components using in situ crack-healing ability and proof-testing.

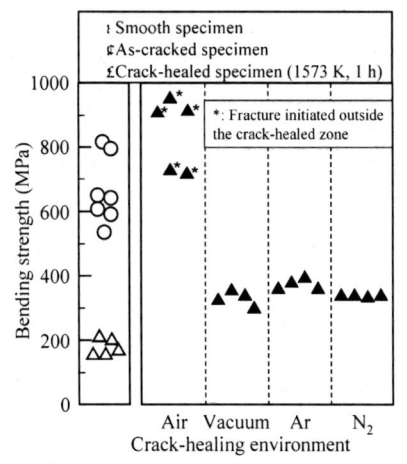

Figure 2 Effect of environment on the crack-healing behavior of alumina/ 15 vol. % SiC particles.

MATERIALS

The summary of chemical compositions, sintering conditions of the materials used is listed in Table 1. The detail of the above information of AS15P used in this section was as follows: The quantity of SiC powder added was 15 vol.% of Al_2O_3 powder. Alcohol was added to the mixture was thoroughly blended for 48 h. The mixture was then placed in an evaporator to extract the solvent and next, in a vacuum desiccator to produce a dry powder mixture. The hot-pressing condition for the material was as follows, temperature: 1873 K; time: 4 h and pressure 35 MPa, in nitrogen environment.

The sintered plates were cut into bar specimens. The specimens were polished to a mirror finish according to the JIS standards on a tensile surface[14]. All fracture tests were performed on a three-point bending system.

Table 1 Chemical composition of ceramics and sintering condition

Sample	Composition	Sintering conditions			
		Temperature (K)	Time (h)	Environment	Hot-pressing (MPa)
AS15P(M)	Al_2O_3 + 15vol.% SiC particle	1973	2	N_2	35
AS20P	Al_2O_3 + 20vol.% SiC particle	1923	2	N_2	35
AS15P	Al_2O_3 + 15vol.% SiC particle	1873	4	N_2	35

RESULTS AND DISCUSSIONS
Strength recovery of machined Al₂O₃/SiC composite ceramics by crack-healing
Figure 3 demonstrates that the non-acceptable cracks introduced by heavy machining can be crack-healed completely. The machining cracks were introduced at the bar specimens surface of alumina/ 15 vol.% SiC particles composite, AS15P(M), by ball-drill grinding. A semicircular groove was made at the center of the smooth specimens. The grinding conditions are listed in Table 2. The varied parameters were cut depth by one pass (d_c). A semicircular groove was shown in Fig. 4. The grinding direction was perpendicular to the long side of the specimens.

The horizontal variable in Fig. 3 is cut depth by one pass, and means the machining efficiency. For example, 33 and 100 cycles needed to fabricate the semi-circular groove by the grinding with the cut depth by one pass of 15 μm and 5 μm, respectively. Alternatively, the vertical axis in Fig. A indicates the local fracture stress of the machined specimens and crack-healed machined specimens. From the load as specimens fractured, P_F, section modulus, Z, and stress concentration factor, α, the local fracture stress at the bottom of the semi-circular groove, σ_{LF}, was evaluated as following,

$$\sigma_{LF} = \frac{\alpha \cdot P_F \cdot l}{4 \cdot Z} \qquad (2),$$

where l is span length. Under this geometry, the values of Z and α were 4.2 and 1.4 from the data book [15], respectively.

Figure 3 shows the strength of the machined specimens healed with semicircular groove as a function of cut depth by one pass (d_c). The symbol △ shows the local fracture stress (σ_{LF}) of the as-machined specimen. The average σ_{LF} of the as-machined specimens were 53 % and 63 % less compared to the smooth specimens healed by machining under d_c = 3 μm and d_c= 5 μm - 20 μm, respectively. The whole cracks introduced by the ball drill grinding employed in this study were too large not to differ the strength each other.

In Fig. 3, the symbols ○ and ▲ show the fracture stress of the smooth specimen healed and the σ_{LF} of the machined specimen healed at 1573 K for 1 h, respectively. Also, the symbols ◇ and ◆ show the σ_{LF} of the machined specimens healed at 1673 K for 1 h and 10 h, respectively. The σ_{LF} of these specimens increased significantly by crack-healing. The machined specimens healed have the same σ_{LF} without varying d_c and the constant value varied with crack-healing temperature and time. The average fracture stress of the specimens crack-healed at 1673 K for 10 h were found to reach a maximum value and the σ_{LF} was almost equal to the fracture stress of the smooth specimen healed which must have no surface cracks. However, as the crack-healing conditions were 1573 K or 1673 K for 1 h, the average σ_{LF} of the machined specimen healed were not almost the same level as that of the smooth specimen healed. The machined specimens healed also show a large scatter of σ_{LF}.

Thus, the optimized crack-healing conditions for the machined specimen are defined as heating at 1673 K for 10 h. These conditions are quite different from the optimized crack-healing conditions for the indented crack healed at 1573 K for 1 h.

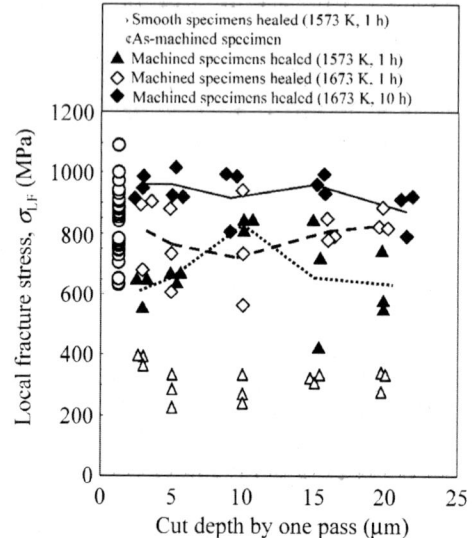

Figure 3 The strength of the machined specimens healed with semicircular groove as a function of cut depth by one pass.

Table 2 Grinding conditions of semicircular groove specimens.

Shape of machined	Grindstone	Number of rotation (rpm)	Table feed speed (mm/min)	Cut depth by one pass (μm)
Semicircular groove	ϕ4 #140 diamond electrocoated	5000	100	3, 5, 10, 15, 20

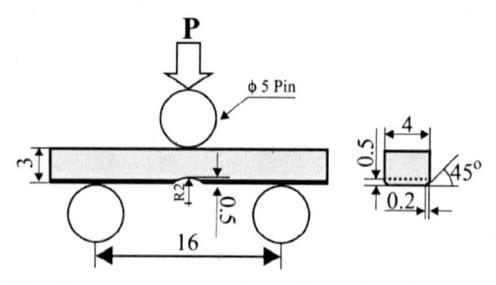

Figure 4 Three-point loading system, test specimen size and semicircular groove (unit: mm).

A new methodology to guarantee the structural integrity of ceramic components using crack-healing

Figure 5 shows comparison between the minimum fracture stress guaranteed (σ_G) and the experimental minimum fracture stress (σ_{Fmin}). Fig. 5 shows the case that σ_p^R = 435 MPa for the crack-healed specimen. The minimum fracture stress guaranteed (σ_G) was estimated by using Eq. (1). The ◇ show the fracture stress of the crack-healed specimens, and the ◆ show the fracture stress of the crack-healed and proof-tested specimens (CHPT specimen). The solid line shows the temperature dependence of σ_G. The minimum fracture stress guaranteed (σ_G) indicates negative temperature dependence. At 873 K, the minimum fracture stress of the crack-healed specimens was 278 MPa, and the value was considerably smaller than that of the CHPT specimen (388 MPa). This fact indicates clearly that the specimens including large embedded flaws could be removed by a proof test at room temperature.

The minimum fracture stress of the proof-tested specimens (σ_{Fmin}) also shows negative temperature dependence up to 1373 K and shows a little higher value than that of the minimum fracture stress guaranteed (σ_G) generally. However, σ_{Fmin} at 1373 K is 289 MPa and showed a little lower value than σ_G=310MPa. The difference is only 6.8%, which is not a serious difference. This difference was assumed to occur because the values of temperature dependence of fracture toughness K_{IC} and intrinsic bending strength σ_0 have dispersion. From the test results, it can be concluded that Eq. (1) is applicable successfully to Al_2O_3/SiC composite, which exhibits large temperature dependence of intrinsic fracture stress (σ_0).

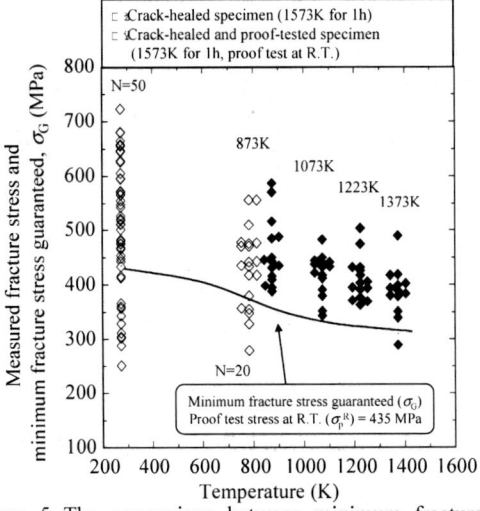

Figure 5 The comparison between minimum fracture stress guaranteed and measured fracture stress as a function of test temperature.

Threshold stress during crack-healing under stress (In situ crack-healing)

Figure 6 shows the bending strength at the crack-healing temperature of AS15P having the pre-crack, which is semi-elliptical (aspect ratio ~ 0.9) and has surface length of 100 μm, crack-healed at 1473 K for 24 h in air under stress. Based on previous studies[16], the crack-healing under stress conditions was confirmed. The applied cyclic stress was sinusoidal wave with a stress ratio, maximum stress/minimum stress, of 0.2 and frequency of 5 Hz. The value of $\sigma_{ap,c}$ was indicated as that of the maximum stress. The open and the closed triangles indicate the bending strength of the specimen crack-healed under static stress ($\sigma_{ap,s}$) and cyclic stress ($\sigma_{ap,c}$), respectively. The bending strength 0 MPa indicates the specimen fractured during crack-healing. The specimens crack-healed under static stresses below 150 MPa were never fractured during crack-healing treatment, and had the same bending strength as the specimens crack-healed under no-stress at 1473 K. A few specimens crack-healed under static stress above 180 MPa were fractured during crack-healing. Therefore, the threshold static stress during crack-healing of AS15P having the pre-crack $\sigma^c_{ap,s}$ was found to be 150 MPa. Also, the threshold cyclic stress, $\sigma^c_{ap,c}$, was found to be 180 MPa. It was found that the crack-healing occurs although the pre-crack growth by the applied stress.

As shown in this figure, the $\sigma^c_{ap,s}$ was found to be smaller than $\sigma^c_{ap,c}$. The threshold stress imposes an upper limit to the crack growth rate, thereby limiting the crack length to less than the critical crack length before crack-healing starts. This implied that the crack growth behavior of all specimens is time dependent rather than cyclic dependent at high temperature. Therefore, applying static stress could be confirmed to be the easiest condition to fracture during crack-healing under stress, and then the threshold stresses of every condition during crack-healing have been found to be the threshold static stress, 150 MPa. The strength of as-cracked specimen (surface length ~ 100 μm) was about 225 MPa. Thus, it was found that ceramic components having the adequate crack-healing ability can crack-heal under stress below 66 % the strength of as-cracked specimen.

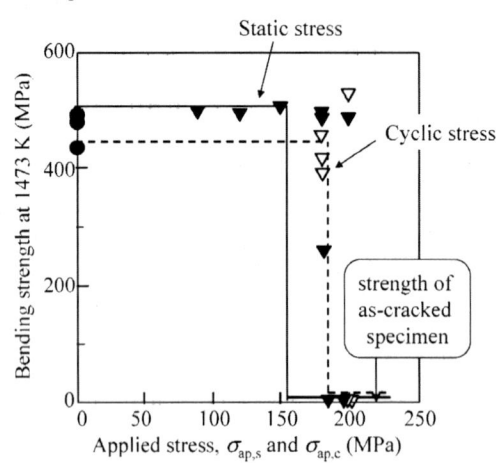

Figure 6 Bending strength of AS15P crack-healed at 1473 K for 24 h under static (closed triangle), cyclic (open triangle) stresses and no-stress (closed circle).

Static fatigue test at high temperature

In the previous section, the crack-healing occurs although the pre-crack growth by the applied stress. In this section, static fatigue test at high temperature of the crack-healed and proof tested specimens was conducted. The specimen used in this study was AS20P, and the proof test and fracture test conditions were same as the previous section. The static fatigue strength of the CHPT specimens was investigated at 1373 K.

The test results are shown in Fig 7 with respect to the correlation with applied stress (σ_{app}) and time to failure (t_f) at the healing temperature of 1373 K. Monotonic bending strengths at 1373 K of the CHPT specimen (▼) were shown on the left-hand side in Fig. 7. The static fatigue tests were stopped at $t = 3.6 \times 10^5$ s (100 h) according to the Japan Industrial Standard [17]. The specimens did not fracture in the tests are marked by arrow symbols (→). The maximum applied stress at which a specimen did not fracture up to $t_s = 3.6 \times 10^5$ s was defined as static fatigue limit and is denoted as σ_{ff}.

From Fig. 7, the value of the static fatigue limit (σ_{ff}) for the CHPT specimen at 1373 K is about 310 MPa. This value is almost equal to the value of the minimum fracture stress guaranteed (σ_G) at 1373 K. Moreover, the bending strength of the specimens that survived static fatigue tests were investigated at 1373 K. The bending strengths are shown on the right-hand side of Fig. 7. The CHPT specimens that survived the static fatigue test showed bending strengths similar to monotonically tested specimens. Moreover, all survived specimens fracture occurred outside of the crack-healed zone. As the results, it can be easily understood that the crack-healed zone exhibited higher bending strength than that of the base material. Thus, Al_2O_3/SiC composite can heal a crack even under static stress. Assuming that very small fatigue crack initiated during fatigue testing, the crack will be healed during fatigue testing, and the embedded flaws were not sensitive because these flaws were blocked from environment.

It can be concluded that a small crack initiated during fatigue test can easily be healed during fatigue testing, and the fatigue limit (σ_{ff}) becomes almost equal to the minimum fracture stress guaranteed (σ_G). Thus, from the results and previous sections, for the whole lifetime, a new concept which may be called [crack-healing + proof test + in situ crack-healing] will be very desirable.

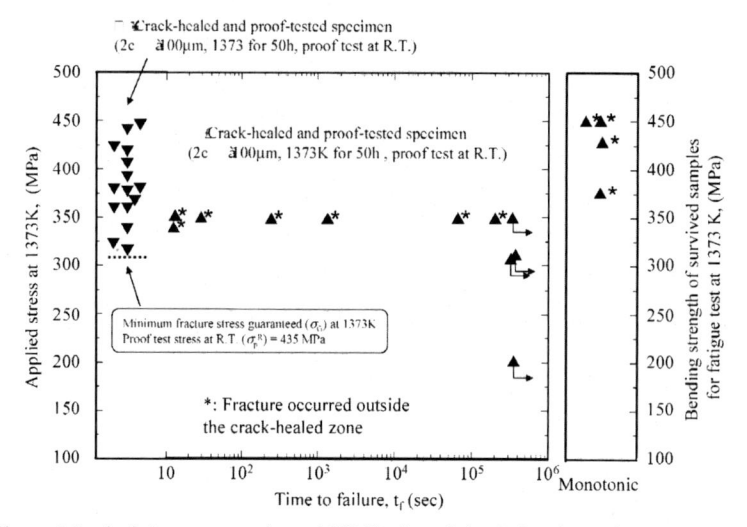

Figure 7 Static fatigue test results at 1373 K of crack-healed and proof-tested specimen (AS20P).

CONCLUSIONS

A new methodology to guarantee the structural integrity of ceramics components which may be called "[crack-healing + proof test + in situ crack-healing]" was proposed and the flow chart was shown. The main conclusions of this work are as follows:
1. The cracks created under the semicircular groove bottom introduced by machining were healed completely, and the σ_{LF} of the machined specimen increased from approximately 300 MPa to 855 MPa by healing at 1673 K for 10 h.
2. The temperature dependence of the minimum fracture stress guaranteed (σ_G) is calculated using temperature dependence of K_{IC} and intrinsic bending strength (σ_0). The σ_G showed very good agreement with the σ_{Fmin} from R.T. to 1373 K.
3. The crack-healing under stress occurred although the pre-crack is grown by applied stress.
4. The fatigue limit (σ_{ff}) becomes almost equal to the minimum fracture stress guaranteed (σ_G) because a small crack initiated during fatigue test can easily be healed during fatigue testing.
5. For the whole lifetime, a new concept which may be called [crack-healing + proof test + in situ crack-healing] will be very desirable.

REFERENCES

[1] K. Ando, M.C. Chu, S. Sato, F. Yao and Y. Kobayashi, "The study on crack healing behavior of silicon nitride ceramics", *Jpn. Soc. Mech. Eng.*, **64-623A**, 1936-1942 (1998) (in Japanese).

[2] F.F. Lange and T.K. Gupta, "Crack-healing by heat treatment", *J. Am. Ceram. Soc.*, **53**, 54-55 (1970).

[3] F.F. Lange and K.C. Radford, "Healing of surface of surface cracks in polycrystalline Al$_2$O$_3$", *J. Am. Ceram. Soc.*, **53**, 420-421 (1970).

[4] T.K. Gupta, "Crack-healing and strengthening of thermally shocked alumina", *J. Am. Ceram. Soc.*, **59**, 259-262 (1976).

[5] K. Ando, T. Ikeda, S. Sato, F. Yao and Y. Kobayashi, "A preliminary study on crack healing behavior of Si3N4/SiC composite ceramics", *Fatigue Fract. Wng. Mater. Struct.*, **21**, 119-122 (1998).

[6] M.C. Chu, S. Sato, Y. Kobayashi and K. Ando, "Damage healing and strengthening behavior in intelligent mullite/SiC ceramics", *Fatigue Fract. Wng. Mater. Struct.*, **18-9**, 1019-1029 (1995)

[7] K. Takahashi, M. Yokouchi, S.K. Lee, and K. Ando, "Crack-healing behavior of Al2O3 toughened by SiC whisckers", *J. Am. Ceram. Soc.*, **86-12**, 2143-2147 (2003).

[8] Y.W. Kim, K. Ando and M.C. Chu, "Crack-healing behavior of liquid-phase sintered silicon carbide ceramics", *J. Am. Ceram. Soc.*, **86**, 465-470 (2003).

[9] K. Ando, K. Furusawa, M.C. Chu, T. Hanagata, K. Tuji, S. Sato, "Crack-healing behavior under stress of mullite/SiC ceramics and the resultant fatigue strength", *J. Am. Ceram. Soc.*, **84-9**, 2073-2078 (2001).

[10] K. Ando, M.C. Chu and S. Sato, "Fatigue strength of crack-healed Si3N4/SiC composite ceramics", *Fatigue Fract. Wng. Mater. Struct.*, **22**, 897-903 (1999).

[11] K. Ando, Y. Shirai, M. Nakatani, Y. Kobayashi and S. Sato, "[Crack-healing + Proof test]: A new methodology to guarantee the structural integrity of ceramics component", *J. Euro. Ceram. Soc.*, **22**, 121-128 (2002).

[12] K. Ando, K. Furusawa, K. Takahashi and S. Sato, "Crack-healing ability of structural ceramics and a new methodology to guarantee the structural integrity using the ability and proof-test", *J. Euro. Ceram. Soc.*, **25**, 549-558 (2005).

[13] B.S. Kim, K. Ando, M.C. Chu and S. Sato, "Crack-Healing Behavior of Monolithic Alumina and Strength of Crack-Healed Member", *J. Soc. Mat. Sci., Japan*, **52**, 667-673 (2003) (in Japanese).

[14] Japan Standard Association, "Testing method for flexural strength (modulus of rupture) of fine ceramics", JIS, R 1601 (1993).

[15] M. Nishida, "Stress Concentration", *Morikita Pub. Co., Ltd.*, 572-574 (1967) (in Japanese).

[16] K. Ando, K. Furusawa, K. Takahashi, M.C. Chu and S. Sato, "Crack-healing behavior of structural ceramics under constant and cyclic stress at elevated temperature", *J. Ceram. Soc. Jpn.*, **110**, 741-747 (2002) (in Japanese).

[17] Japan Standard Association, "Test methods for static bending fatigue of fine ceramics", JIS R1632 (2003).

Joining and Brazing

JOINING METHODS FOR CERAMIC, COMPACT, MICROCHANNEL HEAT EXCHANGERS

C.A. Lewinsohn, J. Cutts, M. Wilson, and H. Anderson
Ceramatec Inc.
Salt Lake City, UT, 84119

Ceramic microchannel, heat exchangers can be used to increase the efficiency of numerous power generation cycles as well as microreactors for chemical synthesis. The Sulfur-Iodide (SI) process has been investigated extensively as an alternate process to generate hydrogen through the thermo-chemical decomposition of water.[1] The commercial viability of this process hinges on the durability and efficiency of heat exchangers/decomposers that operate at high temperatures in corrosive environments. In cooperation with the DOE and the University of Nevada, Las Vegas (UNLV), ceramic based micro-channel decomposer concepts are being developed and tested by Ceramatec, Inc.. The performance benefits of a high temperature, micro-channel heat exchanger are realized from the thermal efficiency due to improved effectiveness of micro-channel heat and mass transfer and the corrosion resistance of the ceramic materials. A critical component in the fabrication of economically viable heat exchangers is ceramic to ceramic joining technology. This paper will compare two types of joining methods suitable for fabricating high temperature, micro-channel heat exchangers for use in a heat exchanger/decomposer for the Sulfur-Iodide (SI) process.

INTRODUCTION

Several versions of thermo-chemical water splitting processes were developed in the 1970's[2,3]. The motivation for these processes is that water can be thermo-chemically split generating hydrogen with an estimated thermal efficiency greater than 50%[4]. Key to these processes is the decomposition of sulfuric acid that occurs in multiple reactions. These endothermic reactions are driven by utilizing high temperature waste heat (as in a nuclear power plant or solar collector) ranging from 450C to 900C[5]. The realization of these processes requires the implementation of high temperature, corrosion resistant heat exchangers.

Over the last 30 years significant advancements have been made in heat exchanger technology. Compact heat exchanger technology that reduces the length scale at which heat and mass transfer occur has enabled high efficiency heat exchangers[6]. In metal alloys these compact heat exchangers are used in car radiators[7], petrochemical processing[8] and HVAC[6]. The introduction of super-alloys into these compact heat exchangers has defined the current state of the art for high temperature heat exchangers.

Compact heat exchangers made from super-alloys, however, do not have sufficient material and mechanical properties for the sulfuric acid decomposition process[9,10,11,12]. Sandia National Labs has been chartered to develop an experimental test loop wherein materials and sub-scale components can be evaluated for use in the Sulfur-Iodide process. Results from Sandia showed that the super-alloy components exhibited rapid corrosion[13]. On the other hand, a Japanese energy consortium that also is investigating the Sulfur-Iodide process has selected high temperature ceramics (silicon carbide) for their demonstrations[14]. Due to well know issues with machining and joining ceramics[15,16,17], the Japanese heat exchanger design is large and expensive. Ceramatec, however has developed ceramic, compact, microchannel heat exchangers to serve as heat exchangers/decomposers.

463

Joining Methods for Ceramic, Compact, Microchannel Heat Exchangers

The Ceramatec compact, microchannel heat exchanger is based on a shell and plate design (Figure 1) consisting of modular stacks of microchannel containing plates that form the primary heat exchange surfaces in a compact arrangement (Figure 2). These modules enable scaling to commercial-scale processes; the microchannels enhance the heat transfer while maintaining low pressure drops within the system. The ceramic materials provide for long-life applications. An enabling feature of this design is the method of joining the microchannel plates. In this paper two types of joining methods suitable for fabricating high temperature, micro-channel heat exchangers for use in a heat exchangers/decomposer for the Sulfur-Iodide (SI) process will be compared: diffusion bonding and pyrolysis of preceramic polymers.

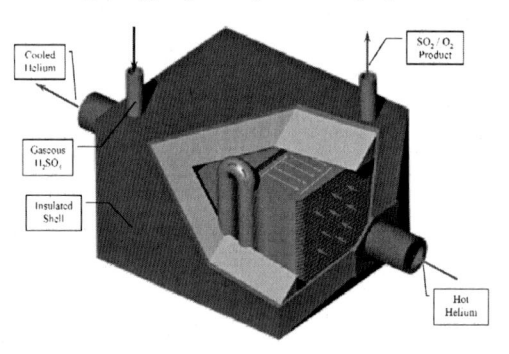

Figure 1. Shell and Plate Design of a Compact Heat Exchanger

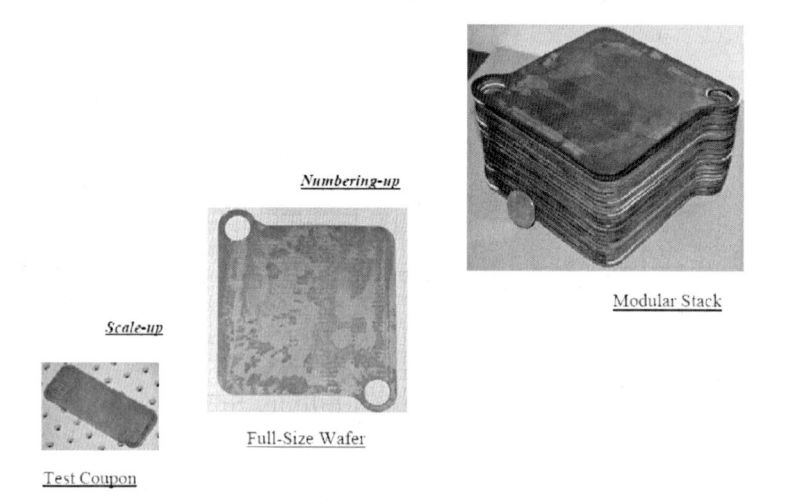

Figure 2. Modular stack design of ceramic, microchannel devices.

METHODS

Two types of joining were evaluated: diffusion bonding and pyrolysis of preceramic polymers. The strength of joints was evaluated using sandwich type specimens of three 25.4 or 12.5 mm^L x 12.5 mm^W x 5 mm^T SiC coupons. The shear strength of the joints was measured, using the electromechanical test frame, by applying a compressive load to only the middle coupon of the sandwich specimen and supporting only the outer two pieces of the bottom of the specimen (see Figure 3). Four to nine specimens joined by each of the two methods were tested. The test configuration did not follow a standard, but provided a simple way to measure the effect of joint compositions and process treatments. The shear strength was calculated by dividing the load at fracture by the joining area. Edge effects and residual stresses due to differential shrinkage would be expected to induce shear stresses at the edges of the joints, but finite element, or other, analysis was not performed to estimate these stresses.

Diffusion bonding involved a novel method using material in the form of a sheet, formed by tape casting, of organic binders and ceramic powders as an interlayer between densified substrates. The interlayers can be featured to make gaskets for manifolds. A diffusion bond between the substrates is accomplished by a high temperature sintering step at 2150°C with 517-1034 kPa applied load in an argin environment.

To avoid deformation of components during joining, as well as the cost of high temperature fabrication steps, an alternative method also was evaluated. In this method, pyrolysis of preceramic polymers was used to form a bond between densified components. Mixtures of powder formed by partial pyrolysis of a commercially available preceramic polymer, allyl-hydridopolycarbosilane (aHPCS, Starfire Systems, Watervliet, NY), silicon carbide powder, and unpyrolysed polymer were mixed and applied onto the surfaces to be joined. Loads between 345-1379 kPa were applied to the specimens during joining at 1200°C in an argon environment.

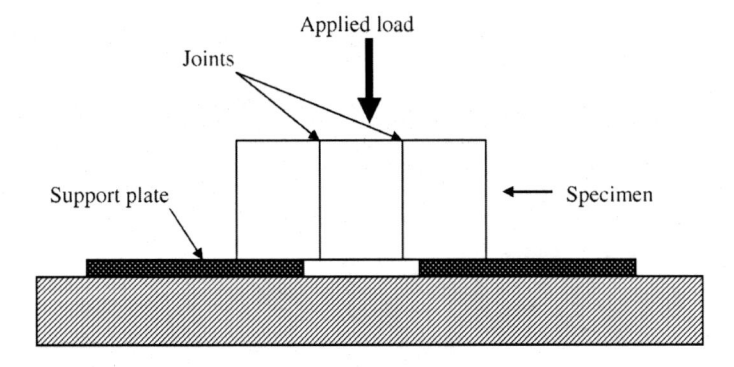

Figure 3 Schematic illustration of joining specimens and testing method.

RESULTS AND DISCUSSION

Images of joints fabricated by the diffusion bonding and polymer pyrolysis approaches are shown in Figure 4 and 5, respectively. It is clear that interdiffusion occurred between the joint material and the substrates. A slight density variation, however, exists in the joint as revealed by the brightness difference in Figure 4. Excellent bonding between the joints made from pyrolysed polymer and the substrates was also seen although there was also evidence of residual porosity (Figure 5) The leak rates of the joints was not measured. The shear strength of the joints made by diffusion bonding was around 77 MPa (95% confidence interval = 68.5 – 85.5 MPa); pyrolysis of preceramic precursors, 41 MPa (95% confidence interval = 23.5 – 58.5 MPa). Microchannel plates were bonded together using the precursor pyrolysis method and these plates could not be separated by hand. These results indicate that both methods of joining show promise for bonding ceramic, microchannel heat exchanger plates however each method has various advantages and disadvantages as shown in Table II. Evaluation of both methods of joining continues.

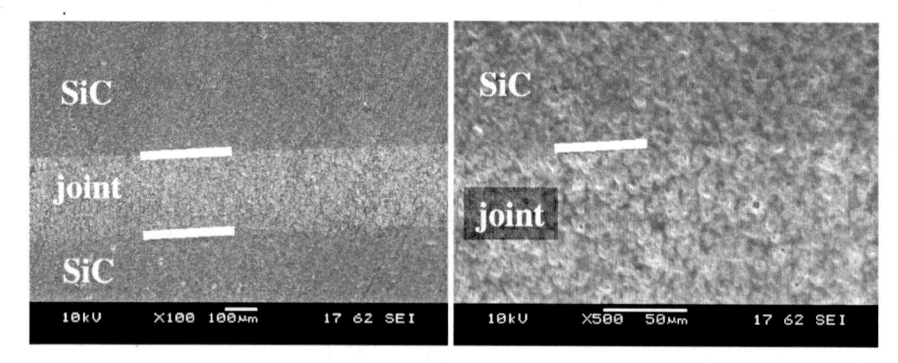

Figure 4. Micrographs of diffusion bonded joint

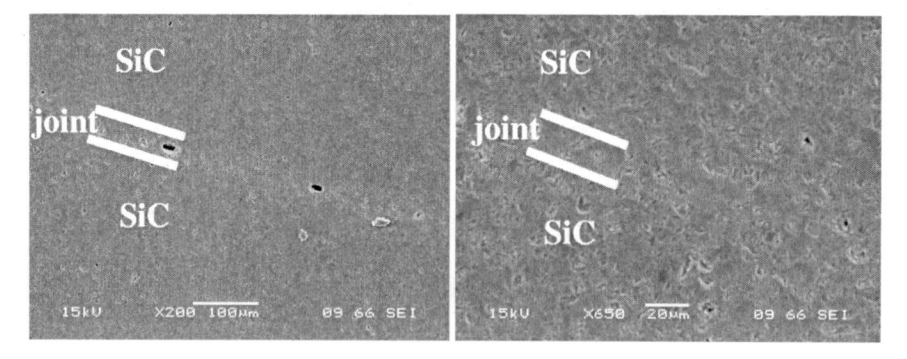

Figure 5. Micrographs of polymer-derived joint

Table II

Method	Advantages	Disadvantages
Diffusion Bonding	High strength, low leak rate, matched thermomechanical properties	High processing temperature: part deformation, high capital and operating costs
Polymer Pyrolysis	Low temperature processing, flexible forming techniques	Lower strength, higher leak rate

SUMMARY

In summary, ceramic microchannel components are candidates for application in high temperature heat exchangers and micro reactors. A cost-effective method of joining heat exchanger plates must provide strong, leak-tight joints without causing deformation of components during fabrication. Two methods of joining are currently being investigated: diffusion bonding and pyrolysis of preceramic polymers. Each method shows promise, however additional development and testing is required.

REFERENCES

[1] *Nuclear Hydrogen R&D Plan*. Department of Energy, Office of Nuclear Energy, Science and Technology. (2004, March). Retrieved January 17, 2006, from http://www.hydrogen.energy.gov/pdfs/nuclear_energy_h2_plan.pdf
[2] Caprioglio, G.; McCorkle, K.H.; Besenbruch, G.E.; Rode, J.S.; "Thermochemical water-splitting cycle, bench-scale investigations and process engineering. Annual report, October 1, 1978 – September 30, 1979." DOE annual Report GA-A-15788 (OSTI ID 5416940). General Atomic Co., San Diego, CA (USA).
[3] Forsberg, Charles; et. al.; "Nuclear Thermochemical Production of Hydrogen with a Lower Temperature Iodine-Westinghouse-Ispra Sulfur Process.", OECD Nuclear Energy Agency, Second Information Exchange Meeting on Nuclear Production of Hydrogen; Argonne, Illinois; October 2-3, 2003. http://www.ornl.gov/~webworks/cppr/y2001/pres/118529.pdf.
[4] Besenbruch, G.E.; "General Atomic Sulfur-Iodine Thermochemical Water-Splitting Process." Am Chem Soc, Div Pet Chem, 271, pp 48-53, American Chemical Society Annual Meeting; 27 Mar 1982; Las Vegas, NV, USA.
[5] Shultz, K.R.; "Use of the Modular Helium Reactor for Hydrogen Production"; World Nuclear Association Annual Symposium; 3-5 September, 2003; London, England. http://www.world-nuclear.org/sym/2003/pdf/schultz.pdf.
[6] Kays, W. M., London, A. L.; "Compact Heat Exchangers,"2d ed., McGraw-Hill Book Company, New York, 1964.
[7] Kelly, Kevin W, et. al; "Crossflow Micro Heat Exchanger." US Patent 6,415,860. Filed 9 Feb 2000.
[8] Bowdery, Tony; "LNG Applications of Diffusion Bonded Heat Exchangers"; AIChE Spring Meeting, 23-27 April 2006. Orlando, FL.
[9] Tiegs, T. N. (1981, July). *Materials Testing for Solar Thermal Chemical Process Heat*. Metals and Ceramics Division, Oak Ridge National Laboratory. Oak Ridge, Tennessee. ONRL/TM-7833, 1-59.
[10] Irwin, H.A., Ammon, R. L. *Status of Materials Evaluation for Sulfuric Acid Vaporization and Decomposition Applications*. Adv. Energy Syst. Div., Westinghouse Electric Corp., Pittsburg, PA, USA. Advances in Hydrogen Energy (1981), 2(Hydrogen Energy Prog., Vol. 4), 1977-99.
[11] Coen-Porisini, Fernanda. *Corrosion Tests on Possible Containment Materials for H2SO4 Decomposition.* Jt. Res. Counc., ERATOM, Ispra, Italy. Advances in Hydrogen Energy (1979), 1(Hydrogen Energy Syst., Vol. 4), 2091-112.

[12] Ishiyama, Shintaro and Maruyama, Shigeki. *Hot Corrosion Resistant Ceramics for Compact Heat Exchanger.* (Japan Atomic Energy Research Institute, Japan; Toshiba Corp.). Jpn. Kokai Tokkyo Koho (2005), 17pp.

[13] Gelbard, Fred; "Sulfuric Acid Decomposition Status Report." UNLV HTHX Quarterly Review Meeting, 17 Mar 2005, Ceramatec, Inc. Salt Lake City UT. http://nstg.nevada.edu/heatpresentations/031705/Gelbard%202005%203%2017%20UNLV%20presentation.pdf.

[14] Ishiyama, Shintaro, et al.; "Compact Heat Exchanger Made of Ceramics Having Corrosion Resistance at High Temperature", US Patent Application Publication, US 2005/0056410 A1.

[15] J.M. Fragomeni and S.K. El-Rahaiby, Review of Ceramic Joining Technology, Rept. No. 9, Ceramic Information Analysis Center, Purdue University, Indiana (1995).

[16] C.A. Lewinsohn, C.H. Henager Jr., and M. Singh, "Brazeless approaches to joining silicon carbide-based ceramics for high temperature applications," pp. 201-208 in Advances in Joining of Ceramics, (C. Lewinsohn, M. Singh, and R. Loehman [Eds.]), Ceramic Transactions Vol. 138, The American Ceramic Society, Westerville (OH, USA), 2003.

[17] M. Singh, "Joining of Sintered Silicon Carbide Ceramics for High Temperature Applications" Journal of Materials Science Letters, **17** [6] 459-461 (1998).

GLASS-TO-METAL (GTM) SEAL DEVELOPMENT USING FINITE ELEMENT ANALYSIS: ASSESSMENT OF MATERIAL MODELS AND DESIGN CHANGES

Rajan Tandon, Michael K. Neilsen, and Timothy C. Jones
Sandia National Laboratories
P. O. Box 5800
Albuquerque, NM 87185

James F. Mahoney
Honeywell FM&T
2000 East, 95th Street
Kansas City, MO 64141

ABSTRACT

Glass-to-metal (GTM) seals maintain hermeticity while allowing the passage of electrical signals. Typically, these seals are comprised of one or more metal pins encapsulated in a glass which is contained in a metal shell. In compression seals, the coefficient of thermal expansion of the metal shell is greater than the glass, and the glass is expected to be in compression. Recent development builds of a multi-pin GTM seal revealed severe cracking of the glass, with cracks originating at or near the pin-glass interface, and propagating circumferentially. A series of finite element analyses (FEA) was performed for this seal with the material set: 304 stainless steel (SS304) shell, Schott S-8061 (or equivalent) glass, and Alloy 52 pins. Stress-strain data for both metals was fit by linear-hardening and power-law hardening plasticity models. The glass layer thickness and its location with respect to geometrical features in the shell were varied. Several additional design changes in the shell were explored. Results reveal that: (1) plastic deformation in the small-strain regime in the metals lead to radial tensile stresses in glass, (2) small changes in the mechanical behavior of the metals dramatically change the calculated stresses in the glass, and (3) seemingly minor design changes in the shell geometry influence the stresses in the glass significantly. Based on these results, guidelines for materials selection and design of seals are provided.

INTRODUCTION

Hermetic glass-to-metal (GTM) seals are used in a large number of applications: in various engine controls, altimeters, and flight data recorders for aerospace applications, air bag initiators, multi fuel sensors, and feedthru's for fuel cell in automotives applications, in high performance lithium batteries for consumer electronics, in MEMS and photonic devices, in feedthrus for medical implantable devices, in pyrotechnic initiator seals, and guidance components for military applications, and in many pressure, flow and temperature sensors.[1] A commonly used classification for the large variety of GTM seals is based on the thermal expansions (α) of the shell, the pins and the glass. This classification is depicted in Fig. 1, which is adapted from information found in Ref. 2. In matched seals, the α's of the glass and metal shell are nearly equal. These seals rely on forming a chemical bond at the glass-shell interface in order to maintain hermeticity. Matched seals are further subdivided into hard and soft glass seals, based on the α's of the glass utilized. In the unmatched seal category, the two α's are unequal, with

Glass-to-Metal (GTM) Seal Development Using Finite Element Analysis

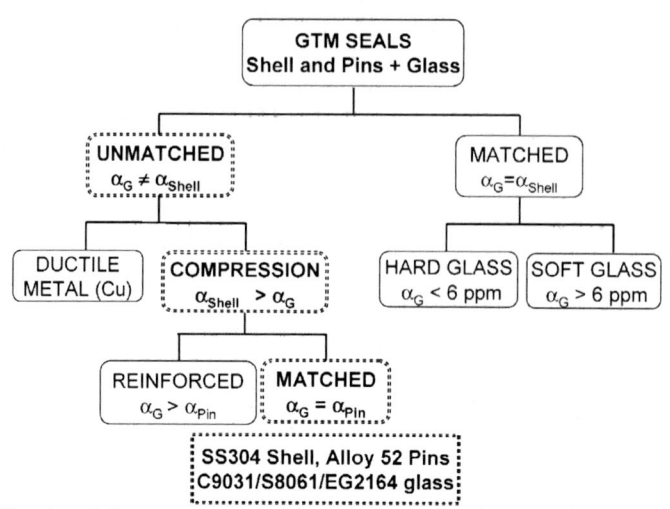

```
                    ┌─────────────────────┐
                    │     GTM SEALS       │
                    │ Shell and Pins + Glass │
                    └──────────┬──────────┘
                               │
               ┌───────────────┴───────────────┐
        ┌ ─ ─ ─┴─ ─ ─ ─ ┐              ┌──────┴──────┐
        : UNMATCHED    :               │   MATCHED    │
        : α_G ≠ α_Shell:               │ α_G = α_Shell│
        └ ─ ─ ─┬─ ─ ─ ─ ┘              └──────┬──────┘
               │                              │
        ┌──────┴──────┐               ┌───────┴────────┐
   ┌────────┐ ┌ ─ ─ ─ ─ ─ ─ ┐   ┌──────────┐ ┌──────────┐
   │DUCTILE │ :COMPRESSION  :   │HARD GLASS│ │SOFT GLASS│
   │METAL(Cu)│ :α_Shell > α_G:  │α_G < 6 ppm│ │α_G > 6 ppm│
   └────────┘ └ ─ ─ ─┬ ─ ─ ─ ┘  └──────────┘ └──────────┘
                    │
             ┌──────┴──────┐
       ┌────────┐  ┌ ─ ─ ─ ─ ─ ┐
       │REINFORCED│ : MATCHED  :
       │α_G > α_Pin│ :α_G = α_Pin:
       └────────┘  └ ─ ─ ─┬ ─ ─ ┘
                          │
               ┌ ─ ─ ─ ─ ─┴ ─ ─ ─ ─ ─ ─ ┐
               : SS304 Shell, Alloy 52 Pins :
               : C9031/S8061/EG2164 glass   :
               └ ─ ─ ─ ─ ─ ─ ─ ─ ─ ─ ─ ─ ─ ┘
```

Fig. 1 Classification of glass-to-metal seals based on the relative thermal expansions of the components. The seals described in this study are depicted in the boxes with dotted lines. The specific material set to be discussed is shown in the square box at the bottom.

that of the metal shell being higher. The ductile metal seal (also referred to as Housekeeper seal) relies on extensive plastic deformation of the thinned metal sheet to form a joint.[2] In compression seals, the higher expansion shell exerts force on the glass upon cooling from the sealing temperature, and stress in the glass is generated below its glass transition temperature, T_g. The compressive stress is usually enough to form a hermetic seal; however, a small degree of chemical bonding at the glass-metal interface is beneficial. In a reinforced compression seal, the α of the glass is larger than that of the pin, creating additional compression along that interface, and further reducing the likelihood of leakage. The seals of interest in this study are compression seals, with α's of the pin and glass being nearly equal (matched), please refer to dotted boxes in Fig. 1.

We describe finite element analysis (FEA) results for an electrical connector that has a compression GTM seal. During early development and prototyping, it was expected that the glass surface would be in compression. However, as shown in Fig. 2, severe circumferential cracking at the pins was observed in the components as manufactured. These cracks negatively influence the electrical characteristics of the connector, and there is a possibility that they may propagate through-the-thickness during service and cause a loss of hermeticity. FEA revealed that the shell geometry and glass thickness and placement would significantly change the stress in the glass. It was found that small-strain plastic deformation in the metals (pins and the shell) induced during low temperature thermal cycling could create high tensile stresses in the glass. The stresses predicted are very sensitive to the material model chosen to represent the plastic deformation behavior of the metals. The results from this work were used to assist in the design of a connector with reduced stress, and lowered likelihood of cracking. They also point out the need for a more complete characterization of materials properties, and their incorporation into constitutive materials models so that the fidelity of the FEA can be improved. Guidelines for materials selection for future designs are also provided.

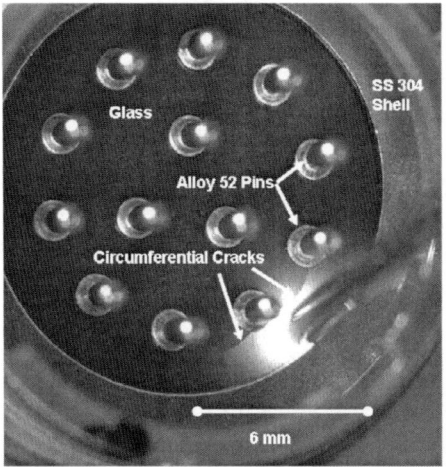

Fig. 2 A connector showing the 304 stainless steel (SS304) shell, Alloy 52 pins, and the (blue-colored) alkali-barium-silicate glass. A circumferential crack in the glass, between two of the pins can be observed under light wand illumination. (Photo courtesy of Saundra Monroe at Sandia National Laboratories.)

ANALYSIS

Materials Set and Thermo-Elastic Properties:
    The material set is a fully annealed 304 stainless steel shell, fully annealed Alloy 52 (50.5% Ni, 0.3% Mn, 0.1% Si, balance Fe by weight) pins, and an alkali-barium-silicate glass (either Schott 8061, Corning 9031, or Electroglas 2164). The metals are considered fully annealed because the entire system is exposed to temperatures > 900°C for a period of 2-4 hrs. during the glass seal formation, and this is expected to relieve any effects of work-hardening. After the seal is formed, the system is cooled to room temperature. For the purposes of the FEA, the three glasses from the different manufacturers are considered nominally equivalent, although there are very slight compositional differences between them. Since the T_g of the Schott 8061 glass is reported to be ~467°C,[3] the stress-free temperature of the glass in the analysis was assumed to be 450°C to try and capture any stress-relaxation effects in the lower range of T_g. The thermal contraction of the three materials from 500°C to room temperature was measured using a dilatometer[*] and these results were linearly extrapolated to -55°C. The measured thermal strains for each material were input into the FEA analyses. Over the temperature range of interest, the α for SS304 was ~ 16 ppm, which is twice that of the glass and the Alloy 52. The glass was modeled as isotropic, linear elastic below the stress-free temperature, while four constitutive

---

[*] Dilatometer 402ED, Netzsch Instruments, Inc., Burlington, MA 01803

Glass-to-Metal (GTM) Seal Development Using Finite Element Analysis

models for the SS304 were used to fit the range of experimentally available true stress-true strain data for the material. These experimental data were gathered for different lots of material over the course of several years of production, and all these materials would have been deemed acceptable for production. The four models fit to the data, a linear hardening model and 3 different power-law representations, are shown in Fig. 3a. Fig. 3b shows the two models, linear hardening and power law, that were utilized for the Alloy 52 pins. Based on experimental data, the linear hardening model for the SS304 shell included a temperature-dependent yield criteria such that the yield stress at any temperature (σ_{ysT}) ranged from ~$0.55\sigma_{ysRT}$ (room temperature yield stress) at 450°C to ~1.2 σ_{ysRT} at -55°C. The von-Mises yield criterion for the shell and the pins was used. The Young's modulus of the glass was taken as 67.6 GPa, and that of the Alloy 52 as 207 GPa, and both were assumed to be independent of the temperature. The Young's modulus of steel at and below room temperature was taken as 195 GPa, and was assumed to be temperature dependent, decreasing to ~ 0.84 times this value at 450°C. It may be noted that most of these materials assumptions have been used in analyses at Sandia Labs for design purposes for several years, and when used with glass-strength based failure criteria, have yielded a fair indication of the presence or absence of cracks.

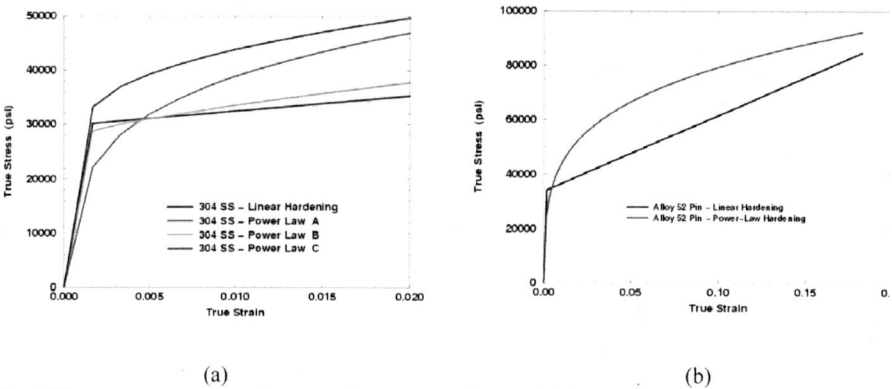

(a)   (b)

Fig. 3 The true stress-true strain curves that were used to model the metal behavior. (a) The four constitutive models fit to data for SS304 and (b) the two models for Alloy 52 are shown.

FEA Approach and Models Run

The FEA runs were conducted with ADAGIO in the SIERRA multi-physics code framework[4] using 8-node brick elements. The number of elements in a typical model was 500,000. The size of the mesh elements ranged from ~ 20 to 500 microns. The meshes in the glass and at both the metal interfaces (pin and shell) were tied together; this approach is equivalent to assuming that the shear and tensile strength of the interfaces is infinite. The materials combination, overall size, number of pins, and many other features of the manufactured component were dictated by the system design and manufacturing constraints; however, when cracking was observed, leeway in the design to vary some parameters was obtained. These parameters are shown in a cut-out view of the connector in Fig. 4. The glass offset (location of the top of the glass disk from the metal connector wall), could be varied from 0-0.76 mm and the glass thickness itself could be varied from 2.4-4.2 mm. The design also

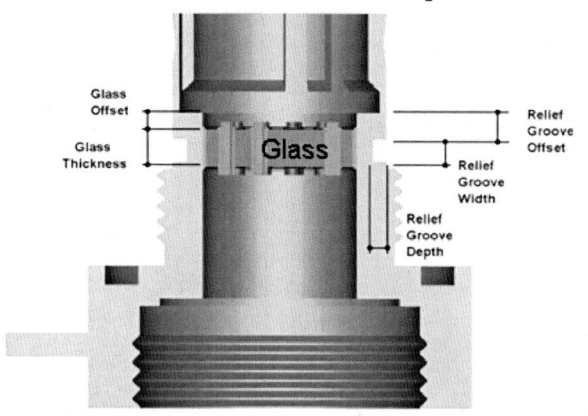

Fig. 4 A cut-out view of the component showing the geometrical features that could be varied.

allowed the incorporation of a groove (or slot), and varying its offset from -0.25 to +0.25 mm, its width from 1.27-1.78 mm, and its depth from 0.64-1.46 mm. As will be shown in the results, this slot is appropriately called the "stress-relief" groove. In addition to varying the sizes and locations of these geometrical features, computations using various material models for the metals were performed for select geometries. Analyses were conducted to determine stress at room temperature, in the sub-ambient at -55°C, and then when the component is cycled to temperatures in excess of room temperature, up to 200°C.

RESULTS AND DISCUSSION

Fig. 5 is a highly exaggerated view of the deformation (deformation magnification 50x) of the three constituent materials for a representative geometry, shown at room temperature after the glass sealing operation. In this matched compression glass seal, we expect the glass to be held in compression; however, at least two geometric and two materials effects may lead to localized tensile stresses. The geometric effects are due mainly to the large thermal expansion difference between the shell and the glass. As the connecter cools from the stress-free temperature, the entire shell contracts much more than the glass in both the radial direction and axial directions. In the radial direction, the shell region where the glass is in contact with the steel is constrained, and tensile hoop stresses develop in the steel. The rest of the shell contracts more or less independently of the constrained region and gets "pulled-in" towards the axis. Because the lower portion of the connecter has a larger volume than the steel above the glass, a net bending moment develops in the glass, such that the top surface is under tension. This effect is illustrated in Fig. 5 as bending. The other purely geometric effect, called pinching, is also illustrated in Fig. 5. The axial difference in contraction between the steel and the glass leads to an axial tensile stress in the steel during cooling, and a local axial compression in the glass adjoining the metal. The higher stiffness of steel leads to the glass conforming to the strain in the steel, i.e., the strain in the glass is higher than what it would have been if the elastic stiffness of

Glass-to-Metal (GTM) Seal Development Using Finite Element Analysis

the two materials was matched. The axial compression in the glass near the interface then sets up a local radial tensile strain radially inwards, slightly away from the interface. This radial tensile

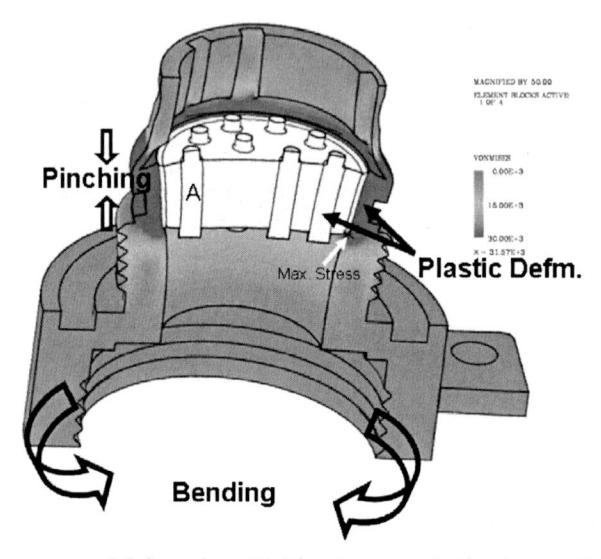

Fig. 5 View of the exaggerated deformations (50x) in a connecter that has been cooled to room temperature after fabrication. The two geometric (bending and pinching) and two materials effects (plastic deformation of the shell and pin) that might be leading to tension in the glass are indicated. The color gradation metal corresponds to the von Mises stress (in psi).

strain may be visualized by imagining a deck of cards held tightly along one edge, and can be seen in Fig. 5 near the surface of the pin marked A. Now since the pins are also significantly stiffer than the glass, this localized residual strain is further magnified by as much as a factor of 1.2-1.4.[5]

One of the materials behavior effects is the small-strain plastic deformation of the steel due to the tensile hoop stress, and the compressive radial stress in the steel created due to the thermal expansion mismatch. As seen in Fig. 5, the maximum von-Mises stress occurs at the location marked by the white arrow at the bottom right, near the glass-steel interface. The magnitude of the stress is ~31,500 psi (= 220 MPa), and this stress would cause plastic deformation in the steel if the linear-hardening or power laws A or B models (see Fig. 3a) were the accurate descriptions of the constitutive behavior. Cooling the connecter to below room temperature (as required by qualification protocols) would further intensify the stress, especially in the region where the glass and steel are in contact (region in red in Fig. 5). Yielding allows the steel to follow the contraction behavior (strain) of the glass, and the steel shell around the glass now has a much larger diameter than if yielding had not occurred. When the connector is heated back to room temperature, the steel expands and places the glass in radial tension. So in effect, the impact of yielding of the steel on the connecter behavior is to decrease the amount of

Glass-to-Metal (GTM) Seal Development Using Finite Element Analysis

compressive loading placed on the glass. The other material effect of importance is caused by the plastic deformation of the pins. The calculations show that the in-plane compression in the glass leads to a small plastic strain (~0.3% equivalent strain) in the pins. Since plastic deformation is a volume-conserving process, the material in the pins is pushed up and outwards, towards the free surface. This leads to an additional local stress near the pins, and could be another reason why cracks often nucleate at or near the pins (Fig. 2).

The important effect of metal yielding on the stress generation in glass described above is highlighted by the results of the calculations in which the various constitutive models for metal behavior (Fig. 3) were exercised. Calculations utilizing all the combinations of metal models at room, sub-ambient and elevated temperatures were performed but only results for two of the stainless steel power law models (A and C) in combination with the linear and the power law model for the pins will be discussed. The thermal cycle simulated was the cooling of the connecter to -55° C, and then reheating to 200° C. The maximum principle stresses in the glass at 200° C are shown in Fig. 6a-6d.

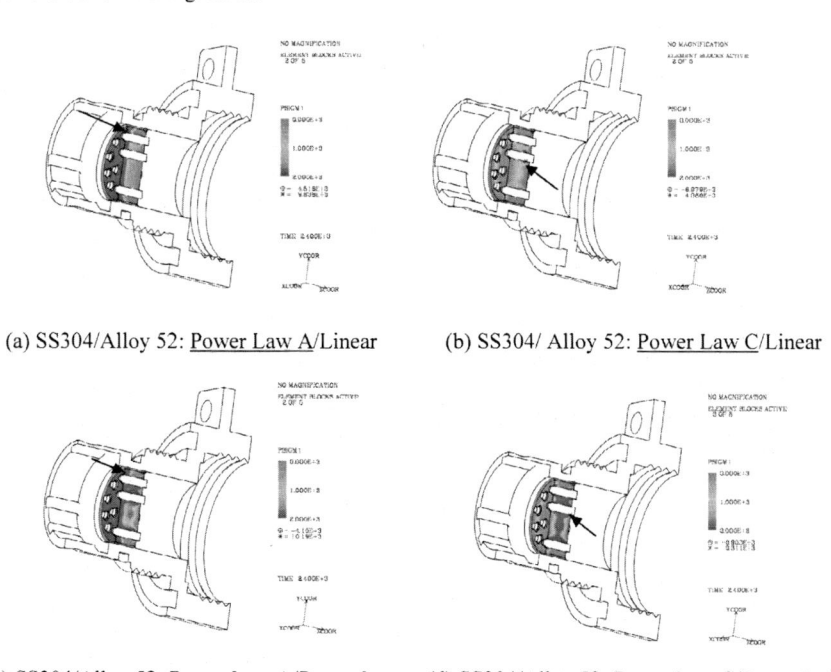

(a) SS304/Alloy 52: Power Law A/Linear   (b) SS304/ Alloy 52: Power Law C/Linear

(c) SS304/Alloy 52: Power Law A/Power Law   (d) SS304/Alloy 52: Power Law C/Power Law

Fig. 6 Calculated maximum principal stresses (in psi) in glass at 200°C with Power Law models for the steel shell (A and C), and the linear hardening and power law model for the pins. The color scale used for the magnitude of the stress is identical for all cases, with red regions having stress is excess of 14 MPa. The maximum stress location is at (x), and indicated by an arrow.

Glass-to-Metal (GTM) Seal Development Using Finite Element Analysis

Comparison of Fig. 6a and 6b reveals that if the pin yield behavior is captured by linear hardening, Power Law A for the steel with its lower yield strength results in much higher stress on the glass surface. The highest stress is predicted to be ~66 MPa at the top glass surface near a pin with an orientation in the radial direction, consistent with the circumferential cracking observed. However, if the steel has a much higher yield stress, as for Power Law C, the magnitude of the stresses is significantly lower, with the highest stress being ~ 32 MPa, in the mid-plane. Based on the arguments presented previously, this indicates that if the steel follows Power Law C behavior, the entire cross-section does not yield when the connecter is exposed to the cold cycle. Hence the amount of plastic flow is minimal, and on heating the tensile stresses generated in the glass are low. The effect of the pin models on the predicted stress can be seen by comparing Fig. 6a with Fig. 6c. If the pin has a lower yield stress (Power Law), the stresses on the glass surface are higher, and the region of high tensile stress extends through the thickness of the glass. The maximum stress in this case is near a pin at the top glass surface, with a magnitude of ~71 MPa. Similarly, comparing Fig. 6b with Fig. 6d, it is apparent that the lower yield strength model for the pin (Power Law) leads to a high stress through the mid-plane, with the peak stress being ~65 MPa at the bottom glass surface. Comparing Fig. 6c and 6d, it can be seen that using the higher yielding steel model (Power Law C) still has an effect of mitigating the tensile stress in the glass.

Figure 7 shows the maximum stress in the top surface of the glass as a composite function of the glass offset and the "stress-relief" groove (slot) depth (see Fig. 4 for a definition of these geometrical parameters). For this calculation, linear hardening models for both the steel and the pins were used. The thermal cycle simulated was fabrication at the sealing temperature, cooling to -55°C, and then heating to 200°C. The stresses at 200°C are shown. The glass thickness was fixed at 2.9 mm, the slot width at 1.5 mm and slot offset at 0 mm.

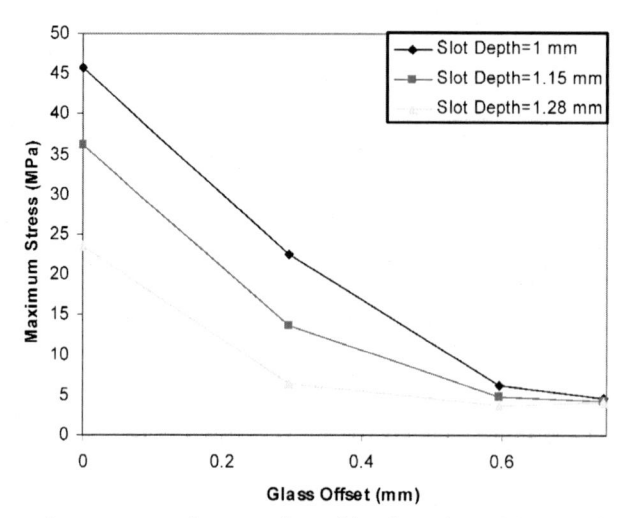

Fig 7 The maximum stress on the top surface of the glass after a -55°C to 200°C thermal cycle, as a function of the glass offset and of the slot depth. The lines are aids to observation.

For any given glass offset, the maximum stress at the top of the glass surface decreases with increasing slot depth. As the slot depth increases, the amount of steel compressing the glass decreases; however, the amount of steel axially pinching and transmitting the bending moment created by the massive bottom portion of the steel connector to the top surface of the glass is also reduced. This has a net effect of decreasing the tensile stress experienced by the glass due to the thermal mismatch at elevated temperatures. Because this slot has such a major effect on stress reduction, it was used in the design to reduce the stresses and the failure probability of the glass. For all slot depths, the maximum stress on the top surface of the glass decreases as a function of increasing glass offset. Increasing the glass offset has a beneficial effect on the stress for two possible reasons: (1) the effect of the bending stress imposed by the massive bottom portion of the steel connecter on the top surface is reduced and (2) Moving the glass downwards has the effect of making the glass thickness more co-planar with the slot. This increased symmetry of the glass about the slot reduces the pinching effect, and leads to reduced stress in the glass. The other geometrical features of the slot (slot width and slot offset) were found to have a negligible effect on the stresses in the glass. An increase in the glass thickness led to an increase in the stress; however, reducing the glass thickness to a level where the predicted stresses were low was not advisable because the shock load bearing capacity of the glass would be compromised. Utilizing the results of this study, geometrical changes were made to the design, and components that met the qualification requirements were produced. In particular, an intermediate glass thickness, a large glass offset, and a high relief groove offset was chosen for the final design.

CONCLUSION

The FEA calculations revealed that small-strain plastic deformation in the metals (shell and pins) cause local and global tensile stresses in the glass, in a geometry where compression was expected. Based on this extreme sensitivity to the plastic deformation, it would be prudent to make detailed measurements of the stress-strain behavior of the metals over the temperature range of interest, and to use improved constitutive materials models in the FEA. Because there can be significant lot-to-lot variability in the metal stress-strain behavior, specifying that only lots having acceptable (high) yield stresses be used in production would help in improving seal reliability. Since the "pinching" effect appears to localized near the glass-SS304 interface and the stiffer pins magnify this stress, placing the pins as far from the metal shell as possible might also help reduce the fracture probability of glass. Having the option to use a glass with higher strength and fracture toughness would also help in improving the reliability of the glass-to-metal seal.

ACKNOWLEDGEMENTS

Sandia is a multi-program laboratory operated by Sandia Corporation, a Lockheed Martin Company, for the United States Department of Energy's National Nuclear Security Administration under Contract-DE-AC04-94AL85000. One of the authors (RT) would like to gratefully acknowledge the energy, enthusiasm and technical mentorship by Saundra Monroe at SNL.

REFERENCES
[1] www.teknaseal.com
[2] www.twi.co.uk

[3]http://www.us.schott.com/epackaging/english/all_about_GTMs/tables.html
[4]J.A. Mitchell, "Adagio: Non-linear Quasi-static Structural Response using the Sierra Framework," in K. J. Bathe, editor, First MIT Conference on Computational Fluid and Solid Mechanics. Elsevier Science, 2001
[5]R. W. Davidge, Mechanical Behavior of Ceramics, Cambridge University Press, 1979

INTEGRATIVE DESIGN WITH CERAMICS: OPTIMIZATION STRATEGIES FOR CERAMIC/METAL JOINTS

A. Bezold, H.R. Maier, E.M. Pfaff
Institute for Materials Applications in Mechanical Engineering
Aachen University
Nizzaallee 32
D-52072 Aachen, Germany

ABSTRACT

Major goal of the concept of integrative design with ceramics is the development of an iterative optimization strategy and the effective implementation of FEM based results and empiric knowledge for computer-aided design of functional units with ceramic components. Furthermore new design elements are established and connected to built up an application, material, joining and manufacturing orientated design model including the decision criteria: material alternatives for ceramic and non-ceramic components, loading alternatives (force, heat and mass flow), joining alternatives (material, form and force fit) and manufacturing alternatives (component properties and residual stresses). Thus disadvantageous superposition of loading, joining and residual stresses should be prevented in order to improve reliability and economic efficiency.
Based on the above concept a new tool for structural multi-criteria optimization of functional units with ceramic components has been developed. MatLab was used as a programming environment which allows selecting optimization strategy, algorithms and analysis model. Basics of algorithms are briefly summarized. The general purpose FEA solver ABAQUS and CARES/Life for reliability analysis are linked together and implemented in the optimization toolbox solving sizing as well as shape optimization tasks with ceramic and non-ceramic components. A cylindrical tube ceramic/metal braze joint as a lap and butt version is selected as a case study part to demonstrate the capabilities of the above concept.

INTRODUCTION

The area of design optimization, especially structural optimization, is an active area of research. There are several reasons for the interest, including the need to handle a broader class of engineering problems, to include realistic definitions of design variables, to find techniques to locate the global optimum, and to improve computational efficiency of the numerical procedure. A great variety of optimization algorithms, genetic, stochastic, optimality criteria were already applied to solve sizing, shape, and topology optimization problems.
To make the optimization concept attractive as a design methodology would require that sizing, shape and topology aspects of structural optimization be addressed simultaneously. Topology optimization is regarded not to be universally applicable to the optimization of ceramic components e.g. due to manufacturing constraints. Thus, sizing and shape optimization has been integrated for multi-objective optimization concept of ceramic-containing components.
But what makes a multi-objective optimization concept attractive for ceramic-containing components? Design engineers are used to deal with metal materials and have less experience how to handle statistic strength data of ceramic materials. Furthermore there are few generic rules for designing e.g. ceramic/metal joints and a successful design strongly depends on the

experience of the design engineer. But joining ceramics has become a key engineering technology with particular emphasis on brazing methods e.g. for SOFCS or other gas tight applications like OTMs. Based on this the evaluation of the developed optimization strategy will be demonstrated at a generic ceramic/metal braze joint geometry.

INTEGRATIVE DESIGN CONCEPT
The present paper focuses on a computer aided optimization concept linking existing design rules, numerical stress calculation and reliability analysis. Thus a new optimization approach is developed base on the concept of integrative design with ceramic.
The basic idea of integrative design with ceramics was first introduced by Maier[1,2]. Major goal of the integrative design concept is the enhancement of an iterative optimization strategy to transfer FEA based results and empiric design knowledge into a computer aided design methodology and the appropriate interface design of ceramic-containing components (functional units). Main issue is to built up an application, material, joining and manufacturing oriented design model. The major characteristics of the integrative design philosophy are briefly summarized.

Material oriented design
Characteristics of engineering ceramics are expressed best by the following basic equations (1) – (6). Fracture stress

$$\sigma_{compr,f} \approx 2 \cdot \tau_f \approx 20 \text{ to } 40 \text{ times of } \sigma_{t,f}. \qquad (1)$$

Thus, prefer compressive stresses σ_{compr} or shear stresses τ and limit tensile stresses σ_t. Ceramics are linear elastic up to fracture:

$$\sigma_t = \varepsilon \cdot E. \qquad (2)$$

Strength distribution is the result of statistically distributed defects. Critical fracture stress is given by:

$$\sigma_{t,cr}(x,y,z) = \frac{K_{IC}}{\sqrt{a_{cr}} \cdot Y} = f[\sigma_L(x,y,z), \sigma_J(x,y,z), \sigma_R(x,y,z)]. \qquad (3)$$

K_{IC}: fracture toughness, a_{cr}: critical flaw size, Y: geometry constant, $\sigma_L(x,y,z)$: loading stresses, $\sigma_J(x,y,z)$: residual joining stresses, $\sigma_R(x,y,z)$: macro distributed residual stresses.

"Strength" is substituted by "fracture probability" P_f at a given stress level:

$$P_f = 1 - \exp\left[-\frac{1}{V_0} \int \left(\frac{\sigma_t(x,y,z)}{\sigma_{0V}}\right)^{m_V} dV\right] \qquad (4)$$

with V_0: standard volume and m_V, σ_{0V}: Weibull parameters. Fracture probability/stress is a function of component size:

$$\frac{\sigma_{t1}}{\sigma_{t2}}(P_f = const.) = \left(\frac{V_{t2}}{V_{t1}}\right)^{\frac{1}{m_f}}. \tag{5}$$

Sub-critical crack growth (SCG) does increase long term fracture probability:

$$v = \frac{da}{dt} = A \cdot K_I^n \text{ with A, n: SCG parameters.} \tag{6}$$

Load oriented design
Design with ceramic components must be treated as a whole (functional unit) due to the interaction with their ceramic and/or non-ceramic neighbors. Hence it is the main task of the design engineer to optimize force, heat and mass flow in consideration of the neighborhood to minimize tensile stresses arising from external loading. The design of a heat exchanger for coal combustion power station by Himmelstein and Maier[3] has successfully demonstrated this concept.

Joining oriented design
It is not a primary goal to minimize the resulting tensile stress distribution arising from the joining technique on its own but rather the superposed impact of joining and loading stresses (and residual stresses). There is no only one joining method that yields satisfactory to all loading conditions.

Production oriented design
Material properties like Young's modulus, Poisson's ratio, thermal expansion, thermal conductivity, heat capacity and fracture toughness, Weibull and crack growth parameters as well as the macroscopic residual stress distribution $\sigma_R(x,y,z)$ depend on the selection of the ceramic material and the best suited processing technology. Macroscopic residual stresses can originate from density gradients during shaping, humidity gradients during drying and temperature gradients during sintering or from surface modification during finishing. Macroscopic residual stresses can offer a great potential for compensating resultant tensile stresses ($\sigma_L(x,y,z)$ and $\sigma_J(x,y,z)$).

MULTI-OBJECTIVE OPTIMIZATION

Then aim of structural optimization is the minimization of an objective function, e.g. maximum stresses, maximum deflection or fracture probability, etc., subjected to geometrical and behavioral constraints. Geometrical constraints are classified as restrictions on cross-sectional dimensions, maximum height, etc., whereas behavioral constraints can be restrictions on stresses, displacements, weight, fracture probability, etc. When there are multiple objective functions to be considered, the design problem becomes multi-objective. In this case, the usual design optimization method for a scalar objective function cannot be used. A multi-objective optimization problem can be stated as follows [4,5]:

$$\min \mathbf{F}(\mathbf{x}, \mathbf{p}) = [F_1, F_2 \ldots F_m]^T,$$
$$\text{with } \mathbf{g}(\mathbf{x}, \mathbf{p}) \leq 0,$$
$$\mathbf{h}(\mathbf{x}, \mathbf{p}) = 0 \text{ and}$$

Integrative Design with Ceramics: Optimization Strategies for Ceramic/Metal Joints

$$x_{i,LB} \leq x_i \leq x_{i,UB} \ (i = 1, ..., n). \tag{7}$$

The objective function vector **F** is a function of design vector **x** and a fixed parameter vector **p**; **g** and **h** are inequality and equality constraints, respectively. $x_{i,LB}$ and $x_{i,UB}$ are the lower and upper bounds for the ith design variable, respectively. Stadler[4,5] applied the notion of Pareto optimality to the fields of engineering and science in the 1970s. The most widely used method for multi-objective optimization is the weighted sum method. The method transforms multiple objectives into an aggregated scalar objective function by multiplying each objective function by a weighting factor and summing up all contributors:

$$F_{weighted\ sum} = w_1 F_1 + w_2 F_2 + ... + w_m F_m. \tag{8}$$

w_i (i = 1, ...,m) is a weighting factor for the i-th objective function (which potentially can be divided by a scaling factor, i.e., $w_i = \alpha_i/sf_i$). If

$$\sum_{i=1}^{m} w_i = 1 \text{ and } 0 \leq w_i \leq 1, \tag{9}$$

then the weighted sum is said to be a convex combination of objectives. Each single objective optimization determines one particular optimal solution point on the Pareto front. The weighted sum method then changes weights systematically, and each different single objective optimization determines a different optimal solution. The solutions obtained approximate the Pareto front. Note that if there are non-unique anchor points, weights that have zero values may produce weak Pareto optimal solutions. However, using standard weighted sum method, the way in which the Pareto front is computed is determined a priori, e.g. by predefining weights. The adaptive weighted sum method (AWS) suggested by Kim[6,7] "learns" the shape of the Pareto front iteratively until some desired level of resolution is achieved. AWS method is on the way to be implemented in the optimization procedure.

Algorithms
    In spite of the availability of several optimization techniques, no one method yields satisfactory performance across the spectrum of problems confronting a design engineer. Genetic algorithms, heuristic concepts, mathematical programming, deterministic (gradient based) and stochastic (e.g. simulated annealing) algorithms offer a great potential in numerical optimization. For solving multi-objective structural optimization problems using the integrative design concept latter both simulated annealing and a gradient based quasi-Newton method as well as a hybrid algorithm were applied and are described in the following.
Simulated annealing (SA) is a multivariable optimization technique based on the Monte-Carlo method used in statistical mechanics studies of condensed systems is adapted for solving single and multi-objective structural optimization problems[8,9,10]. This SA procedure draws an analogy between energy minimization in physical systems and objective function minimization in structural systems. The search for a minimum is simulated by a relaxation of the statistical mechanical system where a probabilistic acceptance criterion is used to accept or reject candidate designs. It takes the objective function of an optimization problem as the energy corresponding to a given state, and the candidates (vectors of design variables) in the search space are treated as

the possible states for an equilibrium temperature, which is a control parameter in the process. The general scheme of SA can be stated as follows[11]:
1) Generate randomly a candidate **x**.
2) Choose T > 0 be the initial temperature.
3) Stop, if a stopping criterion is satisfied; otherwise repeat the following steps:
   a) Exit this loop if equilibrium is reached.
   b) Let **x'** be a randomly selected neighbour of **x**.
   c) Generate a uniform random number U in [0, 1].
   d) If $p = \exp\{-[f(\mathbf{x'})-f(\mathbf{x})]/T\} > U$, then **x** =**x'** is the new current optimum design.
4) Let T be a new (lower) temperature value; go to step 3.

A major drawback of SA, compared with other stochastic methods like genetic algorithms is that it searches only from a single point, instead of from a population of points. To overcome the above problems Tzan and Pantelides[12] suggested Simulated Annealing with Automatic Reduction of Search Range (SAARSR). As a second alternative a gradient based algorithm (quasi-Newton method) suggested by Broyden–Fletcher–Goldfarb–Shanno (BFGS) was used to have a faster rate of convergence and global convergence. The quasi-Newton methods that build up an approximation of the inverse Hessian are often regarded as the most sophisticated for solving unconstrained problems. It is known that Newton based algorithms could fail for general nonlinear problems and get stuck at saddle-points or couldn't escape local minima. Thus a hybrid-algorithm was introduced by linking BFGS and SA to overcome above drawbacks.

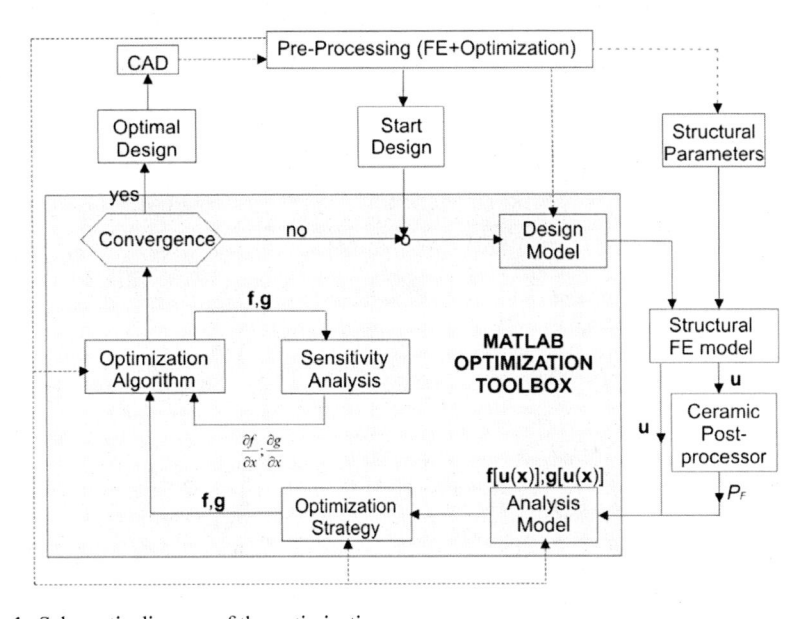

Fig. 1. Schematic diagram of the optimization process.

Optimization methodology for integrative design

ABAQUS[13] finite element code is used for calculation of residual joining stresses after the brazing cycle and superposed loading stresses for metal as well as for ceramic materials of the starting design. The CARES/LIFE[14] program is used for probabilistic failure analysis. On basis of the resulting stress distribution and experimentally determined Weibull data the fast fracture probability of the ceramic components is evaluated. All system responses **u** (stresses, deformations, etc.) and fracture probability P_F are available as objectives for further optimization process as depicted in Fig. 1. Starting the inner loop of the optimization process (grey shaded in Fig. 1) the analysis model is built by initializing system constraints and design variables, choosing an appropriate optimization strategy (single-, bi- or multi-objective) for parameter as well as for shape optimization. Algorithms can be selected manually (default SA).

To enhance computational time a sensitivity analysis should be carried out to identify design variables with strong impact on the objective function. For reading FEA input file and writing the new design variables to a modified input file several interface programs were built using Python script language and implemented in a MatLab[15] optimization environment. This is a major advantage because all variables and results could be easily visualized and processed.

The optimization loop is passed if stopping parameters are satisfied and convergence is achieved. Then the optimum design could be transferred to CAD or to FEA program for further processing.

CASE STUDY – CERAMIC/METALL TUBE JOINT

A generic cylindrical lap and butt geometry of a ceramic/metal tube joint (Fig. 2) has been used as an example to demonstrate the introduced multi-objective optimization concept. Finite element analysis of the component is linked with probabilistic failure analysis of ceramic materials and automated optimization for determining the best design.

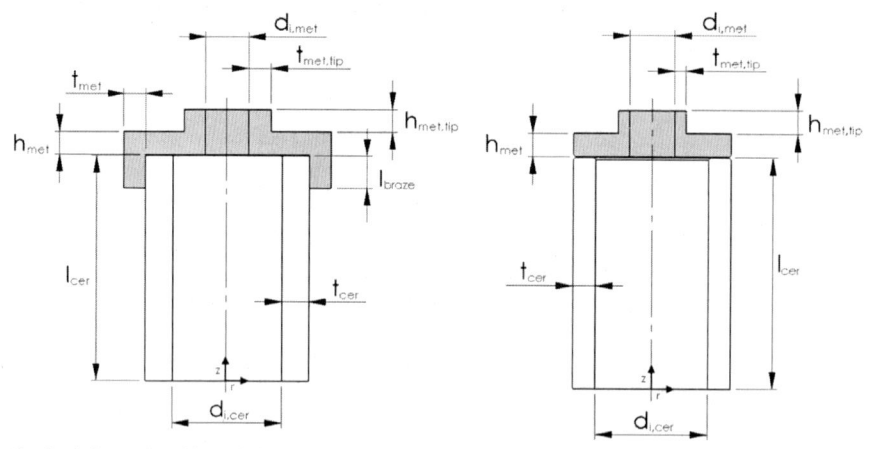

Fig. 2. Schematic of lap (right) and butt (left) joint composed of alumina tube (white), Ti6Al4V cap (grey) and Cusil-ABA braze layer.

Design variables

For the optimization of the lap and the butt joint wall thickness of the ceramic tube t_{cer}, thickness of metal tip $t_{met,tip}$ and axial thickness of metal cap h_{met} have been selected as design variables. Additionally radial thickness of the metal cap cap t_{met} and braze layer length l_{braze} for the lap joint geometry. Length l_{cer} and the inner diameter $d_{i,cer}$ of ceramic tube have been fixed to 20mm and 10mm respectively. Inner diameter of the metal cap $d_{i,net}$ and height of metal tip $h_{met,tip}$ have been fixed to 4mm and 2 mm, and the thickness of the braze layer to 50μm. Limits and starting values of the design variables have been applied by design constraints as summarized in Table I. In practical use the design limits are directed by manufacturing and tooling considerations.

Table I. Design variables of lap and butt joint geometry.

Design variable	Start	Minimum	Maximum
Wall thickness of ceramic tube t_{cer} [mm]	2.5	1.0	4.0
Thickness of metal tip $t_{met,tip}$ [mm]	2.0	1.0	4.0
Axial thickness of metal cap h_{met} [mm]	2.0	1.0	4.0
Radial thickness of metal cap t_{met} [mm]	2.0	1.0	4.0
Length of the braze layer l_{braze} [mm]	2.0	1.0	4.0

Materials

A given Alumina standard has been used as the ceramic partner of the joint and was treated as an elastic solid according equation (2). For fast fracture analysis a Weibull modulus of 12.8 and scale parameter of 379 MPa were applied. Ti6Al4V and Cusil-ABA has been selected as a metal and braze material respectively. For both materials elastic/plastic behavior was assumed. Temperature dependent material properties are summarized in Table II. Time dependent material behavior e.g. sub-critical crack growth for the ceramic and creep of the metal and braze material were not considered yet.

Table II. Temperature dependent material properties (elastic/plastic).

Material		E [GPa]	v	α [10^{-6} 1/K]	R_p [MPa]	E_T [MPa]
Ti 6Al 4V						
	RT	114	0.3	9.0	910	500
	750°C			9.7	565	694
Cusil-ABA						
	RT	93.6	0.36	19.5	330	1050
	550°C		0.38		75	0
	750°C	59.1	0.402		10	
Al$_2$O$_3$						
	RT	410	0.25	7.1		
	750°C			8.9		

Stress analysis

Fast fracture failure probability of the alumina component has been calculated on the basis of the residual joining stresses by cooling the joint from brazing temperature (800°C) to room temperature (case 1 for the lap and case 2 for the butt joint). It is assumed that there is ideal adhesion between the joined materials and the joint is stress free at brazing temperature. Stress

state at room temperature has been used for all following analyses. Stresses arising from external loading were superposed to joining stresses according equation (3). Results of the stress and failure probability analysis are transferred to the optimizer for cost function evaluation. Macroscopic residual stresses originating from manufacturing process or finishing are not included at the moment but offer a great potential for compensating critical tensile stresses. Three external load cases have been investigated:
1. An internal pressure of 60 MPa is applied on the inner surface of the ceramic tube and metal cap (case 3),
2. An axial force of 4kN is applied on the metal tip (case 4) and
3. Internal pressure and axial force are applied simultaneously (case 3+4).

Optimization
Using the above described multi-objective optimization procedure an objective function must be defined. The objective function refers to a quantitative rating given to a particular design. Primary goal is the minimization of fracture probability of the ceramic component. Furthermore the non-ceramic neighbors are recognized by the equivalent von Mises stresses arising in the metal part and braze. Exploiting the full strength potential of the non-ceramic neighbors equivalent von Mises stresses are allowed to be close to the yield stress. Thus an objective function (equation 10) is built with a high weight factor w_{cer} for the fracture probability and low weight factors for the maximum von Mises stresses in the metal (w_{metal}) and braze (w_{braze}) respectively:

$$\min F(X) = w_{cer} P_F + w_{metal} \sigma_{max,mises,metal} + w_{braze} \sigma_{max,mises,braze} \qquad (10)$$

Results of FEA and reliability analysis for a starting set of design variables are used for evaluation of the objective function. Optimization algorithm proceeds searching for a new set of design variables until convergence criteria are satisfied. In the current study convergence was achieved after 100 designs.

RESULTS
Cooling down from brazing temperature to room temperature causes significant tensile stresses due to the mismatch in thermal expansion coefficient of the joint partners. The upper outside corner of the ceramic tube in the butt joint is sensitive to fracture because of tensile stresses arising from the shrinkage of the metal cap during cooling down from brazing temperature. But the stress level still remains uncritical (P_F=0.03%) for the selected geometry. Applying an internal pressure (case 2) fracture probability increases to 1.0% and 1.2% for the axial force (case 3). Optimization procedure determines a new set of design variables reducing fracture probability to 0.1% for both cases. For the combined load case (case 3+4) fracture probability has been decreased from 100% to 1.3%. No further reduction was possible because upper limits of the design variables were reached. Thickness of the ceramic tube has a higher effect on joint quality than thickness of the metal tip and axial thickness of the metal cap. In table IV optimization results of the butt geometry for all load cases are summarized.
Maximum principle stresses for the lap joint occurred at the end of the metal cylindrical metal cap where it is joined to the ceramic tube. These regions indicate the probable crack initiation site because of the highest fracture probability as depicted in Fig. 3-5 for load case 3, 4 and the combined one. A likely cause of this was the tendency for the metal part to pull on the ceramic at

Integrative Design with Ceramics: Optimization Strategies for Ceramic/Metal Joints

Metal cap
max. σ_{Mises} = 634 MPa

Ceramic tube
$\sigma_{I,max}$ = 527 MPa
P_F = 99%

Metal cap
max. σ_{Mises} = 509 MPa

Ceramic tube
$\sigma_{I,max}$ = 415 MPa
P_F = 7%

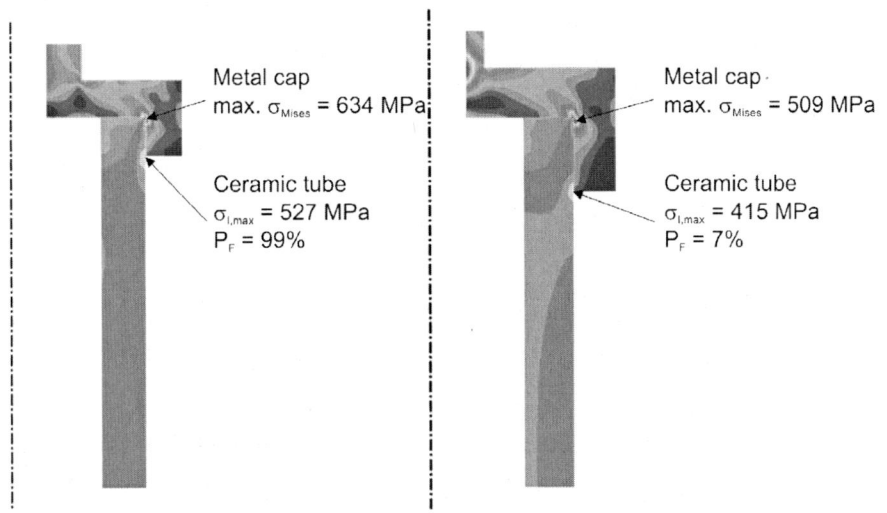

Fig. 3. Maximum principal stress distribution in alumina tube and equivalent von Mises stresses in metal part of initial and optimized lap joint geometry for case 1+3.

Table III. Optimization results of the lap joint.

Criteria/design var.	Case 1+3 (p_i=60MPa)		Case 1+4 (F_{ax}= 4kN)		Case 1+3+4 (p_i+F_{ax})	
	Initial	Optimum	Initial	Optimum	Initial	Optimum
$\sigma_{Imax,cer}$ [MPa]	526.5	414.9	298.4	147.1	679.2	536.3
$\sigma_{mises,metal}$/ $R_{p,met}$ [-]	0.7	0.6	0.4	0.1	1.1	0.7
$\sigma_{mises,braze}$/ $R_{p,braze}$ [-]	1.0	1.0	1.0	1.0	1.1	1.0
Location of $\sigma_{Imax,cer}$	r=r_a	r=r_a	r=r_a	r=r_a	r=r_a	r=r_a
(qualitatively)	z=l_{cer}	z=l_{cer}	z=l_{cer}	z=l_{cer}-l_{braze}	z=l_{cer}	z=l_{cer}
$\Delta\sigma_{Imax,ceramic}$ / Δx [MPa/100μm]	179	94	67	9	231	102
V_{eff}/V_{geo} [10^{-3}]	4.2	18.8	3.7	0.1	-	9.7
P_F [%]	99.1	7.2	1.1	0	100	99.0
Thickness ceramic wall t_{cer} [mm]	-	2.8	-	1.5	-	1.5
Thickness metal tip $t_{met,tip}$ [mm]	-	1.0	-	4.0	-	1.0
Radial thickness metal cap t_{met} [mm]	-	2.6	-	4.0	-	4.0
Axial thickness metal cap h_{met} [mm]	-	2.3	-	4.0	-	1.0
Length braze layer l_{braze} [mm]	-	3.9	-	4.0	-	4.0

this point. Maximum principal stresses arising from cooling down from brazing temperature remain uncritical (101 MPa, P_F=0%) for the starting lap geometry. Applying an internal pressure (case 2) fracture probability increases to 99.1% and 1.1% for the axial force (case 3). Furthermore the gradient of maximum principal stress perpendicular to the direction of the maximum principal stress in these regions dramatically increases from 13 to 169 and 67 MPa/100µm for case 3 and case 4 respectively. This indicates in addition to the fracture probability a high risk for crack initiation and growth. For case 2 the optimizer reduces fracture probability to 7.2% and the stress gradient by a factor of 2 by increasing the length of the cylindrical metal cap. Reducing the weight factor for the contribution of the equivalent von Mises stresses in the metal part the probability of fracture could be further optimized to 1% approximately. Optimization for case 3 results in a fracture probability of 0% and a stress gradient of 9 MPa/100µm. Here length, radial and axial thickness of the cylindrical metal cap has been increased and thickness of the ceramic tube has been reduced to 1.5mm. For the combined load case 3+4 a fracture probability of 100% is predicted. Even by changing optimization strategy no further reduction of the fracture probability is possible using parameter optimization and the given limits of the design variables.

Table III summarizes optimization results of all load cases for the lap joint geometry. Maximum von Mises stresses arising in the metal part and in the braze material are scaled to yield stress indicating the level of loading. For the braze material the ratio is close to 1 indicating that the strength potential of the material is fully exploited. Whereas the ratio for the metal part is between 0.1 and 0.7 for the optimized geometry indicating that there is further potential to increase the quality of the joint.

Metal cap
max. σ_{Mises} = 332 MPa

Ceramic tube
$\sigma_{I,max}$ = 298 MPa
P_F = 1%

Metal cap
max. σ_{Mises} = 119 MPa

Ceramic tube
$\sigma_{I,max}$ = 146 MPa
P_F = 0%

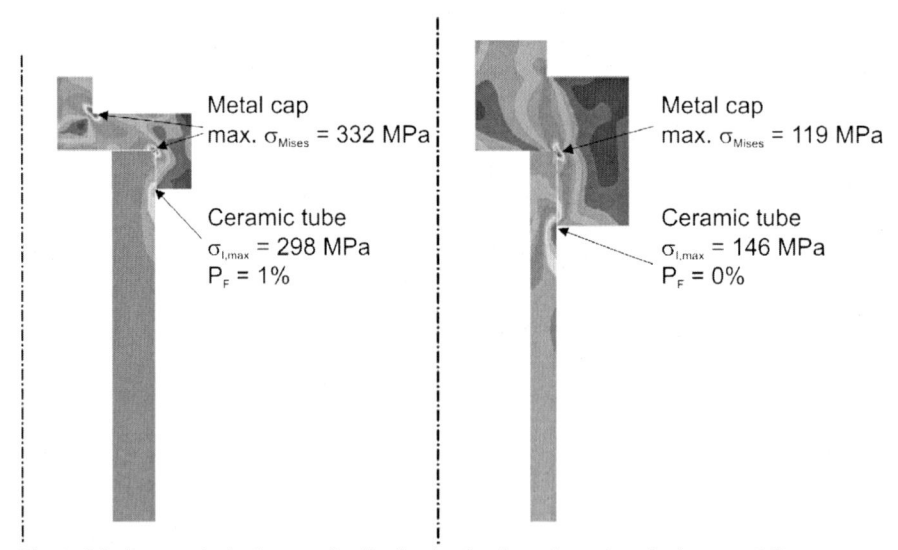

Fig. 4. Maximum principal stress distribution in alumina tube and equivalent von Mises stresses in metal part of initial and optimized lap joint geometry for case 1+4.

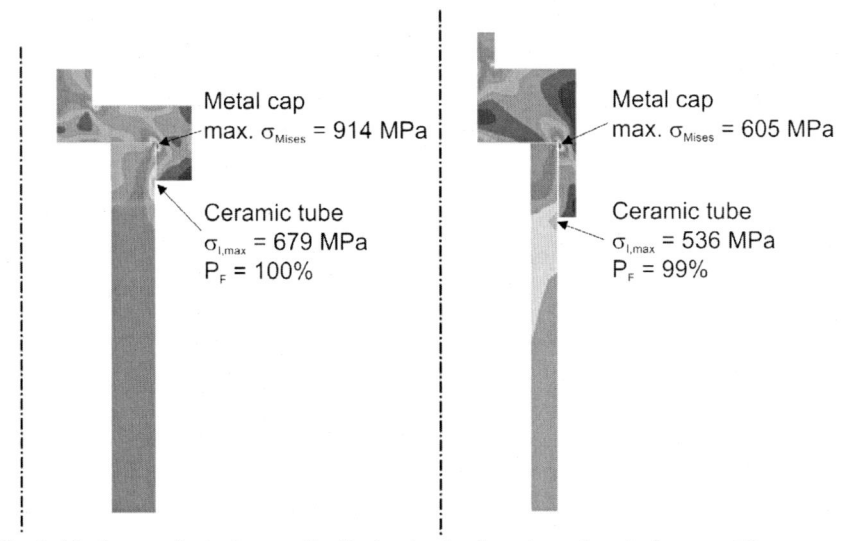

Fig. 5. Maximum principal stress distribution in alumina tube and equivalent von Mises stresses in metal part of initial and optimized lap joint geometry for case 1+3+4.

Table IV. Optimization results of the butt joint.

Criteria/design var.	Case 2+3 (p_i=60MPa)		Case 2+4 (F_{ax}=4kN)		Case 2+3+4 (p_i+F_{ax})	
	Initial	Optimum	Initial	Optimum	Initial	Optimum
$\sigma_{Imax,cer}$ [MPa]	290.8	226.8	292.1	228.6	500.3	295.7
Location of $_{Imax,cer}$ (qualitatively)	$r=r_i$ $z=0.98l_{cer}$	$r=r_a$ $z=0.98l_{cer}$	$r=r_i$ $z=0.98l_{cer}$	$r=r_a$ $z=0.98l_{cer}$	$r=r_i$ $z=l_{cer}$	$r=r_i$ $z=0.98l_{cer}$
$\Delta\sigma_{Imax,cer} / \Delta x$ [MPa/100µm]	71	66	28	66.6	40	30
V_{eff}/V_{geo} [10^{-3}]	1.7	0.3	1.5	0.3	-	0.9
P_F [%]	1.0	0.1	1.2	0.1	100	1.3
Thickness ceramic wall t_{cer} [mm]	-	4.0	-	4.0	-	4.0
Thickness metal tip $t_{met,tip}$ [mm]	-	4.0	-	4.0	-	4.0
Axial thickness metal cap h_{met} [mm]	-	4.0	-	2.9	-	4.0

CONCLUSIONS

The above developed optimization concept based on the synthesis of loading, joining and residual stresses has been applied to the optimization of ceramic/metal braze joints. Case study results show that an optimized set of design variables strongly depend on the joint geometry and the applied loading. Optimization of the butt joint geometry leads to a significantly reduction of fracture probability for all load cases. Whereas for the lap geometry fracture probability could not be reduced for combined load case (internal pressure and axial force). This clearly indicates the limits of pure parameter optimization. However the concept offers the possibility to identify most sensitive design variables. This is of great importance to adjust tolerances in manufacturing and machining process. Future studies will focus on adding shape optimization technique which offers further potential to increase reliability of ceramic to metal joints. Furthermore a discrete-continuous optimization strategy will be integrated by linking a material data base for selection of the best suited joining partners, and interlayer materials.

REFERENCES

[1] Maier, H.R.: "Integratives Konstruieren mit Keramik – Synthese von Eigen-, Verbund- und Lastspannungen", Keramische Komponenten für das Spritzgießen und Extrudieren, Selb (Germany), May 2005 in German.

[2] Maier, H.R.: "Synthese von rechnergestützten und experimentellen Spannungsanalysen am Beispiel von Keramik/Metall- und Keramik/Polymer-Verbunden", Verbundwerkstoffe und Werkstoffverbunde, Oberursel (Germany), October 1995 in German.

[3] Himmelstein, K.; Maier, H. R.: „Design and testing of a prototype heat exchanger for coal combustion power stations", In: 7. International Symposium "Ceramic Materials and Components for Engines" - applications in energy, transportation and environment systems (2001).

[4] Stadler W.: "A survey of multicriteria optimization, or the vector maximum problem". J. of Opt. Theory and Applications **29**, 1–52 (1979).

[5] Stadler W.: "Applications of multicriteria optimization in engineering and the sciences" (A Survey). In: Zeleny M. (ed.) Multiple criteria decision making - past decade and future trends. JAI, Greenwich (1984).

[6] Kim I.Y.; deWeck O.L.: "Adaptive weighted-sum method for bi-objective optimization: Pareto front generation", Struct. Multidisc. Optim., **29**, 149–158 (2005).

[7] Kim I.Y.; deWeck O.L.: "Adaptive weighted sum method for multiobjective optimization: a new method for Pareto front generation", Struct. Multidisc. Optim., **31**, 105–116 (2006).

[8] Jaynes, E.T.: "Information Theory and Statistical Mechanics," Physical Reviews, **106** [6], 620–630 (1957).

[9] Kirkpatrick, S.; Gelatt, C.D.; Vecchi, M.P.: "Optimization by Simulated Annealing," Science, **220**, 671–680 (1983).

[10] Metropolis N.; Rosenbluth A.W.; Rosenbluth M.N.; Teller A.H.: "Equation of state calculation by fast computing Machines", J. Chem. Phys., **21**, 1087-1092 (1953).

[11] Moh J.-S.; Chiang D.-Y.: "Improved Simulated Annealing Search for Structural Optimization", AIAA Journal. **38** [10] (2000).

[12] Tzan S.; Pantelides C.P.: "Annealing strategy for optimal structural design". Journal of Structural Engineering, **122** [7], 815-827 (1996)

[13] ABAQUS v. 6.5.1, Hibbitt, Karlsson & Sorensen, Inc., 2006.

[14] Nemeth, N.N.; Powers L.M.; Janosik, L.A.; Gyekenyesi, J.P.: "CARES/LIFE – Ceramics Analysis and Reliability Evaluation of Structures Life Prediction Program", NASA Lewis Research Center (1993).

[15] MATLAB v. 7.3.0.267, The Matworks, Inc., 2006.

DIFFUSION BONDING OF SILICON CARBIDE FOR MEMS-LDI APPLICATIONS

Michael C. Halbig
U.S. Army Research Laboratory
NASA Glenn Research Center
Cleveland, Ohio 44135

Tarah Shpargel
ASRC Aerospace Corporation
NASA Glenn Research Center
Cleveland, Ohio 44135

Mrityunjay Singh
Ohio Aerospace Institute
NASA Glenn Research Center
Cleveland, Ohio 44135

James D. Kiser
NASA Glenn Research Center
Cleveland, Ohio 44135

ABSTRACT

A robust joining approach is critically needed for a Micro-Electro-Mechanical Systems-Lean Direct Injector (MEMS-LDI) application which requires leak free joints with high temperature mechanical capability. Diffusion bonding is well suited for the MEMS-LDI application. Diffusion bonds were fabricated using titanium interlayers between silicon carbide substrates during hot pressing. The interlayers consisted of either alloyed titanium foil or physically vapor deposited (PVD) titanium coatings. Microscopy shows that well adhered, crack free diffusion bonds are formed under optimal conditions. Under less than optimal conditions, microcracks are present in the bond layer due to the formation of intermetallic phases. Electron microprobe analysis was used to identify the reaction formed phases in the diffusion bond. Various compatibility issues among the phases in the interlayer and substrate are discussed. Also, the effects of temperature, pressure, time, silicon carbide substrate type, and type of titanium interlayer and thickness on the microstructure and composition of joints are discussed.

INTRODUCTION

Silicon carbide based ceramic materials have capabilities that meet the needs for several types of high temperature structural applications. Beneficial properties include high temperature stability, high strength, and corrosion resistance. Monolithic silicon carbide based ceramics are currently being used as seals, heating elements, diffusion furniture in microelectronic industries, mirror and optical components for space applications, and ceramic armor. In addition to these applications, silicon carbide based composites are also being developed for aerospace and ground-based turbine engine and hot structure components as well as nuclear reactor applications. However, limitations in current processing methods create a barrier to the wider utilization of silicon carbide (SiC) based ceramic materials. Complex shaped components can't be easily fabricated through most conventional ceramic processing methods. New designs and/or processing methods are needed. One cost-effective solution for fabricating large, complex-shaped components is through the joining of simple-shaped ceramics.

Several different methods have been investigated and utilized for joining ceramics. These methods have included brazing, adhesive bonding, fusion welding, diffusion bonding, and friction welding[1]. Many of these approaches have been used to join silicon carbide. Reaction formed joints have been fabricated through the use of a carbonaceous mixture reacted with silicon paste to form silicon carbide[2,3]. Other methods use preceramic polymers at the interface that undergo a step to convert the polymer to a ceramic[4] or to a silicon oxycarbide glass[5]. Brazing-like processes in which multiple interlayers are used to form joints through the formation of a thin transient liquid phase layer have been investigated. In this process, temperatures are typically several hundred degrees lower than that required for more traditional joining methods[6]. Diffusion bonding of silicon carbide has also been performed with the use of interlayers of Inconel 600[7], nickel[8], and titanium[8,9].

Diffusion bonding of SiC is a robust joining approach that is well suited for a proposed lean-direct-injector (LDI) application being developed within NASA's Subsonic Foundation Research Program. A diagram of the LDI which consists of joined SiC laminates is shown in Figure 1. Lean-burning and ultra-low emissions are achieved from the laminate designs which consist of fuel and air channels that create intricate and interlaced passages that speed up fuel-air mixing. Additional benefits of the design and of using SiC laminates include: passages of any shape can be created to allow for multiple fuel circuits, thermal protection of the fuel to prevent coking, and low cost fabrication of modules with complicated internal geometries through chemical etching. The requirements for such an application are the ability to bond relatively large geometries (i.e. 10.16 cm diameter disks), leak free operation, high strength, and chemical and thermal stability. Diffusion bonding with the aid of hot pressing is a favorable joining approach for this application since the flat substrate design and the SiC and Ti interlayer materials allow for high temperature bonding under high stresses.

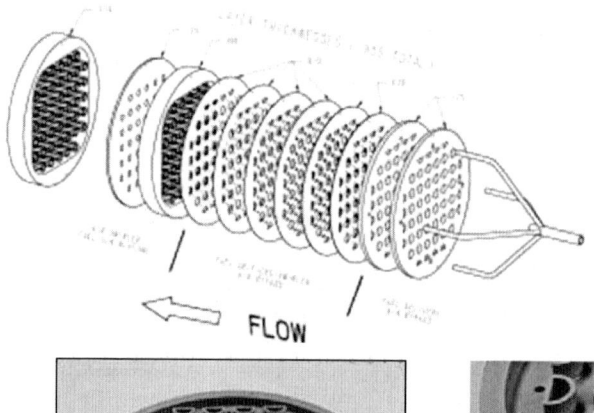

Figure 1. Illustration of the lean direct injector design (LDI)[10].

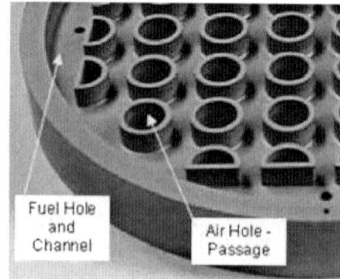

Figure 2. One of the SiC laminates for the LDI (left). Detail of the open channels and air holes (right).

EXPERIMENTAL

Diffusion bonding was conducted on three types of silicon carbide substrates with the aid of two types of titanium interlayers between the SiC substrates. The three types of silicon carbide were α-SiC referred to as CRYSTAR from Saint-Gobain, CVC-SiC (chemical vapor composite processed SiC) from TREX Enterprises, and CVD-SiC (chemically vapor deposited SiC) from Rohm & Haas. The first source of titanium was a 10 micron thick physically vapor deposited (PVD) Ti coating that was applied in-house at NASA GRC on top of the Rohm & Haas CVD-SiC substrates. The Ti coated surface acted as an interlayer. In order to investigate the bonds formed from 10 and 20 micron interlayers, the PVD coated substrate was matched with a second uncoated substrate in one case and with a coated substrate in another case. The second source of titanium was an alloyed titanium foil that was 38 microns in thickness and was used as an interlayer between all three sets of substrates.

Table I. Silicon carbide and titanium material combinations (top) and processing conditions for the select material combinations (bottom).

SiC and Ti Material Combinations:
1. 4.45 cm diameter α-SiC (CRYSTAR from Saint-Gobain) disks joined with a 38 micron Ti foil
2. 4.45 cm diameter CVC SiC (TREX Enterprises) disks joined with a 38 micron Ti foil
3. 2.54 cm x 5.08 cm CVD SiC (Rohm & Haas) joined with a 10 micron PVD Ti coating on one of the mating surfaces
4. 2.54 cm x 5.08 cm CVD SiC (Rohm & Haas) joined with a 38 micron Ti foil
5. 2.54 cm x 5.08 cm CVD SiC (Rohm & Haas) joined with a 10 micron PVD Ti coating on each of the mating surfaces

Processing Conditions for the Material Combinations

Condition	Temp. [°C]	Pressure [MPa]	Time [hr]	Atmosphere	Cooling Rate [°C/min]	Analysis
A (materials 1, 2, and 3)	1250	24, 24, 31	2	vacuum	5	microsopy & microprobe
B (materials 1 and 3)	1300	24, 31	2	vacuum	2	microscopy
C (materials 1 and 3)	1250	50, 50	2	vacuum	2	microscopy
D (materials 1, 4 and 5)	1250	24, 31, 31	2	vacuum	2	microscopy & microprobe
as-received alloyed Ti foil						microprobe

The diffusion bonds were formed using different material combinations and processing conditions as shown in Table I. Processing was conducted at four different conditions. In the baseline processing condition (condition A), the hot pressing was done at a temperature of 1250°C, a minimum clamping pressure (24 or 31 MPa depending on the sample size), a 2 hr hold in a vacuum environment, and a cooling rate of 5°C per minute. The other conditions had a slight variation from the baseline. In condition B, the hot pressing temperature was 1300°C and the cooling rate was 2°C per minute. In condition C, the applied pressure was 50 MPa and the cooling rate was 2°C per minute. In condition D, the cooling rate was 2°C per minute. Two to three samples of the five material combinations of Ti and SiC were processed at each of the four processing conditions as shown in Table I. The objective was to study the bonds between different materials and to identify the optimal bonding materials and processing conditions. Part of the processing matrix also addressed issues of minimizing the formation of microcracks in the diffusion bonds that were formed when Ti foil was used as the interlayer. The minimum clamping stress refers to the minimum hydraulic pressure of 110 psi required to clamp down on a

25.4 cm diameter cylinder that distributes the load to the sample. Since different sized SiC samples were used, the minimum clamping pressures were 24 MPa for 4.45 cm disks and 31 MPa for 2.54 cm x 5.08 cm rectangular samples.

Polished cross sections of the as-received alloyed Ti foil and the diffusion bonds were prepared. All of the microscopy results were obtained through characterization of those polished sections. A scanning electron microscope (SEM) was used during the initial examination of the microstructure of the diffusion bonds. The structure and composition of the foil and the diffusion bonds were further characterized using a JEOL JXA-8200 Superprobe electron microprobe. The quantitative analyses obtained with the electron microprobe made it possible to identify the reaction formed phases in the bond regions.

RESULTS AND DISCUSSION

A previous analysis did not detect vanadium in the alloyed Ti foil and the resulting diffusion bonds[11]. The microprobe results presented here provide a more detailed analysis of the foil and the diffusion bonds. All images labeled as micrographs are backscattered electron images obtained with the electron microprobe, where differences in the compositions of the phases observed resulted in significant differences in contrast between the phases. A micrograph of the polished cross section of the alloyed Ti foil is shown in Figure 3. The primary phase appears dark grey while a secondary phase appears white. Results of electron microprobe analysis for the two phases are shown in Table II. The alloyed Ti foil has a composition of Ti-6Al-4V (weight %) which is an alpha-beta Ti alloy commonly used for aeronautics applications.

Microscopy of the diffusion bonds that were formed with the alloyed Ti foil revealed bonds that were well adhered to the SiC substrates with no delaminations. However, microcracks were observed which traversed across the width of the diffusion bonds perpendicular to the substrate surfaces. A secondary electron image of the diffusion bond formed with the TREX SiC and the alloyed Ti foil is shown in Figure 4. The results are typical for the bonds formed with the alloyed Ti foil. Within the diffusion bond, there are three distinct parallel regions. Examination of the bond near the substrate edge reveals small, scattered pores. Another micrograph of the same diffusion bond is shown in Figure 5. The SiC substrates appear black above and below the diffusion bond. Higher contrast can be seen between the three layers within the bond, and distinct reaction formed phases are much more visible than in the secondary electron image (Figure 4). The corresponding microprobe analysis of the phases in the bond region is given in Table III. The central core region is primarily made up of the constituents of the alloyed foil (Ti, Al, and V). However, some of the Ti has migrated out of this region and alpha and beta type phases (labeled as D and E) have become enriched in Al and V, and more segregated. The two outer regions adjacent to the SiC substrate have high concentrations of Si and/or C. Similar results were obtained for diffusion bonds formed between the other two SiC materials with the alloyed Ti foil as an interlayer. For the material combination of the alloyed Ti foil and the CRYSTAR α-SiC, an image of the diffusion bond is shown in Figure 6 and the results from the microprobe analysis are given in Table IV. An image of the material combination of the alloyed Ti foil and the Rohm & Haas SiC is shown in Figure 7 and the results from the microprobe analysis are given in Table V. Microcracks were present in the diffusion bond whenever the alloyed Ti foil was used as an interlayer. This was despite the use of three different SiC substrates and different processing conditions as shown in Table I. such as a slower cooling rate, a higher processing temperature, and a higher processing stress.

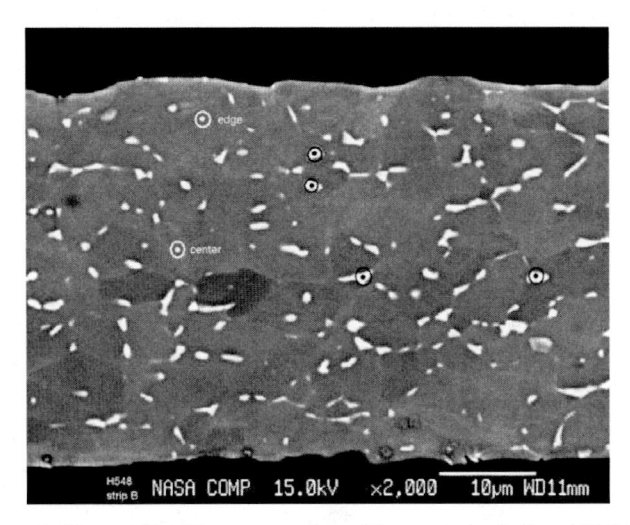

Figure 3. Micrograph of the cross-section of the as-received alloyed Ti foil.

Table II. Microprobe analysis results of the two phases within the as-received alloyed Ti foil as shown in Figure 3 (*values are an average among two data locations for the dark grey phase and three for the white phase*).

Phase			Al	Fe	Ti	V	Total
Atomic Ratio	Grey Phase		10.196	0.042	86.774	2.988	100.000
Weight (%)	Grey Phase		5.999	0.051	90.632	3.318	100.000
Atomic Ratio	White Phase		4.841	1.850	76.507	16.803	100.000
Weight (%)	White Phase		2.748	2.172	77.084	17.997	100.000

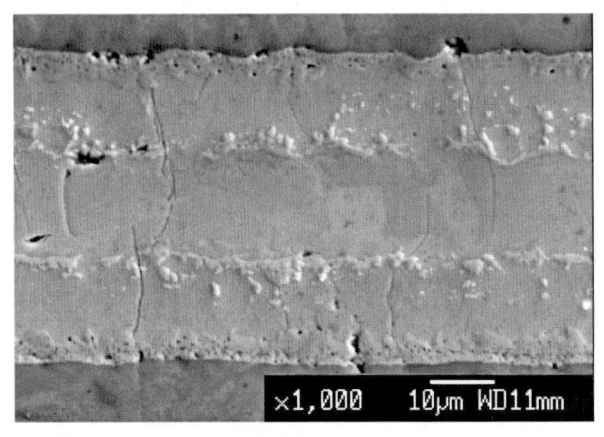

Figure 4. Secondary electron image of the diffusion bond for the alloyed Ti foil and the Trex CVD SiC.

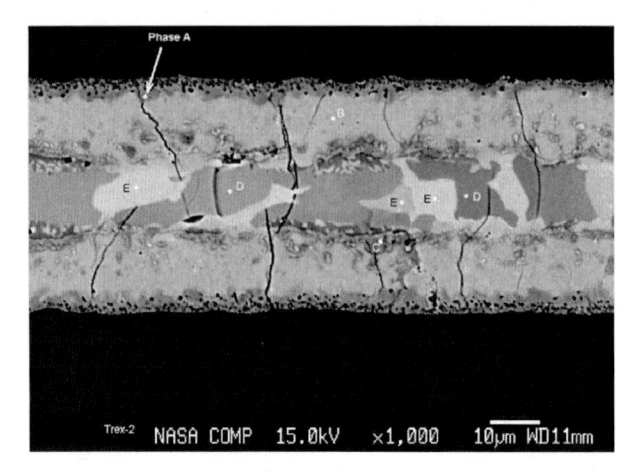

Figure 5. Micrograph of the cross-section of the diffusion bond formed between Trex CVD SiC with the alloyed Ti foil as the interlayer.

Table III. Microprobe analysis of the atomic ratios for the reaction formed phases in the diffusion bond between TREX SiC as shown in Figure 5 (*atomic ratios are an average from five locations for each phase*).

Phase	Al	V	Cr	Si	C	Ti	Fe	Total
Phase A	0.101	0.578	0.026	18.733	24.317	56.243	0.004	100.000
Phase B	0.384	1.220	0.025	34.639	6.904	56.794	0.034	100.000
Phase C	0.604	0.715	0.023	3.148	35.366	60.124	0.020	100.000
Phase D	19.887	1.424	0.026	0.537	13.719	64.381	0.025	100.000
Phase E	26.941	14.935	0.396	0.588	0.747	55.684	0.710	100.000

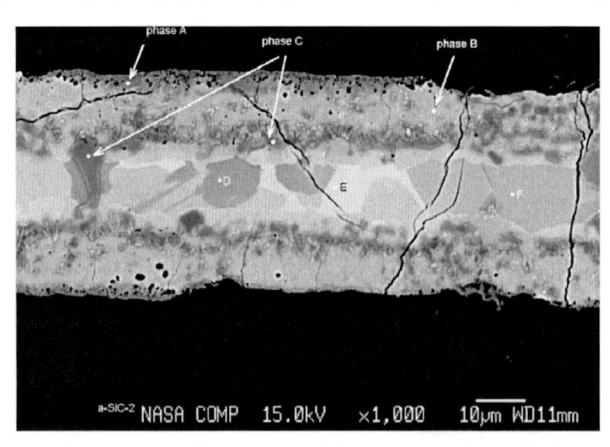

Figure 6. Micrograph of the cross-section of the diffusion bond formed between CRYSTAR α-SiC with the alloyed Ti foil as the interlayer.

Table IV. Microprobe analysis of the atomic ratios for the reaction formed phases in the diffusion bond between α-SiC as shown in Figure 6 (*atomic ratios are an average from five locations for each phase*).

Phase	Al	V	Cr	Si	C	Ti	Fe	Total
Phase A	0.020	0.970	0.023	7.662	34.787	56.531	0.007	100.000
Phase B	0.076	0.923	0.020	34.151	6.502	58.220	0.108	100.000
Phase C	0.223	0.760	0.019	6.015	32.269	60.690	0.025	100.000
Phase D	21.862	2.363	0.050	0.432	10.394	64.865	0.036	100.000
Phase E	27.102	21.318	0.611	0.473	1.901	46.456	2.141	100.000
Phase F	25.975	5.962	0.145	1.186	1.898	64.666	0.168	100.000

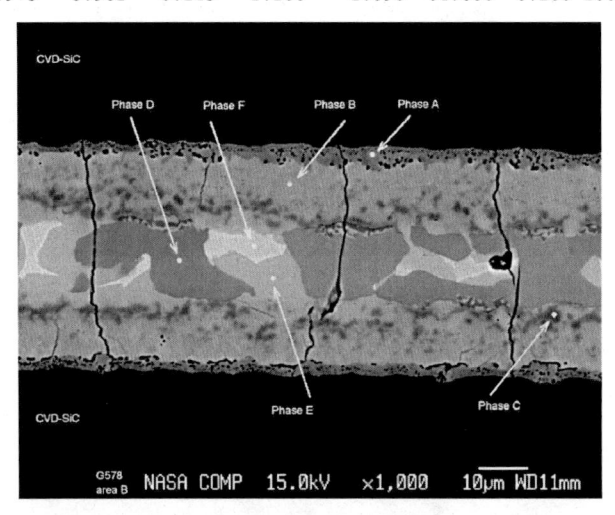

Figure 7. Micrograph of the cross-section of the diffusion bond formed between Rohm & Haas SiC with the alloyed Ti foil as the interlayer.

Table V. Microprobe analysis of the atomic ratios for the reaction formed phases in the diffusion bond between Rohm & Haas SiC as shown in Figure 4 (*atomic ratios are an average from five locations for each phase*).

Phase	Al	V	Cr	Si	C	Ti	Fe	Total
Phase A	0.070	0.848	0.030	17.085	25.802	56.164	0.001	100.000
Phase B	0.303	0.961	0.018	33.766	6.675	58.212	0.064	100.000
Phase C	0.293	0.661	0.010	5.234	35.671	58.122	0.010	100.000
Phase D	19.002	1.392	0.039	0.557	12.707	66.288	0.014	100.000
Phase E	23.244	7.649	0.213	0.758	3.970	64.104	0.062	100.000
Phase F	19.833	20.784	0.634	0.514	4.622	52.302	1.313	100.000

The microcracks observed when forming diffusion bonds with the alloyed Ti foil may be due to a single factor or a combination of factors. However, the primary factor appears to be the formation of the phase $Ti_5Si_3C_x$, which appears to correlate with Phase B in Figure 5 and Table III. This phase appeared in all samples that had microcracking. Titanium silicide, Ti_5Si_3, is

Diffusion Bonding of Silicon Carbide for MEMS-LDI Applications

anisotropic in its thermal expansion[12,13]. The coefficient of thermal expansion (CTE) in the a-direction is 6.11 ppm/K and in the c-direction it is 16.62 ppm/K which gives a ratio of CTE(c)/CTE(a) equal to 2.72. Other researchers have reported the anisotropy to give a ratio as high as 4.39 [14]. This anisotropy could cause the microcracking due to thermal stresses during the cool down phase after hot pressing.

Mixed results were obtained when PVD Ti coatings were used as the interlayer. Figure 8 shows the diffusion bond from when Rohm & Haas CVD SiC was bonded with a 20 micron thick PVD Ti layer. The interlayer was formed from 10 micron coatings on both SiC substrates. Microcracks were observed in the diffusion bond. The corresponding microprobe analysis of the reaction formed phases is given in Table VI. Similar to the bonds formed using the alloyed Ti foil, $Ti_5Si_3C_x$ (Phase B in Figure 8 and Table VI) and microcracking was observed when a relatively thick layer of PVD Ti (20 micrometers) was used as the interlayer. Naka et al.[15] suggested that $Ti_5Si_3C_x$ is an intermediate phase that will not be present when phase reactions have gone to completion. This appears to be the case as seen in the bond formed from the thinner 10 micron PVD Ti interlayer. The micrograph in Figure 9 shows a well formed diffusion bond. The bond is much less complex than those from the alloyed Ti foil which had five to seven phases or the thicker PVC Ti coating which had three phases. Only two phases have formed as seen from the microprobe analysis in Table VII. The absence of microcracking in the diffusion bonds formed with a 10 micron PVD Ti interlayer is because the detrimental phase of Ti_5Si_3 was not present. The source of the dark pores in the bond still has to be determined. It may be due to the formation of more dense phases during the diffusion bonding process. The presence of the pores may not have a significant effect on the mechanical and leakage properties of the bond, since the pores are very small and isolated. Future tests and analysis will determine if the small, isolated pores have an effect on leakage through the bonds.

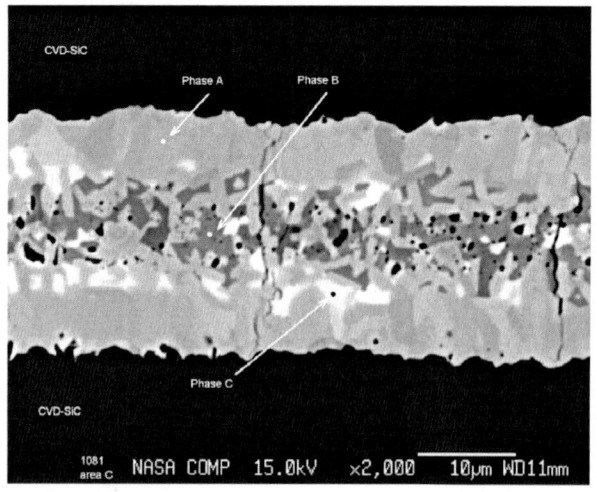

Figure 8. Diffusion bond when a double layer of PVD Ti (20 microns) was used as the interlayer between the Rohm & Haas SiC substrates.

Table VI. Microprobe analysis of the atomic ratios for the reaction formed phases in the diffusion bond as shown in Figure 5 (*atomic ratios are an average from five locations for each phase*).

Phase	Al	Fe	Ti	Si	C	Cr	Total
Phase A	0.011	0.001	56.426	17.792	25.757	0.014	100.000
Phase B	0.007	0.005	35.794	62.621	1.570	0.003	100.000
Phase C	0.027	0.153	58.767	33.891	7.140	0.023	100.000

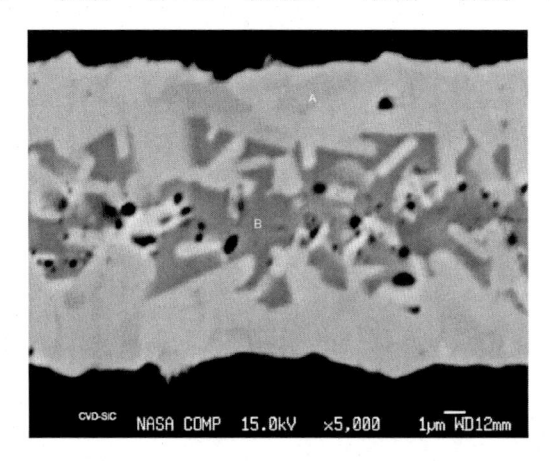

Figure 9. Diffusion bond when a single layer of PVD Ti (10 microns) was used as the interlayer between the Rohm & Haas SiC substrates.

Table VII. Microprobe analysis of the atomic ratios for the reaction formed phases in the diffusion bond as shown in Figure 6 (*atomic ratios are an average from five locations for each phase*).

Phase	C	Si	Ti	Al	Cr	Total
SiC	45.890	54.096	0.011	0.000	0.004	100.000
Phase A	24.686	18.690	56.621	-	0.003	100.000
Phase-B	3.028	61.217	35.752	-	0.003	100.000

CONCLUSIONS

Robust silicon carbide joining technology was developed through the use of titanium interlayers to form diffusion bonds. The diffusion bonds were well adhered to the SiC substrate. No delaminations were observed between the SiC layers. In the case of diffusion bonding with alloyed Ti foil or relatively thick PVD Ti coatings, microcracking was observed. This was due to the formation of the detrimental phase of Ti_5Si_3 which has high anisotropy in its thermal expansion. The best diffusion bonds were obtained when 10 micron thick PVD Ti coatings were used as interlayers between CVD SiC. The bonds were uniform, formed preferred phases, and did not contain microcracking. The preliminary results presented here suggest that diffusion bonding is a good processing approach for fabricating joints. Future analysis and subcomponent tests will seek to confirm that the diffusion bonds are leak free, strong, and stable at elevated temperatures.

ACKNOWLEDGEMENTS

This effort was supported by the NASA Glenn Research Center under the Intelligent Propulsion Systems Foundation Technology task – Ultra-Efficient Engine Technology Project – Vehicle Systems Program. The authors would like to thank James Smith for conducting microprobe analysis and Dr. Robert Okojie for applying PVD Ti coatings on substrates.

REFERENCES

[1] B. Gottselig, E. Gyarmati, A. Naoumidis, and H. Nickel, "Joining of Ceramics Demonstrated by the Example of SiC/Ti," *Journal of the European Ceramic Society*, **6**, 153-160 (1990).

[2] M. Singh, "A Reaction Forming Method for Joining of Silicon Carbide-Based Ceramics," *Scripta Materialia*, **37**, 8, 1151-1154 (1997).

[3] J. Martínez Fernández, A. Muñoz, F. M. Valera-Feria, and M. Singh, "Interfacial and Thermal Characterization of Reaction Formed Joints in Silicon Carbide-Based Materials," *Journal of the European Ceramic Society*, **20**, 2641-2648 (2000).

[4] P. Colombo, A. Donato, B. Riccardi, J. Woltersdorf, E. Pippel, R. Silberglitt, G. Danko, C. Lewinsohn, and R. Jones, "Joining SiC-Based Ceramics and Composites with Preceramic Polymers," *Ceramic Transactions*, **144**, 323-334. (2002).

[5] E. Pippel, J. Woltersdorf, P. Colombo, and A. Donato, "Structure and Composition of Interlayers in Joints Between SiC Bodies," *Journal of the European Ceramic Society*, **17**, 1259-1265 (1997).

[6] M. R. Locatelli, B. J. Dalgleish, K. Nakashima, A. P. Tomsia, and A. M. Gleaser, "New Approaches to Joining Ceramics for High-Temperature Applications," *Ceramics International*, **23**, 313-322 (1997).

[7] J. Li and P. Xiao, "Fabrication and Characterization of Silicon Carbide/Superalloy Interfaces," *Journal of the European Ceramic Society*, **24**, 2149-2156 (2004).

[8] K. Bhanumurthy and R. Schmid-Fetzer, "Solid-State Bonding of Silicon Carbide (HIP-SiC) Below 1000°C," *Materials Science and Engineering*, **A220**, 35-40 (1996).

[9] B. V. Cockeram, "The Diffusion Bonding of Silicon Carbide and Boron Carbide Using Refractory Metals," Proceedings from the Material Solutions Conference '99 on Joining of Advanced and Specialty Materials, 1-4 November 1999, Cincinnati, Ohio.

[10] R. Tacina, C. Wey, P. Laing, and A. Mansour, "A Low Lean Direct Injection, Multipoint Integrated Module Combustor Concept for Advanced Aircraft Gas Turbines," NASA/TM-2002-211347, April 2002.

[11] M. C. Halbig, M. Singh, T. P. Shpargel, J. D. Kiser, "Diffusion Bonding of Silicon Carbide Ceramics Using Titanium Interlayers," Proceedings of the 30th International Conference & Exposition on Advanced Ceramics & Composites, Cocoa Beach, FL, Jan. 22-27, 2006.

[12] J. H. Schneibel, C. J. Rawn, E. A. Payzant, and C. L. Fu, "Controlling the Thermal Expansion Anisotropy of Mo_5Si_3 and Ti_5Si_3 Silicides," *Intermetallics*, **12**, 845-850 (2004).

[13] J. H. Schneibel and C. J. Rawn, "Thermal Expansion Anisotropy of Ternary Silicides Based on Ti_5Si_3," *Acta Materialia*, **52**, 3843-3848 (2004).

[14] L. Zhang and J. Wu, "Thermal Expansion and Elastic Moduli of the Silicide Based Intermetallic Alloys $Ti_5Si_3(X)$ and Nb_5Si_3," *Scripta Materiallia*, **38**, 2, 307-313 (1998).

[15] M. Naka, J. C. Feng, and J. C. Schuster, "Phase Reaction and Diffusion Path of the SiC/Ti System," *Metallurgical and Materials Transactions A*, **24A**, 1385-1390, (1997).

EFFECT OF RESIDUAL STRESS ON FRACTURE BEHAVIOR IN MECHANICAL TEST
FOR EVALUATING SHEAR STRENGTH OF CERAMIC COMPOSITE JOINT

Hisashi Serizawa and Kazuaki Katayama
Joining and Welding Research Institute, Osaka University
11-1 Mihogaoka
Ibaraki, Osaka 567-0047, Japan

Charles A. Lewinsohn
Ceramatec Inc.
2425 South 900 West
Salt Lake City, Utah 84119, USA.

Mrityunjay Singh
MS 106-5, Ceramics Branch, NASA Glenn Research Center
21000 Brookpark Road
Cleveland, OH 44135-3191, USA

Hidekazu Murakawa
Joining and Welding Research Institute, Osaka University
11-1 Mihogaoka
Ibaraki, Osaka 567-0047, Japan

ABSTRACT
As examples of the most typical methods to determine the shear strength of ceramic composite joints, the tensile test of lap joined SiC/SiC composite and the asymmetrical four point bending test of butt joined SiC/SiC composite were analyzed by using finite element method with the interface element. From the computational results, it was revealed that the shear strength in the tensile test was strongly influenced by the residual stress as the increase of the joint layer thickness. In the case of the asymmetrical bending test, it was found that the crack initiation point would move due to the residual stress and the strength was also affected by the joint layer thickness.

INTRODUCTION
Silicon carbide-based fiber reinforced silicon carbide composites (SiC/SiC composites) are promising candidate materials for high heat flux components because of their high-temperature properties, chemical stability and good oxidation resistance[1-3]. For fabricating large or complex shaped parts of SiC/SiC composites, practical methods for joining simple geometrical shapes are essential. To establish useful design databases, the mechanical properties of joints must be accurately measured and quantitatively characterized, where the shear strength of joints is one of the most basic and important mechanical properties. Although, as examples of the most typical methods to determine the shear strength of the composite joints, the tensile test of lap joined composite and the asymmetrical four point bending test of butt joined composite were widely used, it was revealed that the fracture behavior of those tests was different and the apparent shear

Effect of Residual Stress on Fracture Behavior in Mechanical Test

strength of the composite joint obtained experimentally was affected by the combination of the surface energy and the shear strength at the interface[4]. Therefore, the types of fracture behavior for each test configuration have to be precisely studied to measure the shear strength.

On the other hand, the strength of the bonded joint is largely influenced by the geometry of the joint and the test method for evaluating the strength[5,6]. Also, the residual stress is likely to be induced in the joints because of the difference in the coefficient of thermal expansion between the base material and the adhesive layer[7]. In order to study these influences, the level of stress and the order of the singularity in stress field are commonly employed for the relative evaluation of strength. Although detailed information on the stress field is provided, little information on the criteria of the fracture is obtained from these types of study[5-7]. This comes from the fact that, the physics of failure itself is not explicitly modeled in these analyses. To describe deformation and fracture behavior more precisely, new and simple computer simulation methods have been developed[4,8-13]. The methods treat the fracture phenomena as the formation of new surface during crack opening and propagation. Based on the fact that surface energy must be supplied for the formation of new surface, a potential function representing the density of surface energy is introduced to the finite element method (FEM) using cohesive elements[8] or interface elements[4,9-13]. These methods may have a potential capability not only to give insight into the criteria of the fracture but also to make the quantitative prediction of strength itself.

So, in this research, in order to examine the effect of the residual stress on the fracture behavior in the mechanical test for evaluating the shear strength of SiC/SiC composite joints, the tensile test of lap joined composite[4,13] and the asymmetrical four point bending test of butt joined composite[4,7,12,14] were analyzed by using FEM with the interface element. Also, the influence of the joint layer thickness on the shear strength of the composite joint was examined.

INTERFACE ELEMENT

Essentially, the interface element is the distributed nonlinear spring existing between surfaces forming the interface or the potential crack surfaces as shown by Fig.1. The relation between the opening of the interface δ and the bonding stress σ is shown in Fig.2. When the opening δ is small, the bonding between two surfaces is maintained. As the opening δ increases, the bonding stress σ increases until it becomes the maximum value σ_{cr}. With further increase of δ, the bonding strength is rapidly lost and the surfaces are considered to be separated completely. Such interaction between the surfaces can be described by the interface potential[15]. There are rather wide choices for such potential. The authors employed the Lennard-Jones type potential because it explicitly involves the surface energy γ which is necessary to form new surfaces. Thus, the surface potential per unit surface area ϕ can be defined by the following equation.

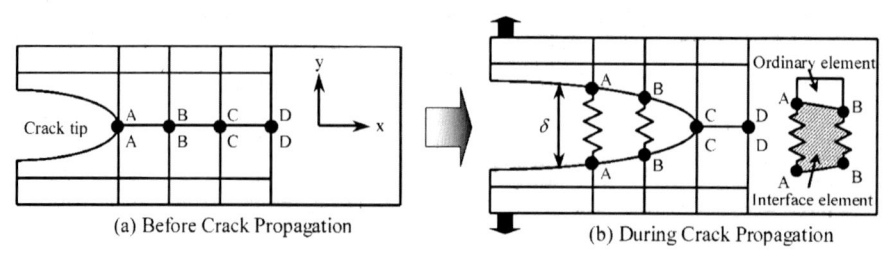

Fig.1 Representation of crack growth using interface element.

(a) Before Crack Propagation

(b) During Crack Propagation

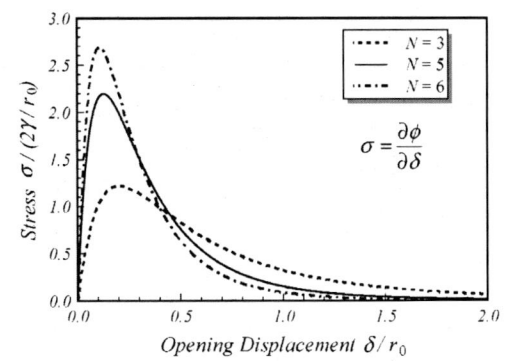

Fig.2 Relationship between crack opening displacement and bonding stress.

$$\phi(\delta_n,\delta_t) \equiv \phi_a(\delta_n,\delta_t) + \phi_b(\delta_n) \tag{1}$$

$$\phi_a(\delta_n,\delta_t) = 2\gamma \cdot \left\{ \left(\frac{r_0}{r_0+\delta}\right)^{2N} - 2 \cdot \left(\frac{r_0}{r_0+\delta}\right)^{N} \right\}, \quad \delta = \sqrt{\delta_n^2 + A \cdot \delta_t^2} \tag{2}$$

$$\phi_b(\delta_n) = \begin{cases} \frac{1}{2} \cdot K \cdot \delta_n^2 & (\delta_n \leq 0) \\ 0 & (\delta_n \geq 0) \end{cases} \tag{3}$$

Where, δ_n and δ_t are the opening and shear deformation at the interface, respectively. The constants γ, r_0, and N are the surface energy per unit area, the scale parameter and the shape parameter of the potential function. In order to prevent overlapping in the opening direction due to a numerical error in the computation, the second term in Eq.(1) was introduced and K was set to have a large value as a constant. Also, to model an interaction between the opening and the shear deformations, an interaction parameter A was employed in Eq.(2). From the above equations, the maximum bonding stress, σ_{cr}, under only the opening deformation δ_n and the maximum shear stress, τ_{cr}, under only the shear deformation δ_t are calculated as follows.

$$\sigma_{cr} = \frac{4\gamma N}{r_0} \cdot \left\{ \left(\frac{N+1}{2N+1}\right)^{\frac{N+1}{N}} - \left(\frac{N+1}{2N+1}\right)^{\frac{2N+1}{N}} \right\} \tag{4}$$

$$\tau_{cr} = \frac{4\gamma N \sqrt{A}}{r_0} \cdot \left\{ \left(\frac{N+1}{2N+1}\right)^{\frac{N+1}{N}} - \left(\frac{N+1}{2N+1}\right)^{\frac{2N+1}{N}} \right\} \tag{5}$$

By arranging such interface elements along the crack propagation path as shown in Fig.1, the growth of the crack under the applied load can be analyzed in a natural manner. In this case, the decision on the crack growth based on the comparison between the driving force and the resistance as in the conventional methods is avoided.

From the results of our previous researches using the interface elements, it was found that the failure mode and the stability limit depend on the combination of the deformability of the ordinary element in FEM and the mechanical properties of the interface element as controlled by the surface energy χ the scale parameter r_0 and the interaction parameter A in Eq.(2); furthermore, the fracture strength in the failure problems of various structures might be quantitatively predicted by selecting the appropriate values for the surface energy χ the scale parameter r_0 and the interaction constant A [4,9-13].

MODEL FOR ANALYSIS

Since, as a result of R & D efforts, an affordable, robust ceramic joining technology (ARCJoinT™) has been developed as one of the most suitable methods for joining SiC/SiC composites among various types of joining between ceramic composites [16,17], SiC/SiC composite ceramic joints joined by ARCJoinT™ were selected for this study. Figure 3 shows schematic models of the beveled lap joint for the tensile test and the asymmetrical four point bending test. The lap joint shown in Fig.3(a) was made from two SiC/SiC composite plates, whose dimensions were 57.5 mm-long, 12.5 mm-wide and 2.125 mm-thick. The angle of the edge, θ, was assumed to be 161 degree according to our ongoing experiments. To prevent a rotation of the lap joint due to a bending moment, the tabs were also joined to the ends of the joint via the ARCJoinT™ method. L_1 and L_2 in Fig.3(b) are the inner and outer span lengths, respectively. According to our previous experimental results for $50^l \times 4^b \times 4^h$ mm³, L_1 and L_2 were chosen to be 12 and 44 mm, respectively [18]. The thickness of the joint layer was varied from 1 to 200 mm for typical examples of ARCJoinT™ [16,17].

(a) Beveled Lap Joint

(b) Asymmetrical Four Point Bending Test

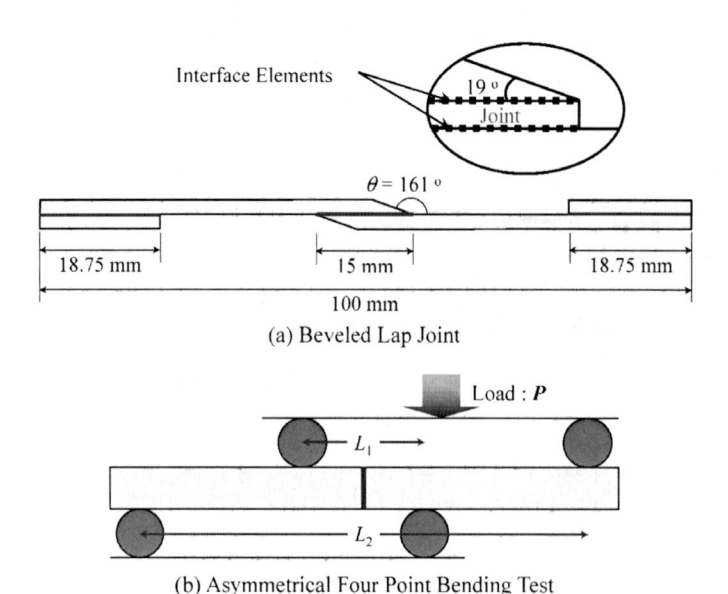

Fig.3 Schematic illustrations of test methods for measuring shear strength.

Young's moduli and Poisson's ratios of SiC/SiC composite and the joint were assumed to be 300 GPa, 350 GPa, 0.15 and 0.20, respectively[13,16-18]. Although the mechanical properties of SiC/SiC composites should be affected by SiC fiber direction, Young's moduli were assumed to be isotropic since the difference between the elastic properties of the composite and the joint material is relatively larger than those due to the composite anisotropy. Because of the brittleness of the ceramic materials, FEM calculations were conducted assuming linear elastic behavior in two-dimensional plain strain. Since the fracture started from the interface between SiC/SiC composite and the joint where the load had a maximum value in the experiments[18], the interface elements were arranged along both the interfaces between the composite and the joint layer and the mesh division near the edges was set to fine. Because the coefficient of thermal expansion of the joint material in ARCJoinT™ is larger than that of SiC/SiC composite, the residual stress was induced by applying the positive inherent strain to the joint layer. The element sizes were decided by continuously refining the mesh until approximate convergence of the numerical solution was achieved.

From the previous studies about the four point and the asymmetrical four point bending tests of a butt joined SiC/SiC composites via the ARCJoinT™ method by using FEM with the interface element, the surface energy γ and the interaction parameter between the opening and the shear deformations A in Eq.(2) were estimated to be 30 N/m and 2.47×10^{-2}, respectively[12]. In this research, a constant K was set to 5.0×10^4 N/m. Then, by changing the scale parameter r_0 and the inherent strain in the ranges from 1.0×10^{-4} to 100 μm and from 5.0×10^{-4} to 1.0×10^{-3}, respectively, the test methods for evaluating the shear strength of the composite joint were analyzed. The shape parameter N was assumed to be 4 according to our previous researches[4,10-13].

RESULTS AND DISCUSSIONS
The tensile load was applied to the beveled lap joint through the horizontal displacement given on both the ends of the composite joint. According to the experimental results, the maximum load obtained was defined as the fracture load. The effects of the scale parameter and the inherent strain on the fracture load under the tensile test of lap joint with the joint layer thickness of 1, 10, 100 μm are summarized into Figs.4-6, respectively. In all cases, the curves can be divided into

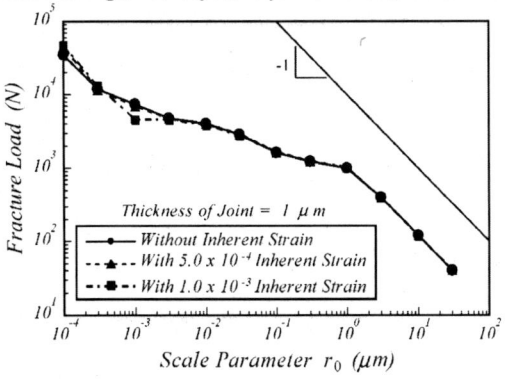

Fig.4 Effect of scale parameter and inherent strain on fracture load of lap joint.
(1 μm joint layer thickness).

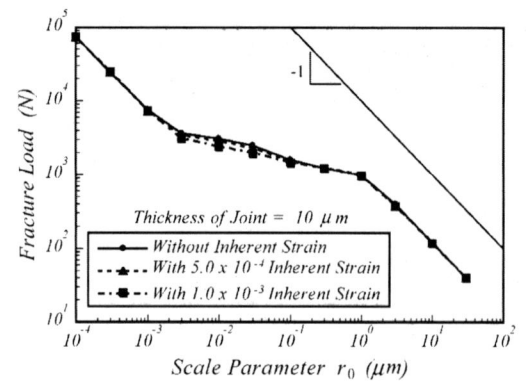

Fig.5 Effect of scale parameter and inherent strain on fracture load of lap joint.
(10 μm joint layer thickness).

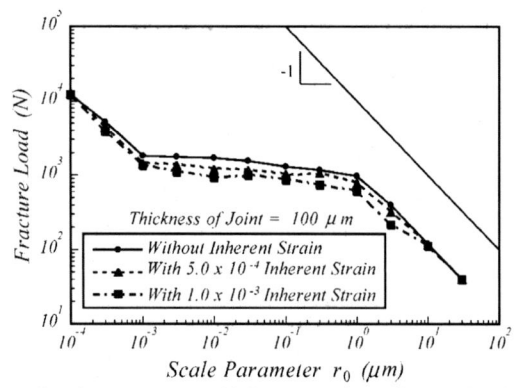

Fig.6 Effect of scale parameter and inherent strain on fracture load of lap joint.
(100 μm joint layer thickness).

three parts with respect to the size of the scale parameter. Since the maximum shear stress of the interface element τ_{cr} is inversely proportional to the scale parameter as shown in Eq.(5), it was found that the fracture behavior in the lap joint was affected by the shear strength at the interface. When the scale parameter were in the middle range, fracture load decreased with increasing the applied inherent strain and these changes could be more clearly appeared in the larger joint layer thickness.

From the close examination of the failure process, it was found that the composite and the joint layer were simply separated without significant deformation when the scale parameter was larger than 1 μm (Zone-III). In the middle range where the scale parameter was from 1.0×10^{-3} to 1 μm (Zone-II), crack like localized sliding was observed at the edge of the interface. The joint broke suddenly without significant shear deformation of the interface when the scale parameter was smaller than 1.0×10^{-3} μm (Zone-I). These failure processes can be related to the shear strength of the interface element in the following way. Since the shear deformation of the interface was dominant, the fracture load of the joint was almost the same as the maximum shear strength τ_{cr} in the Zone-III. According to Eq.(5), both the stiffness and the shear strength of the interface were small in Zone-III. Thus, the joint broke in the simple separation mode. On the other hand, when the scale parameter r_0 was small as in the Zone-I, the fracture strength became larger than the stress induced at the crack tip in FEM model. In this case, the crack like localized sliding was not formed and the failure occurred when the computed stress at the edge of the interface reached the critical shear stress τ_{cr}. Since this phenomenon was caused by the coarseness of the mesh, it could be eliminated by using small enough mesh division. Therefore, it was found that the practical failure process was limited in the Zone-II and IIL.

Because the order of the stress singularity at the edge of the interface becomes to be 0.5 with increasing the joint layer thickness, it was considered that the fracture load with 100 μm joint layer thickness in the middle range was mainly controlled by not the shear strength at the interface but the surface energy. Moreover, since the stress intensity factor (K value) is affected by the residual stress from the view point of the fracture mechanics[19], the fracture toughness (the critical energy release rate) at the edge of the interface was considered to decrease with increasing the applied inherent strain. Then, as a result, the effect of the residual stress on the fracture load in the tensile test would become larger as the increase of the joint layer thickness.

In the case of the asymmetrical four point bending test without the residual stress, it was obtained that the fracture load was not affected by the joint layer thickness. Since the order of the stress singularity at the crack initiation point, which is the center of the composite joint, was in the range from 0.11 to 0.13, the fracture load was considered to be governed by both the surface energy and the shear strength at the interface. So, in the case without the inherent strain, the fracture load became to be nearly proportional to the scale parameter as shown in Fig.7. On the other hand, in the cases with the inherent strain, the effect of the scale parameter on the fracture load can be divided into three parts as same as the results in the lap joint and the fracture load in the middle range decreased with increasing the joint thickness. From the close examination of the failure process and the stress distributions, it was found that the residual stress near the surface of

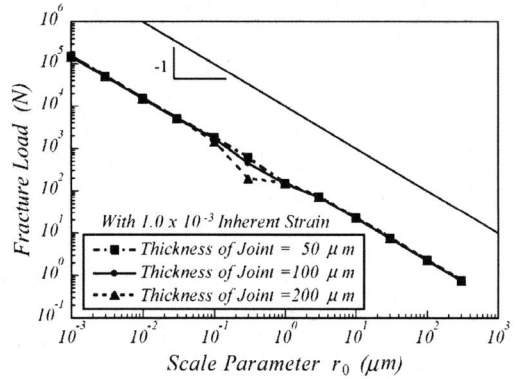

Fig.7 Effect of scale parameter and joint layer thickness on fracture load of asymmetrical bending test.

the composite joint became larger than the stress at the center when the applied load achieved to the maximum value. Namely, it was revealed that the crack initiation point would move from the center of the composite joint to the surface due to the inherent strain, and then the fracture load in the middle range might decreased with increasing the joint layer thickness.

CONCLUSIONS

In order to examine the effect of the residual stress on the fracture behavior in the mechanical test for evaluating the shear strength of SiC/SiC composite joints, the tensile test of lap joined composite and the asymmetrical four point bending test of butt joined composite were analyzed by using finite element method with the interface element. Also, the influence of the joint layer thickness on the shear strength of the composite joint was examined. The conclusions can be summarized as follows.

(1) From the computational results, it was found that the fracture load under the tensile test was strongly influenced by the residual stress as the increase of the joint layer thickness.

(2) In the case of the asymmetrical bending test, it was revealed that the crack initiation point would move due to the residual stress and then the fracture load was also affected by the joint layer thickness as same as the tensile test of the lap joint.

REFERENCES

[1] H. Serizawa, C.A. Lewinsohn, G.E. Youngblood, R.H. Jones, D.E. Johnston and A. Kohyama, "High-Temperature Properties and Creep Resistance of Near-Stoichiometric SiC Fibers," *Ceramic Engineering and Science Proceedings*, **20** [4], 443-450 (1999).

[2] A. Kohyama and Y. Katoh, "Overview of Crest-Ace Program for SiC/SiC Ceramic Composites and Their Energy System Applications," *Ceramic Transactions*, **144**, 3-18 (2002).

[3] B. Riccardi, L. Giancarli, A. Hasegawa, Y. Katoh, A. Kohyama, R.H. Jones and L.L. Snead, "Issues and advances in SiCf/SiC composites development for fusion reactors," *Journal of Nuclear Materials*, **329-333**, 56-65 (2004).

[4] H. Serizawa, D. Fujita, C. A. Lewinsohn, M. Singh and H. Murakawa, "Finite Element Analysis of Mechanical Test Methods for Evaluating Shear Strength of Ceramic Composite Joints Using Interface Element", *Mechanical Properties and Performance of Engineering Ceramics II, Ceramic Engineering and Science Proceedings*, **27** [2], 115-124 (2006).

[5] R.D. Adams, J. Comyn and W.C. Wake, "Structural Adhesive Joints in Engineering", Chapman & Hall, London (1997).

[6] R W. Messler, Jr., "Joining of Materials and Structures", Elsevier Butterworth-Heinemann (2004).

[7] H. Serizawa, C.A. Lewinsohn and H. Murakawa, "FEM Analysis of Experimental Measurement Technique for Mechanical Strength of Ceramic Joints", *Ceramic Engineering and Science Proceedings*, **22** [4], 635-642 (2001).

[8] A. Needleman, "An Analysis of Decohesion Along An Imperfect Interface," *International Journal of Fracture*, **42**, 21-40 (1990).

[9] H. Murakawa, H. Serizawa and Z.Q. Wu, "Computational Analysis of Crack Growth in Composite Materials Using Lennard-Jones Type Potential Function," *Ceramic Engineering and Science Proceedings*, **20** [3], 309-316 (1999).

[10] H. Serizawa, H. Murakawa and C.A. Lewinsohn, "Modeling of Fracture Strength of SiC/SiC Composite Joints by Using Interface Elements," *Ceramic Transactions*, **144**, 335-342 (2002).

[11]H. Murakawa, H. Serizawa, K. Miyamoto, I. Oda, "Strength of Joint Between Dissimilar Elastic Materials," *Proceedings of 2003 International Conference on Computational & Experimental Engineering & Sciences (ICCES'03)*, **6**, (2003) (CD-ROM).

[12]H. Serizawa, H. Murakawa, M. Singh and C.A. Lewinsohn, "Finite Element Analysis of Ceramic Composite Joints by Using a New Interface Potential," *High Temperature Ceramic Matrix Composites 5*, 451-456 (2004).

[13]D. Fujita, H. Serizawa, M. Singh and H. Murakawa, "Numerical Analysis of Single Lap Joined Ceramic Composite Subjected to Tensile Loading", *Ceramic Engineering and Science Proceedings*, **26** [2], 417-424 (2005).

[14]Ö. Ünal, I.E. Anderson and S.I. Maghsoodi, "A Test Method to Measure Shear Strength of Ceramic Joints at High Temperatures," *Journal of American Ceramic Society*, **80** [5] 1281-1284 (1997).

[15]A. Rahman, "Correlations in the Motion of Atoms in Liquid Argon", *Physical Review*, **136** [2A], A405-A411 (1964).

[16]M. Singh, "Design, Fabrication and Characterization of High Temperature Joints in Ceramic Composites," *Key Engineering Materials*, **164-165**, 415-419 (1999).

[17]M. Singh and E. Lara-Curzio, "Design, Fabrication, and Testing of Ceramic Joints for High Temperature SiC/SiC Composites", *Proceedings of ASME TURBOEXPO 2000*, 69-74 (2000).

[18]C.A. Lewinsohn, R.H. Jones, M. Singh, T. Nozawa. M. Kotani, Y. Katoh and A. Kohyama, "Silicon Carbide Based Joining Materials for Fusion Energy and Other High-Temperature, Structural Applications," *Ceramic Engineering and Science Proceedings*, **22** [4] 621-625 (2001).

[19]W. Soboyejo, "Mechanical Properties of Engineered Materials", Marcel Dekker, Inc., New York (2003).

Author Index

Ahmad, J., 135, 145, 155
Ahmad, K., 245
Alexander, D. J., 179
Allen, S., 277
Anderson, H., 289, 463
Ando, K., 399, 443, 449
Ayala, A. A., 361

Baek, S. S., 119, 163
Bansal, N. P., 179
Barbero, E., 347
Basini, V., 319
Bellosi, A., 327
Bezold, A., 479
Boussuge, 319
Byun, J.-H., 191

Calomino, A., 135
Chen, J., 145, 155
Choi, S. R., 179
Corral, E. L., 361
Costabile , A., 237
Cotterell, B., 379
Cutts, J., 463

da Rocha Caffarena, V., 3

Falk, L. K. L., 55
Foucaud, S., 65

Gauthier, W., 423
Gelebart, L., 319
Genson, A., 337
Gerdes, T., 15
Golubov, S., 297
Goursat, P., 65
Gowayed, Y., 135, 145, 155

Halbig, M. C., 491
Halloran, J. W., 327
Hampshire, S., 55, 337
Hanifi, A. R., 337
Hatta, H., 213
Hbaieb, K., 379, 411
Hemrick, J. G., 347
Hetrick, G., 119
Hinoki, T., 207
Hirao, K., 391, 433
Hyuga, H., 391, 433

IsikawaT., 213

Jensen, J. A., 371
John, R., 135, 145, 155
Johnson, A., 289
Jones, T. C., 469

Kang, T. J., 191
Karandikar, P., 101

Author Index

Karlsdottir, S. N., 327
Katayama, K., 503
Katoh, Y., 91, 223, 297
Kiser, J. D., 491
Kobayashi, R., 111
Kohyama, A., 91, 207
Komeya, K., 111
Kondo, S., 91
Kotani, M., 213
Kowalik, R. W., 179
Kumagai, T., 41

Lamon, J., 423
Lange, F. F., 411
Lara-Curzio, E., 223
LaSalvia, J. C., 257
Leixas Capitaneo, J., 3
Lee, J.-K., 199
Lee, J. Y., 191
Le Flem, M., 319
Lewinsohn, C. A., 289, 463, 503
Li, H., 33
Liu, X., 347
Loehman, R. E., 361
Lopes Cosentino, P. A. de S., 3

Mahoney, J. F., 469
Maier, H. R., 479
Marchant, D. D., 101
Marshall, A., 101
McCuiston, R. C., 257
McMeeking, R. M., 411
Meguro, T., 111
Menard, M., 319
Menke, Y., 55
Miller, R., 135, 145, 155
Miyazaki, H., 391, 433
Monnet, I., 319
Monteverde, F., 327
Morscher, G., 135
Moser, B., 257
Murakawa, H., 503

Naganuma, M., 199
Nakao, W., 399, 443, 449
Negahdari, Z., 15
Newman, R. A., 277
Neilsen, M. K., 469
Nozawa, T., 223

Ogasawara, T., 3, 213
Ohji, T., 391, 433
Ojard, G., 135, 145, 155
Ono, M., 449
Osada, T., 399

Pan, W., 245
Park, J.-S., 207
Peters, K.-M., 347
Pfaff, E. M., 23, 479
Phillips, N. S. L., 371
Pomeroy, M. J., 337
Pyzik, A. J., 277

Ritts, A., 33
Ruggles-Wrenn, M. B., 119, 163

Salamone, S., 101
Salehi, S., 269
Salles, V., 65
Santhosh, U., 135, 145, 155
Schneider, N., 55
Sennett, M., 101
Serizawa, H., 503
Sglavo, V. M., 237
Shimoda, K., 207
Shinavski, R. J., 223
Shpargel, T., 491
Siegert, G. T., 163
Silveira Pinho, M., 3
Singh, A. K., 79
Singh, M., 491, 503
Singh, R. P., 79
Snead, L., 297
Sun, J. G., 371

Takahashi, K., 399, 443, 449
Tandon, R., 469
Tatami, J., 111
Teyssandier, F., 307
Trethewey, J. S., 371
Tsutagawa, Y., 443

Van der Biest, O., 269
Vanderspiegel, N. N., 371
Verhelst, J., 269
Vleugels, J., 269

Wakihara, T., 111

Wery, S., 307
Willert-Porada, M., 15
Wilson, M., 289, 463

Xu, J., 347

Yoshizawa, Y., 391, 433
Yu, Q., 33

Zwick, M., 23
Zunjarrao, S. C., 79